Fluid Power Troubleshooting

FLUID POWER AND CONTROL

A Series of Textbooks and Reference Books

1. Hydraulic Pumps and Motors: Selection and Application for Hydraulic Power Control Systems, *by Raymond P. Lambeck*
2. Designing Pneumatic Control Circuits: Efficient Techniques for Practical Application, *by Bruce E. McCord*
3. Fluid Power Troubleshooting, *by Anton H. Hehn*
4. Hydraulic Valves and Controls: Selection and Application, *by John J. Pippenger*
5. Fluid Power Design Handbook, *by Frank Yeaple*

Other Volumes in Preparation

Fluid Power Troubleshooting

ANTON H. HEHN
Hehn and Associates
Skokie, Illinois

MARCEL DEKKER, INC. New York and Basel

Library of Congress Cataloging in Publication Data

Hehn, Anton, H., [date]
Fluid power troubleshooting.

(Fluid power and control; 3)
Includes Index.
1. Fluid power technology. 2. Oil hydraulic machinery
—Maintenance and repair. 3. Pneumatic machinery—Maintenance and repair. I. Title. II. Series.
TJ843.H44 1984 621.2'0288 83-20934
ISBN 0-8247-7048-X

MARCEL DEKKER, INC.
270 Madison Avenue, New York, New York 10016

Current printing (last digit):
10 9 8 7 6 5 4 3 2 1

PRINTED IN THE UNITED STATES OF AMERICA

To my wife
Violet Hofmann Hehn

Foreword

The servicing of hydraulic systems, including maintenance, troubleshooting, and repair, has been one of the major problems facing users of hydraulic equipment.

Although many component manufacturers have prepared and published excellent data on the diagnostics and repair of specific components which may be malfunctioning, little has been published on the diagnostics of entire systems which may include similar but subtly differing products designed and manufactured by several competing manufacturers.

In this work the author has addressed the problem on a broad scale. The material is organized to enable the hydraulic maintenance technician to determine the probable cause of malfunction in even a relatively complex system. While many authors have addressed this problem, few have made adequate reference to or use of the specific maintenance instructions and maintenance suggestions of component manufacturers whose products are included in a malfunctioning circuit.

The author has taken great care to have much of his material reviewed by other fluid power experts, especially those best qualified to deal with fluid circuits which may be malfunctioning for obscure reasons often difficult to pinpoint. He has incorporated many of the suggestions received into his narrative. The net result is a volume which will be a most useful tool for the circuit designer, providing the insight needed to devise circuits and define the installation and mounting of the various components to minimize application problems. Maintenance experts will find the author's practical, down-to-earth troubleshooting hints and suggested remedies a welcome change from the more usual obscure approaches to circuit problems.

Z. J. Lansky

Preface

Industrial fluid power is a relatively young field of energy transmission and control. The manufacture of hydraulic and pneumatic components for various industrial applications has expanded very rapidly since the end of World War II. Modern hydraulic and pneumatic equipment can economically convert mechanical energy into fluid energy, and with simple components this energy may be regulated to provide direction, speed, and force control. No other type of power transmission provides the range of control of force, speed, and direction that is possible with fluid power transmission. The ever-increasing use of industrial robots and automated machinery in plants is made possible in part by continuing development in the field of fluid power systems and controls. Today, fluid power equipment is used in nearly every branch of industry: petrochemical, mining, machine tool, plastics processing, material handling, packaging, pharmaceuticals, and others. Increased use of fluid power equipment has resulted in an expanded need for technicians, engineers, managers, and mechanics to become acquainted with the operation, care, and troubleshooting of this equipment.

This book has been prepared to provide designers and users of fluid power systems with a reference source in the area of fluid power troubleshooting. The material presented is practical in character; theory is kept to a minimum. The information presented will provide the reader with answers to many of the most common questions pertaining to locating and isolating fluid power problems within a system.

The author has drawn on the knowledge obtained and information gathered during many years of designing, maintaining, and troubleshooting equipment actuated and controlled by fluid power. The arrangement of the material follows that developed during the conduct of numerous seminars dealing with this subject matter.

Chapter 1 presents a review of fluid power principles and a comparison of pneumatic, hydraulic, and electric systems. Chapters 2 and 3 deal with the operation, maintenance, and troubleshooting of hydraulic and pneumatic systems. Chapter 4 discusses the sources, effects, and control of contaminants in a hydraulic system. Chapter 5 reviews compressed air filtration, lubrication, and moisture control. Chapters 6 through 10 deal with hydrostatic transmissions, the generation and control of heat, selection and care of hydraulic fluids, fluid conductors and connectors, and seals used in static and dynamic applications. Chapters 11 and 12 concern themselves with maintenance and troubleshooting of systems, malfunction detection, and diagnostic instrumentation. The book concludes with a brief discussion of fluid power system noise and a review of electrohydraulic servosystems. Some helpful technical information and a detailed list of fluid power symbols are provided in the appendixes. It is hoped that this material will enable the reader, when troubleshooting, to spot quickly a component that is not working properly.

The author wholeheartedly thanks Mr. John J. Pippinger for the editorial work, revisions, and additions to the text and Mr. Ray Lambeck, Dr. Zeke Lansky, Dr. J. Slater, Mr. Frank Yeaple, and Mr. Bruce McCord for their review of the manuscript and the many fine suggestions received from each. The author extends many thanks to the fluid power equipment manufacturers for their generous contributions of technical information and illustrations.

Anton H. Hehn

Contents

Fluid Power Troubleshooting

1

Fluid Power Principles: A Review

1.1 HYDRAULIC PRINCIPLES

Proper use and maintenance of hydraulic systems is dependent on a thorough knowledge of the fluid and of the functions of the mechanical components. To operate and maintain a hydraulic system, people in the industrial field must equip themselves with a knowledge of the basic physical laws of fluid power and should be familiar with the seven basic components of a hydraulic system.

Many hydraulic systems seem exceedingly complicated. However, their basic design is quite simple. Regardless of the complexity or simplicity of a hydraulic system, each system contains seven basic components: (1) a reservoir to hold the fluid supply, (2) connecting lines to transmit the fluid power, (3) a pump to convert input power into fluid power, (4) a pressure control valve to regulate pressure, (5) a directional control valve to control the direction of fluid flow, (6) a flow control device to regulate speed or fluid flow, and (7) an actuator to convert hydraulic power into mechanical motion.

1.1.1 Basic Laws of Fluids

Although hydraulics is one of the oldest branches of science, it was only in recent years that knowledge of the physical laws of fluids was put to practical use in industry. Ninety years ago it would have been impossible to apply hydraulics as we do today. Machine tools for producing the required precision parts did not exist. Moreover, oil technology had not progressed far enough to produce a satisfactory hydraulic oil.

One reason for the ever-increasing use of hydraulic apparatus is the fact that properly constructed hydraulic systems possess a number of favorable characteristics: (1) they can eliminate complicated mechanisms such as cams, gears, and levers; (2) the liquid used is not subject to breakage as are mechanical parts; (3) the components of a hydraulic system are not subject to great wear; (4) forces can be rapidly generated and transmitted over considerable distances with very little loss and can also be carried up and down and around corners; and (5) hydraulic systems make possible widely variable motions in both rotary and straight-line transmission of power. It should also be noted that the increase in automation in industry has produced a tremendous expansion in the use of hydraulic and electronic equipment.

The fluid most commonly used in hydraulic systems is petroleum-based, although, water has found application in a limited number of hydraulic installations. Synthetic fluids and fire-resistant fluids are also used in the hydraulic field, and their use is almost mandatory in operations where fire hazards exist.

The primary functions of a hydraulic fluid are to transmit force and to reproduce quickly any change in direction or magnitude in the force (Fig. 1.1). To accomplish this, the fluid must be relatively incompressible. Liquids possess physical properties which give them the ability to transmit power and multiply force.

Liquids are nearly incompressible. Even when liquids are subject to extreme pressures of thousands of pounds to the square inch, the volume of change is negligible. The practical significance of this fact is that force can be transmitted undiminished in all directions when the force is applied on one surface of a confined field. Pascal discovered this principle in 1620, thus providing the fundamental law underlying the whole science of hydraulics. *Pascal's law* states that "pressure exerted on a confined liquid is transmitted undiminished in all directions and acts with equal force on all equal areas." This means that pressure applied to a liquid at one point will be transmitted to any point reached by the liquid.

Because liquids are incompressible, they possess a distinct advantage in the transmission of power. Pressure exerted on one surface of a confined liquid is transmitted undiminished to other surfaces. However, when a blow is struck on the end of a metal bar, the direction of thrust cannot be altered. The main force of the flow is carried straight through the bar to the other end. This happens because the bar is rigid. In this case, only the use of gears and other complex mechanisms can change the direction of the force.

When a force is applied to a confined liquid, the liquid exhibits the same effect of rigidity as a solid. However, the force is transmitted not only straight through to the other end, but in every direction throughout the confined fluid. The force is transmitted equally forward, backward, and sideways. The pressure of a liquid can be directed around corners. It can be bent back on itself. It can transmit force to any point reached by the confined liquid.

A tremendous force can be effected depending on the difference in area sizes where the force is applied and where the work is effected. This hydraulic advantage makes it possible for a small force to lift a great weight. The hydraulic jack demonstrates this principle.

Pascal's law can be used to multiply force since it states that equal force is exerted on equal surfaces. This means that a pressure of 10 lb applied on 1 in.^{2} of surface would be applied equally to each square inch on all surfaces. Thus a pressure of 10 lb on a 1-in.^{2} area would act against a 10-in.^{2} surface with a total force of 100 lb.

Force can be multiplied successfully by the use of hydraulic principles. However, it must be noted that to increase pressure, distance is sacrificed. For example, a 10-in. piston will move one-tenth the distance of a 1-in piston. This ratio of pressure and distance is continually employed in various hydraulic components. It is well to remember that when pressure is increased, distance is sacrificed, and vice versa. The basic rule for two pistons in a hydraulic system is that the distances moved are inversely proportional to their areas.

In addition to Pascal's law, there is another important principle governing the behavior of fluids, *Bernoulli's theorem*. It states: "As liquid flows through a restriction it gains speed but loses pressure; as it emerges from the restriction it loses speed but gains pressure." This is the second important basic law of hydraulics; it deals with fluids that are in motion. There is a proportionate relationship between the cross-sectional area of a passageway and the velocity of flow. In a system, the volume of flow will always be the same through any line. A given volume will pass through a given distance in the same time regardless of the cross-sectional area of the supply lines. The only variable will be a pressure increase or pressure drop.

Some attention must be given to the flow characteristics of fluids in order to understand better the results of these characteristics on hydraulic operations. There are two types of flow in fluid power operations: laminar (or straight line) and turbulent. *Laminar flow* is a smooth movement of fluid through the system as a result of proper control of fluid velocity and pressure drop. *Turbulent* or *"rough flow"* is caused by excess velocity, rough internal pipe surfaces, obstructions, excessive pressure drop, and the use of excessive elbows or bends in the piping system.

In addition to laminar and turbulent flow, it is necessary to mention other factors that directly affect hydraulic fluid performance. These are:

1. *Specific gravity (or weight of the fluid)*: This is important when using fire-resistant hydraulic fluids, because some of these fluids are heavier than conventional petroleum-base fluids.

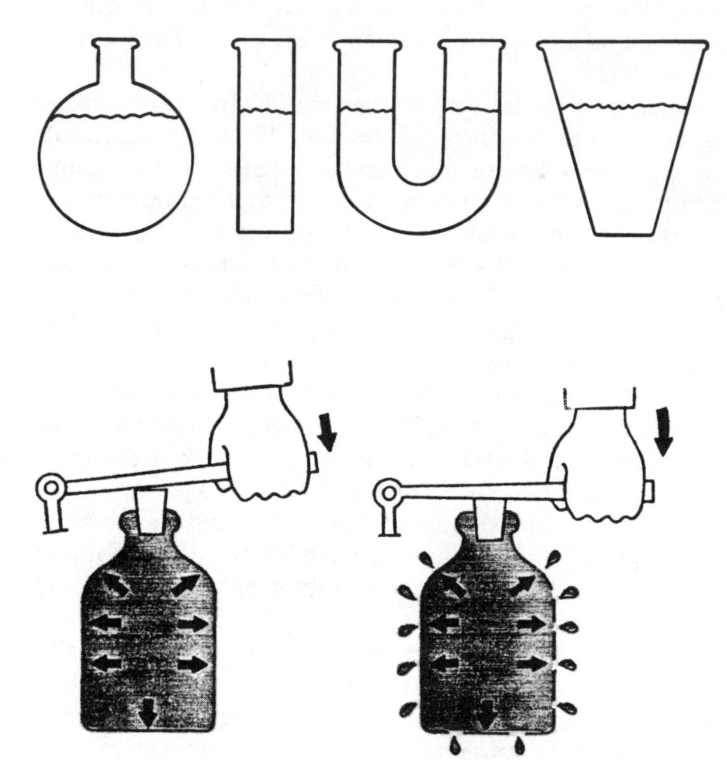

Figure 1.1 Basic principles of fluids. (Courtesy of John Deere and Co., Waterloo, Iowa.)

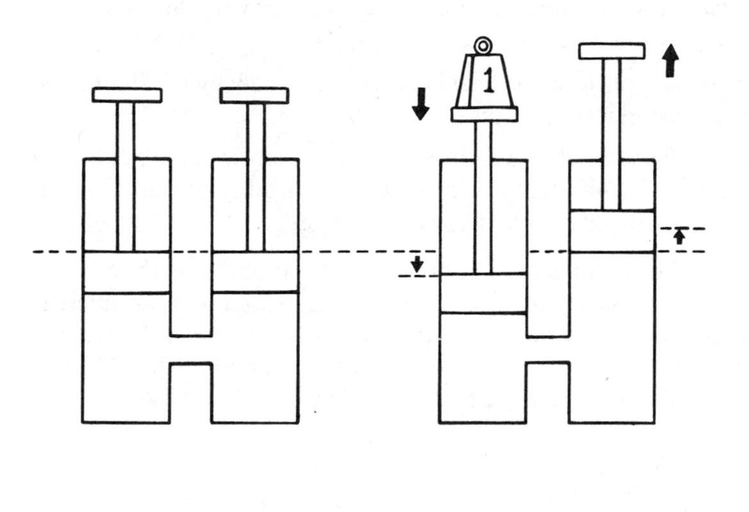

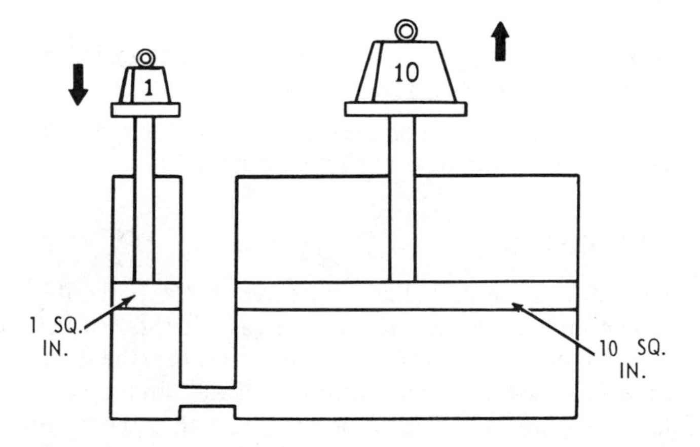

Figure 1.1 (continued)

2. *Atmospheric pressure*: This acts on the fluid whenever it is exposed to air (important for suction inlet problems).
3. *Inertia*: Inertia must be considered when moving the fluid from a rest position to motion.
4. *Friction*: Considerable heat is generated merely by the fluid passing the inside wall of a pipe.
5. *Air entrainment*: Since air is compressible, any mixture of air with the hydraulic fluid will affect the compressibility of the fluid, resulting in soft or spongy hydraulic action.

Before discussing the design and operation of the basic components of a hydraulic system it is necessary to understand the terms used in hydraulics. It is also important to become aware of some misconceptions regarding hydraulic operations. The terms pressure, force, work, power, and torque are continually used in describing hydraulic functions. They are defined as follows:

Pressure = force per unit area (psi)

Force = psi × area (in.^{2})

Work = force × distance (ft-lb)

Power = time rate of doing work (given force moving through a given distance at a given speed) (ft-lb/sec)

Torque = twisting effort developed on the output shaft of a hydraulic motor or input shaft of driving member (lb-ft)

1.2 PNEUMATIC PRINCIPLES

Each time you take a breath, you are inhaling a little over 3/100 of a cubic foot of air—a colorless, odorless, tasteless mixture of gases without which you could not survive for more than 5 or 6 min. The air we breathe is the same air that we compress and use in an endless variety of ways in industry. A great many scientists have worked hard to provide us with a thorough knowledge of the basic laws that govern the behavior of air.

One scientist who was instrumental in extending our knowledge regarding air was Evangelista Torricelli, a student of Galileo. He suspected that the suction of a pump (the vacuum in the cylinder) used to get water out from the bottom of mine shafts was not responsible for *pulling* up the water. Instead, he suspected that the water was *pushed* up into the cylinder by the pressure or weight of the air once a partial vacuum was established by the pump. In other words, he suspected that the air had a finite weight.

To test this, he filled a glass tube about 36 in. long with mercury. One end of the tube was sealed. Holding his thumb over the open end of the tube, he upended the tube into a dish partially filled with mercury. On removing his thumb, he watched the mercury flow out of the tube into the dish and then stop flowing when the column fell to a height of about 30 in. The weight of the air (atmosphere) exerted a force equal to 30 in. of mercury. Since mercury is 13.6 times as dense as water, this was equal to nearly 34 ft of water. If a good enough vacuum could be established, 33.9 ft is the highest the atmosphere would lift a column of water. The experiment resulted in the invention of the barometer.

In another experiment using glass tubes and mercury, Robert Boyle, an English physicist of the latter part of the seventeenth century, established the relationship between the pressure and volume of air. He proved experimentally that when the pressure is doubled, the volume of the air in a confined vessel is reduced by one-half. This discovery became known as *Boyle's law,* which states that "with a constant temperature, volume is inversely proportional to pressure."

From his investigation and experiments, a scientist named Charles gave his name to the other principal law—together with Boyle's—which makes up the "ideal gas law." *Charles's law* states that "with a constant volume, absolute pressure is directly proportional to absolute temperature." Charles's law is involved when you are cautioned against throwing empty aerosol cans into a fire. As the remaining gas in the can is heated, the pressure might increase sufficiently to explode the container.

Because air cannot be compressed without its temperature changing, Boyle's and Charles's laws do not operate separately in a real pneumatic system. They operate together, according to the *Ideal Gas Law,* calculated by the formula

$$\frac{P_1 \times V_1}{T_1} = \frac{P_2 \times V_2}{T_2}$$

This formula uses absolute pressure (psia) and absolute temperatures (°R) in its calculations. The formula also shows that the pressure, volume, and temperature of the second state of a gas are equal to the pressure, volume, and temperature of the first state. However, in actual practice other factors, such as humidity, heat of friction, and efficiency losses, also affect the gas.

Absolute temperature is used when applying Charles's law. Absolute temperature starts at absolute zero, which is −459.67°F, or 0°R (zero degrees Rankine). Rankine was the scientist who discovered absolute zero. On this basis, 0°F equals 460°R and 60°F equals 520°R (460°R + 60°F).

The dynamics of air are determined by the air's pressure, temperature, and water vapor content (relative humidity). Therefore, industry has adopted a standard on air specifying these three factors. Pressure differentials, pressure drop, flow, and performance criteria of various pneumatic equipment and components are based and designed around this "standard" air. A *standard cubic foot* (scf) of air is 1 ft^3 in volume at sea-level atmospheric pressure (14.69 psi), 68°F temperature, and 36% relative humidity (Fig. 1.2). For air to perform some useful work, there must be pressure differentials. Flowing air (a function of pressure difference) is measured by the cubic-foot volume past a given point during some period of time, usually minutes; thus we arrive at the unit *standard cubic feet per minute* (scfm). It is well to make a distinction between the weight of air and the pressure of air. *Weight* is the air's density or mass. *Pressure* is the force that air exerts against a surface (psi).

Although it is well to know the definition of a standard cubic foot of air, it will rarely be encountered under these precise conditions. Normally, we deal with *free* or *"actual" air,* with its constantly changing temperature and relative humidity. The atmospheric pressure also changes as the altitude increases and decreases.

The amount of water vapor in the air is called the *relative humidity.* The capacity for air to hold water in vapor or gaseous form is primarily a function of its temperature. The higher the temperature, the greater the capacity to hold water vapor. Briefly, relative humidity is the amount of water vapor in the air at a given temperature, relative to its total capacity to hold water at that temperature. By withdrawing heat energy from air, moisture can be removed; that is, a portion of the water vapor can be condensed. When the relative humidity of air is 100%, the point where it can hold no additional water has been reached. This is called the *dew point* of the air. At this point the water vapor will begin to transform, or condense, into liquid "dew."

Air is "springy" or *compressible.* This compressibility of air means that more and more air can be squeezed into a constant volume (or storage tank).

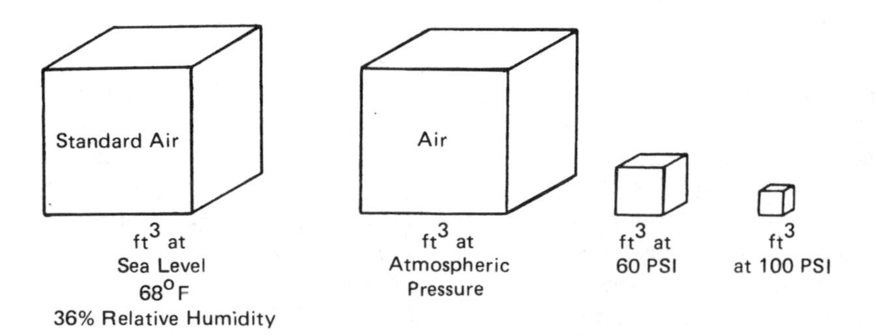

Figure 1.2 Relationship of air volume and pressure.

As more air is forced into this constant volume, the pressure rises (Boyle's law). This ratio of free air required to achieve elevated pressure in a constant volume is expressed by the formula

$$\text{compression ratio} = \frac{\text{atmospheric pressure} + \text{gauge pressure}}{\text{atmospheric pressure}}$$

Therefore, if the number of cubic feet of free air that must be pumped into a 1-ft^3 container to get a 100-psi gauge reading are to be determined, the equation above can be used. At sea level the atmospheric pressure is 14.69 psi and the compression ratio to obtain a 100-psig container pressure is

$$\frac{114.69}{14.69} = 7.8$$

At a 5000-ft elevation with an atmospheric pressure of 12.22 psi, the ratio would be

$$\frac{112.22}{12.22} = 9.18$$

This indicates that, at sea level, it takes approximately 8 ft^3 of free air to obtain 1 ft^3 of air at 100 psig pressure.

If the volume of compressed air used or needed by a pneumatic actuator is known and it becomes necessary to convert this value into standard cubic feet of air, the following formula applies:

$$\text{free air (scf)} = \text{compressed air } (ft^3) \times \frac{\text{psig} + 14.7}{14.7}$$

Pressure gauges normally show only elevated pressures. *Gauge pressure* is designated as *psig*—pounds per square inch gauge. The sum of gauge pressure and atmospheric pressure is called *absolute pressure,* designated as *psia*—pounds per square inch absolute. Knowing and using the absolute pressure is also required in calculating flow and pressure drop or losses for various components, such as valves, fittings, and piping.

In a dynamic system, friction is present between parts that are moving in relation to one another. In a pneumatic system, air is moving but the pipe is not. The faster air travels through a system, the more pressure energy is changed into heat. Transformation of pressure energy into heat occurs in any dynamic fluid power system as frictional resistance is being overcome. However, we do not encounter pneumatic systems running hot. The reason for this is that air cools as it expands to a point downstream. Cooling from air expansion has more temperature effect on a pneumatic system than does frictional resistance.

The volume of air flowing through a pipe in a period of time is a rate of flow. *Flow rate* in pneumatic systems is usually measured in cubic feet per minute (cfm). We know that a cubic foot of air can be under various pressures. A cubic foot of air can be at 90 psi, 100 psi, or 60 psi. But the cubic foot of air referred to in "cfm" generally indicates a cubic foot of air as it enters a compressor. After this air is compressed, it will have a smaller volume, depending on how much it has been compressed. As this reduced volume passes through system piping, it is still referred to as a cubic foot of air since that is what it is under normal conditions.

Streamline or *laminar flow* is the ideal type of airflow in a pneumatic system because the air layers move in nearly parallel lines. Like all fluids, the layer of air next to the surface of the pipe moves the slowest because of the friction between the fluid and the pipe. The layer of moving fluid (air) next to the outermost layer moves a little faster, and so on, until the fluid layers nearest the center of the flow passage move the fastest.

Turbulent flow conditions usually occur because the flow passage is too small for the desired flow velocity of the air. The density and viscosity of the air also affect turbulent flow, but not as much as the flow passage and the flow velocity. Rough or irregularly formed air passages, sudden enlargement or reduction in the diameter of the flow passages, and sudden changes in the direction of flow should all be avoided. When air must pass through a passage of reduced size, the restriction should be smooth and gradual. Turbulent flow heats the air, wastes power by requiring higher air pressure, and can damage the flow passages and ports in the pneumatic equipment.

When air and liquids or other gases come into contact, they can intermix either via diffusion or dispersion. *Diffusion* can be described as the rapid intermingling of the molecules of one gas with another. This should not be confused with *evaporation,* which is the changing of a liquid to a gas. To prevent compressed gases from rapidly diffusing into the surrounding air, they must be stored in closed containers.

Dispersion can be described as the temporary mixing of liquid particles with a gas. When air is compressed, compressor lubricating oil is picked up by the moving air in the compressor and dispersed in fine particles that remain suspended in the air for a time. If enough heat is generated in the compressor, some of the oil evaporates and is diffused in the air.

Although diffusion and dispersion should be kept to a minimum, liquids (especially water) are continually evaporating into the surrounding gases. Because liquids are heavier than gases, they do not mix readily. Therefore, when air and suspended water or oil are put in the same tank or flow through the air lines, the water or oil will settle out and flow to the lowest places. That is why water usually collects at the bottom of a vertical air line, where it should be removed through a drain valve.

The types of energy used in pneumatic systems include the following:

Electrical energy: operates the compressor motor

Pneumatic energy: produced by the compressor

Kinetic energy: produced when the compressed air is lifting an object

Potential energy: which the lifted object now has

Heat energy: produced by friction in the compressor motor, the compressor, the moving air, and the moving piston

1.3 HYDRAULIC SYSTEM OPERATION

Now that the basic principles of a hydraulic system have been described, let us look at the function that each component performs in the operation of the complete system. First, the electric motor is started; it drives the pump shaft that in turn actuates the pump. Once the pump is in operation, it brings into play three direct actions on the fluid:

1. A partial vacuum is created in the pump inlet pipe.
2. Atmospheric pressure acting on the surface of the fluid in the reservoir forces the fluid into the inlet pipe to fill the partial vacuum.
3. The mechanical action of the pump forces the fluid from the inlet to the outlet of the pump and through the hydraulic system.

At this step there is fluid flow through the outlet pipe at a pressure created by the resistance to flow. It is usually desirable to install a relief valve at the outlet of the pump (close to the pump) with no other valving between the relief valve and the pump outlet. The relief valve is set to pass fluid to a lower pressure area (usually the reservoir) when a predetermined pressure level is encountered due to resistance to flow of the fluid entering the power transmission system. Fluid entering the power transmission system (hydraulic circuit) may encounter flow rate control devices. The flow may be directed to the inlet of one or more directional control valves with the rate of flow being governed by the directional control valve. Flow control valves can be in actuator lines or installed in a tee connection to divert a selected flow back to the reservoir or to another lower pressure area.

Fluid pressure is affected by the design of the directional control valve. There are two basic types of directional control valves that will be frequently encountered: closed center and open center in the neutral position. In the *closed center* type, fluid flow is blocked in the neutral position of the directional control valve. The restriction causes a rise in pressure in the line from the outlet of the pump to the inlet of the directional control valve. The pressure will continue to increase until it reaches the pressure setting of the relief valve. In this situation, fluid is directed through the relief valve to the reservoir. However, high pressure will be maintained in the line from the pump to the directional control valve because of the relief valve action of creating only that passageway size needed to pass the quantity of fluid delivered from the pump back to the tank at the pressure level established by the spring setting in the valve. In the *open-center* type of directional valve, fluid will pass through the directional valve and return to the reservoir at low pressure when the valve is in the neutral or center position reflecting the energy needs to pass the fluid through the valve and piping.

For the system to perform work, the directional control valve must be connected to a cylinder or motor. Using a closed-center valve, there will usually be high pressure on the inlet side of the control valve and no pressure value of any significance on the outlet side, which is connected to the cylinder. When the control valve is shifted to either extreme position, the restriction to flow has been reduced and the system pressure will drop to whatever pressure is necessary to move the piston in the cylinder and the restriction caused by the interconnecting fluid lines. The fluid in the rod end chamber of the cylinder will flow through the directional control back to the reservoir usually only meeting atmospheric pressure. When the valve is shifted to the reverse position, the system pressure will drop to whatever pressure is necessary to reverse movement of the cylinder piston assembly. The fluid in the unpressurized chamber of the cylinder will return to the reservoir through the directional valve. Figure 1.3 reviews the operation of such a system, which is a simple hydraulic system. In industrial applications there are many components that may be added to the basic system, such as accumulators, filters, flow metering valves, check valves, and heat exchangers.

Various pressures exist during the operation of a fluid power system. These pressures are directly related to one of the laws of fluid power and can be described as follows:

1. To cause flow, there must be a pressure differential across an orifice or restriction. To stop flow, there must be no pressure differential. When the cylinder moves, there must be a pressure differential (commonly referred to as a *pressure drop*) in order to move the piston.
2. With a closed-center-type directional valve centered, pump delivery from a constant-displacement-type unit is to the tank over the relief valve at its pressure setting. Shifting the valve controlling direction directs pump flow to one side or the other of the piston and opens the other side to tank. Pressure drops to whatever is required to move the piston. Under stall conditions or at the end of

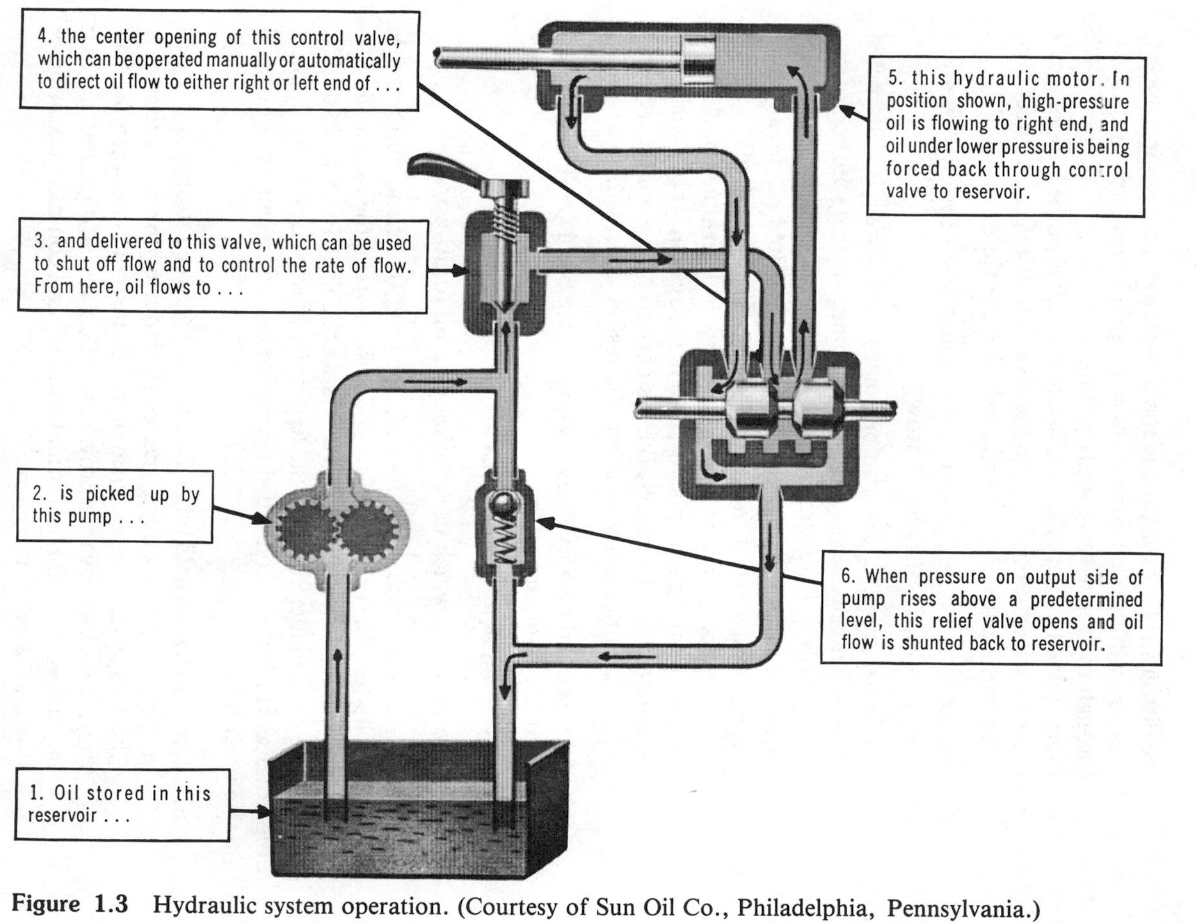

Figure 1.3 Hydraulic system operation. (Courtesy of Sun Oil Co., Philadelphia, Pennsylvania.)

the piston stroke, even though the directional valve is not centered, pressure will rise and pump flow will again be diverted to the tank through the relief valve at its setting.

3. Fluids take the path of least resistance. Fluid is pushed out of a cylinder because the pipe open to the reservoir through the directional valve is the path of least resistance for the fluid.
4. Pumps do not create pressure, they cause fluids to flow. Pressure is the result of resistance to flow.
5. The maximum load that can be placed on a system is determined by the relief valve setting. When the cylinder reaches the end of travel, the flow of fluid is restricted and pressure builds up to the relief setting. It cannot go any higher because the fluid can escape over the relief valve when the relief valve setting is reached.
6. During operation of the system, pressure is determined by the load imposed on the piston plus friction in the system. If external loads were placed on the piston, the system pressure would increase to the pressure necessary to move the piston and load.

Next, let us build up a hydraulic system, piece by piece. The basic hydraulic system has two parts:

1. The *pump,* which moves the oil.
2. The *cylinder,* which uses the moving oil to do work.

In a hand pump, when you apply force to the lever, the hand pump forces oil into the cylinder. The pressure of this oil pushes up on the piston and lifts the weight. In effect, the pump converts a mechanical force to hydraulic power, while the cylinder converts the hydraulic power back to mechanical force to do work. But for continued operation of the system, we must add some new features:

3. *Check valves* to hold oil in the cylinders between strokes and to prevent oil from returning to the reservoir during the pressure stroke. Check-type valves open when oil is flowing but close when the flow stops. Check-type valves provide free flow in one direction (neglecting bias springs and/or normal friction) and complete cessation of flow in the other direction.
4. A *reservoir* to store the oil. If you keep on stroking the pump to raise the weight, a supply of extra oil is needed. The reservoir has an air vent which allows oil to be forced into the pump by gravity and/or atmospheric pressure when the piston is retracted.

Notice that the pump is usually smaller than the cylinder. This means that each stroke of the pump would move only enough oil to move the piston a small amount. However, the load lifted by the cylinder is much greater than the force applied to the pump piston. If you want to raise the weight faster, you must work the pump faster, increasing the volume of oil to the cylinder.

Let us now complete the circuit and add some new features. We will first add a gear-type pump. This is one of many types of pumps that transform the rotary force of a motor or engine to hydraulic energy.

5. The *directional control valve* directs the oil. This allows the operator to control the constant supply of oil from the pump to and from the hydraulic cylinder. When the directional control valve (open center, tandem type) is in the neutral position, the flow of oil from the pump goes directly through the valve to a line which carries the oil back to the reservoir. The *tandem-type* designation indicates a center spool condition which directs pressure to tank and blocks both cylinder ports. Thus, the valve has trapped oil on both sides of the hydraulic cylinder in the neutral position of the flow directing member preventing hydraulic cylinder movement in either direction.

   When the control valve spool is moved, the pump oil is directed at the option of the signal source to the cavity on one end of the cylinder piston; if this is the blind end it could be raising a weight. At the same time, the line at the other end of the cylinder is connected to the return passage, thus allowing the oil forced from the one side of the piston to be returned to the reservoir. When the valve control spool is moved in the opposite direction, oil is directed to the end of the piston, causing a reversal of the load, and in the case of a vertical load, the load would provide additional energy to push the oil through the line and valve to return the piston assembly to the rest position, at a rate determined by the flow passage through the directional valve.

6. The *relief valve* protects the system from excessive pressures. If the pressure required to lift the load is abnormal, this valve opens and relieves the pressure by dumping some of the oil back to the reservoir to establish a maximum pressure as set on the relief valve adjustment. The relief valve may also be required when the piston reaches the end of the stroke. If the piston bottoms out, then the relief valve is essential. If the piston stops before bottoming out, the valving may provide the needed pressure relaxation through the center position of the directional valve.

7. A *flow control valve* can be used in a hydraulic system to regulate the volume of oil directed to or from the actuator. This valve can be a variable restrictor or orifice.

   Reducing the orifice size will lessen the volume of oil permitted to flow to or from the actuator. This will reduce actuator speed. Thus flow control valves are used to regulate the speed of an actuator.

This completes our basic hydraulic system. To summarize:

pump = generating force

cylinder = working force

valve = oil control
reservoir = oil storage

As you have seen in the simple hydraulic system that we have just developed, the purpose is to transmit power from a source (engine or motor) to the location where this power is required for work.

To look at the advantages and disadvantages of the hydraulic system, let us compare it to the other common methods of transferring this power. These would be mechanical (shafts, gears, or cables) or electrical.

Advantages of Hydraulic System

1. *Flexibility:* Unlike the mechanical method of power transmission, where the relative positions of the engine and work site must remain nearly constant with the flexibility of hydraulic lines, power can be moved to almost any location.
2. *Multiplication of force*: In the hydraulic system, very small forces can be used to move very large loads simply by providing suitable cylinder or hydraulic motor sizes.
3. *Simplicity*: The hydraulic system usually has fewer moving parts and fewer points of wear, and it lubricates itself.
4. *Compactness:* Compare the size of a small hydraulic motor with an electric motor of equal horsepower. Then imagine the size of the gears and shafts that would be required to create the forces that can be attained in a small hydraulic press. The hydraulic system can handle more horsepower for its size than either of the other systems.
5. *Economy*: This is usually the natural result of the simplicity and compactness, which provide relatively low cost for the power transmitted. Also, power and frictional losses are comparatively small.
6. *Safety:* There are fewer moving parts such as gears, chains, belt, and electrical contacts than in other systems. Overloads can be more easily controlled by using relief valves than is possible with the overload devices on the other systems.

Disadvantages of Hydraulic Systems

1. *Efficiency*: Although the efficiency of the hydraulic system is usually much better than that of the electrical system, it may be lower than for the mechanical transmission of power. An overall assessment of power transmission needs must be completed to determine true machine needs and best power transmission choice.
2. *Need for cleanliness*: Hydraulic systems can be damaged by rust, corrosion, dirt, heat, and breakdown of fluids. Cleanliness, proper design, and maintenance are usually as critical in the hydraulic system as in the other methods of power transmission.

1.3.1 Comparing Hydraulic Systems

Two major types of hydraulic systems are used today: open-center systems and closed-center systems. The simple hydraulic system that we developed earlier is what we call an *open-center system* (Fig. 1.4). This system requires that the control valve spool be open in the center to allow pump flow to pass through the valve and return to the reservoir. The pump we have used supplies a constant flow of oil and the oil must have a path for return when it is not required to operate a function. In the *closed-center system* shown in Fig. 1.5 the pump is capable of "taking a break" when oil is not required to operate a function. This is a variable-displacement-type pump. Flow is automatically reduced to a value just sufficient to maintain desired pressure level. Therefore, the control valve is closed in the center, which stops (dead-ends) the flow of oil from the pump—the "closed-center" feature. If the pump does not "take a break" or reduce pumping rate when the control valve is in the center position, the oil will recirculate to the tank through the relief valve. Passing flow across a relief valve creates a great deal of heat in the system because all the energy put into the oil flow is wasted.

To summarize, in the open-center system the pump runs constantly, with the valve opened in the center to allow oil to return to the reservoir. In the closed-center system the valve spool is closed in the center to dead-end the pump oil supply in neutral.

Let us look at a closed-center system with a variable-displacement pump. In neutral, the pump forces oil into the system until the pressure rises to a predetermined level. Then a pressure-regulating mechanism allows the pump to automatically reduce the volume of fluid delivered into the circuit to maintain pressure to the valve. When the control valve is operated, oil is diverted from the pump to the bottom of a cylinder or any other destination as required in the machine cycle.

The drop in pressure caused by connecting the pump pressure line to

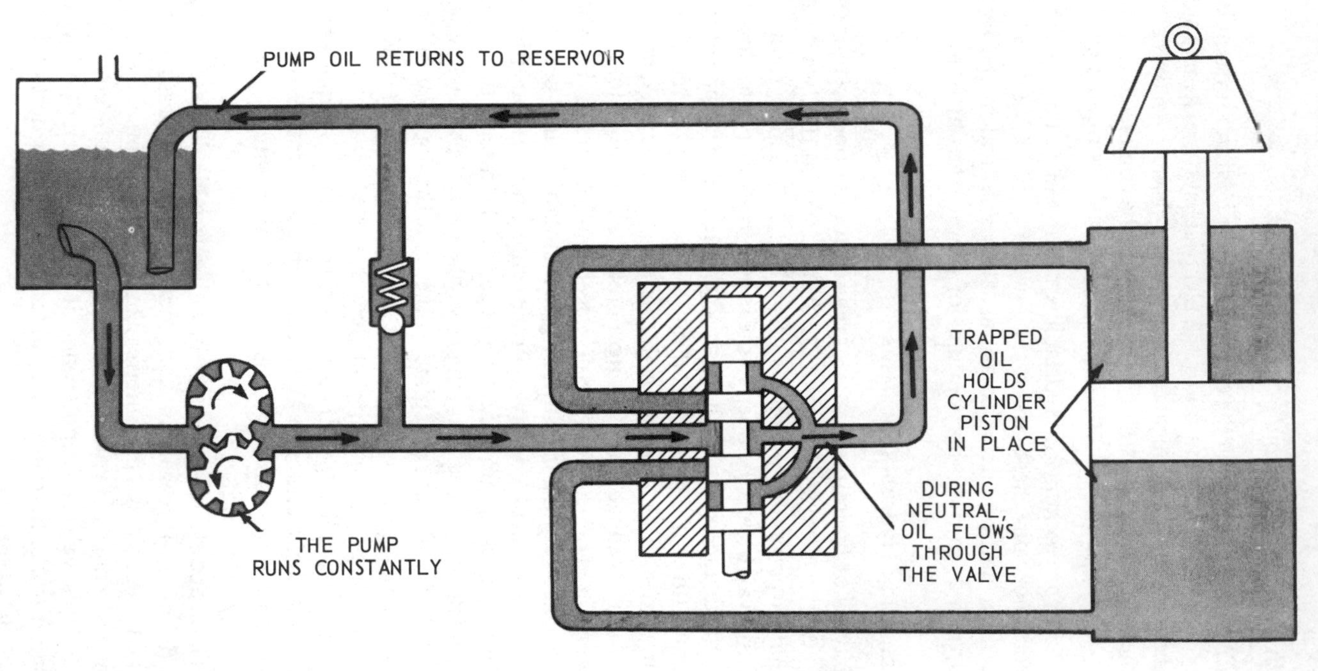

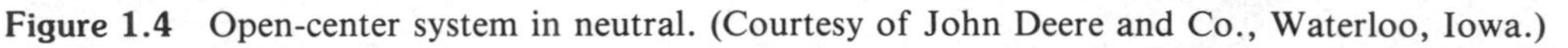

Figure 1.4 Open-center system in neutral. (Courtesy of John Deere and Co., Waterloo, Iowa.)

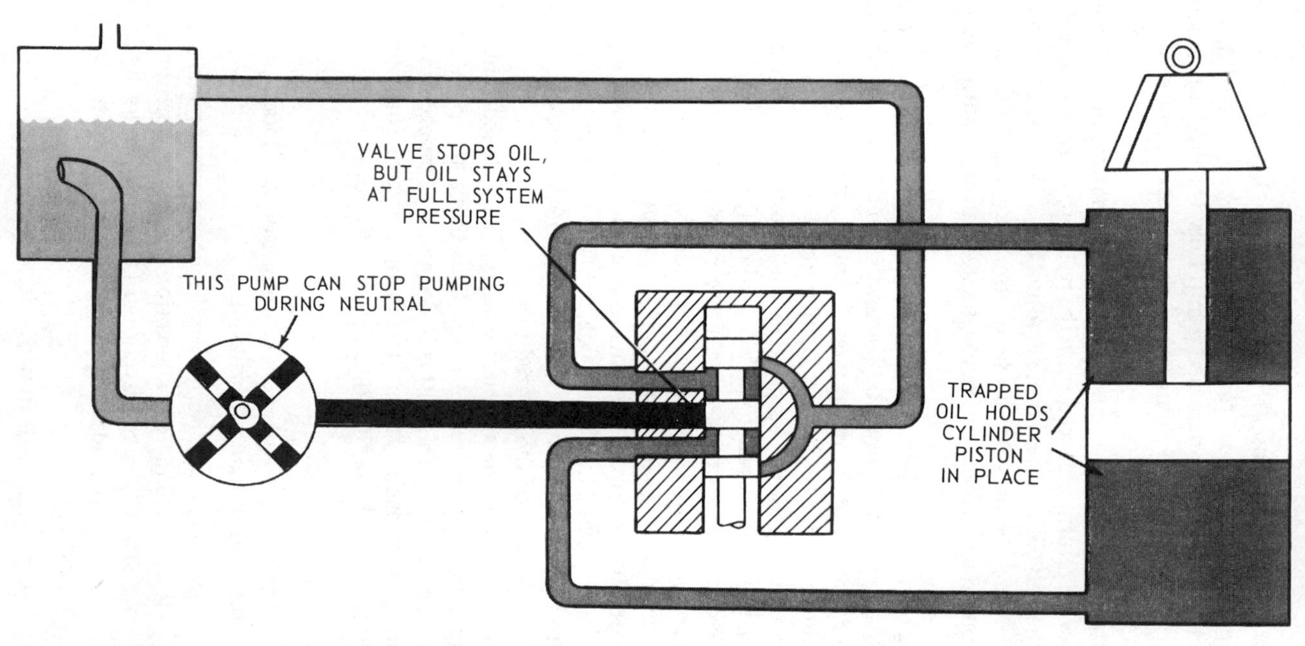

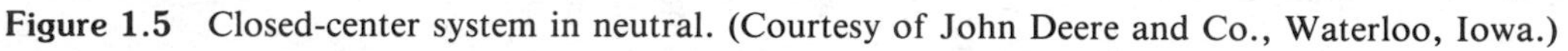

Figure 1.5 Closed-center system in neutral. (Courtesy of John Deere and Co., Waterloo, Iowa.)

the bottom of the cylinder causes the pump to go back to work, pumping oil to the bottom of the piston and raising the load. The mechanism that converts this pressure drop into increased pump flow is called a *pressure compensator* and is built into the pump assembly. Its operation is described in detail in Chapter 2.

When the valve is moved, the top of the piston is connected to a return line, thus allowing return oil forced from the piston to be returned to the reservoir. When the valve is returned to neutral, oil is again trapped on both sides of the cylinder and the pressure passage from the pump is dead-ended. At this time the pump again takes a break as it automatically reduces output flow to the valve inlet.

Moving the spool in the downward position directs oil to the top of the piston, moving the load downward. Then oil from the bottom of the piston is sent into the return line. With the closed-center system, if the load exceeds the predetermined standby pressure or if the piston reaches the end of its stroke, the pressure buildup simply tells the pump to take a break, thus eliminating the need for a relief valve to protect the system. A relief valve may be in the circuit as a safety valve in the event of a pump compensator malfunction or as a means to minimize shock conditions which are faster than the response time of the compensator. As an example, if an accident occurs which causes an abrupt stop of the moving machine member, a high pressure shock wave can be created in the fluid column. The compensator cannot react fast enough to cancel the adverse effects of the shock wave. The relief valve can, particularly if it is of the piloted type.

We have now built the simplest of open- and closed-center systems. However, most hydraulic systems require their pump to operate more than one function. Let us look at how this is done and compare the advantages and disadvantages of each system. To operate several functions at once, hydraulic systems may have the following connections:

1. Open-center systems
   a. Open-center with series connection (cylinder ports usually blocked in neutral position).
   b. Open-center with series-parallel connection (cylinder ports usually blocked in neutral position—pressure port is connected in parallel with each directional valve inlet and a separate series bypass is provided. When any spool is shifted it blocks the bypass).
   c. Open-center with flow divider.
2. Closed-center systems

a. Closed-center with fixed-displacement pump and accumulator.

b. Closed-center with variable-displacement pump.

Next we discuss each of these variations of the basic closed- and open-center systems.

Open-Center System with Series Connection

In this system, oil from the pump is routed to the three control valves in series (Fig. 1.6). The return from the first valve is routed to the inlet of the second, and so on. In neutral, the oil passes through the valves in series and returns to the reservoir (valve cylinder ports are normally blocked in neutral). When a control valve is operated, incoming oil is diverted to the cylinder which that valve serves. Return oil from the cylinder is directed through the return line and on to the next valve.

This system is satisfactory as long as only one valve is operated at a time. In this case, the full output of the pump at full system pressure is available.

Figure 1.7 depicts the flow path that exists when raising a load with an open-center type control valve. When several such valves are connected, it

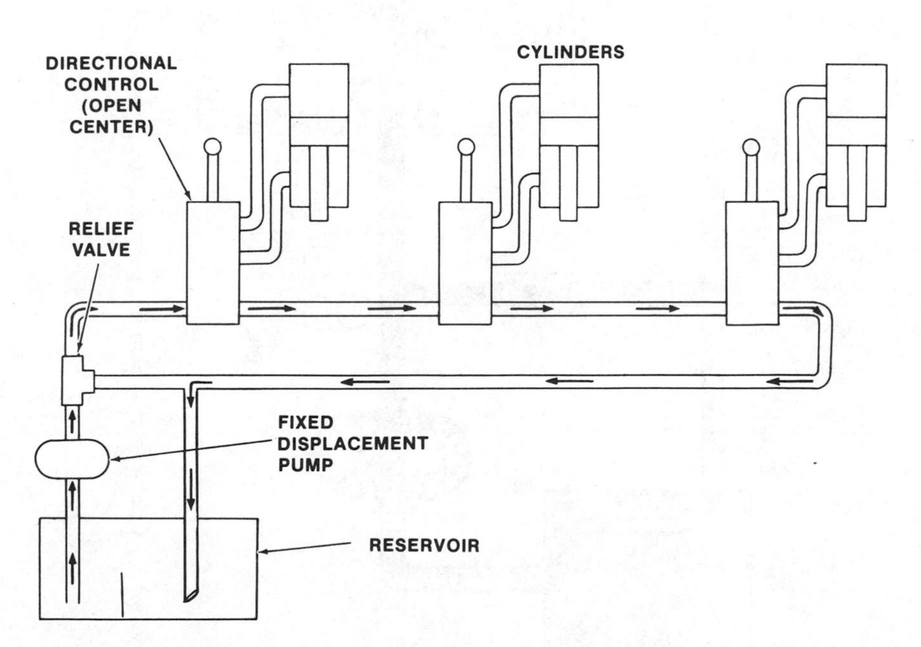

Figure 1.6 Open-center system with series connection. (Courtesy of John Deere and Co., Waterloo, Iowa.)

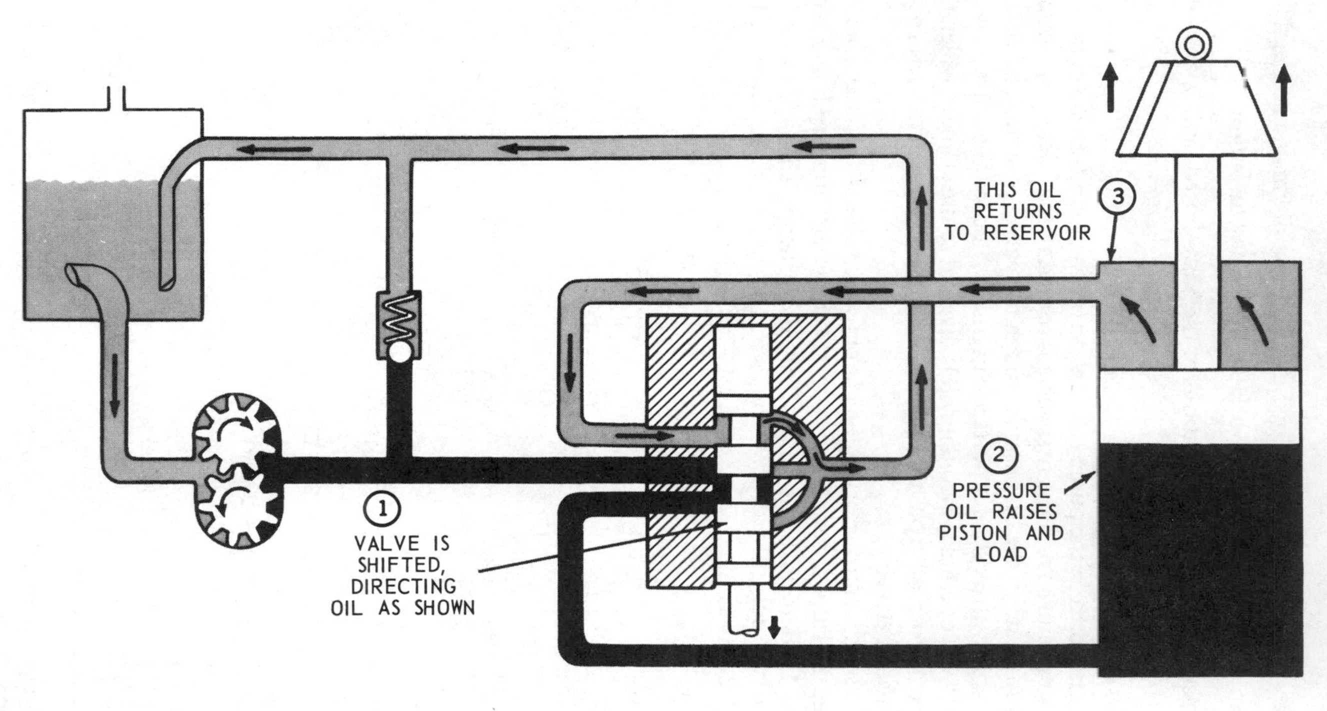

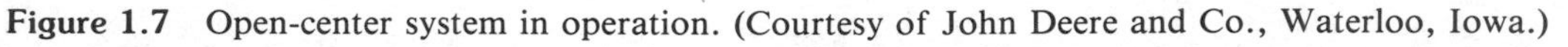

Figure 1.7 Open-center system in operation. (Courtesy of John Deere and Co., Waterloo, Iowa.)

can be seen that flow must pass through the actuator (cylinder) in order to reach other valves.

Open-Center System with Series-Parallel Connection

This system is a variation of the series-connected type (Fig. 1.8). Oil from the pump is routed through the control valves in series—but also in parallel. The valves are sometimes "stacked" to allow for the extra passages. In neutral, the oil passes through the valves in series. But when any valve is operated, the return is closed and oil is available to all valves through the parallel connection. When two or more valves are operated at once, the cylinder that needs least pressure will operate first, then the next least, and so on. However, this ability to satisfy two or more functions at once is an advantage over the series connection.

Open-Center System with Flow Divider

This system is comprised of a fixed displacement pump, relief valve, control valves, and actuators (Fig. 1.9). A flow divider is connected between the outlet line of the pump and the control valves. The flow divider splits the flow output from the pump to provide equal flow to each control valve. This device performs the flow-splitting function regardless of the pressure in each of its outlet ports. If a flow divider were not used, then, when both control valves were actuated simultaneously, the actuator with the least resistance to motion would receive the full pump output since this would represent the path of least resistance for the fluid. The other actuator would receive fluid when the lower resistance one was satisfied—when it had stopped moving. The flow divider negates this action. It directs an equal amount of fluid to each actuator regardless of the resistance to motion that is encountered by each. Thus the actuators move in unison—they are synchronized within the capabilities of the valve and system.

Closed-Center System with Fixed-Displacement Pump and Accumulator

This system is comprised of a pump of small but constant volume which charges an accumulator (Fig. 1.10). When the accumulator is charged to full pressure, the unloading valve diverts the pump flow back to the reservoir. The check valve traps pressure oil in the working circuit. When a control valve is operated, the accumulator discharges its oil and actuates the cylinder. As pressure begins to drop, pump flow is directed by the unloading valve to the accumulator to recharge it.

This system, using a small-capacity pump, is effective when operating oil is needed only for a limited time. However, when the functions need a lot of oil for longer periods, the accumulator system cannot handle it unless

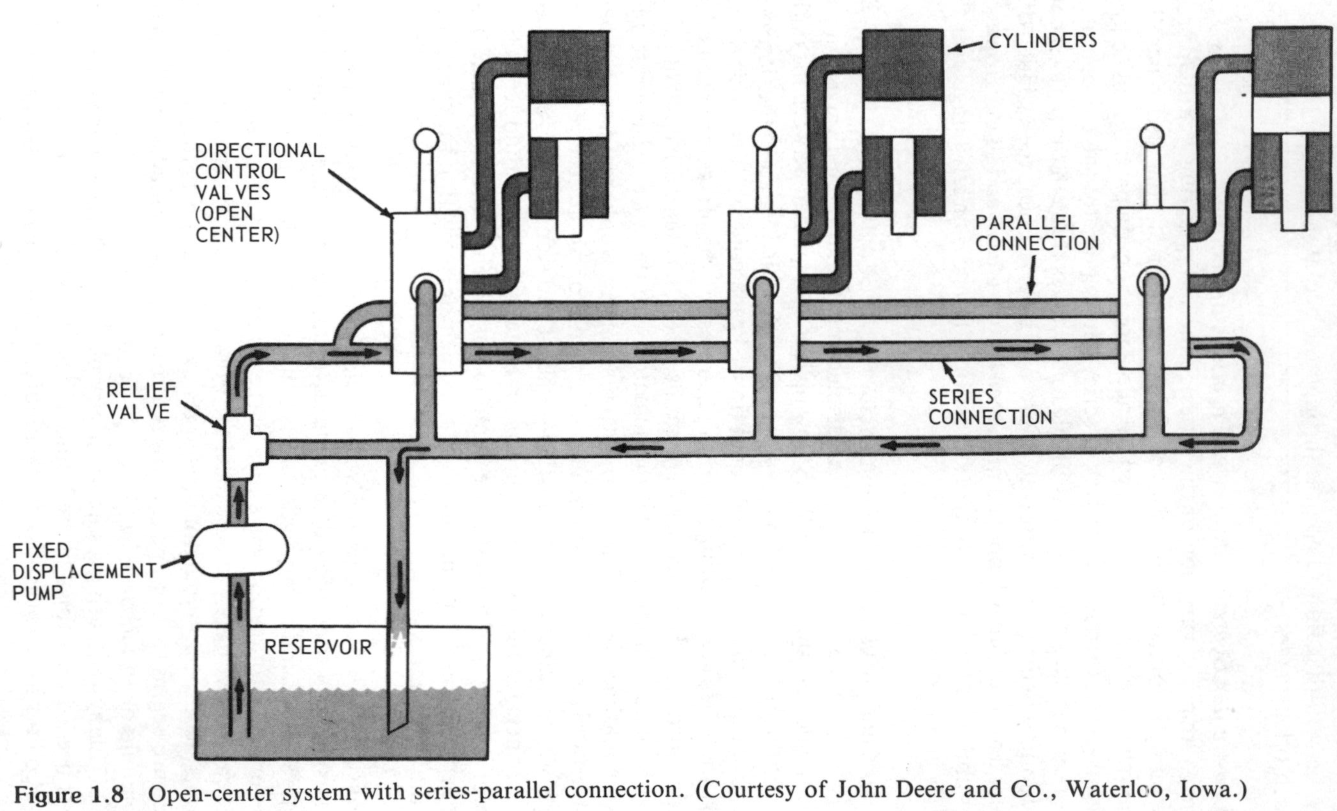

Figure 1.8 Open-center system with series-parallel connection. (Courtesy of John Deere and Co., Waterloo, Iowa.)

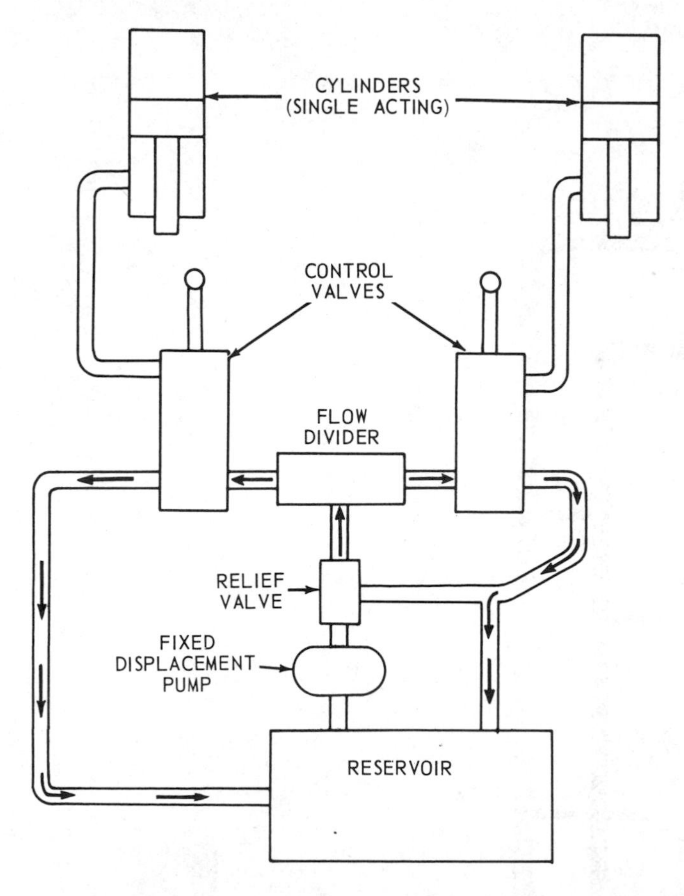

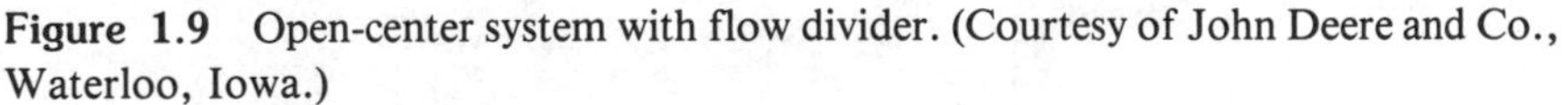
Figure 1.9 Open-center system with flow divider. (Courtesy of John Deere and Co., Waterloo, Iowa.)

the accumulator is very large, and then only for a time limited by the accumulator capacity.

Closed-Center System with Variable-Displacement Pump

This system is very much like the one discussed immediately above, but now we are adding a charging pump (Fig. 1.11). This pumps oil from the reservoir to the variable-displacement pump. The charging pump supplies only the makeup oil required in the system and provides some inlet pressure to make the variable-displacement pump more efficient. Return oil from the system functions is sent directly to the inlet of the variable-displacement pump, as shown. Design considerations must make provisions for cylinder areas created

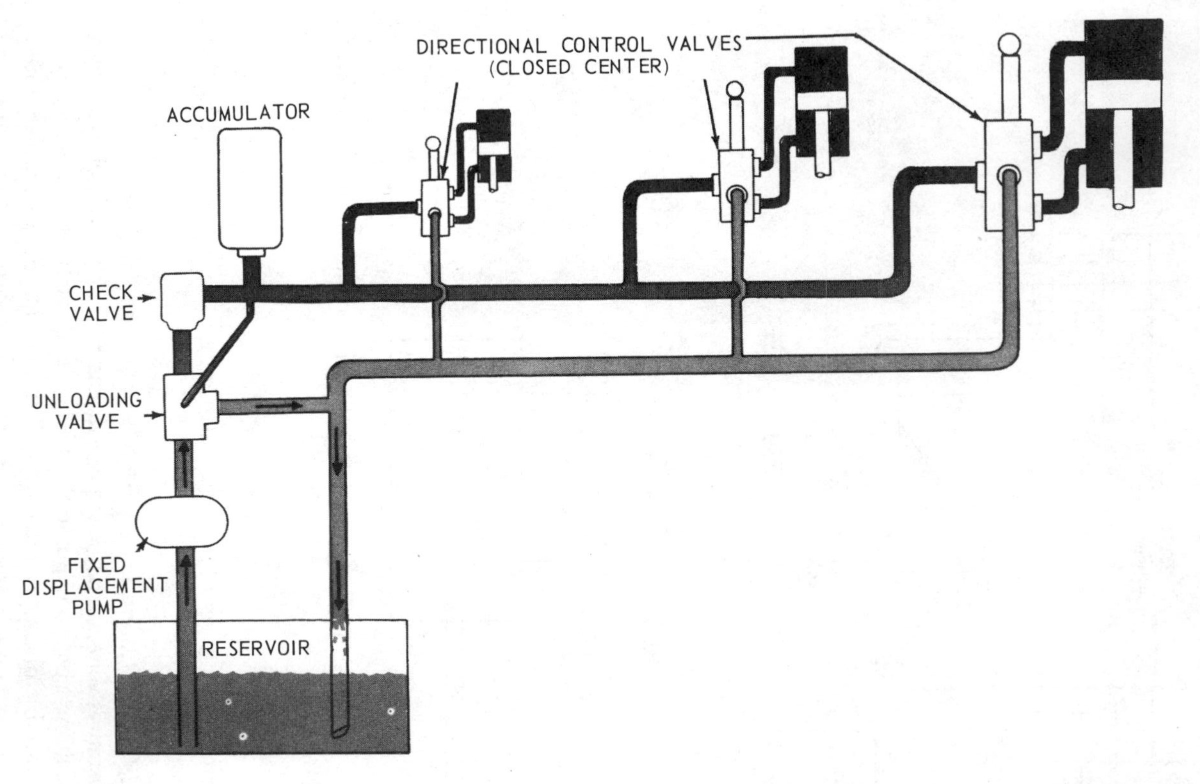

Figure 1.10 Closed-center system with fixed-displacement pump and accumulator. (Courtesy of John Deere and Co., Waterloo, Iowa.)

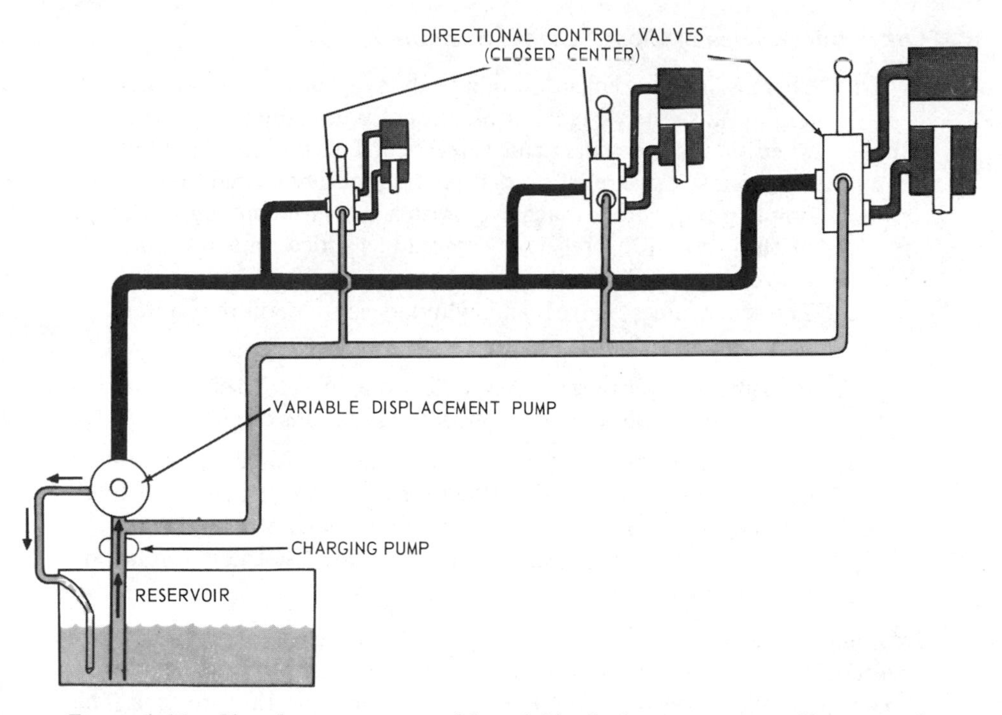

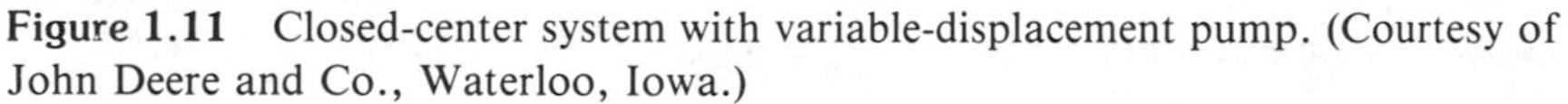
Figure 1.11 Closed-center system with variable-displacement pump. (Courtesy of John Deere and Co., Waterloo, Iowa.)

by rod displacement and make certain that capacities of the reservoir and charge pump are adequate to accept the extra fluid as the cylinders retract and the makeup fluid as the rods extend.

We saw earlier that the open-center system is the simplest and least expensive for hydraulic systems which have only a few functions. But as more functions are added with varying demands for each function, the open-center system requires the use of flow dividers to proportion the oil flow to these functions. The use of these flow dividers in an open-center system reduces efficiency, with resulting heat buildup.

Today's machines need more hydraulic power and the trend has been to the closed-center system because the machines tend to have mulitiple functions. In most cases, each of these functions has a requirement for different quantities of oil. With the closed-center system, the quantity of oil to each function can be controlled by line size, valve size, or by orificing with less heat build-up when compared to the flow dividers necessary in a comparable open-center system.

Other Advantages of Closed-Center Systems

1. There is no requirement for relief valves in a basic closed-center system because the variable displacement pump simply reduces flow when standby pressure value is reached. This prevents heat buildup in systems where relief pressure is frequently reached. The reduction in pump output is achieved with a pressure-compensating type of pump control. A relief valve should be added for safety purposes if pressure shock is anticipated.
2. The size of lines, valves, and cylinders can be tailored to the flow requirements of each function.
3. By using a larger pump, reserve flow is available to ensure full hydraulic speed at low engine rpm. More functions can also be served.
4. On functions such as brakes, which require force but very little movement on a piston, the closed-system is very efficient. By holding the valve open, standby pressure is constantly applied to the brake piston with no loss of efficiency because the pump has returned to standby.

In a similar open-center system, the pump would operate at maximum system pressure, with the fluid that is not required by the system for motion passing through the relief valve back to the reservoir to maintain this pressure. This causes an unacceptable amount of heat to be generated in the system.

1.4 PNEUMATIC SYSTEM OPERATION

So far, we have reviewed the physical discoveries made to help us understand the behavior of air. Let us now review what a typical pneumatic system consists of (Fig. 1.12). First, an air compressor assembly driven by an electric motor or internal combustion engine is needed to supply a source of pressurized air. A *pressure control* is used to *limit* the maximum air pressure. A *filter* screens out the dirt and grit from the air to protect the sensitive control components. A pressure *regulator* provides a means for adjusting the pressure to any desired value. The *lubricator* adds a clean mist of oil into the airstream to reduce friction and wear. The *directional control* and *functional control* components provide a method for controlling the speed and movement of the *actuator*.

The air compressor assembly is made up of a number of components—not just a compressor. A typical system consists of a compressor, drive motor, heat exchanger, filter, safety valve, and storage tank. Piping is also part of the unit.

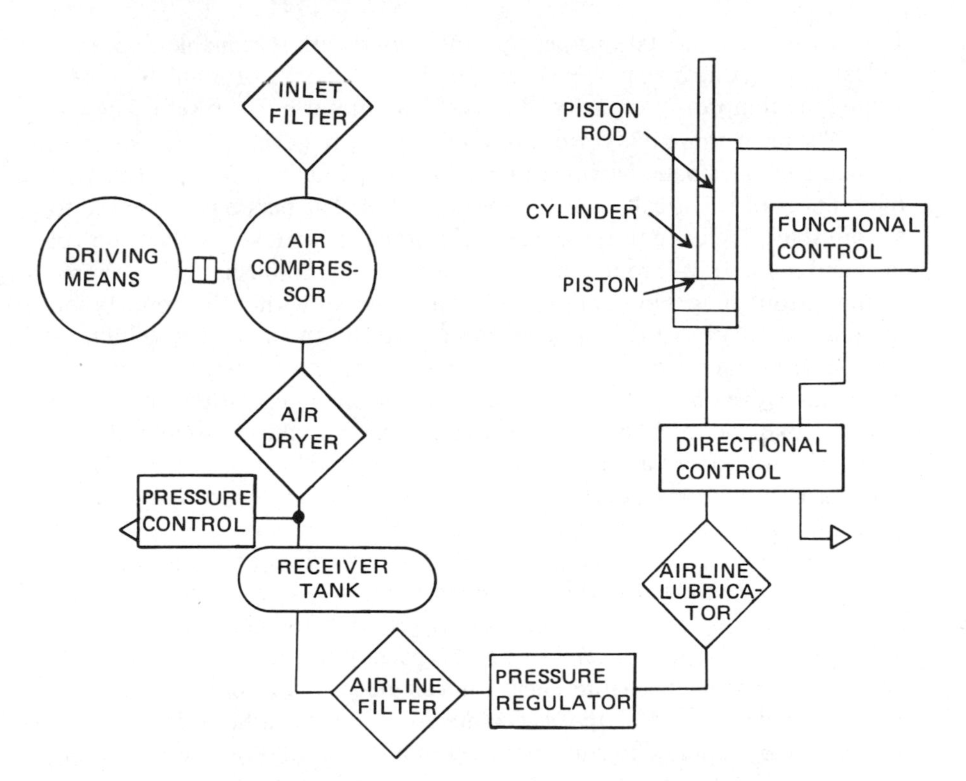

Figure 1.12 Typical pneumatic system.

1.4.1 Compressors

Compressors convert the mechanical energy transmitted by a prime mover (i.e., electric motor, internal combustion engine) into potential energy of compressed air. To perform any appreciable amount of work with a pneumatic system, a device is needed which can supply a receiver tank with a sufficient amount of air at a desired pressure. This device is a *positive-displacement* compressor.

The volumetric efficiency of a compressor is the ratio of the actual air delivered to the total displacement. For the compressors under consideration, this efficiency is from 4 to 4.2 scfm/hp. In other words, the electric motor-driven compressors will draw free air at the intake, pressurize it to a nominal 100 psig, and discharge it at the rate of approximately 4 scfm per input horsepower. A 50-hp compressor, for example, has a 200-scfm capacity at 100 psig.

The figure 4 scfm at 100 psig per hp can be used with reasonable assurance for calculating the drive power needed for a compressor or for quickly determining the compressor capacity when the drive motor horsepower is known.

The most common type of positive-displacement compressor found in an industrial pneumatic system is a *reciprocating piston* compressor. This type of compressor is basically a piston insidc a bore. The piston is connected to a crankshaft, which in turn is connected to the prime mover. Valves control inlet and outlet flow through the compressor. As the crankshaft of a reciprocating piston compressor rotates, the piston moves within the bore. When the crankshaft pulls the piston in one direction, an increasing volume is formed within the bore. With the resulting less-than-atmospheric pressure and the intake valve open, atmospheric air fills the chamber. At the end of the piston stroke, the chamber is filled with air and the intake valve closes. Starting its upward travel, the piston compresses the air. When air pressure in the bore reaches a high level, the outlet valve is opened. Compressed air discharges from the compressor into an air receiver tank.

The operation described above is that of a *single-stage* piston compressor, that is, a compressor which compresses air in a single stroke before it is discharged. Single-stage compressors are generally used in systems that require compressed air between 40 and 100 psig. When a gas is compressed, as in a compressor, it generates heat. In a single-stage compressor, when air is compressed above 80 psi, the heat created (heat due to air being compressed) and the energy required to compress the hot air become excessive. However, many industrial systems operate above 80 psi. For this reason, two-stage compressors are usually found in industrial pneumatic systems.

A *two-stage* piston compressor consists basically of a large and a small piston each in its own cylinder bore and each connected to the same crankshaft, associated inlet and outlet valves, and an intercooler. The piston with the large diameter performs the first stage of compression. The smaller piston compresses air in the second stage. As the crankshaft is turned by the prime mover, the large-diameter piston strokes downward. Air enters the chamber from the atmosphere through the open inlet valve. When the piston starts its upward movement, the inlet valve closes. Air is compressed (and heated) until a preset pressure is reached. The outlet valve then opens, discharging hot, compressed air. This air is directed by means of a finned tube, called an *intercooler,* to the smaller, second-stage piston. During the travel time from first to second stage, the air is cooled by means of air blowing over the tube or water flowing across the tube. By the time the air reaches the second-stage piston, a great portion of the heat of first-stage compression has been dissipated. The air is now cooler and ready to be compressed a second time. With compressed air at its inlet, the smaller-diameter piston is pulled downward. Compressed air fills the chamber and inlet valve closes. The piston is stroked upward, compressing the air further. The compressed air, as it discharges from the compressor, is at an elevated temperature. But this excess temperature above the ambient is not nearly as great as if one stage were used in the compressing process. Two-stage compressors do not use or waste as much energy in compressing air as do single-stage units.

Another type of positive-displacement compressors is the *vane*-type compressor. Vane compressors generate a pumping action by causing vanes to track along a circular housing. The compressing mechanism of a vane compressor basically consists of housing, rotor, and a series of floating vanes. The vanes are typically made of carbon, cloth impregnated with phenolic resin, or a similar rigid, low-friction material. For oil-free models, bronze and carbon vanes are often used.

The rotor of the vane compressor houses the vanes and the rotor is attached to a shaft which is connected to a driver motor. As the motor turns the rotor, the vanes are thrown out by centrifugal force and track along the housing. As the vanes make contact with the housing, a seal is formed between the vane tip and ring. The rotor is positioned off center with respect to the housing. As the rotor is turned, an increasing and decreasing volume is formed within the housing, thereby creating the pumping and compression action.

A more complex positive displacement compressor is the *helical* or *screw* type. A helical compressor generates compressed air by running two meshing rotors on one another like two screws. The compressing mechanism consists of two helical screws. Air is drawn in from one end and exits the other. There are two basic types: oil flooded and dry.

The *dry* helical compressor used the timing gears to turn the two screws. This maintains a constant clearance, which means that lubrication is usually not needed. These units have a high efficiency when run at high speed. In the *oil-flooded* type of compressor timing gears are not needed. However, lubrication is needed because the screws are run on each other. Efficiency is higher because the lubricating oil provides a good air seal between the two rotating members and the housing. However, oil separators are needed to remove the oil from the air downstream of the compressor.

Still another type of compressor is the *lobed-rotor* compressor. This constant displacement compressor is very much like the dry helical type. A timing mechanism is employed to eliminate problems with lobe and housing clearances. Because the rotors do not touch, a certain amount of "slip" exists. The slip increases as output pressure increases. This device should be operated at maximum speed for the highest efficiency.

Centrifugal compressors are not constant displacement types. Centrifugal types are best suited for moving large volumes of air at relatively low pressures. The compressing mechanism consists of an impeller, a diffuser section where velocity energy is converted to pressure, followed by a collector where air velocity is further reduced, again increasing pressure. With centrifugal compressors, air demand should never be allowed to drop much below the rated flow. If this is allowed to happen, the compressor will be unloaded and will surge. In this condition, the pressure at the outlet becomes unstable. If continuous operation in this condition occurs, bearings, blades, and even the housing may be damaged.

1.4.2 Heat Exchangers (Aftercoolers)

A common type of heat exchanger used to cool the compressor discharge air is the type using water as the cooling medium. These are water-to-air heat exchangers. Smaller, tank-mounted compressors use air-to-air heat exchangers almost exclusively, but air-to-air exchangers are available for large compressors also.

1.4.3 Storage Tanks

The size of the storage tank is usually commensurate with the compressor capacity. Tank capacities are usually specified in liquid gallons rather than cubic feet; if you wish to convert from gallons to cubic feet, multiply the gallons by 0.1337.

Storage tanks perform two basic functions in the air systems: (1) as a energy accumulator, and (2) to dampen severe pulsations. Storage tanks are constructed in accordance with strict safety codes and are rated to withstand safely the system pressure. As a further safeguard against a catastrophic explosion, these tanks are fitted with safety relief valves. To establish the size of the storage tank for a given compressor capacity, the following formula can be used:

$$\text{receiver size (ft}^3\text{)} = \frac{\text{compressor output (cfm} \times \text{psia)}}{\text{output pressure (psig} + \text{psia)}}$$

With a variable demand application, where the compressor is not able to meet the intermittent high flow demands, the formula is modified as follows:

$$V = \frac{t(Q_{out} - Q_{in}) \times P_a}{P_1 - P_2}$$

where

V = receiver tank size, ft^3
t = time to charge or fill tank to maximum pressure, min
Q_{out} = maximum flow rate out of tank, scfm
Q_{in} = maximum flow rate from compressor, scfm

P_a = atmospheric pressure, psia
P_1 = maximum tank pressure, psig
P_2 = minimum working pressure, psig

Always be sure that the receiver tank has a drain, preferably an automatic one, at its lowest point. If the receiver tank is mounted horizontally, it is a good practice to tilt it slightly to ensure good drainage.

1.5 COMPARISON OF AIR SYSTEMS TO ELECTRIC AND HYDRAULIC SYSTEMS

1.5.1 Air Versus Electric

Air-powered tools and equipment have four major advantages over electric-powered equipment.

1. *Flexibility:* Due to its ability to provide a source of power by means of portable compressors, air tools can be operated in remote areas where other power sources are unavailable. Air tools run cooler and have the advantage of variable speed and torque—no damage from overload or stalling.
2. *Light weight:* The air tool is lighter in weight and lends itself to a higher rate of production per hour with less worker fatigue.
3. *Safety*: Air equipment reduces the danger of fire and explosion hazard in such industries as painting and mining; there are no electrical connections and it is both spark- and shockproof.
4. *Low-cost operation and maintenance*: Because they have fewer moving parts, accessibility, and simple design, air-powered tools usually have good service records and can be inexpensive to maintain.

1.5.2 Air Versus Hydraulic

On many jobs it will be immediately apparent which medium, air or hydraulics, is more suitable. But some jobs could be done with either medium, or with vacuum, and the following points should be considered to help make the best decision.

1. *Power level:* Branch air circuits usually operate in the range ¼ to 1½ hp, while most hydraulic systems operate from 1½ hp up. This is a general rule, as there are higher-powered air systems and lower-powered hydraulic systems. But a 100-hp air compressor in a large plant is usually feeding numerous branch circuits which are

operating independently of one another, and a 100-hp hydraulic system is usually operating only one machine, although this machine may have several branch circuits dependent on one another.

2. *Noise level:* If the air exhaust noise is properly muffled, an air system usually operates much quieter than a hydraulic system of the same horsepower rating. In between cycles, an air system is completely silent, while in a hydraulic system the pump remains rotating in an unloaded condition. Where a hydraulic system must be used, and where noise must be reduced to a minimum, the pump should be operated at the minimum speed (rpm) that will deliver sufficient oil, and the pressure in the system should also be kept as low as possible.

3. *Cleanliness:* Normally, an air system is very clean provided that the air line oiler is not feeding an excessive amount of oil into the air, later to be blown out through the control valve exhaust. A hydraulic system that is carefully designed and constructed can be clean, but there may come a time when a cylinder rod packing may start to leak, or a line must be disconnected to replace a component, or a filter element must be replaced. At these times there is a possibility of oil being spilled around the machine.

4. *Speed:* Lightweight mechanisms can usually be operated faster with compressed air because a large volume of air can be drawn from the storage tank, for short periods, to give a very fast cylinder speed. A hydraulic system, to match the speed of an air system, might have to have a very large pump, large valving, and large plumbing, since the power is usually generated by the pump at the rate of use, with no reserve supply. Of course, accumulators can be used in a hydraulic system, but these make a low power system expensive.

5. *Operating cost:* A hydraulic system usually costs considerably less to operate than a compressed air system with the same mechanical power from the cylinder, because hydraulics is more efficient or can be. As air is compressed it is also heated. This heat of compression radiates from the walls of the compressor, the storage tank, and the plumbing system; or it is removed by an aftercooler. This is an escape of power from the fluid system which can never be recovered. We have heat losses, too, in a hydraulic system, but these are not as serious and can be minimized by good design, but the heat losses from air compression cannot be avoided or minimized.

6. *First cost:* If the cost of the compressor is considered, a compressed air system is more costly to build than a hydraulic system.

But some times a low power air circuit can be added to an existing compressor which has reserve capacity, and this can be done at less cost than a hydraulic system.

7. *Rigidity:* Some applications must have rigidity in the fluid stream, and require the use of hydraulics rather than compressed air. Examples are lifts and elevators to be stopped at an intermediate point in the cylinder stroke for adding or removing a part of their load; the feeding of a cutting tool at a slow speed; slow movement of a machine slide or other mechanism having a large area of sliding friction; press applications where two or more cylinders are attached to an unguided or inadequately guided platen, particularly where the load is not distributed evenly over all cylinders; any system where close control of cylinder speed or position must be maintained, or where the cylinder must be accurately stopped at a precise position in midstroke. These and similar applications, whether high or low power, can be accomplished satisfactorily only with hydraulics.

2

Hydraulic Systems: Operation, Maintenance, and Troubleshooting

2.1 HYDRAULIC PUMPS

Hydraulic pumps convert the mechanical energy transmitted by a prime mover (e.g., electric motor, internal combustion engine) into hydraulic working energy or hydraulic horsepower. The initial flow of fluid in a hydraulic system is from the reservoir to the pump. When a hydraulic pump is operated, its mechanical action creates a partial vacuum at the pump inlet that enables atmospheric pressure in the reservoir to force liquid through the inlet line into the pump. In addition, its mechanical action forces the fluid into the hydraulic system.

Basically, the function of the pump is to move the fluid. Pumps do not pump pressure. The pressure encountered in a hydraulic system is created by resistance to flow, the load resistance, and so on. Another form of resistance that creates pressure is the work load imposed on the system. Pumps rated as high-pressure pumps are so rated because of their ability to withstand high pressures or loads.

A pump can produce the flow necessary to the development of pressure, but a pump cannot provide resistance to its own flow. Resistance to flow is caused by a restriction or obstruction in the path of the flow. Without some form of resistance in a system there can be no pressure. It must be emphasized that a pump cannot, itself, produce pressure.

Pumping action is the same for every pump. All pumps generate an increasing volume at the suction side and a decreasing volume at the pressure side. However, the elements that perform the pumping action are not the same in all pumps. Based on their design, pumps may be classified into *positive-*

and *nonpositive-displacement* categories. In industrial hydraulics, the positive-displacement types are prevalent. Examples of nonpositive-displacement pumps are centrifugal types. In the positive-displacement category, three types of designs are in predominant use. They are: (1) gear type, (2) vane type, (3) piston type. The performance characteristics of these pumps are very adaptable to industrial applications, which accounts for their predominance. They are available in a variety of designs and construction features from many manufacturers. Another reason for their use is that they are economical to maintain, repairable, durable, and relatively low in cost.

2.1.1 Gear Pumps

There are two types of *gear pumps*: external gear and internal gear. The *external gear* type consists of two gears enclosed in a closely fitted housing. The gears rotate in opposite directions and mesh at a point in the housing between the outlet and inlet ports. This type of pump may use spur, helical, or herringbone gear tooth forms (see Figs. 2.1 and 2.2). As the gears rotate, a vacuum is formed as the teeth unmesh, which causes liquid to be forced in through the inlet port by atmospheric pressure on the surface of the oil in the reser-

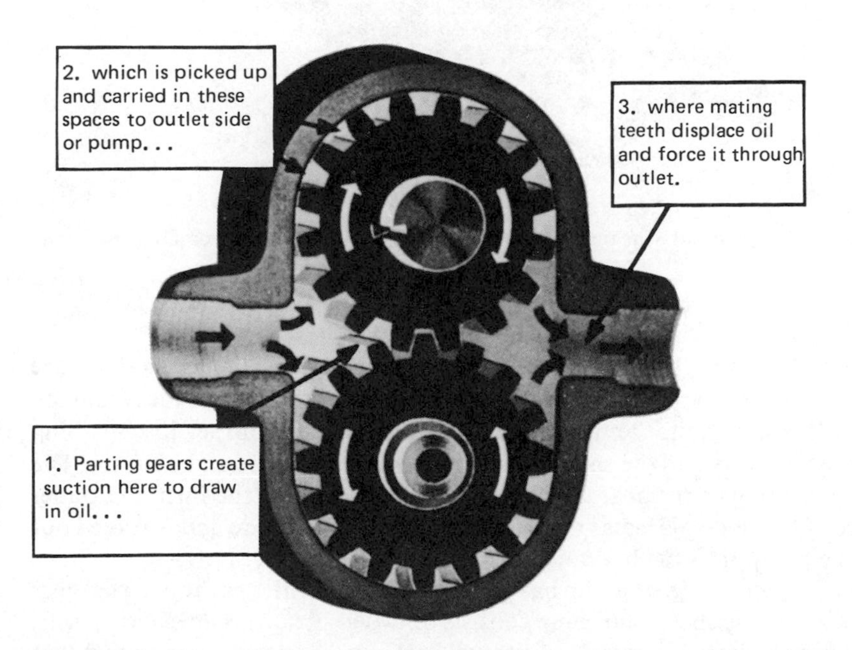

Figure 2.1 External gear pump operation. (Courtesy of Sun Oil Co., Philadelphia, Pennsylvania.)

Figure 2.2 External gear pump construction. (Courtesy of Hydreco Div., Kalamazoo, Michigan.)

voir with possible help from gravity. The liquid is then trapped in the space between the teeth of the two revolving gears and the housing. Fluid from the discharge side cannot return to the intake side because of the close meshing of the two gears and the small clearance between the gears and housing. The close meshing of the gear teeth provides a seal between the inlet and outlet ports. Fluid is displaced as the teeth mesh at the outlet side and is forced out of the pump into the hydraulic system.

The *internal gear* pump has the same characteristics as the external gear pump just described. The pump consists of two gears; one is an external gear, the other an internal gear. The internal gear gets its name from the fact that the gear teeth point in toward the axis instead of away from it. The external gear is mounted on a shaft and is positioned within the internal gear. As the pump operates, the inner gear drives the outer gear; each gear rotates in the same direction. The teeth of both gears engage at only one point in the pump. As the teeth separate, pockets are formed by the gear teeth; it is the change in volume of these pockets that creates a pumping action. Immediately after meshing of the teeth, the volume of the pockets increases, causing fluid to be drawn through the inlet port. Thus the fluid is trapped in the pockets. As the gears rotate, the volume of the pockets decreases and the fluid is forced out of the pump. The inlet side is in the area where the pockets increase and the discharge side is in the area where the pockets decrease (see Fig. 2.3).

A *gerotor pump* is an internal gear pump with an inner gear and an outer gear. The inner gear has one less tooth than the outer gear. If the inner gear is turned by a prime mover, it rotates the larger outer gear. On one side of the pumping mechanism, an increasing volume is formed as gear teeth unmesh. On the other half of the pump, a decreasing volume is formed. A gerotor pump has an unbalanced design similar to a spur gear pump.

The output volume of a gear pump is determined by the volume of fluid each gear displaces, and by the rpm rate at which the gears are turning. Consequently, the output volume of gear pumps can be altered by replacing the original gears with gears of different dimensions or by varying the rpm rate. Gear pumps, whether of the internal or external variety, do not lend themselves to a change in displacement while they are operating.

2.1.2 Vane Pumps

A *vane pump* makes use of centrifugal force and it makes use of variable-volume chambers. Centrifugal force moves the pump vanes into position to form the chambers. The expanding and diminishing of these chambers develops the flow of the fluid in and out of the pump. Vane pumps are used extensively in industry to provide hydraulic power. Wear does not appreciably decrease efficiency, because the vanes always maintain close contact with the cam ring within which they rotate.

The principal parts of the pump are a slotted rotor, vanes, a cam ring with an elliptical inner contour, end caps, and bearings. The rotor is slotted and is driven by a shaft. The slots in some designs may be cut at a slight angle to the outer circumference of the rotor. If so, they are slanted toward the direction of rotation. Current design has generally eliminated tilted slots. Each slot of the rotor serves the purpose of holding a flat, rectangular vane. The vanes are free to move radially in the slot. Specifically, they can move in and out from the axis of rotation. However, they are prevented from moving sideways by end caps that retain the pump assembly.

The vanes are flat rectangular pieces of material. Each vane has three

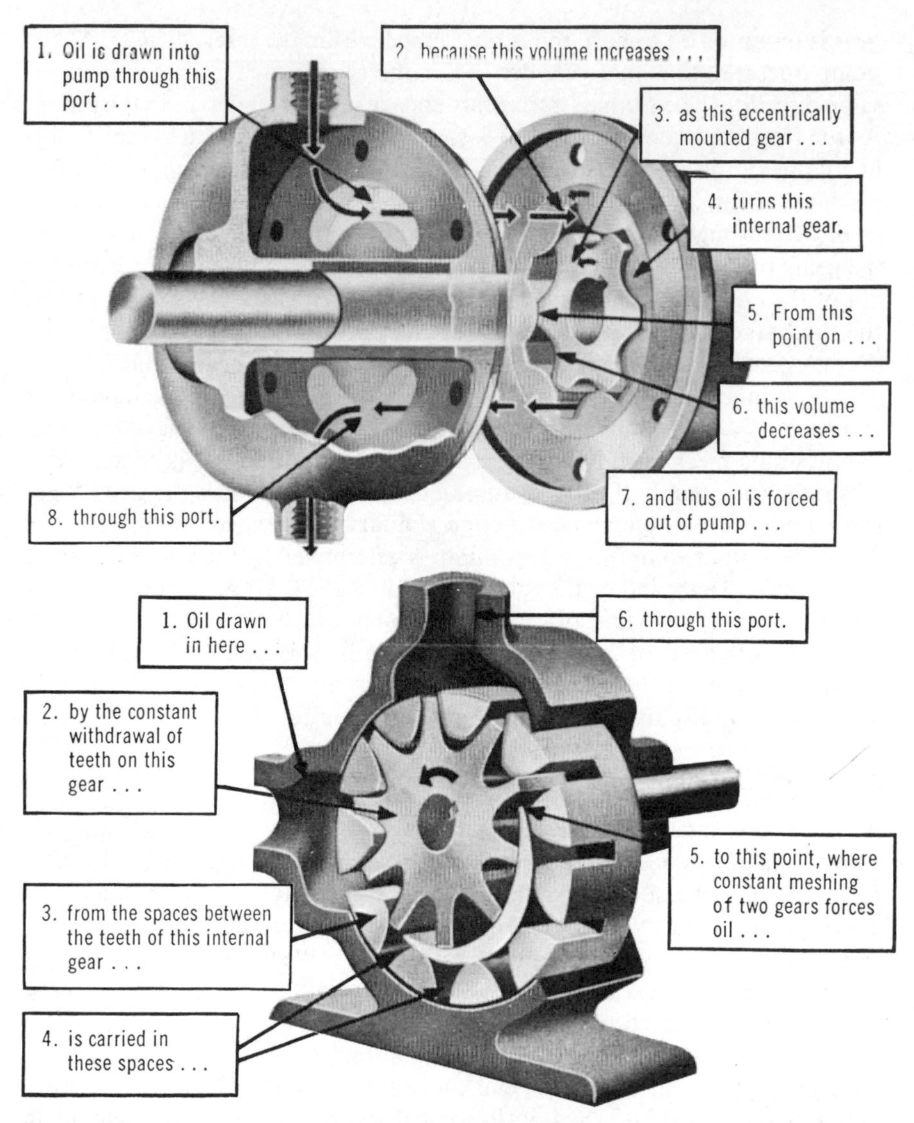

Figure 2.3 Internal gear pump operation. (Courtesy of Sun Oil Co., Philadelphia, Pennsylvania.)

squared edges and one beveled edge. The beveled edge rides along the elliptical-contoured surface of the cam ring casing, with the beveled edge trailing. As the rotor turns, centrifugal force ejects the vanes outward to contact and follow the elliptical cam ring. By this action, the vanes divide the area between the rotor and the ring into a series of chambers. These chambers vary in volume according to their respective positions about the rotor.

The ring that encloses the rotor and the vanes is circular in its outer contour, whereas the inner contour is oval or elliptical. The ring is anchored securely in the housing of the pump; it does not move. Since wear must be minimized, the cam-shaped ring is made of hardened and ground steel. It is the ring contour that makes it possible for each chamber to change in volume as it moves around the ring. Because of the ring contour, each chamber formed by the vanes changes in volume as it moves around the ring. Fluid is drawn into the pump as each chamber increases in volume, and as each chamber decreases in volume the fluid is ejected from the pump. Inlet and outlet ports are located at points to accommodate the intake and discharge areas of the pump (see Figs. 2.4 and 2.5).

Various arrangements of vane pumps are possible. They may be arranged in parallel or in series. When mounted in parallel or series the pumps are driven by a common shaft; each pump may be contained in its own housing. Each pump can have its own inlet and outlet ports, which may be combined by use of manifolds or piping to satisfy the flow requirements of a single circuit, or double pumps in parallel may be used to provide flow for two separate circuits. Many double pumps have a common intake and independent discharge ports.

The pumping mechanism of industrial vane pumps is often an integral

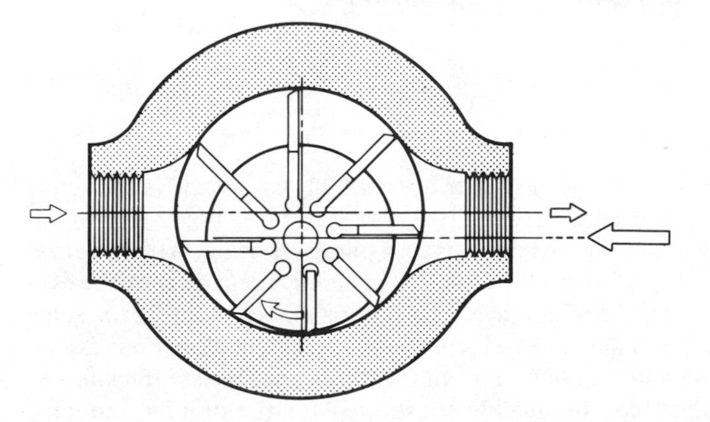

Figure 2.4 Unbalanced vane pump. (Courtesy of Sperry Vickers, Troy, Michigan.)

Figure 2.5 Balanced vane pump. (Courtesy of Sperry Vickers, Troy, Michigan.)

unit called a *cartridge assembly*. A cartridge assembly consists of vanes, rotor, and a cam ring sandwiched between two port plates (see Fig. 2.6). An advantage of using a cartridge assembly is easy pump servicing. After a period of time when pump parts naturally wear, the pumping mechanism can be easily removed and replaced with a new cartridge assembly. Also, if for some reason the pump's volume must be increased or decreased, a cartridge assembly with the same outside dimension, but with the appropriate thickness to change the volume, can be quickly substituted for the original pumping mechanism.

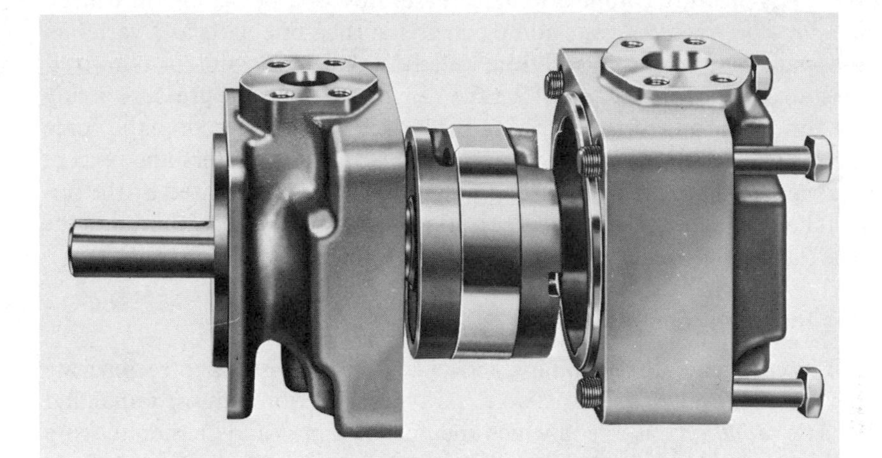

Figure 2.6 Cartridge vane pump. (Courtesy of Sperry Vickers, Troy, Michigan.)

Variable-Volume Vane Pumps

The amount of fluid that a vane pump displaces is determined by the difference between the maximum and the minimum distances the vanes are extended and the width of the vanes. While the pump is operating, nothing can be done to change the width of a vane. But a vane pump can be designed so that the distance the vanes are extended can be changed. This is known as a *variable-volume* vane pump. Variable-volume vane pumps are unbalanced pumps. Their rings are circular and not cam-shaped. However, they are still referred to as cam rings.

The pumping mechanism of a variable volume vane pump basically consists of a rotor, vanes, a cam ring which is free to move, a port plate, a thrust bearing to guide the cam ring, and something to vary the position of the cam ring. When the rotor is turned, an increasing and then a decreasing volume is generated and thus pumping occurs. With the screw adjustment turned out slightly, the cam ring is not as off-center to the rotor as before. Increasing and decreasing volume is still being generated, but not as much flow is being delivered by the pump. The exposed length of the vanes at full extension has decreased.

With the screw adjustment backed completely out, the cam ring naturally centers with the rotor. No increasing and decreasing volume is generated. No pumping occurs. With this arrangement a vane pump output flow can be changed anywhere from full flow to zero flow by means of the screw adjustment.

Generally, variable-volume vane pumps are pressure compensated. A pump that is pressure compensated decreases flow at a preset pressure level. A *pressure-compensated* vane pump consists of the same parts as a variable-volume vane pump. But, in addition, an adjustable spring and/or pilot control piston assembly is used to offset the cam ring. When the pressure acting on the inner contour of the cam ring is high enough to overcome the force of the spring and/or the pilot piston assembly, the ring centers and, except for leakage, pumping ceases. System pressure is therefore limited by the setting of the compensator spring and/or pilot assembly. In effect, this takes the place of a system's relief valve for pressure level control.

2.1.3 Piston Pumps

Piston pumps generate a pumping action by causing pistons to reciprocate within a piston bore. There are two main types of piston pumps: radial and axial. The *radial* type is one in which the pistons and cylinders radiate from the axis of a circular cylinder block like the spokes of a wheel. The *axial* type is one in which the pistons and cylinders are parallel to each other and parallel to the axis.

A piston moving back and forth in a cylinder can draw liquid in and then push it out. One piston in a pump may not be practical; however, a number of pistons can be put in the same cylinder block. The radial piston pump usually consists of a number of pistons arranged around a hub in a metal block, which is bored with accurately machined cylindrical compartments to accommodate the pistons. All cylinder holes are bored an equal distance apart and they connect with a hole that is bored in the center of the block (Fig. 2.7).

The pump consists of a pintle which remains stationary and is actually a valve, a cylinder block which revolves around the pintle and contains the cylinders in which the pistons operate, a rotor of hardened steel against which the piston heads press, and a slide block which houses and supports the rotor. The slide block does not revolve; but the rotor revolves due to the friction set up by the sliding action between the piston heads and the inner contour of the rotor. The slide block, which in effect is a casing or housing, contains the rotor, which is free to revolve.

The center point of the rotor is different from the center point of the cylinder block. It is that difference that produces the pumping action. If the rotor has the same center point as that of the cylinder block, there will be no pumping action since the piston does not move back and forth in the cylinder as it rotates with the block. Figure 2.8 details the construction of a radial piston pump.

Axial Piston Pumps—Bent-Axis Design

In bent-axis piston pumps the pistons and cylinders are parallel to each other and rotate as a group on the shaft (Fig. 2.9b). The piston barrel revolves as

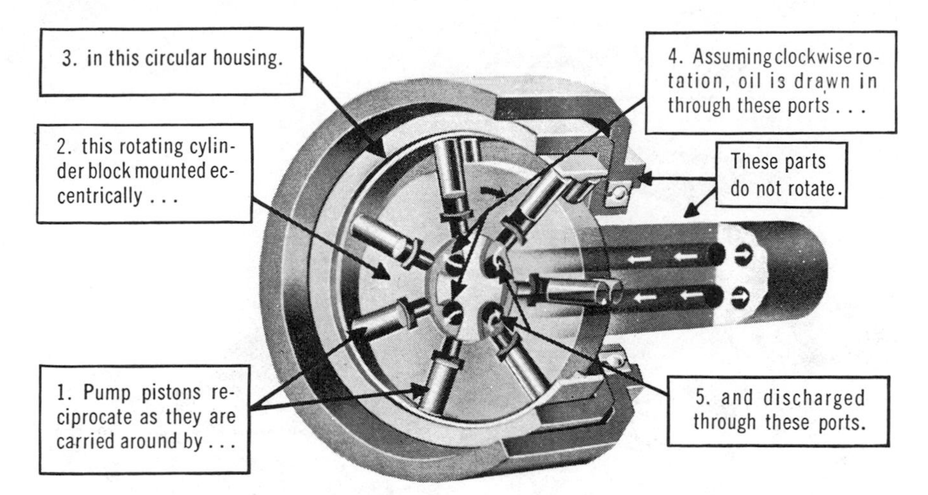

Figure 2.7 Radial piston pump operation. (Courtesy of Sun Oil Co., Philadelphia, Pennsylvania.)

the pistons are driven back and forth in their bore according to the angle of deflection of the barrel assembly (no flow is created when the cylinder barrel is parallel to the drive shaft). The principal parts of the pump are the housing, the control lens, and a rotating group. The assembly, called the *rotating group*, consists of a drive shaft, pistons, and a cylinder barrel. The drive shaft is supported by bearings and is free to rotate when driven by an external force (See Fig. 2.14, the bent-axis motor, for similar construction.)

The valve plate is used to direct fluid in and out of the cylinder barrel (Fig. 2.9c). There are two kidney-shaped passages cut in the plate, one in each half of the plate, through which liquid passes from the intake or to the discharge.

The block that contains the cylinders is called a cylinder barrel. Its face may rest squarely against the face of the valve plate. The high-pressure design shown in Fig. 2.9b–e uses a concave/convex valve plate that improves the pump performance and provides an excellent geometric pattern for most efficient pressure support and good piston thrust characteristics. The barrel is connected to the drive shaft by means of a universal link and knuckles.

Each piston is joined to the socket ring by means of a connecting rod. The rod has a ball end which is the ball part of a ball-and-socket joint (Fig. 2.9b), permitting the piston to move freely in its cylinder. Thus the base of each piston is connected to the drive-shaft flange assembly, which is secured to the drive assembly. The ball ends of the piston rods permit tilting of the pistons, piston rods, cylinder block, and so on. When the rotating group is

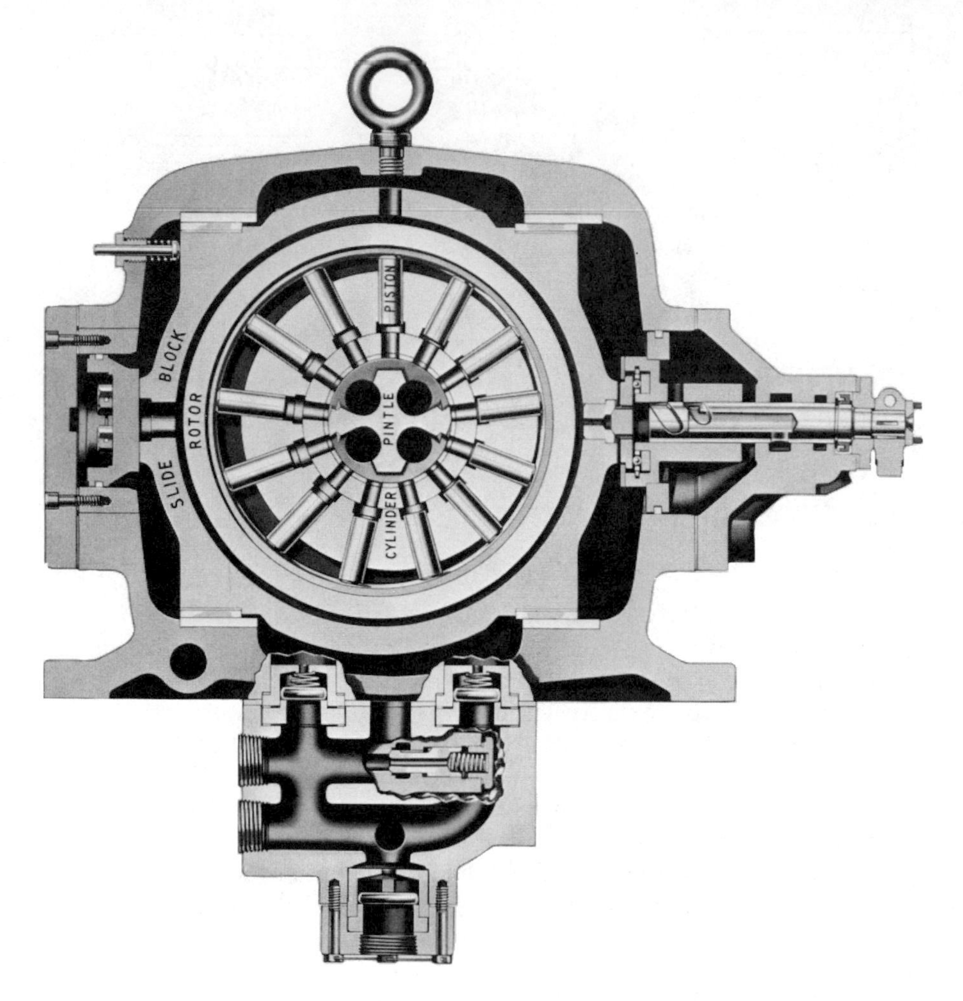

Figure 2.8 Radial piston pump. (Courtesy of Oilgear Co., Milwaukee, Wisconsin.)

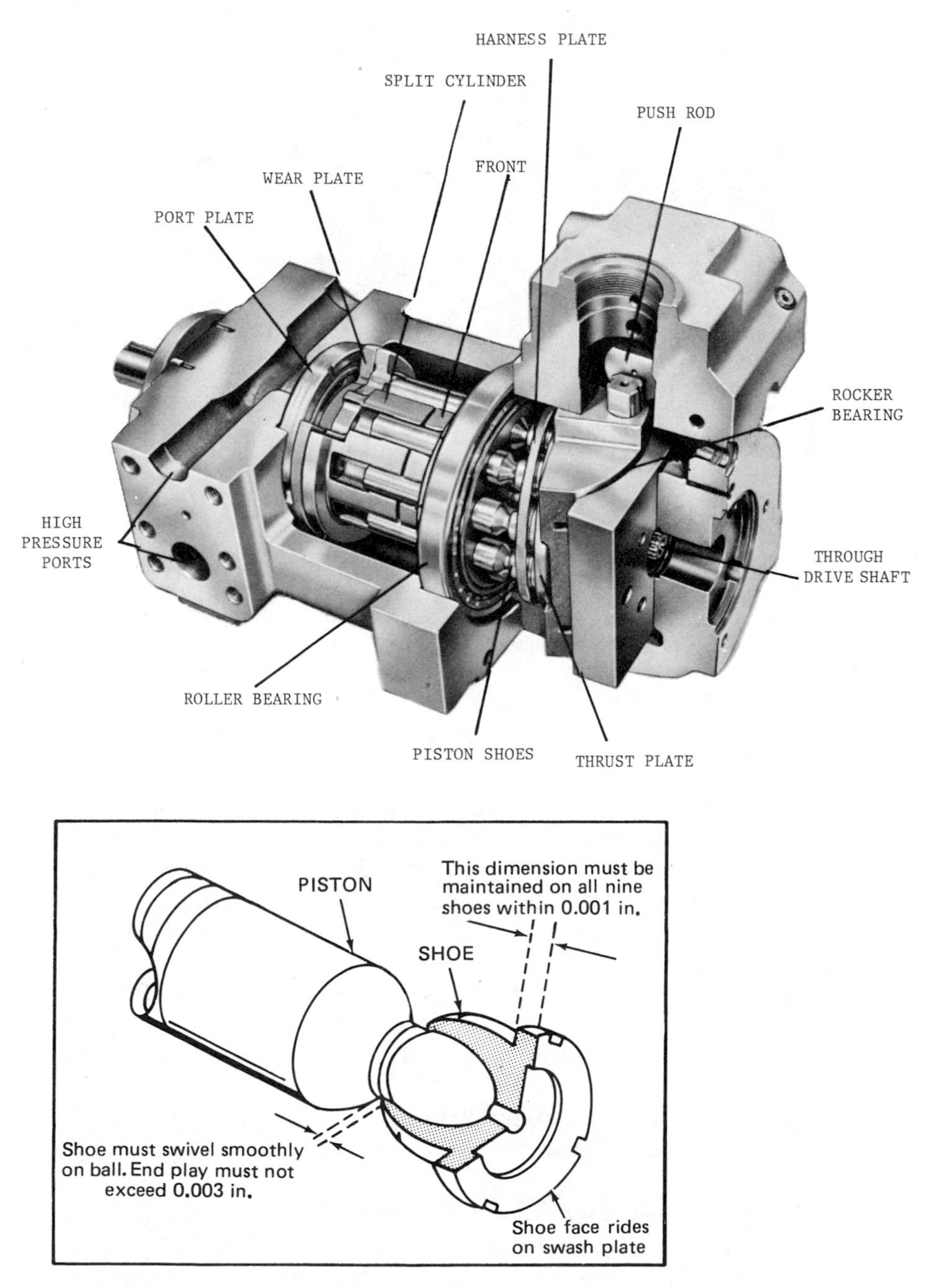

Figure 2.9a Variable-displacement axial piston pump. (Courtesy of Oilgear Co., Milwaukee, Wisconsin.)

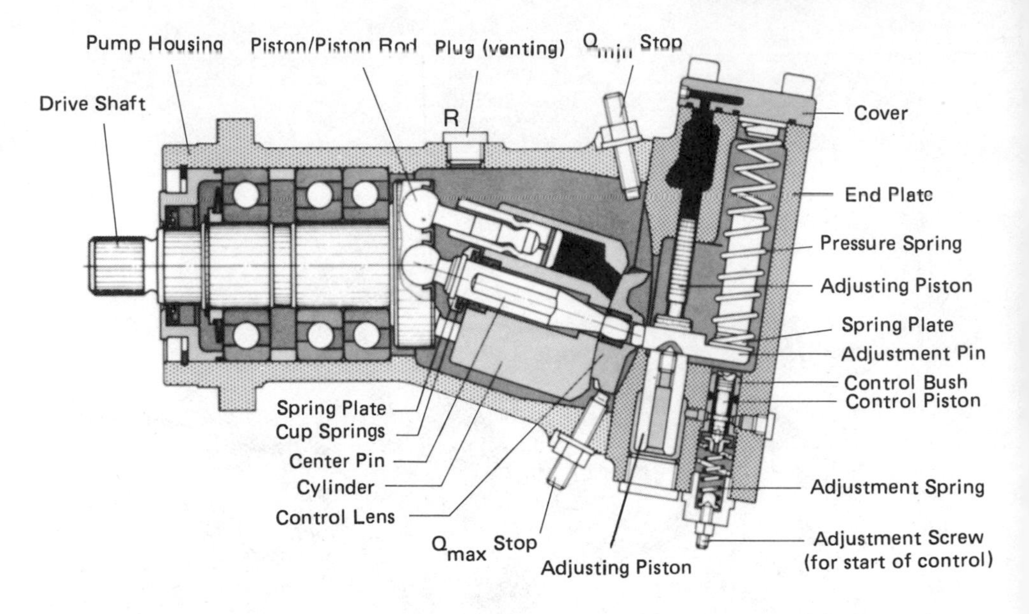

Figure 2.9b Bent axis pump with pressure compensation. (Courtesy of the Rexroth Corporation, Bethlehem, Pennsylvania.)

tilted and held at a fixed angle, the rotation of the flange induces an in-and-out movement of the pistons in their cylinder and creates a fixed flow per revolution. The pump of Fig. 2.9b is structured to react to a pressure signal to reduce flow from an established maximum to an established minimum or to a flow value to maintain pressure within the circuit. The pump of Fig. 2.9d and Fig. 2.9e is structured to permit flow in either direction through the pump by moving the rotating group either direction from the in-line no-flow position. Thus direction of fluid flow can be determined by the pump rotating group, which can control both direction and rate of flow passing through the pump assembly.

Many different control assemblies can be fitted to both the bent-axis type pump and the swash-plate type pump of Fig. 2.9a.

Axial Piston Pump—In-Line Swash Plate Design

Another type of axial piston pump that is widely used in industry is constructed differently from the bent-axis pump just described. This second style is

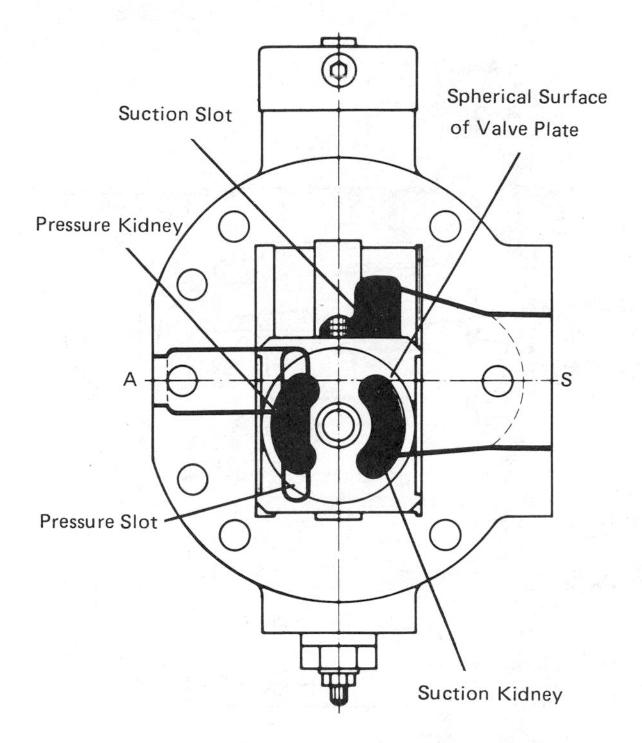

Figure 2.9c View of bent axis pump showing rear cover and valve plate as viewed from section at bottom of cylinder barrel. (Courtesy of the Rexroth Corporation, Bethlehem, Pennsylvania.)

the *swashplate* pump. The pumping mechanism consists basically of a cylinder barrel, piston with shoes, swashplate, shoeplate, shoeplate bias spring, and a port plate (Fig. 2.9a).

The cylinder barrel has several piston bores (usually five or seven), each fitted with one piston. The swashplate is positioned at an angle relative to the piston axis to create a reciprocating motion. If the swashplate is at a right angle to the drive shaft and cylinder barrel, there will be no displacement and consequently no fluid flow through the pump. The shoe attached to the piston rides on the surface of the swashplate. As the cylinder barrel is rotated, the piston shoe follows the surface of the swashplate. Since the swashplate can be positioned at a predetermined angle to provide the desired pump flow, this

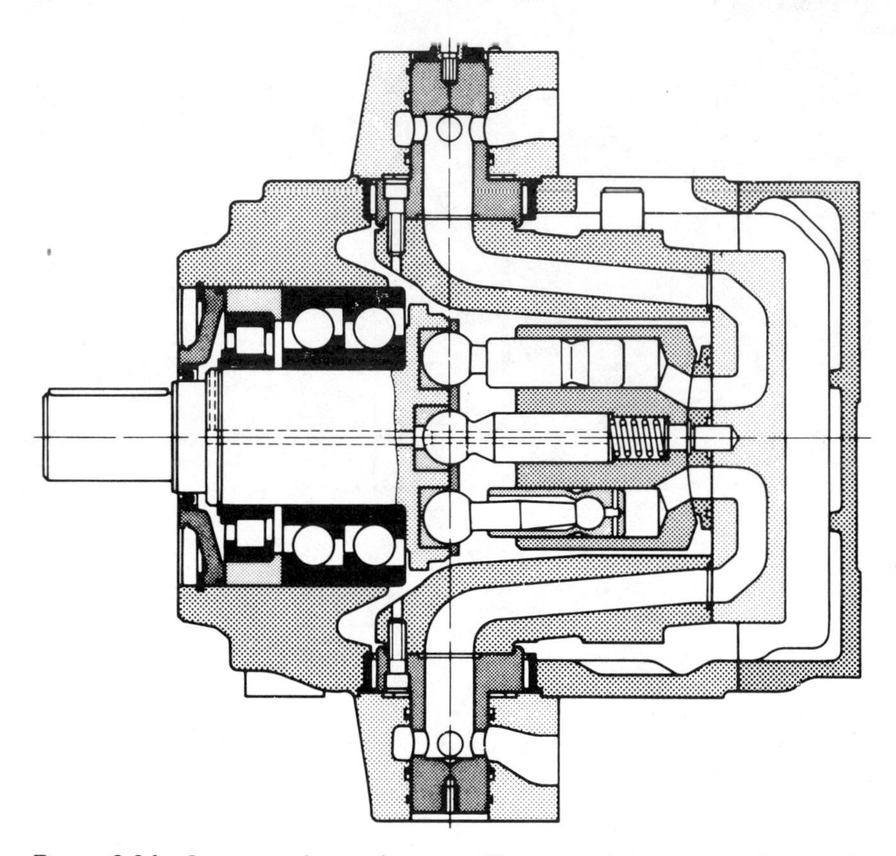

Figure 2.9d Overcenter bent axis pump. (Courtesy of the Rexroth Corporation, Bethlehem, Pennsylvania.)

results in the appropriate piston reciprocation in the bore. In one half of the circle of rotation, the piston moves out of the cylinder barrel and generates an increasing volume. In the other half of the circle of rotation, the piston moves into the cylinder barrel and generates a decreasing volume.

A shaft is attached to the cylinder barrel, which connects it with the prime mover. This shaft can be located at the end of the barrel where the porting is taking place. Or, more commonly, it can be positioned at the swashplate end. In this case, the swashplate and shoeplate have a hole in their centers to accept the shaft. If the shaft is positioned at the other end, the port plate has a shaft hole.

The construction of the swashplate-type pump allows for a through shaft which is widely used to drive a charge pump for a hydrostatic transmis-

Counterclockwise Swivel (Left)

Inlet Port A

Swivel Housing

Rotation

Outlet Port B

Clockwise Swivel (Right)

Figure 2.9e Top view of an overcenter bent-axis pump. (Courtesy of the Rexroth Corporation, Bethlehem, Pennsylvania.)

sion. The hydrostatic transmission usually has the charge pump as an integral part of the input variable delivery pump assembly. Our next subject matter will discuss this pump usage in more detail.

Variable-Volume Axial Piston Pump. The displacement of an axial piston pump, or any piston pump, is determined by the *stroke,* which is the

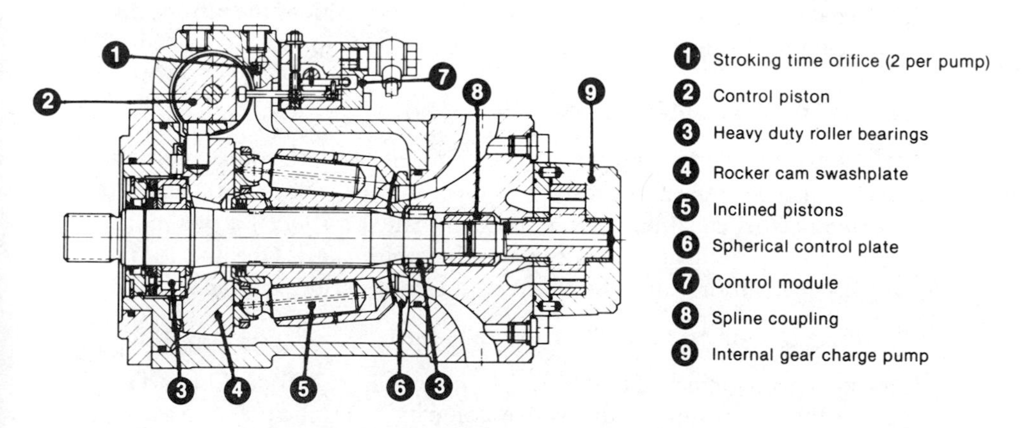

Figure 2.9f Swashplate type axial piston pump with charge pump. (Courtesy of the Rexroth Corporation, Bethlehem, Pennsylvania.)

distance the pistons are pulled in and pushed out of the cylinder barrel. Since the swashplate angle controls this distance in an axial piston pump, we need only to change the angle of the swashplate to alter the piston stroke and pump volume (see Fig. 2.9a). The cross-section view of Fig. 2.9f illustrates a swashplate-type pump with a concave/convex-type valve plate similar to the bent-axis pump. This pump also includes an integral charge pump driven from the input shaft. With a large swashplate angle, the pistons have a long stroke within the cylinder barrel. Conversely, with a small swashplate angle, the pistons have a short stroke within the cylinder barrel. Varying the angle of the swashplate changes the pump's displacement and thus the volume of fluid pumped. Several means of varying the swashplate angle are available from various manufacturers. These range from a simple hand-lever device to a sophisticated servo-valve control.

Some variable-displacement piston pumps have the capability of allowing the swashplate or the cylinder barrel in the bent-axis pump to be moved beyond the centered position. It should be noted that when these elements are centered, the pump will develop no flow. Crossing over the center position results in the increasing and decreasing volumes being generated at opposite ports. Thus flow through the pump reverses. *Over-center* axial piston pumps are used extensively in hydrostatic transmissions to control the direction of rotation as well as the speed of the hydraulic drive motors.

Axial piston pumps also have the ability to be made pressure compensated. The displacement mechanism (swashplate or bent-axis cylinder barrel) of the pump is connected to a piston which senses system pressure. When system pressure becomes higher than the spring setting of the compensator piston, the piston moves the displacement mechanism into the centered position—generally against a mechanical stop. In this position, the pistons do not reciprocate in the cylinder barrel. The result is no flow to the system.

2.2 HYDRAULIC MOTORS

Hydraulic motors convert hydraulic working energy into rotary mechanical energy, which is applied to a resisting object by means of a shaft. The rotary or turning effort capability of the motor is called *torque*. Torque indicates that a force is present at a distance from the motor shaft. One unit for measuring torque is lb-in., and the expression that describes torque is:

$$\text{torque (lb-in.)} = \text{force (lb)} \times \text{distance (in.)}$$

A resisting object attached to a motor shaft generates a torque. This is a resistance for the motor which must be overcome by hydraulic pressure acting on the motor's operating mechanism.

Hydraulic motors operate by causing an imbalance which results in the rotation of a shaft. This imbalance is generated in different ways, depending on the motor type. Hydraulic motors used in an industrial system can be divided into vane, gear, and piston types just as pumps were. Motors used in industrial hydraulic systems are nearly always designed to operate in both directions. Even motors that operate in a system in only one direction are usually reversible motors in design. To protect its shaft seal, vane, gear, and piston, reversible motors are generally externally drained.

2.2.1 Vane Motors

A *vane motor* develops an output torque at its shaft by allowing hydraulic pressure to act on vanes which are extended. The rotating group of a vane motor consists basically of vanes, rotor, ring, shaft, and a port plate with kidney-shaped inlet and outlet ports.

In a vane motor, the imbalance necessary to cause shaft rotation is the result of the difference in vane area exposed to hydraulic pressure. The rotor is positioned off-center in the motor housing or cam ring, which causes the area of the vanes exposed to pressure to increase toward the top and decrease at the bottom. When these unequal areas of the vanes are exposed to pressurized fluid entering through the inlet port, torque is developed at the motor shaft. The larger the exposed area of the vanes, or the higher the pressure, the more torque will be developed at the shaft. If the torque developed is large enough, the rotor and shaft will turn.

In a hydraulic motor, two different pressures are involved: system working pressure at the inlet and tank line pressure at the outlet. This results in side loading the shaft, which could be severe at high system pressures. To avoid shaft side loading, the inner contour of the cam ring can be changed from circular to cam-shaped. With this arrangement, two pressure quadrants oppose each other and the forces acting on the shaft are balanced. Shaft side loading is eliminated. This type of motor is called a *balanced design* and is the one generally used in industrial hydraulic systems (see Fig. 2.10 and Fig. 2.11).

The rotating group of industrial vane motors is usually an integral cartridge assembly. The cartridge assembly consists of vanes, rotor, and a cam ring sandwiched between two port plates or, in certain types, between the pressure plates and the body. An advantage of using a cartridge assembly is easy motor servicing. After a period of time when motor parts naturally wear, the rotating group can be easily removed and replaced with a new cartridge assembly. Also, if the same motor is required to develop more torque at the same system pressure, a cartridge assembly with the same outside dimensions, but with a larger exposed vane area may be available and can be quickly substituted for the original.

Before a vane motor will operate, its vanes must be extended. Unlike a vane pump, centrifugal force cannot be depended on to throw out the vanes

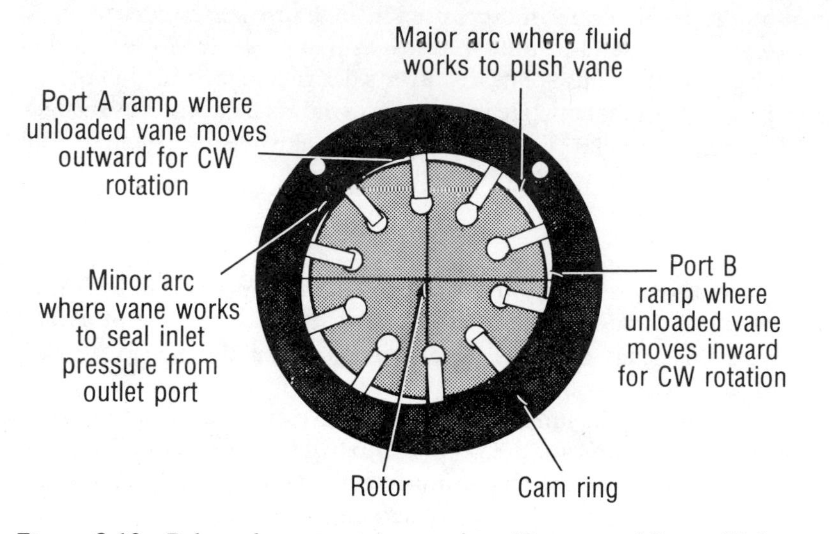

Figure 2.10 Balanced vane motor operation. (Courtesy of Sperry Vickers, Troy, Michigan.)

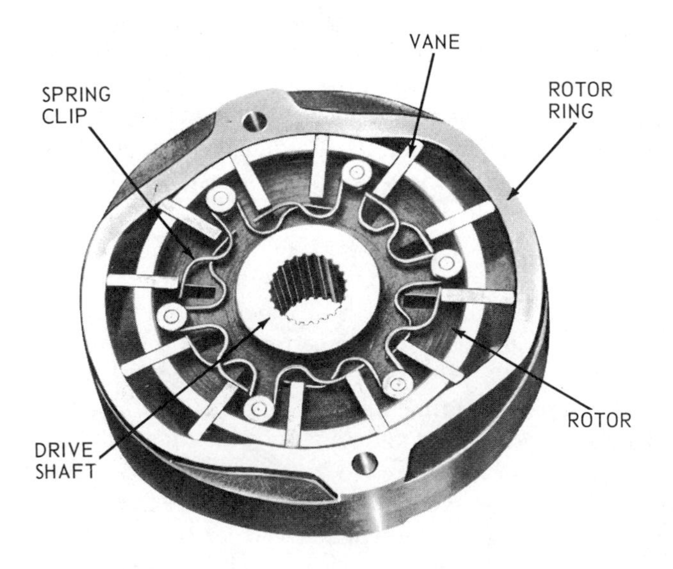

Figure 2.11 Balanced vane motor with spring-actuated vanes. (Courtesy of John Deere and Co., Waterloo, Iowa.)

and create a seal between the cam ring and vane tip. The vanes are spring loaded so that they are extended continuously.

2.2.2 Gear Motors

A *gear motor* develops an output torque at its shaft by allowing hydraulic pressure to act on gear teeth. A gear motor consists basically of a housing with inlet and outlet ports, and a rotating group made up of two gears. One gear is attached to a shaft that is connected to a load. The other gear is the driven gear.

In a gear motor, the imbalance necessary for motor operation is caused by gear teeth unmeshing. The inlet is subjected to system pressure and the outlet is at return line pressure. As the gear teeth unmesh, all teeth subjected to system pressure are hydraulically balanced except for one side of one tooth on one gear. This is the point where torque is developed. Consequently, the torque developed by a gear motor is a function of one side of one gear tooth. The larger the gear tooth or the higher the pressure, the more torque is produced.

An internal gear motor consists of one external gear which meshes with the teeth on the inside circumference of a larger gear. A popular type of internal gear motor in industrial systems is the *gerotor motor*. This motor is an internal gear motor with an inner drive gear and an outer driven gear which has one more tooth than the inner gear. The inner gear is attached to a shaft which is connected to a load. The imbalance in a gerotor motor is caused by the difference in gear area exposed to hydraulic pressure at the motor inlet.

Fluid pressure acting on these unequally exposed teeth results in a torque at the motor shaft. The larger the gear or the higher the pressure, the more torque will be developed at the shaft. Fluid entering the rotating group of a gerotor motor is separated from the fluid exiting the motor by means of a port plate with kidney-shaped inlet and outlet ports.

2.2.3 Piston Motors

A *piston motor* develops an output torque at its shaft by allowing hydraulic pressure to act on pistons. The rotating group of a piston motor consists basically of swashplate, cylinder barrel, pistons, shoeplate, shoeplate bias spring, port plate, and shaft. The pistons fit inside the cylinder barrel. The swashplate is positioned at an angle and acts as a surface on which the shoe side of the piston travels. The piston shoes are held in contact with the swashplate by the shoeplate and bias spring. A port plate separates incoming fluid from the discharge fluid. A shaft is connected to the cylinder barrel (Fig. 2.12).

To describe how a piston motor works, let us observe the operation of one piston in a cylinder barrel. With the swashplate positioned at an angle,

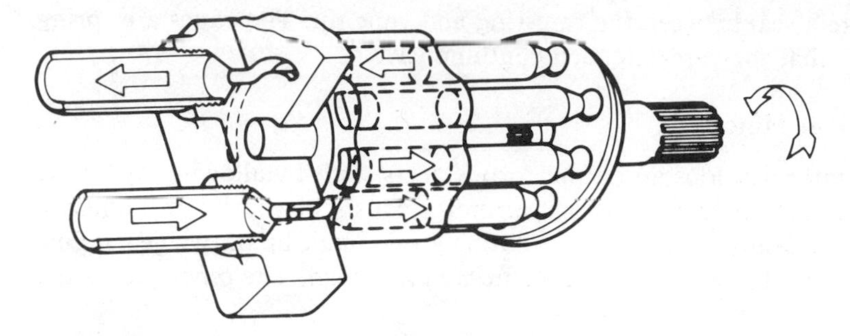

Figure 2.12 Axial piston motor operation.

the piston shoe does not have a very stable surface on which to position itself. When fluid pressure acts on the piston, a force is developed which pushes the piston out and causes the piston shoe to slide across the swashplate surface. As the piston shoe slides, it develops a torque at the shaft attached to the barrel. The amount of torque depends on the angle of slide caused by the swashplate and the pressure in the system. If the torque is large enough, the shaft will turn.

Torque continues to be developed by the piston as long as it is pushed out of the cylinder barrel by fluid pressure. Once the piston passes over the center of the circle, it is pushed back into the cylinder barrel by the swashplate. At this point, the piston bore will be open to the outlet port of the port plate. A single piston in a piston motor develops torque for only half of the full circle of rotation of the cylinder barrel and shaft. In actual practice, a cylinder barrel of a piston motor is fitted with many pistons. This allows the motor shaft to rotate continuously as well as obtain maximum torque.

The displacement of an axial piston motor, or any piston motor, is determined by the distance the pistons are reciprocated in the cylinder barrel. Since the swashplate angle controls this distance in an axial piston motor, changing the angle of the swashplate will alter the piston stroke and motor displacement. With a large swashplate angle, the pistons have a long stroke within the cylinder barrel; and, with a small swashplate angle, the pistons have a short stroke. By varying the angle of the swashplate, the motor's displacement can be changed. A fixed stop is usually provided to establish a minimum stroke to prevent overspeeding and/or mechanical lockup. Major changes of speed and direction are usually the function of the drive pump supplying fluid to the motor. Any change of speed related to the motor is generally associated with an overdrive function. Torque is then sacrificed for an increase in output speed.

Some axial piston motors have the piston and cylinder block assembly tilted or bent with respect to the drive shaft. The angle of the bore compared to the drive shaft determines the stroke or motor displacement. As the angle is changed, the stroke or displacement is varied. Thus the motor speed is variable by changing the tilt angle within a safe working range. Figure 2.13 and 2.14 provide some construction details for the bent-axis piston motor.

2.2.4 Rotary Motion (Fluid Motors and Rotary Actuators)

Torque

Torque is defined as rotary force. In rotary motion it is the exact equivalent of the thrust or force produced by a cylinder. The effect of torque is to produce a force or thrust at the end of its radius arm. But since the motor is prevented from rotating, it is producing nothing but torque. There is no work done or power produced because there is no movement involved.

Torque, unlike linear thrust, can only be measured or used because of the force produced at the end of a radius arm of (so many) pounds at a distance of (so many) inches from the shaft centerline. This is always true no matter whether the motor is driving a fan, a winch, a vehicle wheel, a conveyor, or any other device. The combination of the length of radius arm and the force produced at the end of it is called a *moment*.

T(torque) = F (force) × D (distance)

In a similar manner, if motor torque and length of radius arm are known, the formula above may be transposed to solve for force, F:

Force = torque ÷ distance

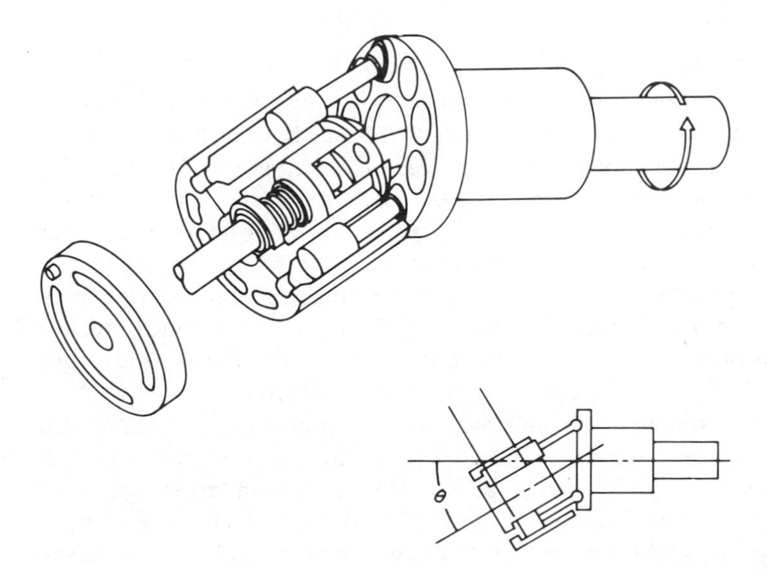

Figure 2.13 Bent-axis motor operation.

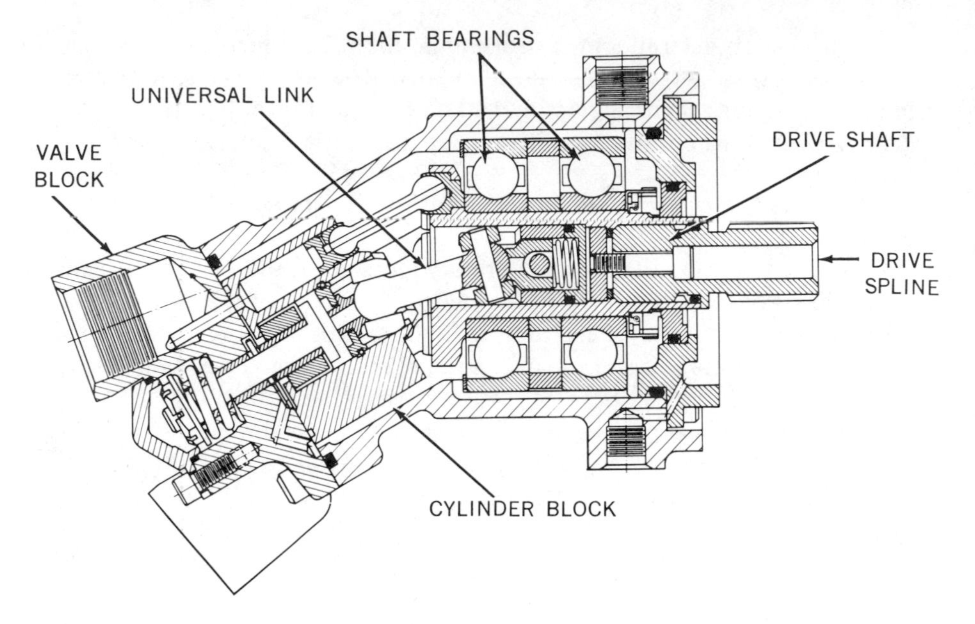

Figure 2.14 Cross section of bent-axis hydraulic motor. (Courtesy of Rexroth Corp.)

If the motor torque and linear force that it produces are known, the length of the radius arm can be solved by transposing the formula to read

distance = torque ÷ force

Fluid Motor Torque

The amount of torque that can be produced with any given motor does not depend directly on its speed, although its speed may have a slight effect due to an unrelated cause known as *porting loss and due to internal friction. The torque depends on only two factors:* (1) *the psi pressure level in the fluid, and* (2) *the number of cubic inches of fluid displaced by the working elements in one revolution.* Only the psi pressure level can be varied by the designer or the machine operator, as the working area or displacement has already been fixed in the motor design. On motors, then, the amount of torque produced is in direct proportion to the psi in the fluid. If we double the psi, the torque doubles, and so on. It is a directly proportional effect.

A quick estimate of the approximate horsepower that will be needed in a system can be made by figuring the power output in mechanical units using pounds of weight or force to be handled, distance in feet that the load has to be moved, then dividing by the length of time in which the work must be done. Then a suitable markup must be added to take care of system losses. If this is done before attempting to select a pump or cylinder size, it will give a good approximate starting point in sizing the system. This formula may be used to calculate the necessary system horsepower in mechanical units:

$$\text{hp} = \frac{\text{distance (feet) that the load is moved} \times \text{load force (pounds)}}{\text{time (minutes needed to move this load this distance)} \times 33{,}000}$$

Relationship Between Horsepower, Torque, and Speed

There is a basic relationship between horsepower, torque, and speed that holds true for all rotary motion of any kind, whether hydraulic motor, hydraulic pump, air motor, oscillating motor, electric motor, waterwheel, or what not. It is expressed by the formula

$$\text{hp} = \frac{\text{torque (ft-lb)} \times \text{speed (rpm)}}{5252}$$

This formula is, of course, valid only for units of feet, pounds, and minutes. If the data are expressed in other units, they will have to be converted.

A close look at this formula should reveal how it may be used in the selection and use of both air and hydraulic motors:

1. Remember that in a fluid motor it is the applied pressure in psi which determines the level of torque. *Example:* A certain motor connected to an oil supply of 1000 psi is developing 75 ft-lb of torque. If the supply pressure is reduced to 250 psi, the torque will reduce in the same proportion, to 18.75 ft-lb).
2. Also remember that the speed is proportional to the flow (cfm or gpm) supplied. Double the cfm or gpm and the speed doubles, and so on. *Example:* If a motor is operating at 900 rpm on a fluid supply of 5 gpm, it will speed up to 2700 rpm if the supply is increased to 15 gpm. This is always true if slippage is neglected.

Using the basic formula given above, the horsepower, the speed, or the torque can be found if the other two are known. We can transpose the formula into these three forms, any one of which may be used:

To find hp:	$\text{hp} = (\text{T} \times \text{rpm}) \div 5252$	T must be in foot-pounds
To find torque:	$\text{T} = (\text{hp} \times 5252) \div \text{rpm}$	T will be in foot-pounds
To find speed:	$\text{rpm} = (\text{hp} \times 5252) \div \text{T}$	T must be in foot-pounds

2.3 STARTUP, OPERATION, AND MAINTENANCE OF HYDRAULIC PUMPS AND MOTORS

To guarantee efficient operation of pumps or motors it is important to review carefully the operating instruction and installation hints supplied with the unit or the information given in the relevant catalog sheet. It is imperative that all internal parts be kept clean during all modes of operation, but particularly during installation. The fluid used in the system must be of a recommended type, and adequate filtration must be provided. The following general instructions apply for most types of hydraulic pumps. However, in some cases additional instructions may be needed. Maintenance and operation are dependent on actual operating conditions.

2.3.1 General Instructions

Installation

It is recommended that the unit be flushed out with the proper hydraulic fluid being used, to remove any anticorrosive agent that may be present. Installation should be carried out in accordance with the drawings and instructions supplied, and particular attention must be paid to the associated pipework system. Unnecessary stresses resulting from incorrectly fitted pipes or misaligned drive assemblies should be avoided. Pump or motor leakage lines must end below the oil level in the tank and be adequately sized to ensure that the back pressure at the housing does not exceed the maximum permissible value. The leakage line must be connected so that the housing always remains filled with oil, so that no siphoning effect is obtained in the tank.

It is essential that suction lines be designed and installed per the manufacturer's instructions. Caution must be exercised to assure that the maximum suction vacuum pressure, or preload pressure, does not exceed the limits specified by the manufacturer. Filters or valves fitted must be given equal consideration. All pipe connections must be sealed (airtight) to prevent entry of air into the pump which will cause noisy operation and possible pump damage.

Initial Startup

When aligning the pump or motor housing mechanically, the possibility of torsion due to uneven foundations or irregular mounting points must be eliminated. Similarly, the alignment of the input/output drive shaft with the connecting part must be checked carefully. Verify that all required voltage and current of electrical control and regulating devices are in accordance with the available supply.

Checking for correct direction of rotation is carried out by loosening the drive coupling either before starting the drive assembly or after filling the unit to be tested with hydraulic oil and quickly switching the drive (for pumps and motors) on and off. This is called *jogging* the unit. If a leakage or drain line is provided, the housing must be filled with hydraulic fluid to give internal lubrication.

In view of the many designs of pumps and motors available, it is always wise first to read the manufacturer's instructions. Most units must be started without load; certain designs, however, are started with load. Others must be started according to a particular procedure. This is particularly the case with boost or auxiliary units on the same shaft. In the case of pumps, prefilling may also be necessary; this depends on the suction head and the frictional losses in the suction line. Before initiating startup, be certain that all valves in the system (particularly on the suction or inlet side) are in the "free-flow" position. The motor should be repeatedly jogged on and off alternately without reaching full speed until it is seen that the unit is operating quietly and efficiently.

At initial startup it may be necessary to prefill the pump with oil or to vent the pump discharge lines to obtain unrestricted inflow of the hydraulic fluid and to reduce the noise level. This can be carried out at a suitable point in the outlet pressure piping or by loosening the outlet pressure piping connection until the flow appears free of air bubbles. If not otherwise indicated, it is customary to set pumps or motors with variable displacement, at approximately half stroke, to facilitate initial suction. When a system is started for the first time (and the system is filled with the hydraulic fluid) the fluid level in the tank must be continuously checked during the startup to ensure that it does not fall below the required fluid level, as shown on the sight gauge.

Certain hydraulic systems contain pressure relief valves which must be set according to the pressure indicated by the manufacturer. In such cases this should be carried out per the manufacturer's recommendations or the details shown on the schematic, for adjustment to be made to the minimum setting. For initial setting, pressure must be gradually increased until the system is operating efficiently. The pressure must not be higher, to prevent unnecessary waste of energy, and overheating of the fluid. After a short run at the recommended operating speed and normal operating pressure, check that the bearings, bushings, and fluid, have not exceeded the normal operating temperature.

Preventive Maintenance

The frequency of maintenance cannot be specified. It is determined purely by experience. The regularity with which it is carried out is important, however. The following points should be checked at regular intervals, as specified by the manufacturer, or based on experience.

Alignment. It is advisable periodically to check the correct alignment of the system components at operating temperature and (if possible) at operating pressure. It is also advisable to check mounting bolts, screws, couplings, pipes, connections, and so on.

Filtration. All filters (particularly suction filters) in the hydraulic system must be checked at suitable intervals and filter elements cleaned or replaced as required.

2.3.2 Troubleshooting

Hydraulic pumps and motors will give many years of trouble-free service when correctly installed and used. If a fault should develop, however, it is important to locate the cause and eliminate it as quickly as possible. Troubleshooting is simplified if a schematic of the hydraulic system is available. Specific pump troubleshooting guidelines are provided in the following chart.

Problem	Cause	Solution
Insufficient system pressure	1. System relief valve not set high enough	Set adjusting screw to obtain desired minimum operating pressure
	2. Oil bypassing to reservoir	Inspect circuit pressure progressively; watch for open center valves or other valves open to reservoir
	3. Pressure being relieved through relief valve	Relief valve is usually not required with pressure-compensated vane pumps; relief valves may create additional heat and present another pressure to set
	4. Vane or vanes stuck in rotor slots	Dismantle pump and inspect for wedged chips or sticky oil
	5. Pump running too slowly	Check minimum pump speed recommendations
	6. Defective pressure gauge or gauge line is shut off; dirt may plug gauge orifice	Install good pressure gauge in a line open to pump pressure
Excessive pump noise	1. Pump-motor coupling misalignment	Realign pump and motor accurately; align to within 0.005 in. total indicator reading
	2. Oil level low	Fill reservoir so that surface of oil is well above end of suction line during all of work cycle; about 1½ pipe diameters minimum
	3. Pump running too fast	Reduce speed; speeds above rating are harmful and cause early failure of pumps; refer

Problem	Cause	Solution
		to pump rating for maximum speed
	4. Wrong type of oil	Use a good, clean hydraulic oil having viscosity in accordance with manufacturer's recommendations; antifoaming additives preferred
	5. Air leak in suction line Air leak in case drain line Air leak around shaft packing	Pour hydraulic oil and/or grease joints around shaft while listening for change in sound of operation; tighten or replace
	6. Direction of pump rotation not correct	Arrow on pump housing must agree with direction of rotation
	7. Reservoirs not vented	Allow reservoir to breathe so oil level may fluctuate as required
	8. Air-bound pump	Air is locked in pumping chamber and has no way of escape; stop pressure line or install special bypass line back to tank so that air can pass out of the pump; an air bleed valve need is indicated
	9. Restricted flow through suction piping	Check suction piping and fittings to make sure full size is used throughout; make sure suction line is not plugged with rags or other foreign material
	10. Pump case drain does not terminate below oil level	Extend slip line piping so that it terminates below the oil surface when oil is at its lowest during any part of one machine cycle
	11. Pressure ring (on vane pumps) is worn	Replace; this condition caused by hot, thin, dirty oil or no oil at all; an air-bound condi-

(continued)

Problem	Cause	Solution
		tion will also contribute to the worn pressure ring
	12. Air bubbles in intake line	Provide reservoir with baffles; all return lines to reservoir must end below oil surface and on opposite side of the baffle from intake lines; check for reservoir design violations
	13. Restricted filter or strainer	Clean filter or strainer
	14. Sticking vane (on vane pumps)	Remove cover assembly and check rotor and vane for presence of metal chips or sticky oil; some pump models have chamfered edges on the vanes; see pump drawings for proper installation
	15. Worn or broken parts	Replace
	16. Reservoir air vent plugged	Air must be allowed to circulate in the reservoir; clean and/or replace breather
System excessively hot	1. Pump operated at higher pressures than required	Reduce pump pressure to minimum required for desired performance
	2. Pump discharging through relief valve	Check for mechanical binding in the machine being driven
		In fixed-displacement pump systems make sure the relief valve is set at a suitable margin above normal system operating pressure levels
		For variable-displacement pump circuits make certain relief is set approximately 25% higher than compensator pressure level
	3. Pump slippage too high	Inspect pumping element; replace if worn or damaged
	4. Cooling inadequate	Check for relief set too high; check for pressure compen-

Problem	Cause	Solution
		sator set too high; look for internal leaks evidenced by "hot spots"; only then add oil cooler and/or increase reservoir capacity
	5. High ambient temperature	Relocate power unit or baffle against heat source
	6. Excessive friction	Internal parts may be too tight; packing may be incorrectly installed or overtightened
	7. Oil in reservoir low	Raise oil level to recommended point; look for external leaks; look for leaks into coolant reservoir
	8. Pump drain or return line too close to pump suction	Separate the drain return and suction lines by a baffle in the reservoir; place the drain line in a location where it must travel the farthest distance practical before the oil reenters the pump
	9. System leakage excessive	Check progressively through the system for losses
Leakage at oil seal	1. Seal installed incorrectly	Correct installation
	2. Pressure in pump case	Observe case drain line for restriction; check drain line circuitry for excessive backpressure arrangement
	3. Poor coupling alignment	Realign pump and motor shafts; align to within 0.005 in. total indicator reading
	4. Packing damaged at installation; damaged or scratched shaft seal	Replace oil seal assembly; slip packing carefully over keyway, avoiding cuts
	5. Abrasives on pump shaft	Protect shaft from abrasive dust and foreign material
Bearing failure	1. Abuse during coupling installation to pump	Most pumps are not designed to handle end thrusts against

(continued)

Problem	Cause	Solution
		the drive shaft; eliminate all end play; couplings should be a slip fit onto the pump shaft
	2. Overhung load	Many pumps are not designed to handle any overhung load or side thrust on the drive shaft; see manufacturer's recommendations
	3. Incorrect fluid	See manufacturer's oil recommendations
	4. Excessive or shock load	Reduce operation pressure if possible; observe maximum rating of operating pressure; make necessary circuit changes; install shock absorber in appropriate area
	5. Chips or other foreign matter in bearings (contamination)	Make sure clean oil is used; essential for efficient operation and long life of bearings
	6. Coupling misalignment	Realign pump and motor
Pump not delivering oil	1. Wrong direction of pump rotation	Observe arrow on pump case or nameplate; direction of rotation must correspond
	2. Oil level low in reservoir	Maintain oil level in reservoir well above bottom of suction line at all times
	3. Air leak in suction line	Apply good pipe compound nonsoluble in oil, and tighten joints Replace line if necessary; consider flexible suction hose
	4. Pump running too slowly	Increase speed; check manufacturer's minimum speed recommendations to be sure of proper priming
	5. Suction filter or plugged line	Filters must be cleaned of lint or dirt soon after first start of unit; periodic checks should be made as a preventive maintenance precaution

Problem	Cause	Solution
	6. Bleed-off in other portion of circuit	Check for open center valves or other controls connected to tank
	7. Oil viscosity too high for proper priming	Thinner oil should be used per recommendations for given temperatures and service
	8. Pump shaft, vanes, rotor, or other parts damaged	Replace broken parts; study for signs of excessive shock, dirt, foreign material, or other probable causes of failure
	9. Sheared key at rotor or coupling	First determine cause and make necessary repairs, then replace sheared key
	10. Pump cover too loose	Tighten bolts on pump cover to recommended torque value

Troubleshooting guidelines for hydraulic motors are indicated below.

Problem	Cause	Solution
Motor turning in wrong direction	1. Piping between control valve is incorrect	Check manufacturer's literature and circuit to determine correct piping
Motor not turning over or not developing proper speed or torque	1. Driven mechanism binding because of misalignment	Remove motor and check torque requirement of driven shaft
	2. Free recirculation of oil to reservoir	Check circuit, valving and valve position
	3. Sticky relief valve (open)	Remove dirt from under pressure adjustment ball or piston
		Clean and polish sticky spool
	4. Motor adjustment mechanism not set at proper angle (on adjustable motors)	Set hand wheel to proper motor displacement angle
		Set to maximum angle to start unit and decrease angle to desired speed and torque

(continued)

Problem	Cause	Solution
	5. Setting overload relief valve not high enough	Check system pressure and reset relief valve
	6. Pump not delivering sufficient pressure or volume	Check pump delivery, pressure, and motor speed
	7. Motor innards bound	Remove stoppage or reshim if applicable
Will not hold load	1. No external brake	Hydraulic motors have inherent internal leakage; an external brake should be considered
	2. External brake not holding	Check for causes of brake slipping or other malfunction

2.4 HYDRAULIC VALVES

There are three different classifications of valves.

1. *Directional control*: to start, stop, and reverse cylinders and motors
2. *Volume control*: to regulate rate of fluid flow or speed of an actuator
3. *Pressure control*: to regulate or limit force

Valves are mechanical devices consisting of a body internally bored with cylindrical chambers and passageways. The chambers may contain pistons, spools, poppets, balls, or springs. The valve may also be equipped with an adjustment screw. The arrangement of the passages in the block together with the movement of the piston or spool causes the diversion of fluid flow. As the spool moves in its chamber, some ports are closed while others are opened. Each position of the spool changes flow direction. Volume of flow and pressure are also affected by the movement of components within the valve block and by the adjustment devices. Various methods or a combination of methods are used to activate a valve. They may be operated manually, mechanically, electrically, pneumatically, and hydraulically.

The name of the individual valve is based on its usual function. For example, a relief valve is one that limits the system pressure and relieves excess pressure. The operation, construction, and application of the various valve types to perform specific functions within a hydraulic system are described next for each valve type.

2.4.1 Directional Control Valves

A *directional control valve* consists of a body with internal flow passages which are connected and disconnected by a movable part. This action results in the control of fluid direction. Various valve configurations are utilized depending on the component to be controlled and the control actions or positions desired. The number of passages in the valve establishes the number of *ways* that the flow can be diverted; the number of *positions* the valve can be shifted to is determined by the configuration as well as the valve actuation mechanism (Fig. 2.15).

A *two-way* directional valve consists of two passages which are con-

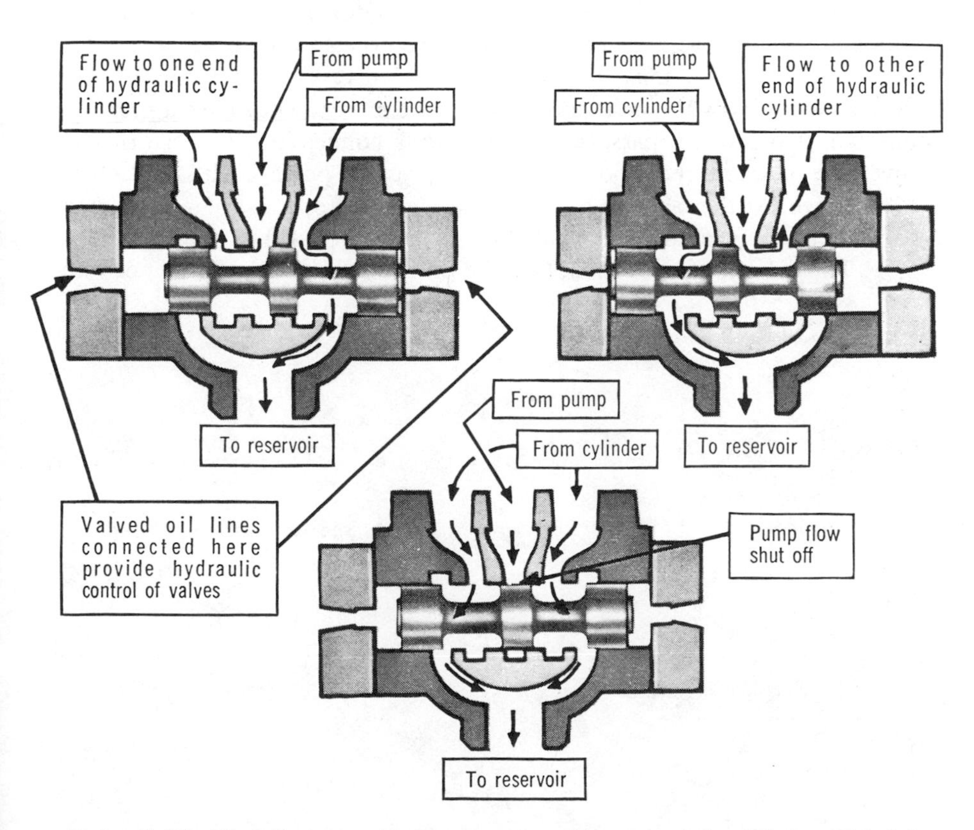

Figure 2.15 Direction control valve operation. (Courtesy of Sun Oil Co., Philadelphia, Pennsylvania.)

nected and disconnected. In one extreme spool position, the flow path through the valve is open. In the other extreme, the flow path is blocked. A two-way directional valve gives an on-off function. This function is used in many systems to serve as a safety interlock and to isolate and connect various system parts (see Fig. 2.16).

A *three-way* directional valve consists of three passages within a valve body: a pressure passage, one actuator passage, and one tank or exhaust passage. The function of this valve is to pressurize and drain one actuator port. When the spool of a three-way valve is in one extreme position, the pressure passage is connected with the actuator passage. When it is in the other extreme position, the spool connects the actuator passage with the tank line (see Fig. 2.17). A three-way directional valve is used to operate single-acting actuators such as rams and spring-return cylinders. In these applications, a three-way valve directs pressurized fluid to one end of the cylinder. When the spool is shifted to the other extreme position, the actuator passage is connected to the tank-exhaust passage. The actuator is returned by a spring or weight.

Four-way directional valves are capable of reversing the motion of a double-acting cylinder or reversible hydraulic motor. They have a pressure passage, two actuator passages, and one tank connection. To perform the reversing function, the spool connects the pressure passage with one actuator passage. At the same time, the remaining actuator passage is connected to the tank connection by the valve spool (see Fig. 2.18).

The directional valve spool can be positioned in one extreme position

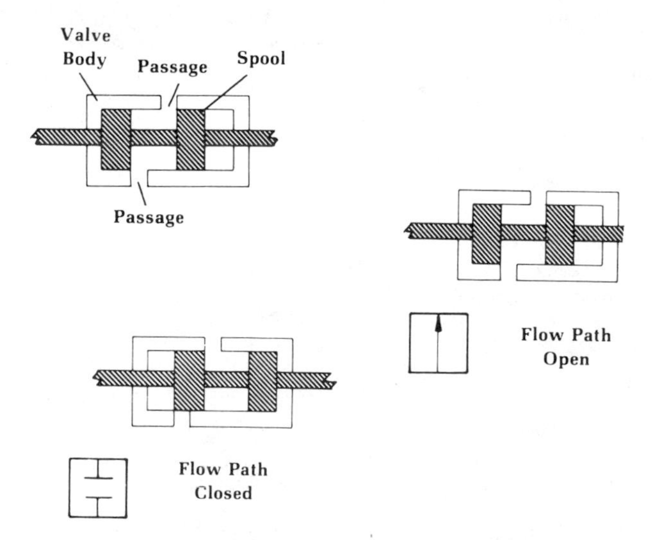

Figure 2.16 Two-way valve configuration.

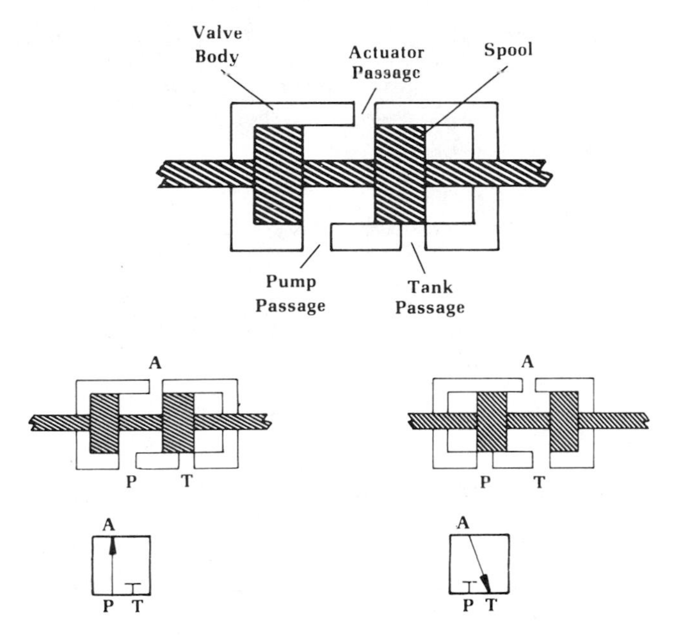

Figure 2.17 Three-way valve configuration.

or the other. The spool is moved to these positions by mechanical, electrical, pneumatic, hydraulic, or manual power. Directional valves whose spools are moved by muscle power are known as *manually operated* or *manually actuated* valves. Various types of manual actuators include levers, pushbuttons, and pedals. A very common type of mechanical actuator is a *plunger*. Equipped with a roller at its top, the plunger is depressed by a cam which is attached to an actuator. Manual actuators are used on directional valves whose operation must be sequenced and controlled at an operator's discretion. Mechanical actuation is used when the shifting of a directional valve must occur at the time an actuator reaches a specific position.

Directional valve spools can also be shifted with fluid pressure either pneumatic or hydraulic. In these valves, pilot pressure is applied to the spool ends or to separate pilot pistons. One form of two-way valve is a *check valve*, shown in Fig. 2.19. This valve allows free flow in one direction and prevents flow in the reverse direction. A spring holds the movable member in the closed position. Fluid pressure must overcome the spring force to cause the valve to open. The strength of the spring establishes the magnitude of pressure necessary for the valve to open.

One of the most common ways of operating a directional valve is with a *solenoid*. A solenoid is an electromagnetic device which consists basically

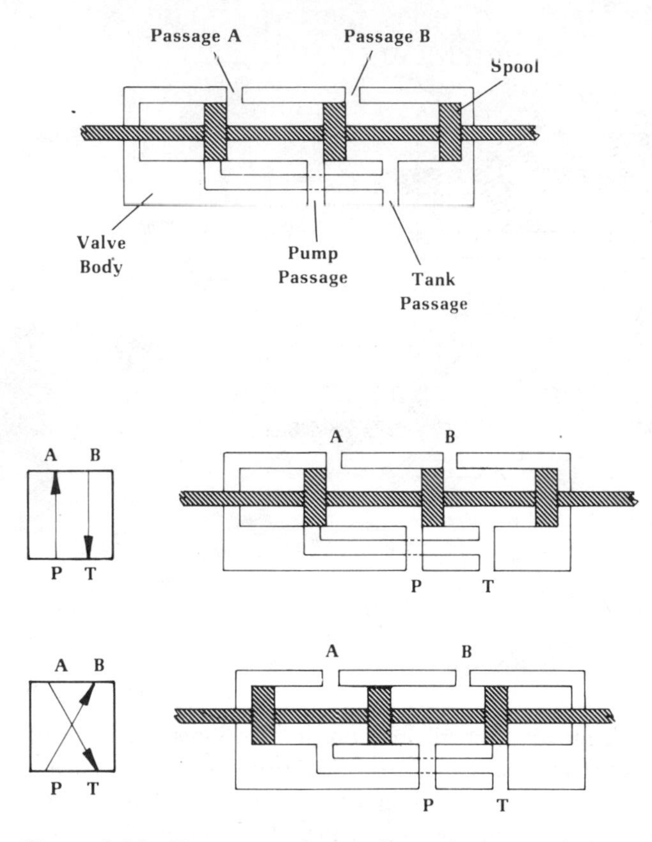

Figure 2.18 Four-way valve configuration.

of a plunger, frame, and wire coil. The coil is wound inside the frame. The plunger is free to move inside the coil. When an electric current passes through a coil of wire, a magnetic field is generated. This magnetic field attracts the plunger and pulls it into the coil. As the plunger moves in, it contacts a push pin and moves the directional valve spool to an extreme position.

A two-position directional valve generally uses one type of actuator to shift a directional valve spool to an extreme position. The spool is generally returned to its original position by means of a spring. Two-position valves of this nature are known as *spring offset valves* in hydraulic systems. Spring offset-returned two-way and three-way valves can be either normally open or normally closed; that is, when the actuator is not energized, fluid flow may or may not pass through the valve. In a two-position three-way valve, since there is always a passage open through the valve, normally closed usually indicates that the pressure passage is blocked when the valve actuator is not energized.

If either of two actuators is used to shift the spool of a two-position valve, *detents* are sometimes used. A detent is a locking device which helps keep a spool in the desired shifted position. The spool is equipped with notches or grooves. Each notch is a receptacle for a spring-loaded movable part. The movable part can be a ball. With the ball in the notch, the spool is held in position. When the spool is shifted, the ball is forced out of one notch and into another notch.

As noted previously, one of the most common means of operating a directional valve is with a solenoid. However, a disadvantage is that the shifting force which can be developed by a solenoid of reasonable size is limited. In large valves, the force required to shift a spool is substantial. As a result, only smaller valves use solenoids for shifting directly. Larger hydraulic directional valves (35 gpm and larger) use pilot pressure for shifting. In these larger directional valves, a small directional valve is often positioned on top of the larger valve. Flow and pressure from the small valve are directed to either side of the large valve spool when shifting is required. These valves are desig-

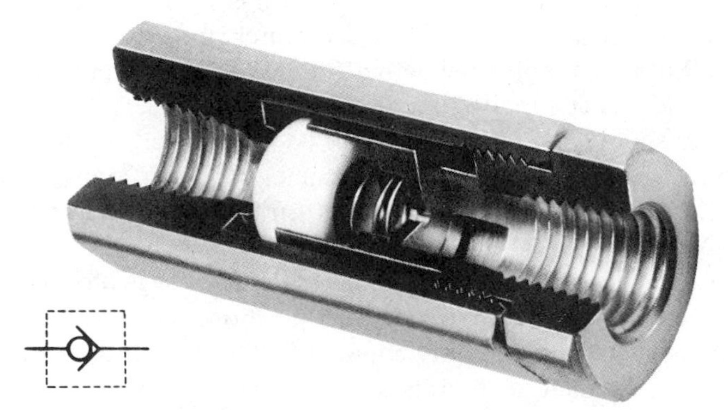

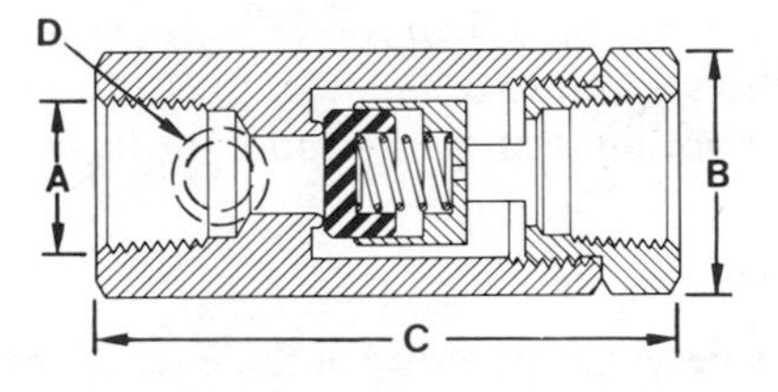

Figure 2.19 In-line check valve. (Courtesy of RegO Corp., Chicago, Illinois.)

nated *solenoid-controlled, pilot-operated* directional valves. They are commonly referred to as *piggy-back valves* (Fig. 2.20). In referring to the various possible flow paths available through a directional valve, only the flow paths as the spool was in either extreme were considered. But there are intermediate spool positions.

Hydraulic four-way valves are quite often three-position valves consisting of two extreme positions and a center position. The two extreme positions of four-way directional valves are directly related to the actuator's motion. They are the power positions of the valve. The center position of the directional valve is designed to satisfy a need of the system. For this reason, a directional valve's center position is commonly referred to as a *neutral condition.*

There are a variety of center positions available with four-way directional valves. Some of the more popular ones are the open center, closed center, tandem center, and float center. These center positions can be achieved within the same valve body simply by using the appropriate spool. Several spool designs are shown in Fig. 2.21. A directional valve with an open-center spool has P, T, A, and B passages all connected to each other in the center position. An *open center* allows free movement of an actuator while pump flow is returned to the tank. The *closed-center* spool has P, T, A, and B passages all blocked in the center position. A closed center stops the motion of an actuator and allows each individual actuator in the system to operate independently from one power supply (Fig. 2.22).

A *tandem-center* spool has P and T passages connected and A and B passages blocked in the center position. A tandem-center condition stops the motion of an actuator, but allows pump flow to return to tank without going over a relief valve.

A directional valve with a *float-center* spool has the P passage blocked and the A, B, and T passages connected in the center position. A float-center condition allows independent operation of actuators tied to the same power source and allows free movement of each actuator.

Directional valves with three positions must have the ability to hold the spool in the center position. *Spring centering* is the most common means of centering a directional valve spool. A spring-centered valve has a spring located at each end of the valve spool. When the valve is actuated, the spool moves from the center position to one extreme, compressing a spring. When the valve is not actuated, the spring returns the spool to the center position.

2.4.2 Flow Control Valves

In any machine system, it is necessary to control the energy that the system delivers. This control must be capable of varying the size of the force, the direction of the force, and the speed at which the force is applied.

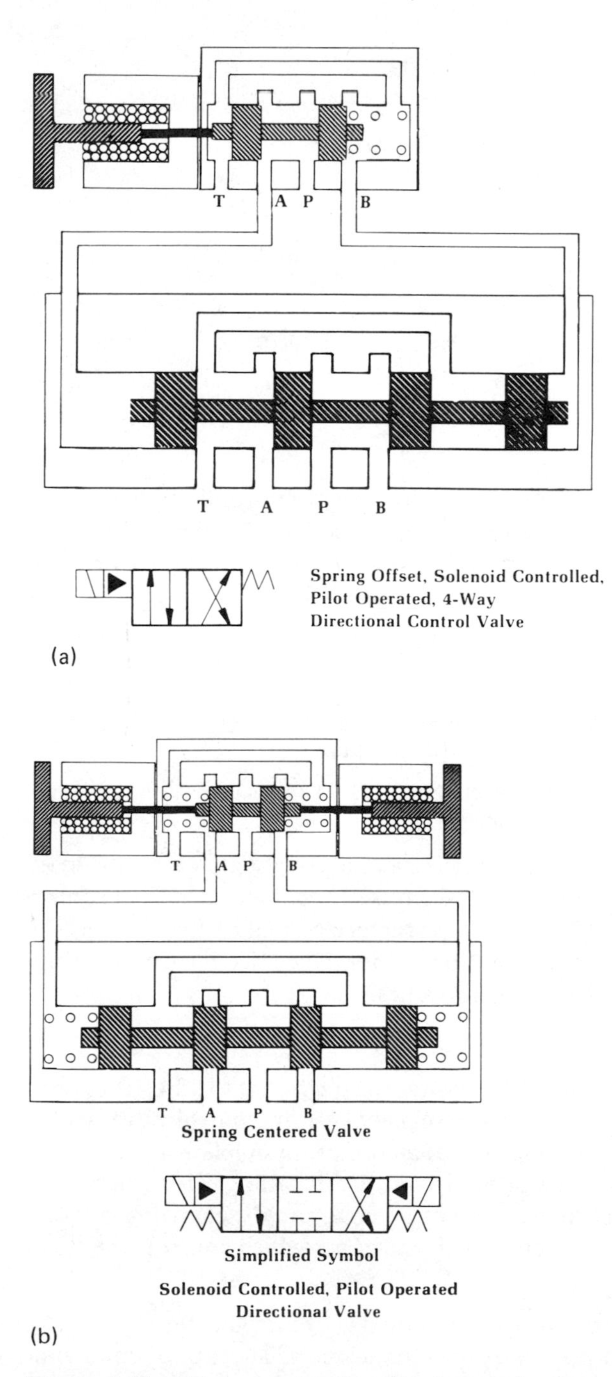

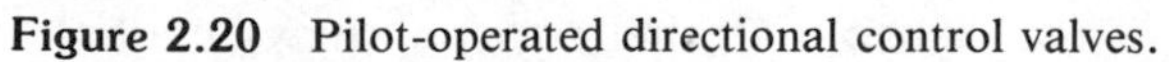

Figure 2.20 Pilot-operated directional control valves.

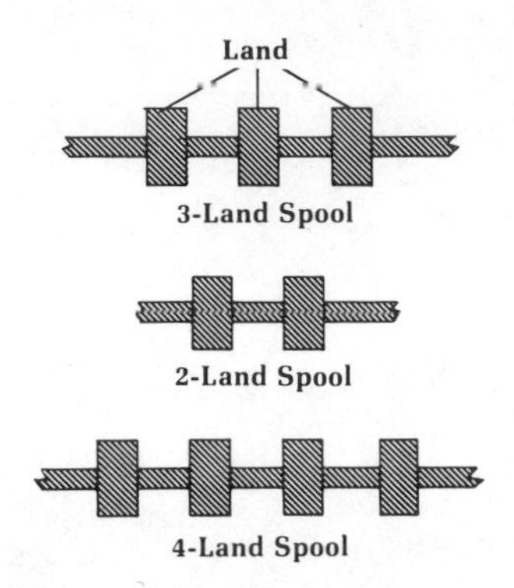

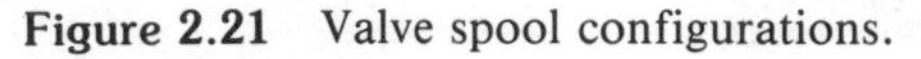

Figure 2.21 Valve spool configurations.

If the system uses the medium of fluid power to transmit its energy, the controls are a function of pressure (force), direction of flow, and volume (speed). A simple hydraulic system could use a relief valve to control pressure and a control valve to control direction of flow. Other methods of fluid volume control are:

Vary the speed of the pump's prime mover.

Use a variable-displacement pump and control system to change pump displacement to match speed requirements.

Depend on the throttling ability inherent in most directional control valves typical of many mobile circuits.

Use a flow control valve for throttling or metering so that actuator speed meets design requirements. Consider use of *compensated* flow control valve, which uses reducing valve structure or relief valve function to maintain a uniform pressure drop across the orifice regardless of upstream or downstream pressure values.

Use a servo or closed-loop control system.

When flow rates must be changed, the simplest method is to add a flow control valve in the system at an appropriate place. A flow control valve is a device that controls flow rate through use of an orifice. In simpler valves, the orifice and its associated pressure drop provide the throttling effect. In others, an orifice is used to detect changes in rate of flow, and the changing pressure drop through the orifice operates a valve member which controls the flow.

Types of Flow Controls

The simplest way to control the flow of a fluid is by introducing a restriction in a line. This restriction may be as uncomplicated as a length of conductor with an inside diameter of reduced size. Or the restriction may be a holed

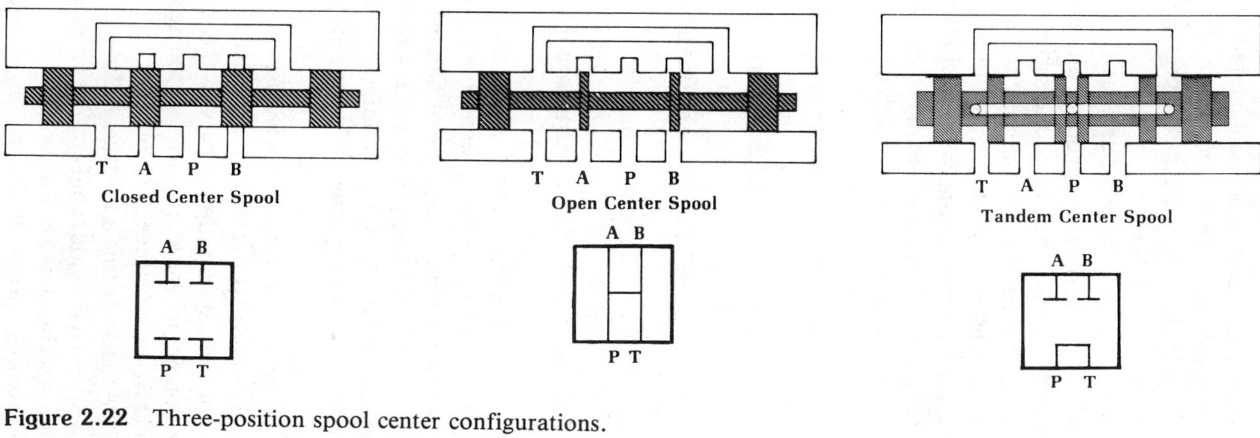

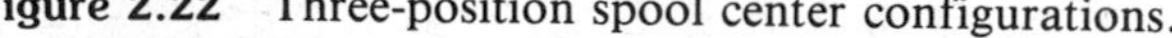

Figure 2.22 Three-position spool center configurations.

plate. A piece of bar stock with a hole in it, machined to fit the conical nose of a 37° or 45° flare fitting will act as an orifice to provide the pressure drop for flow control. The size of hole will determine the pressure drop across this fixed orifice, depending on rate of flow.

These types of orifices cannot be purchased and, within their limitations, function very well. Because they will produce a pressure drop regardless of direction of flow, they cannot be placed indiscriminately throughout the system. In fact, this sort of hardware can easily be lost in a system and steps should be taken to document or mark their locations. A paint mark on the smaller fluid line or on the fitting is one way to "find" them. Most service people prefer to attach a metal tag permanently to the orificed device and make certain it is marked on the original of the machine circuit drawing. Some maintenance directors make it mandatory to include a circuit drawing in the electrical control cabinet so that information of this type is immediately available to service personnel.

Needle Valve. One of the simplest, adjustable flow controls is the needle valve. These valves are available in all shapes and sizes, and can be purchased with or without an integral check valve. Adding an integral check valve permits free flow in one direction and throttled or metered flow in the other. Usually, valves that have free-flow capability have an arrow or some other marking on the valve body to indicate the direction of free flow, and are called flow control valves. Without the check valve, the are usually called needle valves (Fig. 2.23).

An adjustable stem changes the needle position relative to its orifice for a greater or lesser degree of throttling. Some stems have slots that accommodate screwdrivers; other valves have some type of handwheel for needle adjustment. A third variety has a calibrated stem which resembles a micrometer barrel for precise reset capability. Another manufacturer combines the calibrated stem with color codes, for easy reset. All stems have a setscrew or nut which locks the needle in place once it is set.

The geometry of the needle varies widely, depending on the desired change in orifice configuration as the valve is adjusted. Some needles have one "standard" taper; others have multiple tapers on the same stem. Another type of needle has V notches of varying widths on the needle, a V notch on a cylinder, or even a helix-ended sleeve. The purpose of these different configurations is to provide coarse or fine adjustment, depending on the distance of the stem from its seat, or to provide rapid closure or opening; other special configurations provide special flow responses.

Pressure Compensation. Most fluid power applications do not tolerate the wide range of actuator speed adjustment provided by a needle valve. To provide actuator speed regulation within closer limits, pressure-compensated flow control valves were developed. Over a somewhat limited range of

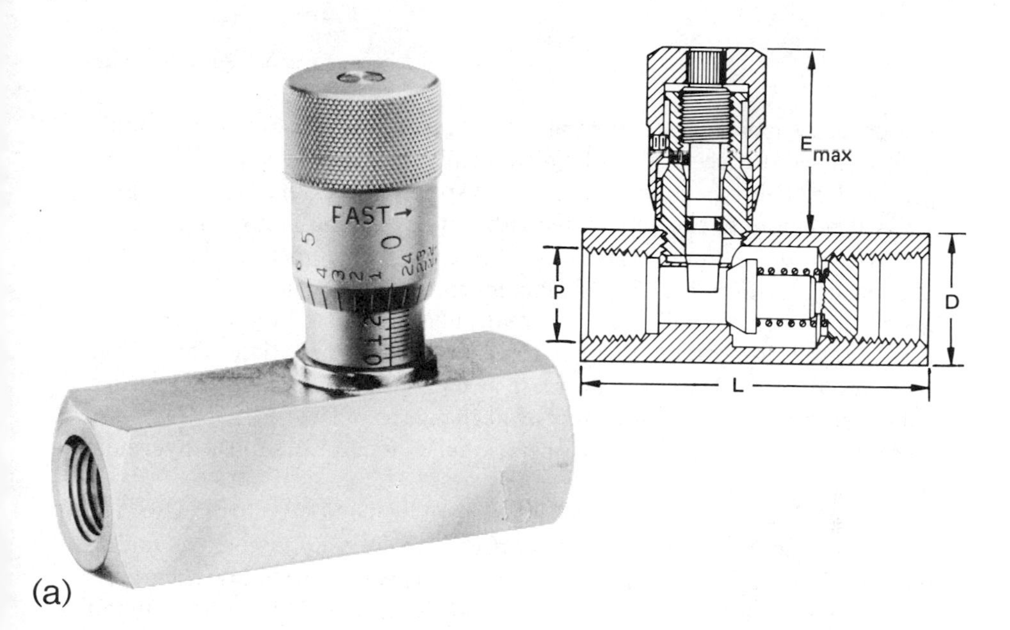

Figure 2.23a Adjustable flow control valve. (Courtesy of RegO Corp., Chicago, Illinois.)

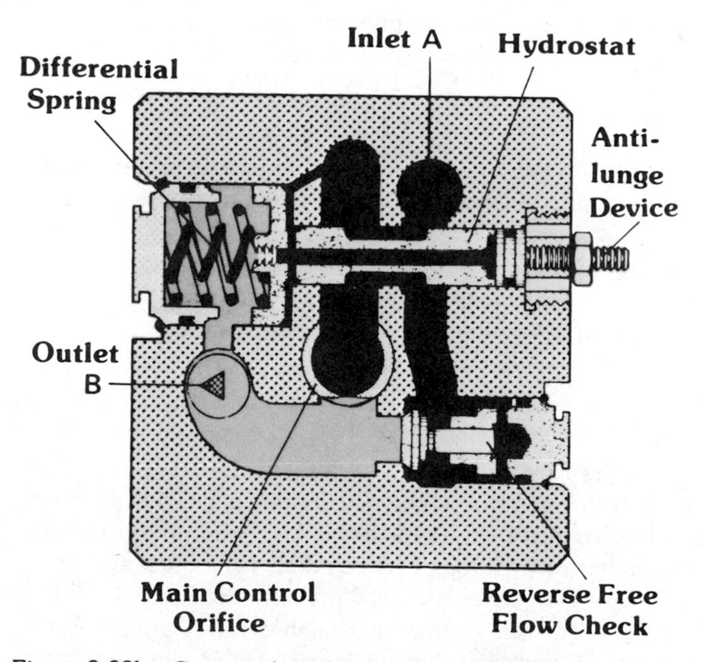

Figure 2.23b Cross section of pressure-compensated flow control. (Courtesy of the Rexroth Corporation, Bethlehem, Pennsylvania.)

pressure drop, a pressure-compensated flow control valve uses a fixed or manually adjusted orifice to control hydraulically the size of a second or throttling orifice. This type of valve provides a specified pressure drop as long as the inlet pressure to the valve falls within its operating limits (Fig. 2.23b).

Temperature Compensation. Changes in oil temperature alter flow rate through orifices because of changes in viscosity which occur as oil is heated or cooled. Usually, when a hydraulic system is first started, fluid temperature is lower than it will be after a period of operation. This means that not only will oil viscosity decrease as temperature increases, but system component clearances and tolerances will change as well. These changes cause flow differences which affect the operational characteristics of the hydraulic system.

To compensate for this flow rate change, some flow control valves use sharp-edged control orifices. These orifices are viscosity insensitive and do not alter flow rates with temperature and viscosity changes. Flow accuracies of 2% are possible with viscosity changes of as much as 250 seconds Saybolt Universal (SSU).

Another compensating method uses the phenomenon of different expansion rates of dissimilar metals. As oil temperature increases, the flow control is made smaller as a temperature-compensating rod changes the size of the orifice opening. This metal expansion must match the viscosity change of the oil so that the net result is a constant flow rate.

2.4.3 Pressure Control Valves

A number of pressure control valves are used in hydraulic systems, each performing a different function. These valves can be listed as follows:

Relief valve

Sequence valve

Counterbalance valve

Pressure-reducing valve

Unloading valve

In their simplest form, pressure controls are two-way valves that are either *normally closed* (no flow through the valves) or *normally open* (flow through the valve). Most pressure control valves have infinite positioning: they can assume an infinite number of positions between their fully open and fully closed positions, depending on flow rates and pressure differentials.

Relief, unloading, sequence, and counterbalance valves are normally closed. They open partially and/or fully, while performing their intended functions. A reducing valve is normally open. It restricts and ultimately blocks flow to a secondary circuit. In either case, a restriction is usually necessary to produce the required pressure control. One exception is the externally piloted unloading valve, which depends for its actuation on an external signal.

Relief Valve

Relief valves are devices installed in a circuit to make certain that system pressure does not exceed safety limits. Relief valves are intended to relieve occasional excess pressures arising during the course of normal operation. Excess fluid is allowed to return to the reservoir through an outlet port in the valve while full adjusted pressure is maintained in the system. For safety, the relief valve is usually installed as close as possible to the pump, with no other valving between the relief valve and pump. Relief valves can be divided in two categories: direct-acting and pilot-operated.

Direct-Acting. A *direct-acting valve* may consist of a poppet or ball held exposed to system fluid pressure on one side and opposed by a spring of preset force on the other. In a fixed, nonadjustable relief valve, when the valve is normally closed, the force exerted by the compression spring exceeds the force exerted by system fluid pressure acting on the ball or poppet. The spring holds the ball or poppet tightly seated. A reservoir port on the spring side of the valve returns leakage fluid to reservoir.

When system pressure begins to exceed the setting of the valve spring, the fluid unseats the ball or poppet, allowing a controlled amount of fluid to bypass to reservoir, keeping system pressure at the valve setting. The spring reseats the ball or poppet when enough fluid is released (bypassed) to drop system pressure below the setting of the valve spring.

Because the usefulness of fixed relief valves would be limited to one setting of its spring, most relief valves are adjustable. This is commonly achieved with an adjusting screw acting on the spring. By turning the screw in or out, the operator compresses or decompresses the spring. The valve can be set to open at any pressure within a desired range. Aside from the adjustable feature, this valve works just like the fixed valve.

The pressure at which a relief valve first begins to open to allow fluid to flow through is known as *cracking pressure.* When the valve is bypassing its full rated flow, it is in a state of full-flow pressure. The difference between full-flow pressure and cracking pressure is sometimes known as *pressure differential,* or *pressure override.*

Direct-acting relief valves can be made in various designs, including poppet, guided piston, and differential piston designs.

Poppet design: Spring-loaded poppet valves are generally used for small flows. They do not leak below cracking pressure and have a fast response, making them ideal for relieving shock pressures. They are often used as safety valves to prevent damage to components from high surge pressures, or to relieve pressures caused by thermal expansion in locked cylinders. If poppet relief valves must operate frequently, chatter caused when the valve relieves may damage the seat.

Guided piston design: Another type of direct-acting relief valve is the guided piston. In this valve a sliding piston, rather than a poppet, connects the pressure and reservoir ports. System pressure acts on the piston and moves it against a spring force. As the piston moves, it uncovers a reservoir port in the valve body. These valves have a fast response but may be prone to chatter. They can be damped to eliminate chatter, but this slows valve reaction time. They are reliable and can operate with good repetitive accuracy if flow does not vary widely. Valves with hardened steel pistons and sleeves have a very long service life. They may leak slightly below cracking pressure unless seals are used to seal the pistons.

Differential piston design: A variation of the guided piston relief valve is the differential piston relief valve. It is so called because the pressure acts on an annular area which is the difference between two piston diameters. The annular area is smaller than the valve's seat area. This permits using a smaller spring than would be needed if pressure acted on the entire seat area. They have a lower pressure differential than poppet or guided piston relief valves.

Pilot-Operated. For applications requiring valves that must relieve large flows with small pressure differential, *pilot-operated relief valves* are often used. The pilot-operated relief valve operates in two stages. A pilot stage which consists of a small, spring-biased relief valve (generally built into the main relief valve) acts as a trigger to control the main relief valve. However, the pilot may also be located remotely and connected to the main valve with pipe or tubing (Fig. 2.24).

The main relief valve is normally closed when the pressure of the inlet oil is below the setting of the main valve spring. An orifice in the main valve permits system pressure fluid to act on a larger area on the spring side of the poppet so that the sum of this force and that of the main spring keep the poppet seated. At this time, the pilot valve is also closed, and pressure in the passage is less than the setting of the pilot valve spring.

As system pressure rises, the pressure in the passage rises as well and when it reaches the setting of the pilot valve, the pilot valve opens. Oil is released behind the main spool or control chamber in which the main spool bias spring is located faster than it can enter through the orifice through the skirt of the main spool and the main spool is no longer balanced. The oil passing through the pilot valve is directed back to tank. Thus the unbalanced condition of the main spool permits excess fluid passage back to tank at a rate suf-

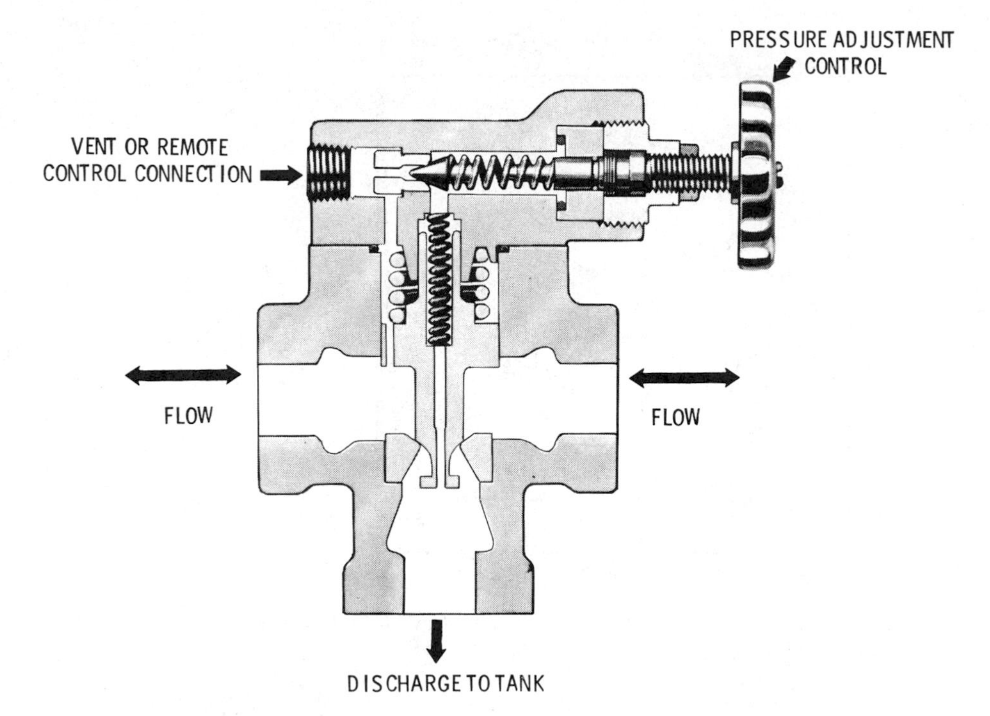

Figure 2.24 Pilot-operated hydraulic relief valve. (Courtesy of Parker Hannifin Corp., Cleveland, Ohio.)

ficient to maintain the desired set pressure. The pilot valve acts as a master adjustment and the piloted main relief valve poppet spool follows in a mirror image pattern to that of the pilot to provide the desired pressure value. The valve closes again when inlet oil pressure drops below the setting of the pilot valve. Pilot operated relief valves have less pressure override than do direct-acting relief valves.

Because these valves do not start opening until the system reaches 90% of full pressure, the efficiency of the system is protected because less oil is released. The operation of a pilot operated relief valve is as fast as a direct-acting relief valve and pilot-operated relief valves maintain a system at a more constant pressure while relieving (Fig. 2.25).

Sequence Valve

Sequence valves control the sequence of operation between two branches of a circuit. They are used to regulate the operating sequence of two separate working components in predetermined order based on pressure signals. Fluid is directed only to that part of the system that is connected with the primary

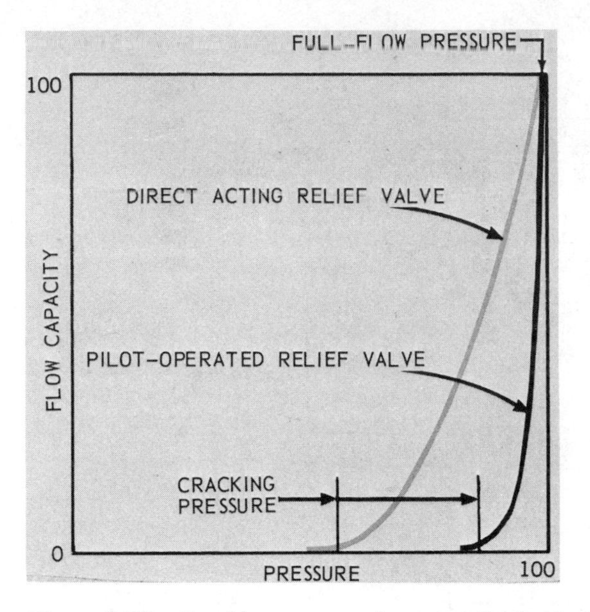

Figure 2.25 Cracking pressure characteristics for hydraulic relief of valves. (Courtesy of John Deere and Co., Waterloo, Iowa.)

port of the valve until the pressure setting is reached. The valve then directs fluid to the secondary branch with no variation in pressure on the primary side of the system.

Sequence valves are normally closed valves. They remain closed until pressure at their inlet port reaches a pressure set by adjusting a spring. At this pressure they open and allow fluid flow through the valve while maintaining pressure at least equal to its setting on the primary side of the system. They are similar to compound relief valves except that their spring chambers are drained externally to the system reservoir instead of internally to the outlet port as in relief valves. A check valve is generally built into the sequence valve to permit free reverse flow through the valve.

Counterbalance Valve

Counterbalance valves are used in a circuit to maintain an adjustable resistance against flow in one direction but permit free flow in the opposite direction. The function of the valve is to prevent uncontrolled movements or to support a weight in one part of the system while fluid is made available for working components in other parts of the same system.

Counterbalance valves have built-in check valves, which allow free reverse flow through the valve. Flow is shut off at the valve until pressure at its inlet port reaches the value set on the valve's spring. Fluid then flows through the valve and returns to the system reservoir. If counterbalance valves are set for too high an operating pressure, the system is inefficient. The correct setting is preferably 10% above the pressure setting required to counterbalance or hold a load.

Pressure-Reducing Valve

Pressure-reducing valves are used to drop the normal operating pressure of a main circuit as it is directed into the branch circuit to the required pressure in the branch circuit. The desired lower pressure can be obtained by adjusting the regulating device on the valve. As soon as the desired pressure is reached in the secondary system, the valve partially closes so that just enough fluid flows through to maintain this desired pressure.

Pressure-reducing valves are normally open valves. An adjustable spring holds the spool open or an integral pilot mechanism quite similar to that used for a pilot-operated relief valve. Downstream or branch pressure acts on the spool to close the valve. Fluid flows unrestricted through the valve until force from the downstream or branch circuit pressure equals the spring or pilot setting. The spool moves, compressing the bias spring assembly and restricting flow through the valve.

Unloading Valves

Unloading valves are often used in two-pump circuits where a large quantity of oil at low pressure is required for one portion of a cycle and a relatively small amount of high pressure during another cycle. Two pumps are used during low-pressure, high-speed operations, the delivery of the large one being diverted to the tank through the unloading valve when the system pressure exceeds its setting.

Remote system pressure is used to open and close the valve automatically at the proper time. Usually, an accumulator, or small pump, supplies the necessary pilot pressure for actuation of the valve. An adjusting screw is provided to regulate pressure. The valve contains a compression spring that acts against the fluid pressure force. It also contains an external pilot connection. Fluid force and spring force move the spool and direct fluid flow according to the requirements of the system.

2.4.4 Servicing, Repair, and Troubleshooting

Valve Servicing

Hydraulic valves are precision made and must be very accurate in controlling the pressure, direction, and volume of fluid within a system. Generally, no packings are used on valves since leakage is slight as long as the valves are carefully fitted and kept in good condition.

Contaminants such as dirt in the oil are the major villains in valve failures. Small amounts of dirt, lint, rust, or sludge can cause annoying malfunctions and damage valve parts extensively. Such material will cause the valve to stick, plug small openings, or abrade the mating surfaces until the valve leaks. Any of these conditions will result in poor machine operation or even complete stoppage. This damage may be eliminated if operators use care in keeping out dirt.

Use only the oils specified for use in the hydraulic system. Follow the recommendations in the operator's manual. Because oxidation produces rust particles, an oil that will not oxidize must be used. The oil should be changed and the filters serviced regularly.

For successful valve service, observe the following precautions.

1. Disconnect the electrical power source before removing hydraulic valve components to eliminate accidental starting or tools shorting out.
2. Move the valve control lever in all directions to release hydraulic pressure in the system before disconnecting any hydraulic valve components.
3. Block up or lower all hydraulic working units before disconnecting any parts.
4. Clean the valve and surrounding area before removing any part for service. *Do not allow water to enter the system.* Make certain that all hose and line connections are tight.
5. As a cleaning agent, use fuel oil or other suitable solvent. Never use paint thinner or acetone as a cleaning agent. Plug port holes immediately after disconnecting lines.

Valve Disassembly Hints

1. Do not perform hydraulic valve internal service work in the shop, on the ground, or where there is danger of dust or dirt being blown into parts. Use only a clean area. Make certain that all tools are clean and free of grease and dirt.
2. During disassembly, be careful to identify the parts for reassembly. Spools are selectively fitted to valve bodies and must be returned to the same bodies from which they were removed. Valve sections must be reassembled in the same order.
3. When it is necessary to clamp a valve housing in a vise, use extreme caution. Do not damage the component. If possible, use a vise equipped with lead or brass jaws or protect the component by wrapping it in a protective covering.

4. All valve housing openings should be sealed when components are removed during service work. This will prevent foreign material from entering the housing.
5. On spring-loaded valves, be very careful when removing the backup plug, as personal injury may result. When springs are under high preload, use a press to remove them.
6. Wash all valve components in a clean mineral oil solvent (or other noncorrosive cleaner). Dry parts carefully with compressed air and place on a clean surface for inspection. Do not wipe valves with waste paper or rags. Lint deposited on any parts may enter the hydraulic system and cause trouble.
7. *Do not use carbon tetrachloride* as a cleaning solvent, as it causes deterioration of rubber seals.
8. After parts are cleaned and dried, coat them immediately with a rust-inhibiting hydraulic oil. Be sure to keep the parts clean and free of moisture until they are installed.
9. Carefully inspect valve springs during valve disassembly. Replace all springs that show signs of being cocked or crooked, or that contain broken, factured, or rusty coils.

Valve Repair

Directional Control Valve Repair. Directional control valve spools are installed in the valve housing by a select hone fit. This is done to provide the closest possible fit between housing and spool for minimum internal leakage and maximum holding qualities. To make this close fit, special factory techniques and equipment are required. Therefore, most valve spools and bodies are furnished for service only in matched sets and are not available individually for replacement.

When repairing, inspect the valve spools and bores for burrs and scoring. Spools may become coated with impurities from the hydraulic oil. When scoring or coating is not deep enough to cause objectionable leakage, the surfaces can be polished with crocus cloth. Do not remove any of the valve material. Replace the valve body and spool if scoring or coating is excessive. If the action of the valve was erratic or sticky before removal, it may be unbalanced due to wear on the spools or body and the valve should be replaced (Fig. 2.26).

Flow Control Valve Repair (Fig. 2.27)

1. On valve spools with orifices, inspect for clogging by dirt or other foreign matter. Clean with compressed air or a small wire.

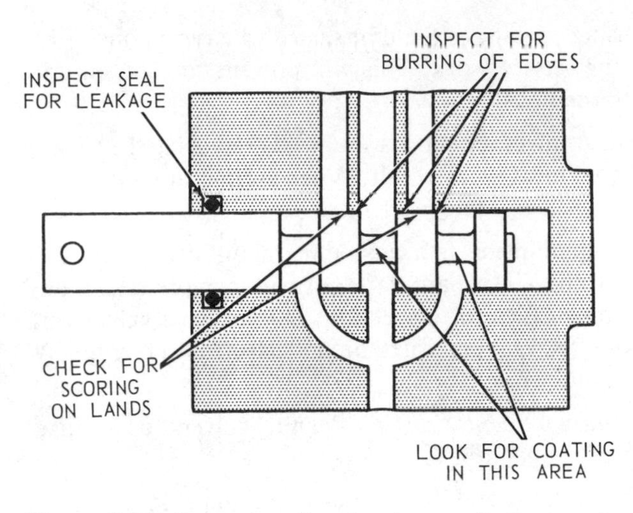

Figure 2.26 Inspecting directional control valve spools and bores.

2. Rewash all parts thoroughly to remove all emery or metal particles. Any such abrasive could quickly damage the entire hydraulic system.
3. Check the valve spool for freedom of movement in the bore. When lightly oiled, the spool should slide into the bore from its own weight.

Pressure Control Valve Repair. Check for weak relief valve spring with spring tester if system checks have indicated low pressure. This may be remedied by replacing the spring (Fig. 2.28).

Valve Seats and Poppets. Check valve seats for evidence of leak-by and scoring. Replace the valve if flat spots appear on the seat or on poppets.

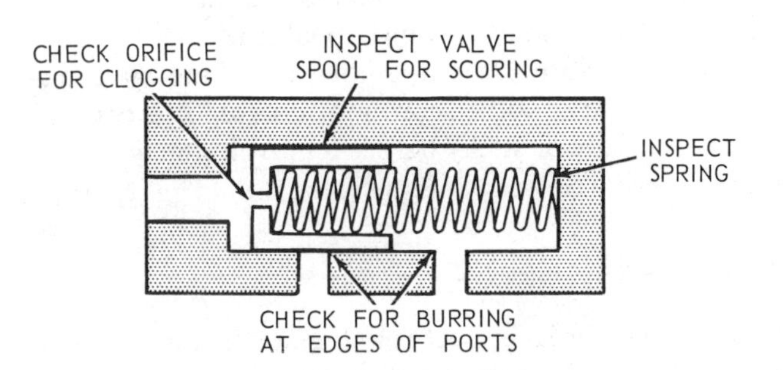

Figure 2.27 Volume control valve repair.

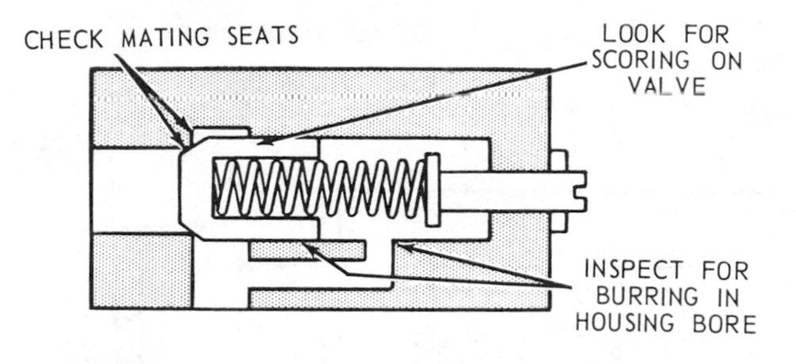

Figure 2.28 Inspecting pressure control valve.

Metal valve seats and poppets may be surface-polished with crocus cloth if scoring is not deep. Do not remove any valve material. Some seats and valve poppets are made of nylon. This material is long-wearing and sufficiently elastic to conform perfectly to mating surfaces, giving a tight seal. The nylon seats on poppet valves will take wear, with no damage to the mating metal point. When repairing these valves, always replace nylon parts with new nylon service parts.

Direction valves can be checked for internal leakage while still assembled provided that they have a spool configuration that blocks the inlet and outlet ports (pressure port from tank port). In such valves, loosening the outlet connection and observing the amount of fluid coming from this connection will give an indication of leakage. The procedure is shown on Fig. 2.29. If more than dripping is observed, the valve is suspect and should be removed from the system for further inspection.

Assembly of Valves

1. When assembling valves, be sure that they are kept absolutely clean. Wash parts in a solvent such as kerosene, blow dry with air, then dip in hydraulic oil with rust inhibitor to prevent rusting. This will also aid in assembly and provide initial lubrication. Petroleum jelly can also be used to hold sealing rings in place during assembly.
2. Double check at this time to make sure that valve mating surfaces are free of burrs and paint.
3. Replace all seals and gaskets when repairing a valve assembly. Soak new seals and gaskets in clean hydraulic oil prior to assembly. This will prevent damage and help seal the valve parts.
4. Be sure to insert valve spools in their matched bores. Valve sections must also be assembled in their correct order.
5. When mounting valves, be sure there is no distortion. This may be

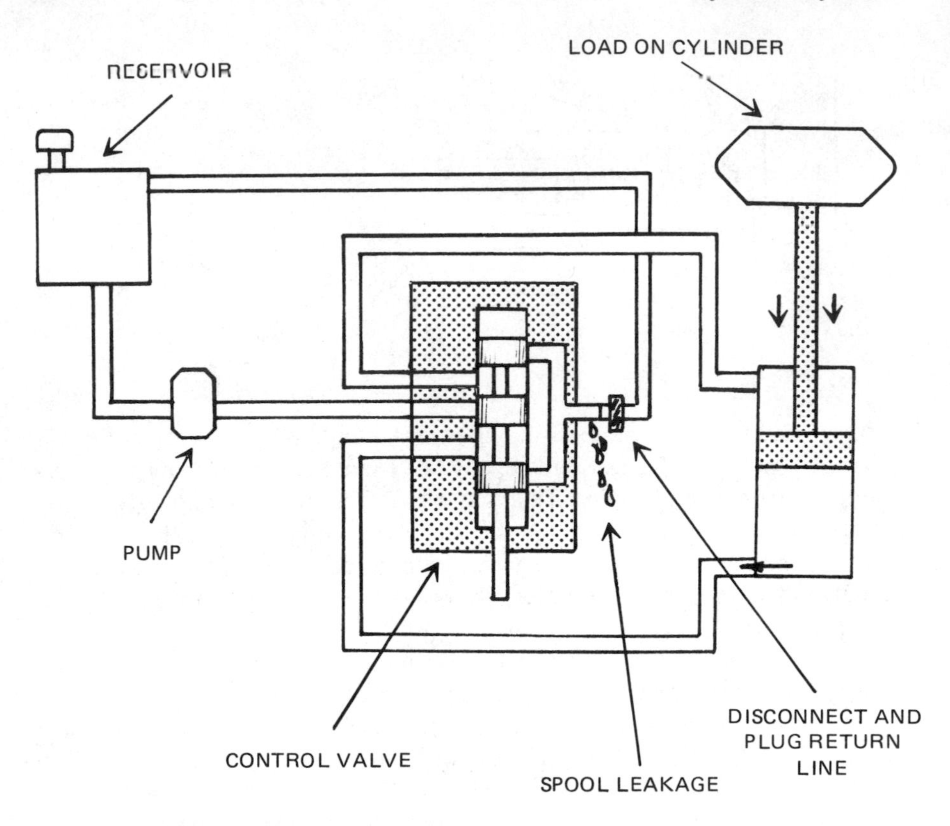

Figure 2.29 Checking control valve for leaks.

caused by uneven tension on the mounting bolts and oil line flanges, uneven mounting surfaces, improper location of the valve, or insufficient allowance for line expansion when oil temperature rises. Any of these may result in valve spool binding.

6. After tightening bolts, check the action of the valve spools. If there is any sticking or binding, adjust the tension of the mounting bolts.

Hydraulic Valve Troubleshooting

The following troubleshooting steps may be used to diagnose most hydraulic valve difficulties. When working on a specific machine, refer to the machine technical manual for more detailed diagnostic information.

I. *Pressure control valves*

  A. *Relief valves*

    1. *Low or erratic pressure*
       a. Incorrect adjustment

b. Dirt, chip, or burr holding valve partially open
c. Worn or damaged poppct or scat
d. Sticking valve piston in main body
e. Weak spring
f. Spring ends damaged
g. Valve cocking in body or on seat
h. Orifice or balance orifice partially blocked

2. *No pressure*
   a. Orifice or balance orifice plugged
   b. Poppet not seating
   c. Loose fit between valve body and spool or poppet
   d. Valve binding in body or cover
   e. Spring broken
   f. Dirt, chip, or burr holding valve partially open
   g. Worn or damaged poppet or seat
   h. Spool or poppet cocked in body or on seat

3. *Excessive noise or chatter*
   a. Oil viscosity too low
   b. Faulty or worn poppet or seat
   c. Fluctuating return line pressure
   d. Pressure setting too close to that of another valve in the circuit
   e. Improper spring used behind valve

4. *Unable to adjust properly without getting excessive system pressure*
   a. Improper spring
   b. Drain line restricted

5. *Overheating of system*
   a. Continuous operation at relief setting
   b. Oil viscosity too high
   c. Leaking at valve seat

B. *Pressure-reducing valves*

1. *Erratic pressure*
   a. Dirt in oil
   b. Worn poppet or seat
   c. Restricted orifice or balance hole
   d. Valve spool binding in body
   e. Drain line not open freely to reservoir
   f. Spring ends not square
   g. Improper spring
   h. Fatigued spring

- i. Valve needs adjustment
- j. Worn spool bore

C. *Pressure sequence valves*

1. *Valve not functioning properly*
   - a. Improper installation
   - b. Improper adjustment
   - c. Broken spring
   - d. Foreign matter on plunger seat or in orifices
   - e. Leaky or blown gasket
   - f. Drain line plugged
   - g. Valve covers not tightened properly or installed wrong
   - h. Valve plunger worn or scored
   - i. Seat of valve stem worn or scored
   - j. Orifices too large, causing jerky operation
   - k. Binding from coating of moving parts with oil impurities (due to overheating or improper oil)
   - l. Bottom cover piston position wrong (upside down)

2. *Premature movement to secondary operation*
   - a. Valve setting too low
   - b. Excessive load on primary cylinder
   - c. High inertia load on primary cylinder

3. *No movement or slowness of secondary operation*
   - a. Valve setting too high
   - b. Relief valve setting too close to that of sequence valve
   - c. Valve spool binding in body

D. *Unloading valves*

1. *Valve fails to completely unload pump*
   - a. Valve setting too high
   - b. Pump failing to build up to unloading valve pressure setting
   - c. Valve spool binding in body
   - d. Oil leakage from accumulator side of circuit

II. *Directional control valves*

A. *Spool valves, rotary valves, and check valves.*

1. *Faulty or incomplete shifting*
   - a. Worn or binding control linkage

b. Insufficient pilot pressure
c. Burned-out or faulty solenoid
d. Defective centering spring
e. Improper spool adjustment

2. *Actuating cylinder creeps or drifts*
   a. Valve spool not centering properly
   b. Valve spool not shifted completely
   c. Valve spool body worn
   d. Leakage past piston in cylinder
   e. Valve seats leaking

3. *Cylinder load drops with spool in centered position*
   a. Pilot operated check stuck open
   b. Counterbalance valve leaking
   c. Leaky piston packing
   d. Safety anti-intensification valve leaking
   e. Leaking directional control valve spool

4. *Cylinder load drops slightly when raised*
   a. Defective check valve spring or seat
   b. Spool valve position improperly adjusted

5. *Oil heats (closed-center systems)*
   a. Valve seat leakage (pressure or return circuits)
   b. Improper adjustment of valves

III. *Volume control valves*

A. *Flow control and flow divider valves*

1. *Variations in flow*
   a. Valve spool binding in body
   b. Leakage in cylinder or motor
   c. Oil viscosity too high or too low
   d. Insufficient pressure drop across valve
   e. Dirt in oil

2. *Erratic pressure*
   a. Worn valve poppet or seat
   b. Dirt in oil

3. *Improper flow*
   a. Valve not adjusted properly
   b. Restricted valve piston travel
   c. Restricted passages or orifice
   d. Cocked valve piston

   e. Circuit relief valve leaking
   f. Oil too hot

4. *Oil heats*
   a. Improper pump speed
   b. Holding hydraulic functions in relief mode
   c. Improper connections

A guide for troubleshooting pressure control valves follows.

Problem	Cause	Solution
Pressure not adjustable	1. Spool stuck open	Free spool
	2. Some other relief valve set lower	Check other relief valve seeing same pressure
Pressure too high	1. Spool stuck closed	Free spool
	2. Incorrect reference pressure	Check line restrictions
	3. Drain line plugged (internal)	Check valve
Reduced pressure too high	1. Incorrect setting	Adjust setting at valve
	2. Stuck spool	Clean valve
	3. Gauge broken	Replace gauge
Reduced pressure	1. Pump pressure too low	Check relief or compensator
	2. Main valve setting too low	Adjust
	3. Vent relief setting too low	Adjust vent relief
System excessively hot	1. Excessive pump slippage	Check clearances and orifice size of pilot section; consider control power losses
	2. Duty cycle too frequent	Reduce duty cycle
	3. Inadequate cooling	Check cooling system or add heat exchanger/cooler

Details of relief valve troubleshooting are given in the following chart.

Problem	Cause	Solution
Lack of pressure	1. Valve stuck open	Free ball, poppet, or spool
	2. Pump not pumping	See troubleshooting guidelines in Sec. 2.3

Problem	Cause	Solution
	3. Incorrect setting or broken spring	Readjust or replace
Valve too slow to open	1. Wrong relief valve type	Change to faster-acting type
Valve noisy or erratic performance	1. Incorrect design	Correct type
	2. Reseat pressure too low	Change valve type
	3. Dirty valve	Clean
	4. Air in system	Remove cause; bleed system

Troubleshooting directional control valves is covered in the following chart.

Problem	Cause	Solution
Valve spool fails to move	1. Solenoids inoperative	Check electrical supply for over- and undervoltage or solenoid burnout
	2. No pilot pressure	Check source of pilot pressure; also check for proper pilot source (internal or external)
	3. Blocked pilot drain	Check plugs, dirt, fittings, and lines
	4. Dirty	Disassemble, clean, and flush
	5. Improper reassembly after overhaul	Review parts drawing for proper assembly
	6. Distortion	Align body and piping to remove strains; check bolt-down torque
	7. Manufacturing burr	Remove spool; check spool and bore
	8. Silted	Remove spool and clean with emery paper; be sure to clean
Valve spool response sluggish	1. Startup oil viscosity too high	Change oil; use tank heater or run pump for oil warmup
	2. Restricted drain	Remove restriction or replace with larger line
	3. Distortion in valve body	Align body piping to remove strains; check bolt-down torque

Problem	Cause	Solution
	4. Malfunctions of solenoids	Check for proper source voltage and frequency; remove solenoid and check coils; look for double-solenoid energization
	5. Dirt in system	Drain and flush system; tear down and clean, if required
	6. Pilot pressure low	Check pilot pressure source
Valve produces undesired response in actuator	1. Improper installation connections	Check drawings for piping and energization
	2. Improper assembly of valves	Compare drawings and parts
	3. Spool end for end	Reverse spool

The following chart gives guidelines for troubleshooting check valves.

Problem	Cause	Solution
Flow stoppage	1. Valve installed backward or free-flow arrow has incorrect direction	Correct installation
	2. Parts broken	Disassemble and check
	3. Pump not pumping	See troubleshooting guidelines in Sec. 2.3
Incorrect pressure drop	1. Valve too small	Change valve
	2. Incorrect spring (in the case of pilot pressure relief)	Correct
Fails to hold pressure	1. Seat is damaged (shock)	Change seat or to correct seat type; eliminate cause of damage
	2. Seat is eroded	Change to larger size
	3. Excessive leakage in component that is being "held"	Check leakage in the cylinder or hydraulic motor; see troubleshooting guidelines in Sections 2.3 and 2.5

2.5 HYDRAULIC AND PNEUMATIC CYLINDERS

2.5.1 Description and Terminology

Cylinders are used more often than any other device for taking the pressure or power in a fluid system and converting it into mechanical force, work, or horsepower. Cylinders produce a straight-line motion, but this can be converted into a limited rotary motion with a lever. Although there are many variations in cylinder construction, the essential parts of a cylinder are shown in Fig. 2.30.

A cylinder consists of a cylinder body, a piston, and piston rod attached to the piston. End caps are attached to the cylinder body with threaded tie rods, or they are welded. As the cylinder rod extends and retracts, it is guided and supported by a bushing called a *rod gland.* The side through which the rod sticks out is called the *head.* The opposite side without the rod is termed the *cap.* Inlet and outlet ports are located in the head and cap ends (Fig. 2.30).

The inside of the barrel is honed to a fine finish to make a tight fluid seal and to preserve the piston packings, which would be worn out in a short time against a rough surface. For efficient operation, a leak-free seal must exist across the cylinder's piston as well as at the rod gland. Hydraulic cylinders often have cast iron piston rings as a piston seal. Piston rings provide a durable seal but they have some leakage flow due to the clearances between the piston ring and cylinder body tube.

For pneumatic systems and hydraulic systems that cannot tolerate any leakage flow, use a resilient piston seal. Resilient seals do not leak under normal conditions, but are less durable than piston rings. Rod gland seals come in several varieties and are generally resilient seals. Some cylinders are equipped with a multilip or U-, V-, or cup-shaped primary seal and a rod wiper which prevents foreign materials from being drawn into the cylinder.

As well as being a bearing, a rod gland bushing is a fulcrum for the piston rod. If the load attached to the piston rod of a long-stroke cylinder is not rigidly guided, then at full extension, the rod will tend to pivot or jackknife at the bushing, causing excessive loading. A *stop tube* is used to protect the rod gland bushing by distributing any loading at full extension between the piston and bushing. A stop tube is a solid, metal collar which fits over the piston rod. A stop tube keeps the piston and rod gland bushing separated when a long-stroke cylinder is fully extended. Cylinders can be mounted in a variety of ways, among which are flange, trunnion, side lug and side tapped, clevis, tie rod, and bolt mounting.

A number of common types of cylinders are used in industry. Following is a list of these together with a brief description.

Single-acting cylinder: a cylinder in which fluid pressure is applied to the movable element in only one direction.

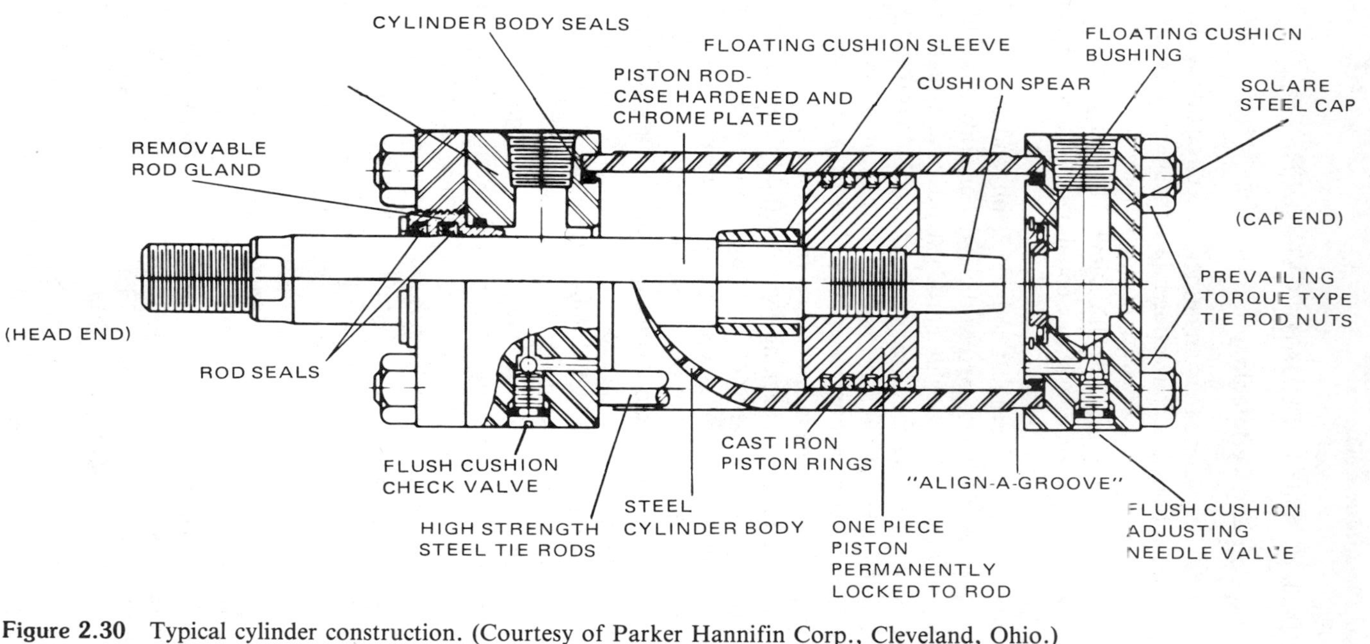

Figure 2.30 Typical cylinder construction. (Courtesy of Parker Hannifin Corp., Cleveland, Ohio.)

Spring-return cylinder: a cylinder in which a spring returns the piston assembly.

Ram cylinder: a cylinder in which the movable element has the same cross-sectional area as the piston rod.

Double-acting cylinder: a cylinder in which fluid pressure is applied to the movable element in two directions.

Single-rod cylinder: a cylinder with a piston rod extending from one end.

Double-rod cylinder: a cylinder with a single piston and a piston rod extending from each end.

Telescoping cylinder: a cylinder with multiple tubular rod segments which provide a long working stroke in a short retracted envelope. The tubular rod segments fit into each other to provide the telescoping action.

Tandem cylinder: two or more cylinders mounted in line with pistons connected by a common piston rod. These cylinders provide increased output force when the bore size of a cylinder is limited but not its stroke.

Duplex cylinder: two cylinders mounted in line with pistons not connected. Duplex cylinders give a three-position capability.

The piston should have two packings facing in opposite directions to seal the fluid in both directions of motion.

On most cylinders the piston rod comes out at only one end. It must be sealed against leakage of the fluid to the outside. A bearing is usually provided to support the rod against possible side thrust of the external load. The piston rod should have a highly polished finish, not only to make a good fluid seal but to avoid excessive wear on the rod seals (packings). The rod should be protected against mechanical damage. Nicks, scratches, dents, and wrench marks will cause the cylinder to leak and may ruin the rod seals. In order to provide an indication of the various construction styles used for differing applications and service environments, Figs. 2.31 to 2.34 are presented.

2.5.2 Cylinder Cushions

Whenever the extension or retraction of the cylinder at a rapid rate causes it to slam into the end, as the piston is stopped, a cushioned cylinder must be considered. The cushion acts as a shock absorber in that it slowly decelerates the piston; it comes to a smooth, gradual stop.

A *cushion* consists of a needle valve flow control and a plug attached to the piston. The plug can be on the rod side, in which case it is called a

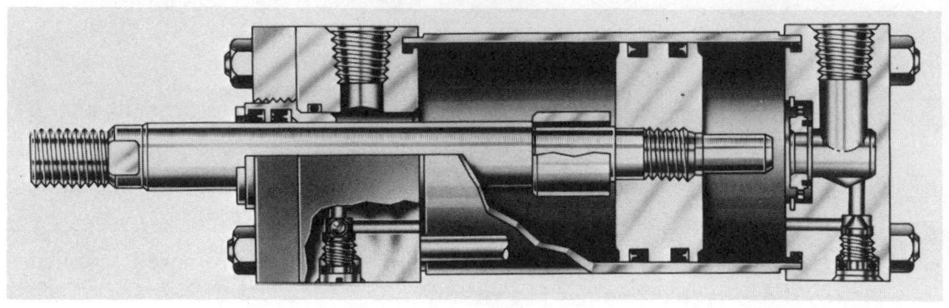

Figure 2.31 Lip seal cushioned cylinder. (Courtesy of Parker Hannifin Corp., Cleveland, Ohio.)

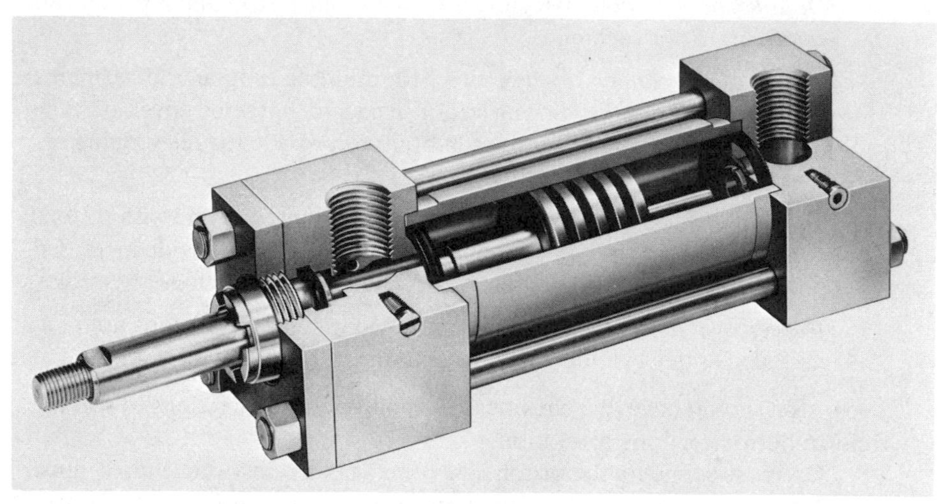

Figure 2.32 Heavy-duty hydraulic cylinder. (Courtesy of Parker Hannifin Corp., Cleveland, Ohio.)

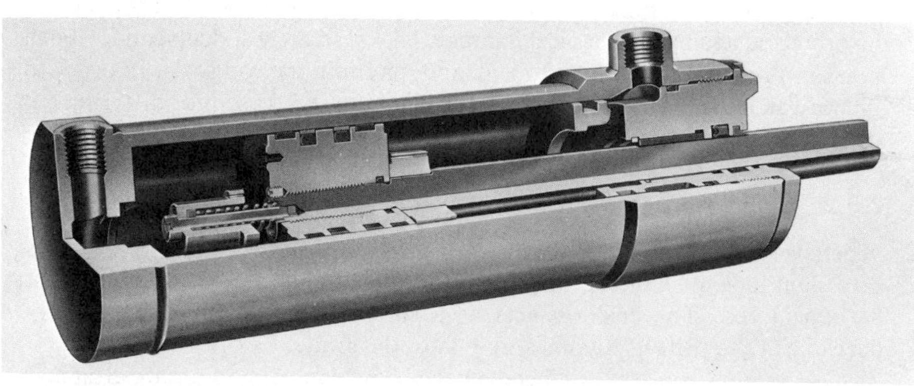

Figure 2.33 Welded construction cylinder. (Courtesy of Parker Hannifin Corp., Cleveland, Ohio.)

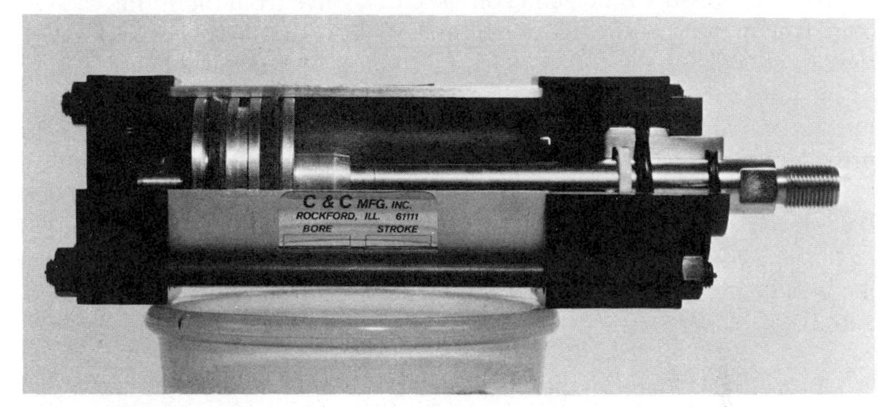

Figure 2.34 Pneumatic cylinder construction. (Courtesy of C&C Mfg., Inc., Rockford, Illinois.)

cushion sleeve. Or it can be on the cap end side, in which case it is called a *cushion spear* or plunger.

In order to understand how a cushion works in a cylinder, let us review the operation. Assume that the cylinder is extending. When the cushion spear enters the cushion cavity, the spear blocks part of the cavity opening; the effect is similar to that of a valve that is shifted and begins to close. As the spear penetrates farther into the cushion cavity, the annular area around the spear decreases, allowing less fluid to flow around the spear from the bore area of the cylinder into the cushion cavity. The piston slows gradually and stops when the spear completely blocks the cavity.

During cushioning, the fluid trapped in the cushion cavity by the spear is pressurized by the force the piston exerts on the fluid. Total force exerted by the piston is the combined force of that generated by the inertia of the load and the force from the fluid. When the pressure of the trapped fluid rises to where the fluid exerts a force equal to these combined forces, a state of equilibrium is reached, and the piston stops. To regulate cushioning (as a function of varying load, for instance) the cylinder is equipped with a needle valve, which can be adjusted to control deceleration rate over the length of the spear.

It is very important that the needle valve be set properly for efficient operation. If set too wide open, there will be little cushioning, if any; the piston will bottom too fast, resulting in shock. If the needle valve is set too closed, the piston will slow suddenly, then move very slowly, unnecessarily prolonging the time required to complete a work cycle, decreasing productivity.

When the piston starts to reverse, fluid flows through the check valve for quick startup. The cushioning process at the retracted end of the cylinder work stroke is identical. The check valve remains seated because force exerted by the pressure fluid acting on the check valve from the spring end exceeds that on the opposite side of the check valve. If there were no check valve, the rate at which the piston would retract would be controlled by the setting of the adjusting needle valve, which would regulate how much oil could flow into the cylinder. Some cylinders are designed without check valves, yet do provide fast reversal.

The shape of the cushion plunger or spear controls the rate at which the opening through which the fluid passes is blocked. Tapered, parabolic, and stepped cushions are available and each type provides a gradual opening area change which smoothly decelerates the cylinder piston (Figs. 2.35 and 2.36).

2.5.3 Cylinder Piston Seals

In the early days, before 1900, hydraulic machinery was used in several European countries: England, France, Germany, Russia, and perhaps others. For example, water power helped England manufacture fine goods for world trade. Steam-driven water pumps, with crude stuffing box packings, pumped water from the Thames River into a central hydraulic system using piston accumulators with weight-loaded pistons. A heavy cast iron pipeline distributed the pressurized water to nearby factories. Simple valving worked the large press ram, and the exhaust water was discharged back into the river. These systems were very temperamental, the packings were leaky and had to be frequently and critically adjusted to prevent excessive leakage and galling of the rams.

Today, cylinder manufacturers carefully select the type and material for piston and rod seals according to design requirements: whether for air or hydraulic use, the kind of hydraulic fluid, pressure and temperature range, tightness of seal required, duty rating, and other factors. Composition and hardness of rubber seals is often important for a particular application, and the mechanic should select replacement seals exactly like the originals. Since various grades of rubber cannot always be correctly identified, original factory

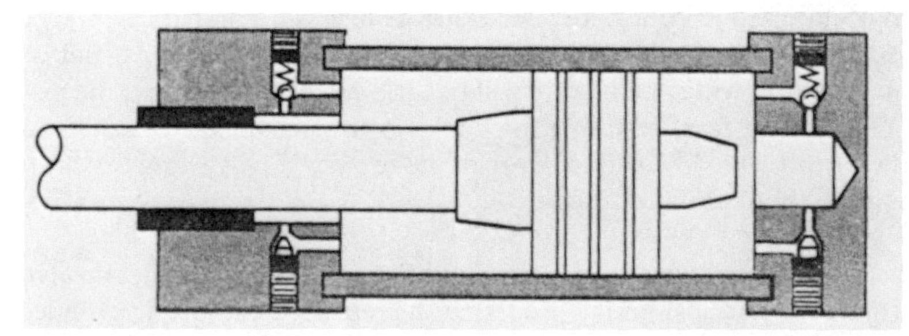

Figure 2.35 Constant taper spear and straight cavity cushion.

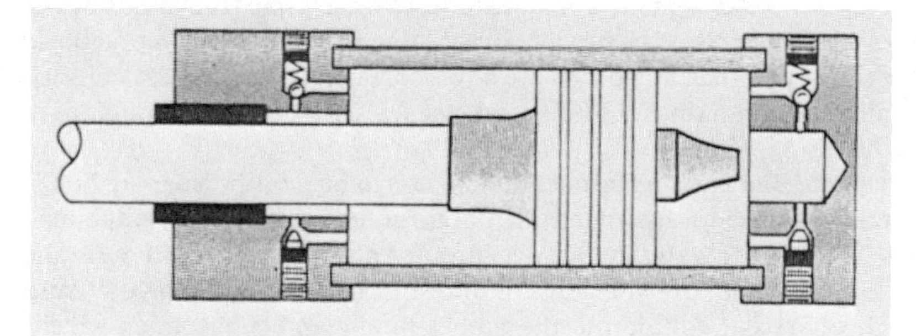

Figure 2.36 Inverted parabola spear and straight cavity cushion.

replacement seals should be obtained if possible. If this is not practical, the mechanic can select substitutes if the advantages and limitations of each packing type are understood.

Seal Materials (Fig. 2.37)

Leather. This is probably the material used in the first self-adjusting packings. It proved to be remarkably good and still has limited use in cylinders, and wide use in devices such as door checks, water pumps, and tire pumps. It will operate in barrels which are too rough for satisfactory life from a rubber seal. It has lower breakaway and running friction than rubber, and requires less lubrication. It hones the barrel on every stroke and the second set of "leathers" often outwears the first. Wax was originally used to fill the leather pores and to stiffen it so that the packing would hold its shape. This limited the operating temperature to 180°F. Polyurethane rubber is used to impregnate the leather for operation to 250°F. The supply of good-quality

Seal material	Recommended for:	Temperature range (°F)
Leather, wax impregnated	Air, oil, water at limited temperature	−65 to +180
Leather, rubber impregnated	Air, oil, water at higher temperature	−65 to +250
Corfam	Air, oil, water	−65 to +250
Buna-N, Hycar	Air, oil, water, water/glycol	−40 to +250
Viton	Air, oil, water, water/glycol, phospate ester	−40 to +250

Figure 2.37 Typical cylinder seal materials.

genuine leather today is limited because most cowhides have numerous blowfly holes, and this makes the production of large-diameter seals impractical. Leather substitutes such as Corfam are now being used, and have less porosity than leather. Backup rings, U seals, and cup seals are the most popular uses for leather.

Synthetic Rubbers. The ones that appear to be most popular are buna-N and polyurethane for hydraulic oil, Viton for high-temperature oil, buna-N for air, buna-S for water, butyl and Viton for phosphate ester fire-resistant fluids, and neoprene for water-/glycol fire-resistant solutions. Polyurethane is a relatively recent addition to the rubber family and is becoming widely used for hydraulic oil because it will outwear buna-N if the barrel has a fine finish and if the oil is kept well filtered. Because of its higher friction it is not the best compound for air or water service.

Natural Rubber. Seals made from natural rubber should not be used in petroleum oil; they are for nonpetroleum-base fluids, for example automotive brake fluid. Not used with air.

Teflon. Because of its hardness it may not be leaktight as a moving seal. Unlike rubber, it may not return to original shape after being distorted. A spring is sometimes used behind it. As a moving seal it has low breakaway and running friction, long life, and will stand high temperature. Used widely in backup rings to protect other seals, and as a thin piston ring with rubber expander to obtain low breakaway friction.

Seal Types

Ring Seals. Automotive-type piston rings are considered the best for long life and ability to withstand high surges and shock loads. They are still used in most cylinders designed for machine tool applications. The rings are usually cast iron working in a steel barrel. Light-duty cylinders may have only two rings, while machine tool cylinders may have four to six rings. Ring seals are not leaktight and are not good for use with pneumatic cylinders.

Multiple-V. Sometimes called *chevrons.* They are used in sets of two up to six or eight rings, depending on the pressure to seal. One ring is used for every 500 to 700 psi, with a minimum of two rings. When used as piston seals on double-acting cylinders, two sets must be used, one set facing each direction, with the lips facing the pressure. They are held in tight contact with the surface to be sealed by preloading and by fluid pressure. For rod seals or for single-acting cylinders, one set is sufficient.

Multiple-V packings, when carefully installed and properly adjusted, are considered to be one of the finest soft seals available. They are valued for their leak-tightness and dependability. Seldom do they fail suddenly. They give sufficient warning, by leaking slightly, long before actual failure. This gives time for replacements to be procured. But in spite of their good performance, cylinder manufacturers are phasing them out as piston seals in favor of single-lip cups, which are cheaper, less critical to install, need no preloading adjustment, and are less likely to leak when the cylinder is repacked.

O-Ring Seals. Although not extensively used on industrial cylinders, O rings are almost universally used on agricultural and mobile cylinders because of their low cost, simplicity, and the simple machining required to mount them. On these applications their high breakaway and running friction, poor sealing at low pressure, and relatively short life are of little concern. Their excellent sealing at high pressures is very desirable, and their life, though short, is adequate for agricultural and mobile cylinders, which are not intended to make as many cycles per day as industrial cylinders.

For reasons of economy O rings may be used as piston seals on low-cost air cylinders, but are not recommended for general air use because of their high friction. If not kept well lubricated they wear out rapidly. Cup seals are preferred for most air cylinders. O rings are not recommended for vacuum applications because of high friction and difficulty of lubrication.

Cup Packings. The base of a cup packing is usually bolted between the head of the piston or backing plate, and a follower plate, with the sealing lip facing the pressure medium. To get a good seal, the packing must be centered and fastened firmly between the follower plate and backing plate with a bolt, or stud and nut, which must be tight enough to seal against leakage. The clearance between the cylinder wall and the back support must be kept at an absolute minimum to prevent extrusion. The back support should have a flat surface with a square corner on the pressure side, because the packing will take the shape of the back support and develop a square shoulder under pressure. The follower plate should have a flat surface on both sides, with the corner on the packing side rounded enough to avoid cutting the packing. The diameter of the follower plate should allow a clearance equal to about 25 or 30% of the packing thickness between the plate and the inside wall of the packing.

Flange Packings. The design of the packing and gland recesses greatly influence the effectiveness of a flange seal. The glands may be either bolted or threaded. Differences between requirements for bolted and threaded glands are minor, but should be considered when using a flange packing. For an effective seal with bolted gland, the clearance between the outside diameter of the gland nose and the gland nose diameter should be comparatively small. A tight fit is not necessary for a threaded gland because the threads center the gland. The fit between the rod or shaft and the gland should be close enough to prevent extrusion. A backup ring may be necessary in some applications. The circumferential edge on the gland nose inner diameter, directly over the shoulder of the packing, should be kept square to prevent further extrusion. The gland nose should be squared off with a smooth, flat surface at the base. It should be long enough to develop a seal on the base of the packing when the gland is tightened, but compression should not exceed 25% of packing thickness. To simplify installation, the gland nose recess should be deep enough to allow for a chamfer. The straight section below the chamfer should be as deep as the packing thickness, plus enough wall to center the gland accurately.

U Packings. A recess for a U packing may be cut into either a cylinder wall or a piston head. Most effective sealing will result if the cross-sectional width of the U packing is kept to a minimum. This tends to reduce the area under pressure and lighten the load on the gland and the studs and on the nuts holding the packing. The fit between the shaft and gland should be close enough to prevent extrusion, and the gland nose should be squared off with a smooth, flat surface at the base. If a supporting ring is used, the top edges should be slightly rounded to avoid cutting the packing. The recess should be deep enough to hold both ring and packing, with the packing slightly squeezed between the ring and the gland nose when the gland is fully tightened. In addition, centering lugs should be fastened on the support ring to center it in the packing. These lugs should be heavy enough to center the rings, but should not interfere with packing lips. Figure 2.38 presents a pictorial summary of the various lip-type pressure energized seals used in hydraulic and pneumatic cylinders.

Some Causes of Piston Seal Failure

If a single-acting piston or piston rod seal is fitted in the wrong direction in relation to operating pressure, it will be ruined very quickly. A simple rule to remember for installing a seal in the right direction is:

> The side of the seal which is softer (the unreinforced part of solid seals) or open (the sealing lip side with U rings and V rings) should face the operating pressure.

Under unfavorable conditions, working temperature can exceed the highest temperature for which the system was designed. Also, perhaps the specified hydraulic fluid has been replaced with one of unsuitable quality. The result is that the seal swells and softens. It can then be damaged by extruding into the clearance between the metal parts, where, due to piston rod movement, pieces of the seal are torn away.

The type of failure most difficult to avoid is seal erosion from piston rod damage. Cylinders on earth-moving machinery are particularly prone to piston rod damage from impact by rocks or stones or by contact with electric cables. Slight damage to the piston rod can often be remedied by rub-

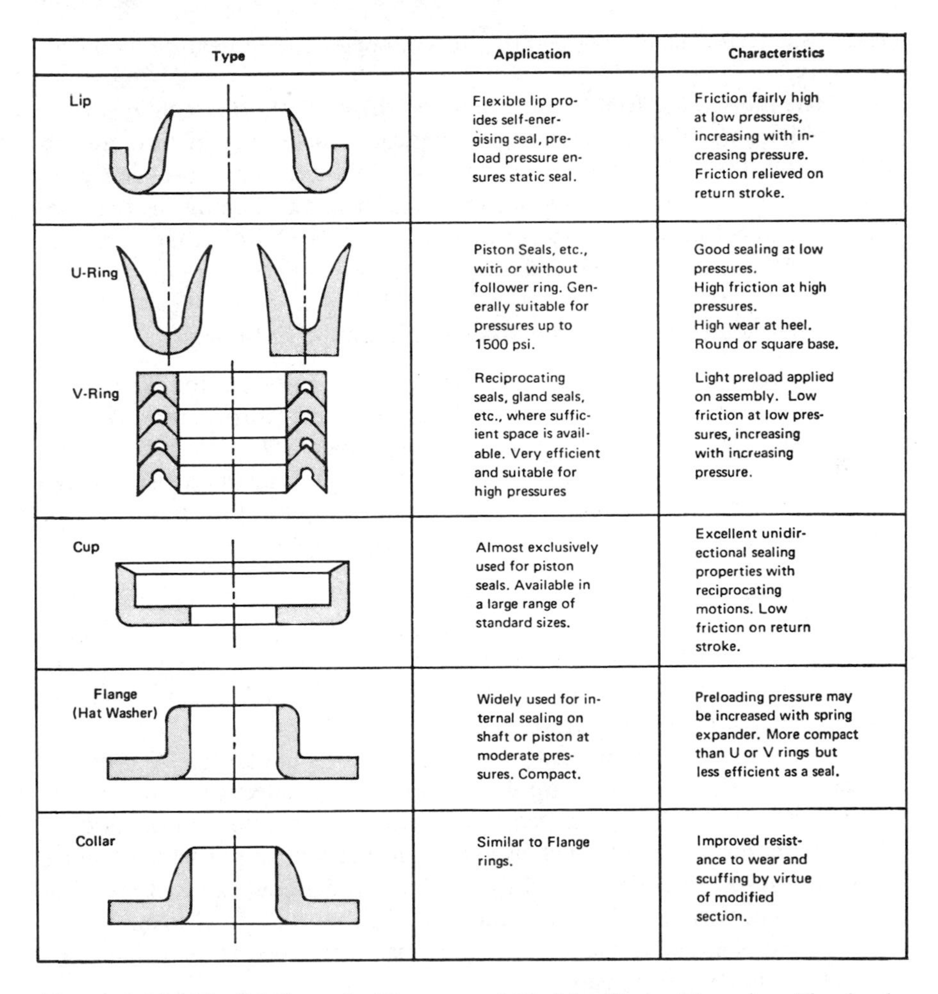

Type	Application	Characteristics
Lip	Flexible lip pro-ides self-ener-gising seal, pre-load pressure en-sures static seal.	Friction fairly high at low pressures, increasing with in-creasing pressure. Friction relieved on return stroke.
U-Ring	Piston Seals, etc., with or without follower ring. Gen-erally suitable for pressures up to 1500 psi.	Good sealing at low pressures. High friction at high pressures. High wear at heel. Round or square base.
V-Ring	Reciprocating seals, gland seals, etc., where suffic-ient space is avail-able. Very efficient and suitable for high pressures	Light preload applied on assembly. Low friction at low pres-sures, increasing with increasing pressure.
Cup	Almost exclusively used for piston seals. Available in a large range of standard sizes.	Excellent unidir-ectional sealing properties with reciprocating motions. Low friction on return stroke.
Flange (Hat Washer)	Widely used for in-ternal sealing on shaft or piston at moderate pres-sures. Compact.	Preloading pressure may be increased with spring expander. More compact than U or V rings but less efficient as a seal.
Collar	Similar to Flange rings.	Improved resist-ance to wear and scuffing by virtue of modified section.

Figure 2.38 Flexible lip seals. (Courtesy of *Machine Design* Magazine, Cleveland, Ohio.)

bing it with fine emery, to allow the machine to continue operating. However, the cylinder should be replaced at the earliest opportunity.

Severe damage to seals can be caused by contamination, such as metal particles or other foreign materials introduced during cylinder maintenance. For example, when screwing the nipples into connectors, fragments of the threads may break loose. The foreign material is trapped between the moving parts and scores the bore sealing surface. The seal is then worn away by the damaged sealing surface. Seal life varies with the quality of the sealing surface finish of the cylinder bore and on the piston rod. Other factors that affect seal performance are operating speed, pressure, temperature, and fluid viscosity.

The lubricating film formed between the seal and the sealing surface tends to become thinner with rising temperature, increasing friction, and reducing seal life. While the piston is at rest, the fluid film is broken by the initial interference pressure. On startup, wear will take place at the seal face before the lubricating film can form. In the ideal condition, there is an unbroken oil film between the seal and the sealing surface.

2.5.4 Installation and Maintenance of Hydraulic Cylinders

To guarantee efficient operation of cylinders it is important to take note of the operating instructions supplied with the unit or the information given in the relevant catalog sheet when undertaking repairs. During assembly and installation the internal parts must be kept clean. The fluid used in the system must be of a recommended type and adequate filtration must be provided.

Installation

In general, the mounting position of hydraulic cylinders is optional. During installation particular attention must be paid to cleanliness. The cylinders must be installed free of tension, and in particular free of radial forces, to prevent functional faults and premature wear.

The pipe connections and thread depths are suitable for all standard connections. The spot faces are designed so that connections with sealing edges or O-ring seals may be used. The thread must not rest on the bottom of the bore. Sealing agents such as hemp and putty must not be used, since they can cause contamination and subsequent functional faults. Before installation pipework must be free of dirt, swarf, sand, chips, and so on. If possible, the pipes should be pickled. Cleaning rags should not be used. The pipes should be laid free of tension.

Startup

For hydraulic cylinders proven hydraulic oils with a mineral oil basis should be used if possible. Before using other types of fluid, prior consultation with the manufacturer is necessary. Before connecting the cylinder, the hydraulic system must be thoroughly flushed. The cylinder connection lines should be shut off during this process. It is recommended to continue flushing for approximately ½ hr. Thereafter the cylinders should be first connected to the pipe system.

Before startup the cylinder should be vented at both ends if necessary. This may be carried out by loosening the connections or alternatively by means of special bleed screws. After thorough venting (the oil must be free of bubbles and there must be no further foaming) the cylinder connections are retightened. A gage connection (a quick coupler half or needle valve) at the cylinder ports is convenient for troubleshooting in the future. Placed at the high point the connection can also serve to bleed air from the entire branch circuit, assuming that it is the high point in the circuit.

Maintenance

Generally speaking, hydraulic cylinders require little maintenance. With heavy impact stress, however, an important point to watch is that the bearing points, such as swivel bearings, are lubricated. Particularly after startup of a new system the cylinders should be checked at frequent intervals for correct function and for leakage. Inspect for proper alignment frequently during startup. If misalignment is suspected, disconnect rod from machine member and cycle the cylinder. Does it freely reenter the interface connection? If not, realign cylinder body or machine member.

It is essential that the hydraulic oil be kept clean. When filling the hydraulic oil into the system a filter of less than 60 μm should be used. Oil should be pumped into the reservoir through a filter. Provisions should be made to tee into a convenient return to tank line with a ball valve dedicated to reservoir fill function. Oil filters incorporated in the system should initially be cleaned at least every 100 hours of operation, thereafter monthly or at least at each oil change. Many filters are equipped with indicators to signal saturation condition and need for change or cleaning. The oil change is dependent on the operating conditions, the condition of the fluid, and the filling capacity of the system. Many large reservoirs are equipped with sophisticated fluid conditioning equipment because of the difficulty of changing fluids and the associated costs. Longer fluid life can be assured with suitable fluid conditioning equipment and an adequate supply of fluid to ensure suitable cooling and filtration. Overworked or aged fluid or contaminated fluids cannot be improved by topping with fresh fluid. It is usually recommended that a complete change be effected. All fluid should be drained from the system while it is at operating temperature if possible. Reservoir should be thoroughly cleaned when oil is renewed.

Storage

It is recommended to store the cylinders in a dry, damp-free place. The storage area must be free of corrosive substances and vapors. Basically, when cylinders are stored as spares they should be provided with a suitable protective oil, as, for example, is used for diesel engines. Specific information can be obtained from the oil recommendation sheets of the individual oil companies. For example, Texaco recommends an engine oil EKM 152 or the more viscous EKM 162 for the protection of hydraulic cylinders. The application of the protective oil is best carried out by running the cylinders briefly on the oil. Thorough cleaning before filling up with the final hydraulic oil is not necessary, since this protective fluid is compatible with pure mineral hydraulic oils. It is, however, recommended that the first oil change be made earlier than usual.

Cylinder ports should be securely plugged when in storage. Rods must bc protected from mechanical damage or oxidation.

When using fluids such as chlorinated or unchlorinated phosphate ester or glycols, it is necessary to wash out the system thoroughly, followed by flushing with appropriate compatible fluids. Normal seals used with petroleum oils are not usually compatible with these fluids.

Care is necessary when changing from one fluid to another. Follow equipment and fluid manufacturers recommendations. Some changes require replacement of all elastomeric seals. Certain fluids will attack unprotected metal. Stainless steel tanks and plumbing may be necessary or epoxy coatings under certain circumstances.

2.5.5 Troubleshooting

Problems with cylinders fall into two general categories: erratic action and cylinder fails to move the load. There are several other problems, such as leakage, seal wear, and drifting. These problems are all interrelated. Following is a description of each problem area, the causes, and the remedy.

Cylinder Drifts

1. For piston seal leaks, pressurize one side of the cylinder piston and disconnect the fluid line at the opposite port. Observe leakage. One to three cubic inches per minute is considered normal for piston ring leakage. Virtually no leak should occur with soft seals on the piston. Replace seals as required. Note that certain types of designs may allow the flat face of the piston to be pressed against the face of the cap or head and thus act as a seal. To avoid errors, the actual leakage past piston seals should be determined with the piston and rod assembly blocked so that the piston faces are not in contact with either end. This procedure is shown in Fig. 2.39.
2. For other circuit leaks, check for leaks through the operating valve and correct if necessary. Correct leaks in connecting lines.
3. A conventional single rod cylinder will creep if restriction on tank port is sufficiently high when used with an open center valve because the rod area will function as a simple ram.
4. A closed center valve can cause similar results except that creep will be according to the amount of clearance flow in the valve. Proper

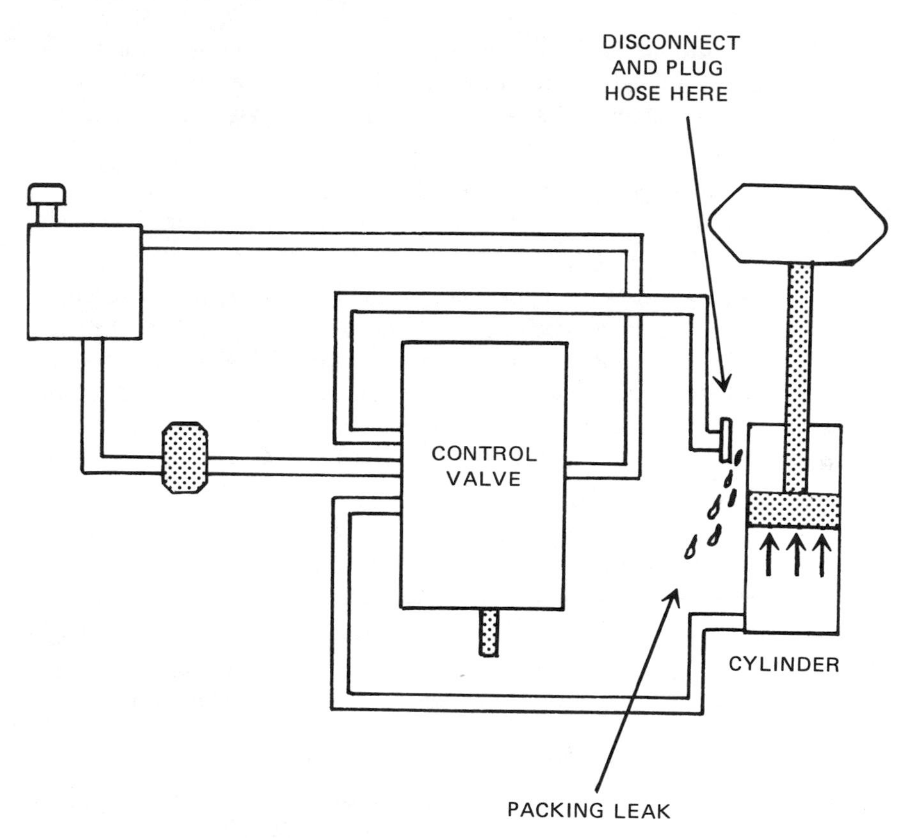

Figure 2.39 Checking double-acting cylinder for leaks.

notching of the valve spool can prevent pressure buildup in the cylinder lines between cycles.

5. Spools with pressure blocked and cylinder ports completely relaxed will also prevent drift if the moving element is not affected by gravity or vibration.
6. Pilot-operated check valves can positively lock the fluid in the cylinder lines. Care must be exercised to ensure adequate pilot pressure when rod differential may cause intensification.

Cylinder Fails to Move the Load When Valve Is Actuated

1. *Binding in machine linkage*: Check linkage to ensure that excessive friction loads are not present.

2. *Pressure too low*: Check the pressure at the cylinder to make certain that it is in accordance with circuit requirements. Review recorded pressure values on machine history chart to see if there has been any change in pressure values.
3. *Piston seal leak*: Operate the valve to cycle the cylinder and observe fluid flow at the valve exhaust ports when the cylinder is at the end of the stroke. Replace seals if flow is excessive.
4. *Cylinder undersized for load*: If this is a new system, recalculate force needs and install appropriate size cylinder to carry the load. Make certain machine members are properly lubricated and machine ways are not tightened excessively restricting movement. *Make sure machine linkage can accept added force.* If the cylinder is on an older machine which worked well in the past, the trouble should be correctable without changing the cylinder size.
5. *Piston rod broken at piston end*: Disassemble and replace piston rod. Check alignment and inspect the rod bearing carefully to make certain excessive side thrust is not present.
6. *Contamination in the hydraulic system, resulting in scored cylinder bore*: Disassemble and replace necessary parts. The contamination must be cleaned from the system before putting it back into service.

Erratic Cylinder Action

1. *Valve sticking or binding*: Check for dirt or gummy deposits. Check for contamination of oil. Check for air in the system. Check for worn parts. Excessive wear may be due to oil contamination.
2. *Cylinder sticking or binding*: Check for overtightened packing on rod seal or piston. Check for dirt, gummy deposits, or air leaks as above. Check for misalignment, worn parts, or defective packings or seals.
3. *Sluggish operation during warm-up period*: Viscosity of oil too high or pour point too high at starting temperature. Change to oil with a lower viscosity or a better viscosity index and a lower pour point. An immersion heater in the oil may help under severe cold conditions. (Provide for suitable oil circulation if heater is installed—oil does not circulate easily from thermal causes and tends to stratify).
4. *Pilot pressure too low*: The control line may be restricted. The metering choke valve may not be adjusted properly. Pilot pressure source may not be adjusted properly. Pilot pressure source may have been tampered with. Check source of pilot pressure and make certain correct level is set as indicated on circuit drawing.

5. *Internal leakage in cylinder*: Repair or replace worn parts and packing. Check the oil to see that the viscosity is not too low. Check for excessive contamination or wear.
6. *Air in system*: Bleed the air and check for leaks. Check to see that the oil intake is well below the surface of oil in the reservoir. Check the pump seals and line connections on the intake side by pouring hydraulic oil over the suspected leak. If the noise stops, the leak has been located. Tighten the joints or change the seal or gasket where necessary. Look for saturated suction strainer.

Cylinder Body Seal Leak

1. *Loose tie rods*: Torque the tie rods to manufacturer's recommendations for that bore size.
2. *Excessive pressure*: Check the maximum pressure rating on the cylinder nameplate. Reduce the pressure to the rated limits. Replace the seal and retorque the tie rods. Use transducer to record potential shock pressures. The transducer should be connected to measure the maximum internal pressures developed in the cylinder. The deceleration of large, fast-moving loads can create internal pressures many times greater than the circuit pressure measured at cylinder ports. Consider use of shock suppressing devices.
3. *Pinched or extruded seal*: Replace the cylinder body seal and retorque the tie rods to manufacturers specifications. Check for transient shock conditions as in item two.
4. *Seal deterioration—soft or gummy*: Check the compatibility of seal material with the fluid used or the lubricant used if an air cylinder is involved. Replace with a seal that is compatible with the operating fluid.
5. *Seal deterioration—hard or loss of elasticity*: Usually, this is due to exposure to elevated temperatures. Protect the seal from heat by shielding the cylinder from the heat source. Replace the seal.
6. *Seal deterioration—loss of radial squeeze due to flat spots or wear on the inside or outside diameter*: Can occur as normal wear due to high cycle rate or length of service. Replace seals.

Rod Gland Seal Leak

1. *Torn or worn seal*: Examine the piston rod for dents, nicks, gouges, or score marks. Replace the piston rod if the surface is rough. Check the gland bearing for wear. If clearance is excessive, replace the gland and seals.

2. *Seal deterioration—soft or gummy*: Repeat the cylinder body seal leak test procedure.
3. *Seal deterioration—hard or loss of elasticity:* Repeat the cylinder body seal procedure.
4. *Seal deterioration—flat spots on inside or outside diameter*: Repeat the cylinder body seal procedure.

Excessive or Rapid Piston Seal Wear

1. *Excessive back pressure due to overadjustment of speed control valves*: Correct the valve adjustment.
2. *Seals incorrectly installed*: Check installation instructions and make necessary corrections.
3. *Wear due to contaminated fluid*: Check for fluid contamination and clean the fluid filters. Replace or recycle fluid if degree of contamination persists with suitable fluid conditioning equipment (i.e., portable filter cart, centrifuge, etc.).

2.6 HYDRAULIC ACCUMULATORS

2.6.1 Types and Applications

An accumulator is a storage device for high-pressure hydraulic oil. The system pump stores oil in the accumulator during periods when it would normally be unloaded or idling. The stored oil is available at a later time either to supplement pump oil or for use when the pump is shut down.

One very popular type of accumulator has an appearance similar to a fluid power cylinder, with the piston acting as a barrier between the stored high-pressure oil and the compressed gas, which is the energy-storing medium. Fluid pressure on both sides of the piston is essentially the same. The *volume rating* of an accumulator (1-gal, 2-gal, 5-gal, etc.) is the approximate cubic capacity of the gas side when all oil is discharged. Volume is slightly less on the oil side due to the piston shape.

Besides the piston-type accumulator, several other configurations are available as shown in Fig. 2.40. Included among all these are the following:

Spring-loaded piston type: Type of piston where a spring is used to preload the piston instead of a gas.

Diaphragm and bladder types (see Fig. 2.41): In these units, an elastomeric (rubber) member is used to separate the gas from the oil. In the bladder type, the gas is inside and the oil surrounds the rubber blad-

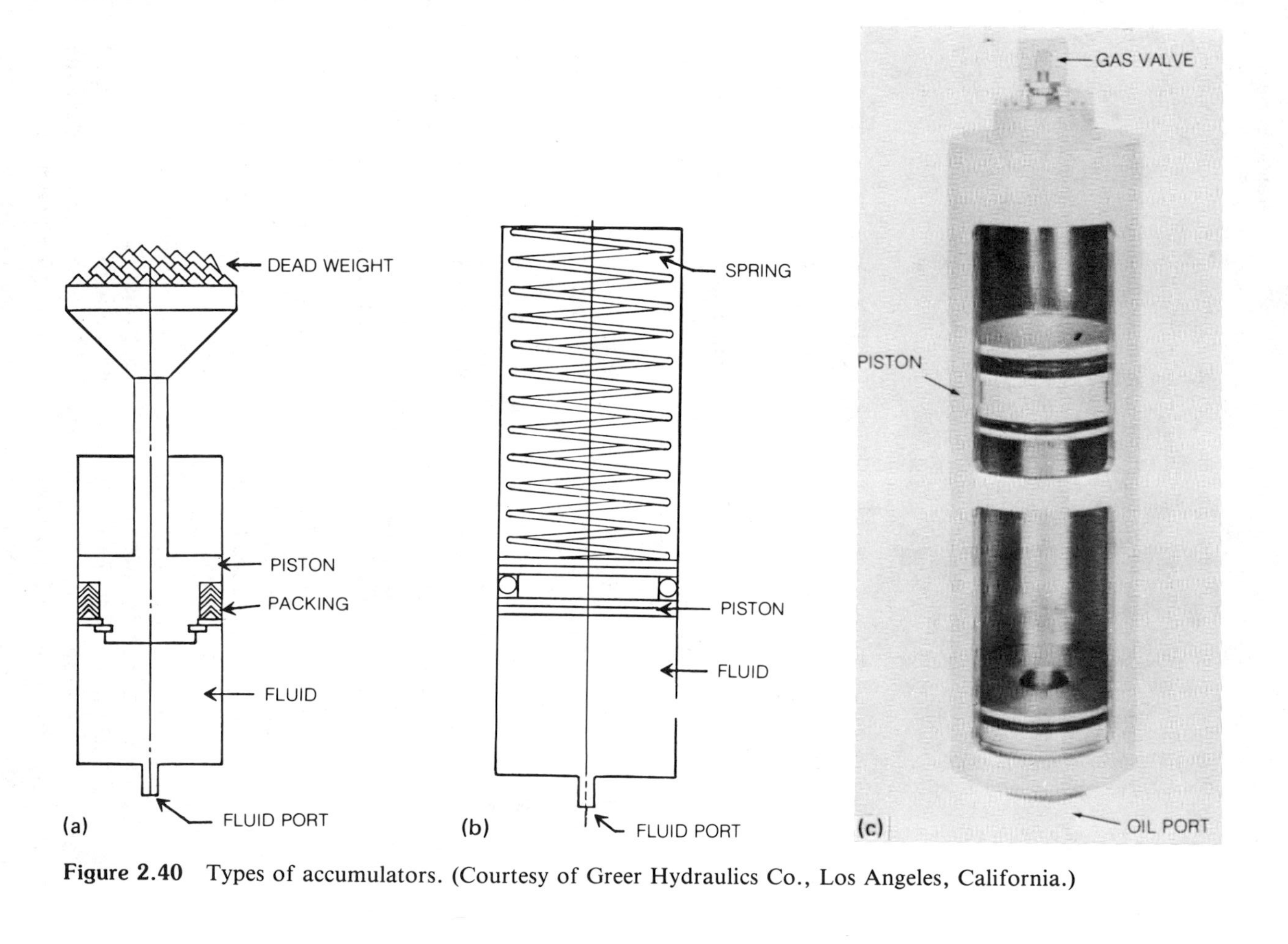

Figure 2.40 Types of accumulators. (Courtesy of Greer Hydraulics Co., Los Angeles, California.)

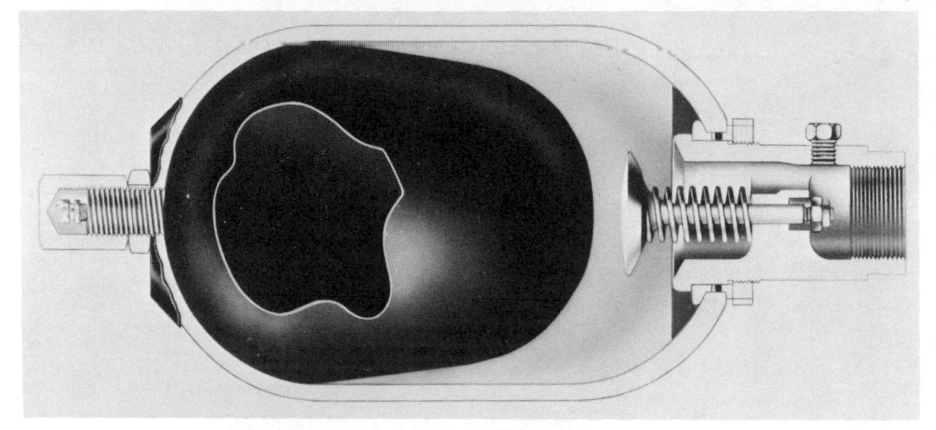

Figure 2.41 Bladder accumulator. (Courtesy of Greer Hydraulics, Los Angeles, California.)

der. In these units, as the oil leaves the accumulator, the rubber member expands and eventually conforms to the internal shape of the accumulator when it is empty of oil.

Weighted type: Physical weight responding to gravity forces is supported by a hydraulic cylinder. The area of the piston in square inches and the mass being supported determines the resulting pressure in pounds per square inch. Frictional forces are normally negligible. Obviously, the weighted accumulator must be installed in a vertical mode. The weighted-type accumulator provides a uniform pressure from completely charged condition to minimum charged condition. The weight and area relationship is a constant value. Consequently, the weighted-type accumulator is well qualified to provide uniform force for holding various types of forming rolls in position and to maintain pressure during long holding cycles such as rubber and plastic molding activities. The weighted accumulator is provided with mechanical signal sources to control charge function. The spring-loaded and gas-charged accumulators depend on a pressure signal to control the charge function.

A piston accumulator may be mounted in any position, although the preferred arrangement where there will be a very heavy momentary discharge is to position it so that the heavy discharge is in a straight line with the work. Potential contamination in the hydraulic fluid may make it desirable to mount the piston-type accumulator vertically to avoid scratching the barrel by solid particles lodged in the piston seals. The preferred mounting position for bag-type accumulators is vertically upright, although they will operate in any position. Contamination in a bag-type accumulator can become embedded in the bladder creating a vulnerable area or prevent the proper operation of the antiextrusion valve, allowing the bag to partially extrude into the oil area when pressure in the hydraulic fluid decreases.

Accumulators can be considered energy-storage devices similar to a receiver tank in a pneumatic system. However, they can be used for other purposes, as shown in Figs. 2.42 and 2.43, including as shock absorbers, as pressure surge dampeners, and for reservoir pressurization where it is undesirable for the tank to be open to the atmosphere through a breather (Fig. 2.43).

Bladder-type devices provide an excellent barrier to entry of dirt and moisture into a hydraulic reservoir. The reservoir can be completely sealed

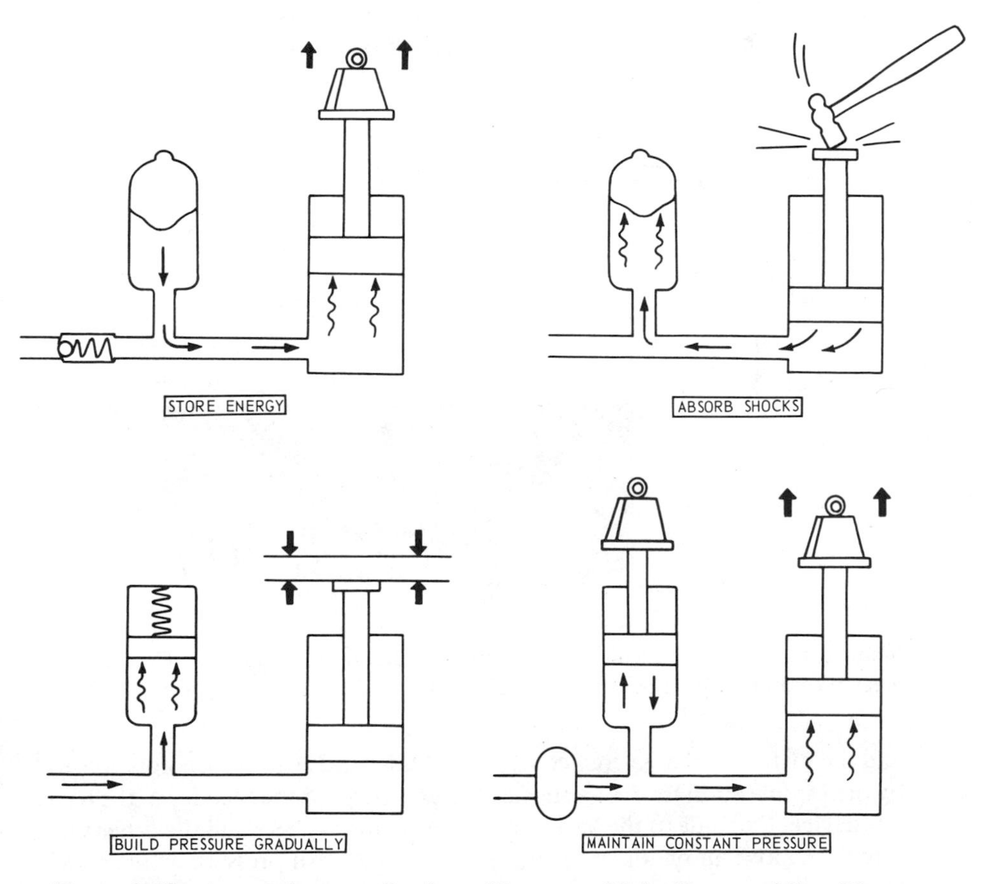

Figure 2.42 Accumulator applications. (Courtesy of John Deere and Co., Waterloo, Iowa.)

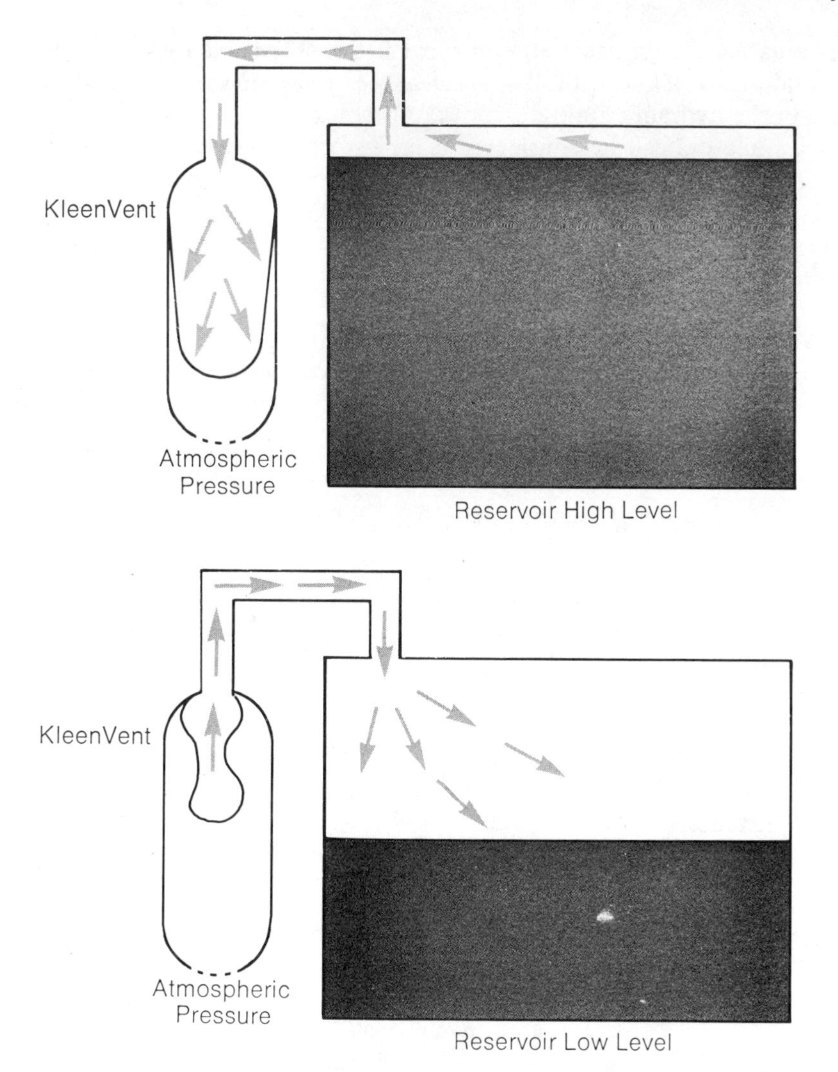

Figure 2.43 Closed reservoir arrangement using bladder accumulator. (Courtesy of Greer Hydraulics Co., Los Angeles, California.)

and the differential area created by the cylinder rods can be accommodated by the captive air in the reservoir plus the excursion of the bladder in the barrier device. One side of the barrier device is to atmosphere and the other side is to the captive air on top of the oil in the reservoir. Air pressure on the reservoir may range up to several psi, depending on the location of the reservoir, type of construction, and design of the barrier assembly. Some hydraulic reservoirs are designed to operate with positive pressure on the captive air above the oil. This positive pressure may be used to provide better pump intake conditions.

2.6.2 Servicing and Repair

Observe the following precautions when working on pneumatic accumulators.

1. *Caution: Never fill an accumulator with oxygen!* An explosion could result if oil and oxygen mix under pressure.
2. Never fill an accumulator with air. When air is compressed, water vapor in the air condenses and can cause rust. This in turn may damage seals and ruin the accumulator. Also, once air leaks into the oil, the oil becomes oxidized and breaks down.
3. Always fill an accumulator with an inert gas such as dry nitrogen. This gas is free of both water vapor and oxygen; this makes it harmless to parts and safe to use.
4. Never charge an accumulator to a pressure more than that recommended by the manufacturer. Read the label and observe the "working pressure."
5. Before removing an accumulator from a hydraulic system, release all hydraulic pressures.
6. Before you disassemble an accumulator, release both gas and hydraulic pressures.
7. When you disassemble an accumulator, make sure that dirt and abrasive material does not enter any of the openings.
8. Install valve to vent pressurized oil from accumulator to tank, preferably automatically when machine shuts down. This may be mandatory in some political areas as a part of a safety code.

Servicing and Precharging Pneumatic Accumulators

Checking Precharged Accumulator on the Machine

1. If you suspect external gas leaks, apply soapy water to the gas valve and seams on the tank at the "gas" end. If bubbles form, there is a leak.
2. If you suspect internal leaks, check for foaming oil in the system reservoir and/or no action of the accumulator. These signs usually indicate a faulty bladder or piston seals inside the accumulator.
3. If the accumulator appears to be in good condition but is still slow or inactive, precharge it as necessary.

Before Removing Accumulator from Machine. First be sure that all hydraulic pressure is released. To do this, shut down the pump and cycle some mechanism in the accumulator hydraulic circuit to relieve oil pressure (or open a bleed screw).

Removing Accumulator from Machine. After all hydraulic pressure has been released, remove the accumulator from the machine for service.

Repairing Accumulator

1. Before dismantling an accumulator, release all gas pressure. Unscrew the gas valve lever very slowly. Install the charging valve first if necessary. Never release the gas by depressing the valve core, as the core might be ruptured.
2. Disassemble the accumulator on a clean bench area.
3. Check all parts for leaks or other damage.
4. Plug the openings with plastic plugs or clean towels as soon as parts are removed.
5. Check bladder or piston seals for damage and replace if necessary.
6. If gas valve cores are replaced, be sure to use the recommended types.
7. Carefully assemble the accumulator.

Charging Accumulators

Energy is stored in the accumulator by compression of an inert gas. Any of the inert gases could be used, but for economic reasons, *nitrogen is normally used because it is nonflammable and is readily available* on the industrial market at a reasonable price. It is obtained in high-pressure bottles of approximately 2200 psi. Where possible, use oil-pumped "dry" nitrogen to reduce the possibility of moisture condensation in the gas chamber, which would seriously damage a piston-type accumulator.

The actual pressure (psi) value of the nitrogen precharge pressure is not highly critical. A range of one-third to one-half of the maximum hydraulic pressure will give good results on most applications. Very little difference in circuit performance will be noticed within this range. Suggested procedure is to precharge initially to the higher value, then do not add more gas until the pressure falls to the lower value.

Caution: Do not use oxygen or any gas mixture containing oxygen, such as compressed air, for precharging accumulators that have rubber seals or separators. In the first place, oxygen deteriorates rubber by oxidation. In the second place, the use of oxygen always creates a possible fire or explosion hazard in any situation, and this is especially dangerous in the vicinity of hydrocarbons such as petroleum oil.

Service notes: On machines using accumulators, test the precharge pressure level if the machine shows symptoms of sudden slowdown before the cycle is completed. If other, more specific information is not available, the precharge level should be one-third to one-half of the maximum system pressure.

One way to test for insufficient accumulator oil capacity is to install two pressure gauges, one on the oil port, the other on the gas port. Observe these two gauges while the machine is going through its cycle. They should have essentially identical readings throughout the entire cycle unless the accumulator runs out of oil. In this case the oil gauge will suddenly drop to near zero as the piston in the accumulator bottoms against the end cap. If the accumulator runs out of gas, or the precharge drops too low, the machine will be slow throughout the cycle, working only at the speed of oil developed by the pump.

Precharging Equipment (Fig. 2.44). A charging and gauging hose assembly should be purchased from the manufacturer of the accumulator brand in use, and should be kept on hand as a service tool. Because of variation in construction of the air valve, a charging hose for one brand may or may not work on a different brand.

Procedure for Precharging. The accumulator may be precharged either while connected into the system or as a loose component lying on the test bench. First, make sure that all oil is discharged from the accumulator. If it is connected into a hydraulic system, open the circuit bleed-down valve to discharge all accumulator oil to the reservoir. Next, connect the charging hose to the accumulator and take a reading of the existing precharge pressure, if any. Connect the hose to the gas bottle. A pressure regulator need not be used. It is not difficult to control the flow rate of gas with the shutoff valve located on the bottle. Slowly open the valving and allow gas to flow into the accumulator. The precharge level cannot be accurately gauged while gas is flowing. Shut off the valve frequently to observe the psi level in the accumulator. After precharging, immediately close all valves and remove the charging hose. For safety, replace the cover on the gas bottle and the cover on the gas valve of the accumulator.

Frequency of Servicing

The gas precharge will gradually leak down on any accumulator. The gas molecules transfer through the pores of the rubber by osmosis. Piston-type accumulators should be tested every month or two, oftener if found to be

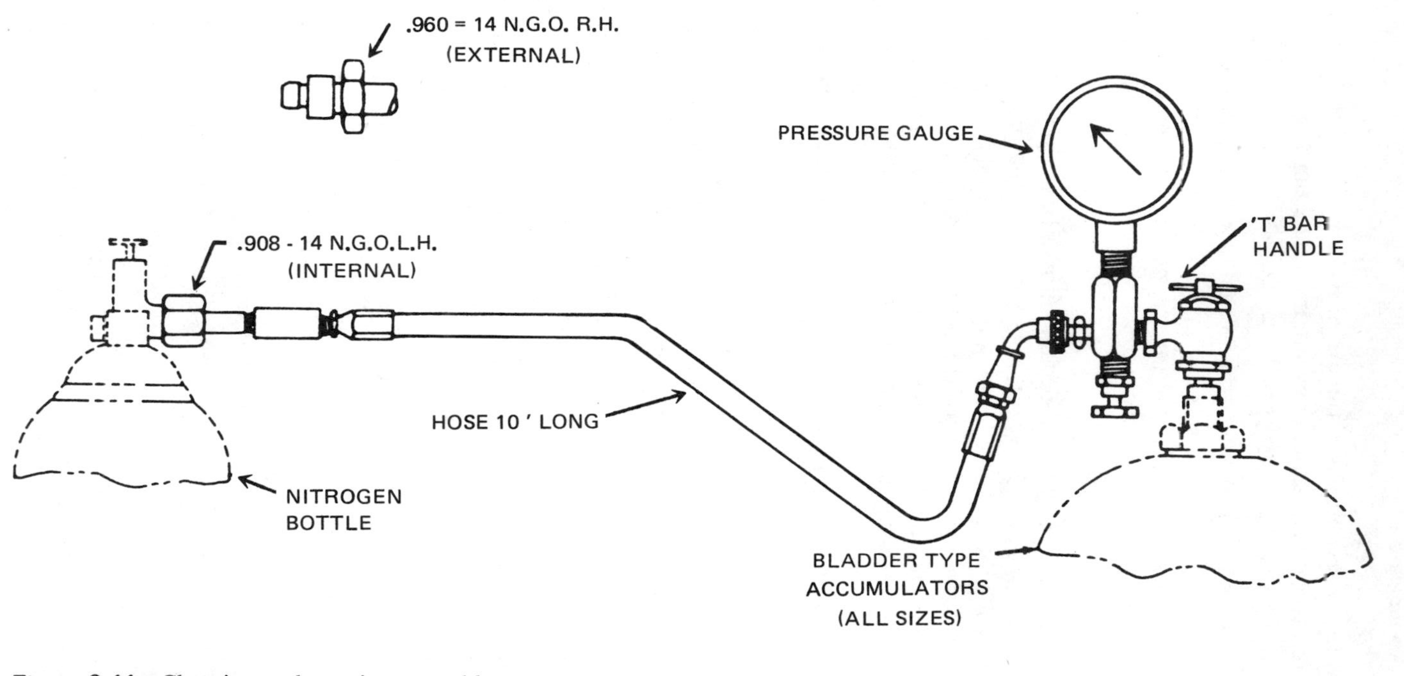

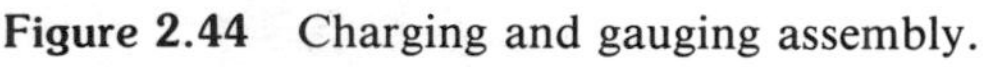
Figure 2.44 Charging and gauging assembly.

necessary. On rapid-cycling applications, precharge psi should be checked every 100,000 cycles. Bag or bladder accumulators have a much greater leak-off rate because of the much greater rubber surface, and should be serviced more often. With piston accumulators, the piston seals (O rings) should be replaced approximately every 1,000,000 cycles because of mechanical wear. Some accumulators use O rings of special cross section, and although it is possible to replace these with standard commercial rings, it is better to obtain exact replacements from the manufacturer.

> *Rule of thumb*: For every 1% an accumulator is oil charged above the minimum system pressure, it will deliver 1 in.^{3} of oil for each gallon of its rated size.

Example of use of rule of thumb: Assume a hydraulic system where the accumulators are to be charged to a maximum pressure of 3000 psi. System pressure will be permitted to fall to 2000 psi after the accumulator has delivered its oil. From 2000 to 3000 psi represents an increase of 50% that the accumulator will be charged above minimum pressure. According to the rule of thumb, a 1-gal accumulator would give out 50 in.^{3} of oil under these conditions. If the circuit required 250 in.^{3} of oil, you would select a 5-gal accumulator.

Installing Accumulator on Machine. Attach the accumulator to the machine and connect all lines. Start the machine and cycle a hydraulic function to bleed any air from the system. Then check the accumulator for proper action.

Details of accumulator troubleshooting are given in the following chart.

Problem	Cause	Solution
Slow reaction	1. Loss of charge or overcharge	Check charge pressure; reset
	2. Unloading valve or pump low pressure set too low	Adjust to higher pressure
	3. Relief valve set too low or stuck open	Reset or clean valve
	4. Pump not pumping	Check pump
	5. Unloading pressure switch set too low	Reset pressure switch
Fails to absorb shock	1. Loss of charge or overcharge	Check and recharge if necessary or reset

2.7 HYDRAULIC SYSTEM RESERVOIRS

An essential part of any hydraulic system is the reservoir. Its primary purpose is to hold the fluid supply for the hydraulic system. Besides holding the fluid supply, a well-designed reservoir serves other useful purposes. For the hydraulic system to operate properly, hydraulic reservoirs must:

Provide a relatively large surface, so heat from the fluid can transfer to the environment by radiation and convection.

Provide a relatively large volume for fluid to slow from its higher velocity through the piping and power components. This allows some of the heavier fluid contaminants to settle out.

Provide space above the fluid for entrained air to escape to the atmosphere.

Provide easy access to remove used fluid and contaminants from the system. Fresh fluid is added here for makeup and changeover.

Leave space for hot-fluid expansion, gravity drain back from a system during shutdown, and storage of large fluid volumes needed only during part of the operating cycle.

Serve as a convenient surface to mount other system components.

2.7.1 Construction and Design (Fig. 2.45)

The reservoir is usually rectangular in shape and is constructed of cold-rolled steel plates with welded joints. A well-designed reservoir is equipped with the following design features and parts:

1. An oil level gauge to give visual indication of the oil level.
2. A filler opening equipped with a strainer used for filling the reservoir. Pump inlet line and all return lines should be extended well below the fluid level to within a few inches of the tank bottom. It is desirable to have the fluid free from air and foaming, thereby reducing pump cavitation problems. Return lines are usually cut at an angle of 45° at the bottom. The flow is directed toward the walls of the tank to allow for maximum heat dissipation.
3. A baffle plate should be placed between the pump intake and the system return line. The baffle plate serves to prevent recirculation of the same fluid.
4. The reservoir should be provided with a breather to allow for air displacememt as the oil level rises and falls. The breather should contain a filter to clean the air. The breather assembly should be large enough to maintain atmospheric pressure in the reservoir.

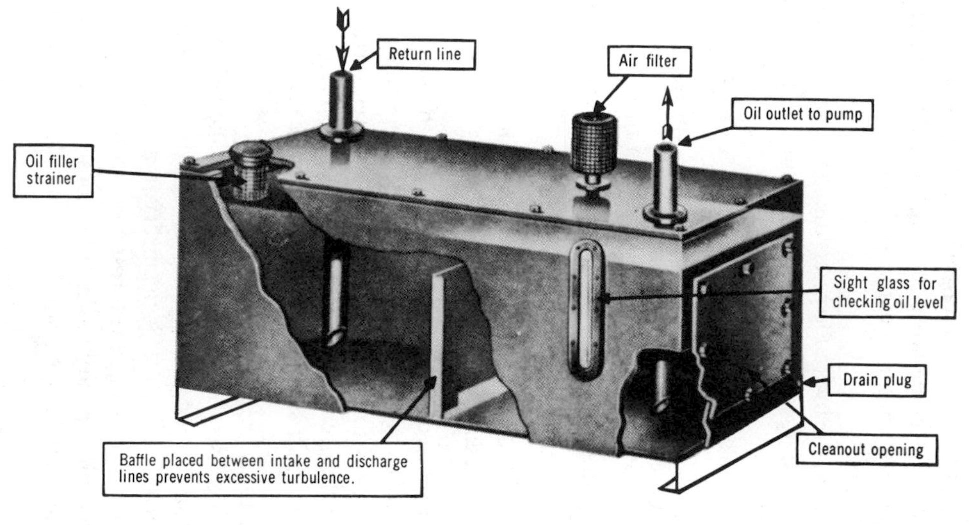

Figure 2.45 Hydraulic reservoir arrangement. (Courtesy of Sun Oil Co., Philadelphia, Pennsylvania.)

Although cooling the fluid is one of the functions of the reservoir, immersion heaters are sometimes used to aid in dissipation of air from the fluid. For example, when a system contains a fire-resistant fluid with a high specific gravity, air entrainment becomes a problem and heat is used to aid in releasing the air. Heat and immersion heaters are also used to thin out heavy, viscous, fire-resistant fluids prior to starting hydraulic systems in cold weather. Because of the poor low-temperature starting characteristics of fire-resistant fluids, immersion heaters are used extensively in reservoirs of systems operating with these fluids.

In some systems, the hydraulic reservoir can be built as an integral part of the equipment which it services. An integral reservoir may cause these particular problems:

Space considerations may limit size. Because heat transfer capacity is a function of size, external oil coolers or heat exchangers may be needed.

Irregular shape may require special baffling to route fluid properly.

Surrounding equipment may limit convectional heat transfer.

Service accessibility may be poor.

Special heat shielding may be needed to isolate components or the operator from reservoir heat.

Geometrically, a square or a rectangular tank has the largest heat transfer surface per unit volume. A cylindrical shape may offer some fabrication economy. If the reservoir is too shallow, the heat transfer surface of the walls will not be used to the greatest advantage. If it is too tall and narrow, there may not be enough fluid surface area for air to escape.

Reservoir Capacity

The functions of the reservoir cannot be accomplished if it is of insufficient size. No set rule can be used to establish the reservoir capacity because there are too many factors involved. It is necessary to analyze the needs of an individual system and then design the reservoir to meet these requirements. A sacrifice in reservoir capacity in the initial design generally results in the eventual addition of auxiliary equipment to supplement the inadequate facility.

When the capacity of a reservoir is to be determined, the following criteria should be satisfied:

1. It should have sufficient volume to allow time for dirt and cuttings to settle and for air to escape.
2. It should be capable of containing all the oil that might drain into the reservoir by gravity as well as the oil from large fluid rams.

3. It should have a surface area large enough to dissipate most of the heat generated in the system (Fig. 2.46).
4. It should be designed to have an operating oil level sufficiently high to prevent "cratering" of the oil or the formation of a vortex around the pump strainers.
5. It should have adequate air space to allow for thermal expansion of the fluid.
6. It should be large enough to provide for future additions of power attachments for the machine it serves.
7. It should have a low center of gravity to give stability to the unit if a motor and pump are mounted on the top surface of the reservoir.
8. It should have adequate dimensions for receiving pump strainers and pipe bosses for auxiliary filters and heat exchangers.
9. It should be designed larger if the fluid circuit contains high-heat-generating components or when the ratio of the working pressure time to the total cycle time is high.
10. It should have a greater capacity if the fluid system is expected to operate continuously. It could be reduced considerably in size for intermittent operation.

Recommended reservoir capacity is three times the flow rate in gallons per minute for fixed-displacement pumps, or three times the mean flow rate for

Total volume (gal)	Dissipating surface (ft²)	Heat dissipation	
		Btu/hr	hp
6½	4½	575	0.23
12	6¾	850	0.34
17	8¼	1,000	0.40
33	13¼	1,700	0.67
80	24	3,000	1.18
120	30	4,000	1.60
250	50	6,000	2.70
500	80	10,200	4.00

Figure 2.46 Heat dissipation capability of hydraulic reservoirs.

pressure-compensated, variable-displacement pumps. These suggested volumes usually allow the fluid to rest between work cycles for heat dissipation, contaminant settling, and deaeration.

If there are circuit components, which require large fluid volumes, such as accumulators or large cylinders, the reservoir may have to be larger so that the fluid level will not drop below the pump inlet regardless of pump flow. Conversely, if external oil cooling is available, if cycling is intermittent, or if ambient conditions permit the reservoir to cool rapidly, it may be possible to use a smaller reservoir.

In any case, the reservoir should contain additional space equal to at least 10% of its fluid capacity. This allows for thermal expansion of the fluid and gravity drain back during shutdown and still provides a free fluid surface for air separation.

Reservoir Construction

The reservoir must not only perform the functions that have been discussed, but should be serviceable and have a neat appearance. In many instances, it is desirable to consolidate the entire pump and reservoir system by constructing the reservoir so that the pump and driving motor can be installed on its top surface. If the pump and motor are mounted directly on the top plate, a smooth-machined surface must be provided which will have adequate strength to support and hold the alignment of the units. The motor can also be attached to a subplate which is mounted above the reservoir top by the use of spacers. This method achieves flexibility and reduces heat transfer to the motor. Except for the flanged-type motor-pump assemblies, a pedestal should generally be provided to support and maintain alignment of the pump.

Inspection and cleanout openings should be large enough to reveal the entire interior of the tank and should be located so that all sections of the reservoir are within easy reach of a person's arm. These openings should be located so they will not be against a wall or part of the machine structure.

Vertical baffles are an important part of the interior of the reservoir. The purpose of the baffles is to separate the fluid entering the reservoir from the fluid leaving and to direct the fluid along a longer path than it would otherwise take. It is difficult to obtain an optimum baffle design that is conducive to good heat transfer, settling of the dirt, and deaeration of the fluid. There are basically two types of baffle configurations used: the circuitous-stream type and the wave-stream type. In the *circuitous* type, the baffle causes the fluid to flow from one end of the tank to the other and then return, which means that the fluid essentially travels twice the length of the reservoir before it can reach the pump-suction strainer. This design, shown in Fig. 2.47, seems to offer good characteristics since it exposes the fluid to the sides of the reservoir, where heat transfer can take place through the walls and also provide

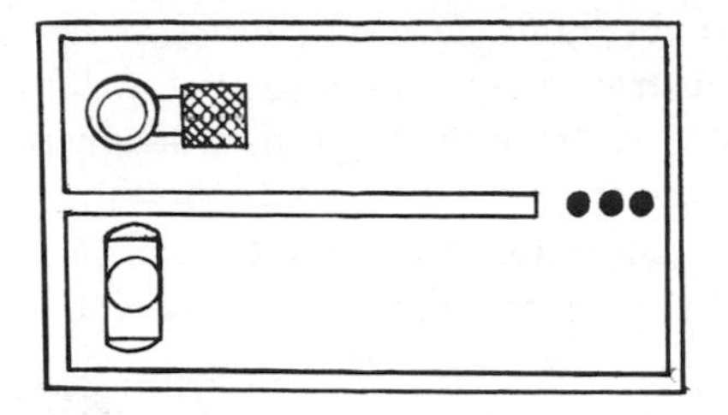

Figure 2.47 Circuitous-stream baffle configuration.

a long path which allows time for the particles to settle and entrained air to escape. At the weir end of the baffle, it is desirable to install magnetic plugs to attract iron particles. These plugs, spaced about 3 in. apart, have been found very effective in eliminating the circulation of such particles.

The *wave-stream* baffle design can have several different forms: single, double, or triple baffles. In the single-baffle form, the fluid travels over the baffle to reach the strainer, whereas in the multiple forms, the fluid must also travel under at least one baffle to reach the pump suction. If a single baffle is employed, it generally divides the reservoir symmetrically; the pump strainer is on one side and the return line is on the other. Since the reservoir top supports a baffle in multiple-baffle designs, some scheme should be used to prevent it from reversing its position, which would change the location of the baffle in the reservoir. Two of the principal advantages of the wave-stream design are that it promotes better deaeration of the fluid and that the stream velocity is lower, which allows particles to settle more effectively (Fig. 2.48).

Two methods used for distributing the return fluid into the tank are to have the ends of the return lines cut at an angle to form a large wedge-shaped

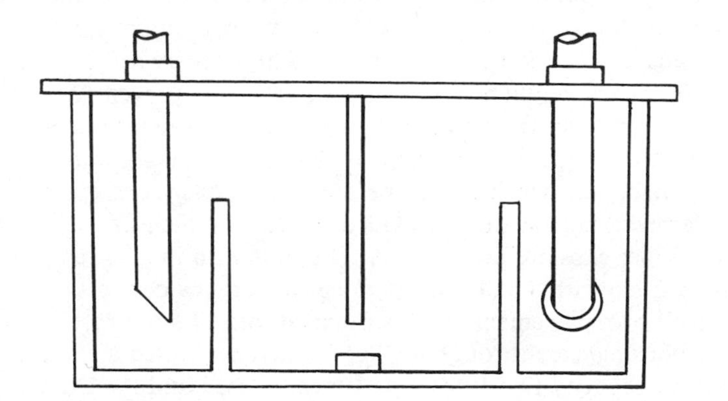

Figure 2.48 Wave-stream baffle configuration.

opening or to have a horizontal pipe section containing several holes to disperse the main stream. The pump-suction strainer should be well below the normal fluid level in the tank and at least 1 in. from the bottom. A strainer located too high will cause a vortex or crater to form, which will permit air to enter the suction line. Air in the system can cause aeration in the oil. The pump may become excessively noisy and the aerated oil usually produces a spongy action in the fluid motor actuator.

In reservoirs where atmospheric pressure is maintained, the air space above the oil should be approximately 4 in. Since the liquid level in the tank may vary with the machine operation, an air breather and filter should be installed. The breather and filter should permit the passage of air in both directions and have a capacity of at least twice the pump rating to compensate for the unequal volumes displaced by single-rod cylinders.

Suction strainers are available which can be installed in the side or top of the reservoir (often fitted in the tank wall). They can be cleaned without the necessity of opening the reservoir. There is considerable concern about suction filters which require major time allocation for service or inspection. The externally serviceable style minimizes this potential down time.

2.7.2 Accessories

Fillers and Filler Breathers

Fluid must be added to the reservoir at startup, after cleanout, and when losses occur. The filler opening should permit reasonably rapid filling, intercept large contaminant particles from the new fluid, and either seal when closed or filter incoming air if vented as a breather.

Standards call for two filler openings, each of which will pass a minimum of 5 gal of fluid per minute. The openings should be on opposite sides or ends of the reservoir. Metal strainer screens of 30-mesh maximum should have internal metal guards and should be attached so that tools are necessary for removal. The filler cover should be permanently attached to prevent loss. If the filler cover does not include a breather, a separate breather should be specified. In either case, 40-μm air filtration is the maximum allowable.

Reservoirs on machines which must operate continuously are often sealed with suitable breathing facilities provided by devices such as those shown in Fig. 2.43. Filler caps are not used. A tee connection in a suitable return to tank line is fitted with a ball shutoff valve and a quick connection terminus. Original fill or replacement fluid is pumped into this return line connection from a portable cart through a filter (sometimes called a filter dolly). Usually a large-capacity fine filtration element is used and the fluid is handled by a suitable transfer pump. Thus all fluid is filtered before entering the tank.

Similarly, a ball valve can be fitted to the lowest drain port in the reservoir. It can be connected to the filter dolly transfer pump to continuously circulate all the fluid in the reservoir on a specific schedule to remove condensate, wear particles from the circuit, and other contaminants that may have loosened from the circuit components.

Fluid-Level Indication

A fluid-level gauge should be provided to indicate the operating level of the oil in the reservoir. The two types most frequently used are the tube type and the button or "bull's-eye" type. Regardless of which type is employed, it should be flush mounted to prevent it from being damaged easily. The tube type has the advantage of always showing the level of the fluid no matter where it is located, whereas the button type has a very limited position range. Two buttons are necessary to show the limits of the high and low fluid level, whereas with the tube type only two marks are needed.

Temperature Indicators

Fluid temperature measurement is not required by standard, but a selection of thermometers is available, many in the same housing as the fluid level indicator. If high fluid temperature is a continuing problem, the heat source in the circuit should be identified and removed.

Fluid Heaters

After shutdown or when the reservoir is exposed to colder temperatures, the fluid may be too cold to perform properly at first. A thermostatically controlled heater can solve this problem.

Usually a circulating pump or propellor-like device is installed in large reservoirs when heating units are used. Often the heater is placed as close as possible to the pump inlet to facilitate the circulation process.

Diffusers

Return line flow diffusers slow down high-velocity fluid entering the reservoir. When correctly installed, they direct the returning fluid away from the pump inlet and along the reservoir walls for cooling.

Magnets

Magnets can be placed in the reservoir to capture and remove metallic particles from the fluid stream. For maximum efficiency, magnets should be cleaned regularly.

Although hydraulic filters are usually not considered reservoir accessories, almost all pump inlet strainers are located within the reservoir, and many other filters mount on or through reservoir surfaces. Because the inlet strainer is out of sight, a pressure gauge will help indicate when cleaning is necessary.

3

Pneumatic Systems: Operation, Maintenance, and Troubleshooting

3.1 AIR COMPRESSORS

Before air can be used to perform the work required in a pneumatic system, it must be given potential energy in the form of pressure. The compressor converts air (at atmospheric pressure) into high pressure by changing its volume. As the air is reduced in volume, its pressure increases.

Many types of air compressors are used in industry. They are classified according to their construction, pressure, and application requirements. An air compressor functions by creating a partial vacuum at its inlet so that air at atmospheric pressure can flow through the inlet filter into the compressor. The compressor reduces the volume while increasing the pressure of the air. When the air has been compressed, it flows through the compressor discharge valves and into a storage tank. Because the compressor operates only when the tank pressure is low, air flows from the compressor into the tank whenever the compressor is running. When tank pressure reaches the desired pressure level, the compressor is usually turned off. Air is prevented from flowing back through the compressor by the discharge valves.

Compressor Classification (Fig. 3.1)
Compressors can be classified into two separate types based on their basic principle of operation: positive displacement and dynamic. *Positive-displacement* compressors confine the air within a closed space and compress it by decreasing the volume of the space. *Dynamic* compressors accelerate the air with rapidly turning rotor blades. This increased airflow also raises the pressure of the air slightly.

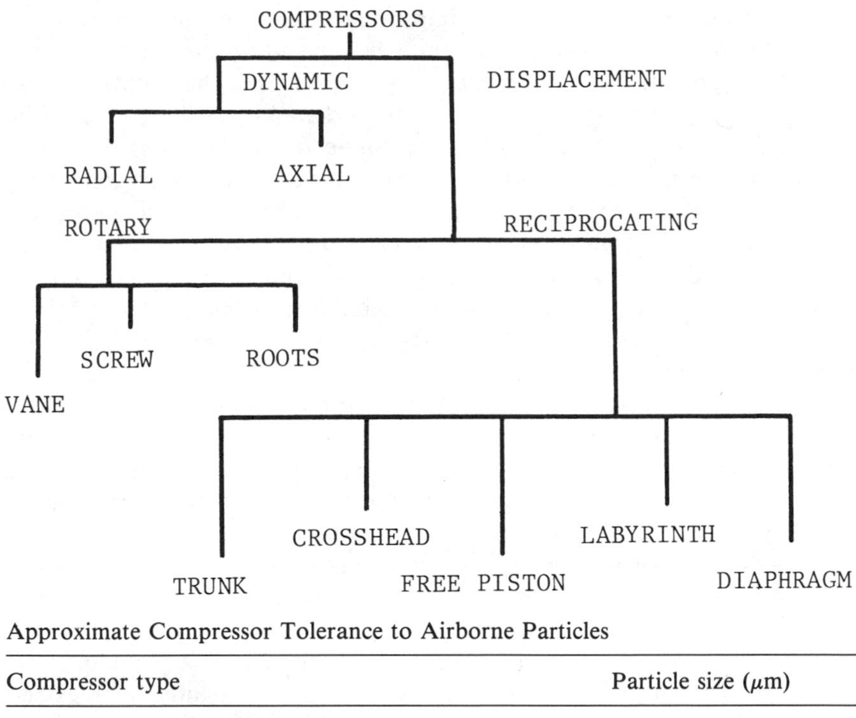

Approximate Compressor Tolerance to Airborne Particles

Compressor type	Particle size (μm)
Reciprocating type (lubricated)	25
Reciprocating type (nonlubricated)	5–10
Screw type (oil-flooded)	10
Screw type (dry)	3–5
Centrifugal	1–3

Figure 3.1 Compressor classification.

Positive-displacement compressors capable of compressing air to comparatively high pressures are commonly used in pneumatic power systems. These may be reciprocating piston, rotary sliding vane, helical screw, liquid piston, impeller, or diaphragm compressors.

3.1.1 Reciprocating Piston Compressors

Reciprocating compressors are manufactured in many varieties, and are usually driven by electric motors. The motors can be connected to the compressor directly or by V belts or reducers. Portable compressors and many large compressors are driven by internal combustion engines. Whatever type of compressor you have in your plant, it is important to keep detailed specifications and test data on file for both the prime mover and the compressor.

A reciprocating or piston compressor operates in the following manner. The compressor crankshaft is rotated or driven by an electric motor or other prime mover. The crankshaft and connecting rod convert this rotary motion into reciprocating motion. The piston is attached to the connecting rod by a crank pin, which permits independent movement of the piston and connecting rod. The piston is moved back and forth in the cylinder by the connecting rod and rotating crankshaft. The rod end of the cylinder is open, allowing the connecting rod to move back and forth while the piston travels in and out. The head end of the cylinder is closed and contains the intake and discharge valves (Fig. 3.2).

As the piston moves back (or toward the crankshaft) it creates enough negative pressure to permit air at atmospheric pressure to open the valves and push air into the expanding space. As the piston completes its stroke and starts returning, the intake valves close. As the piston returns, it compresses the air in the cylinder. When it has almost completed its stroke, it has compressed the air to a pressure high enough to force open the discharge valves. Air discharged from the cylinder at this point is at the discharge pressure of the compressor.

Intake and discharge valves may be of feather, disk or plate, or channel design. The construction of each of the valves differs slightly, but the operation of all is similar. Valve parts include a valve seat, movable valve, and guard or stop plate. The movable valve is frequently held against the seat by a spring. During operation the spring holds the valve in contact with the seat. As the air is compressed in the cylinder, it forces the valve open, and air discharges to the receiver. When the piston reverses its stroke, the spring and

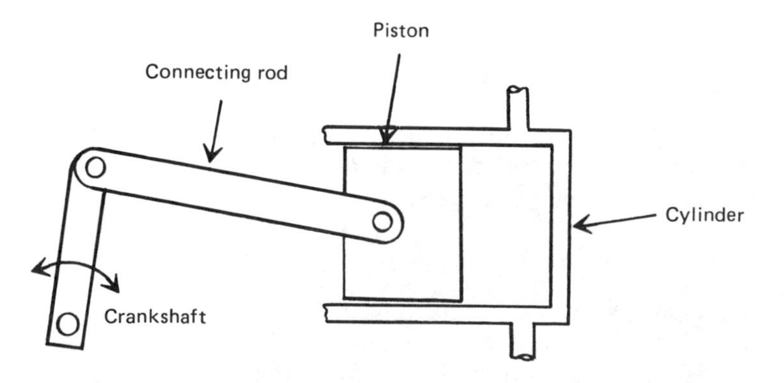

Figure 3.2 Single-stage reciprocating machine.

discharge pressure close the valve, preventing air from returning to the cylinder (Fig. 3.3).

Compressors classified according to their operation include single- or double-acting, and single- or multistage. A compressor that compresses air at only one end of a cylinder is called a *single-acting* compressor. A compressor that compresses air at both ends of a cylinder is called a *double-acting* or *crosshead* compressor.

The terms *single-stage* or *multistage* describe the number of stages or

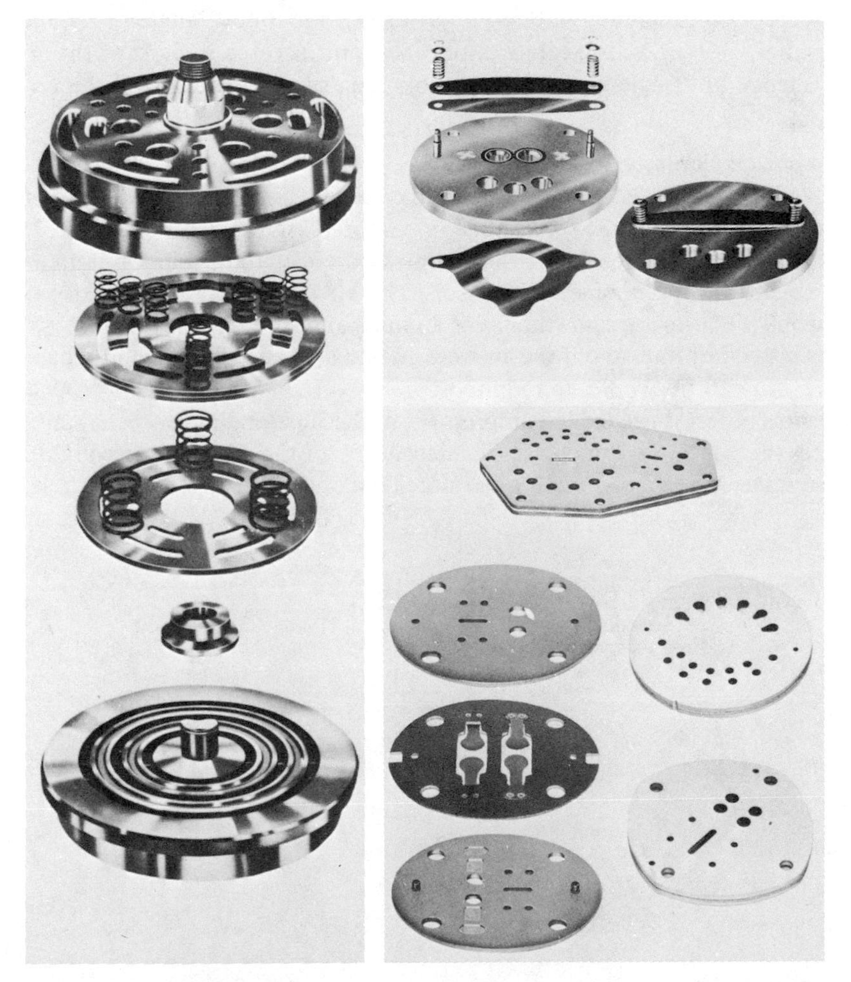

Figure 3.3 Compressor air valve configurations. (Courtesy of Parker Hannifin Corp., Cleveland, Ohio.)

steps that a reciprocating compressor uses to compress air to its final pressure. A compressor that draws in air at atmospheric pressure and compresses it to its final pressure in one stroke is called a single-stage compressor. A compressor that draws in air at atmospheric pressure and compresses it in two or more strokes is called a multistage compressor. A simple multistage compressor can be constructed with two or three cylinders, each with a different piston diameter. (Fig. 3.4).

Compressor Construction

Reciprocating compressors are also identified by the way their cylinders are arranged. These include: vertical, V or Y, W, horizontal, angle or L, and semiradial. Double-acting compressors are manufactured in most of the arrangements, but single-acting compressors are usually made only in the vertical, V, or Y setup.

Horsepower Cooling Requirements

When air is compressed, its temperature increases because of friction between the air molecules and the work done in compressing the air. More than 25 hp is needed to produce 100 ft^3 of compressed air at 100 psi in 1 minute in a one-cylinder single-stage compressor. The air temperature would rise to about 400°F. If, in the same time, this amount of air were compressed in two stages instead of one, and if the air were cooled between the first and second stages, only 10 hp would be required.

Less power is needed to compress the air because cooling between stages reduces the volume and increases the density of air that is compressed in the second stage. When a compressor has more than one stage, air is usually cooled

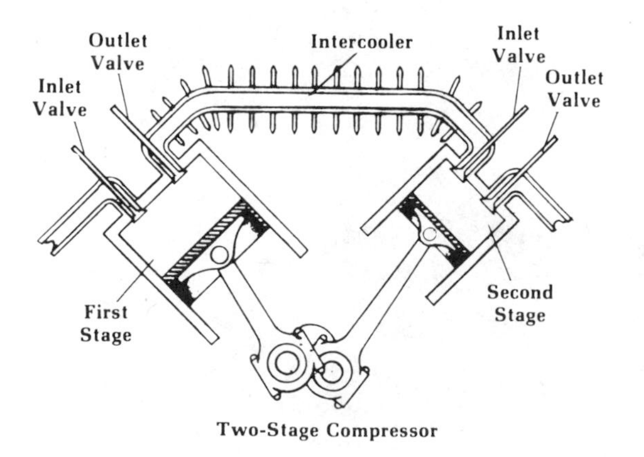

Figure 3.4 Two-stage single-acting machine. (Courtesy of Parker Hannifin Corp., Cleveland, Ohio.)

after it has been discharged from the first stage and before it is compressed in the next stage. This is called *intercooling*. If the air is cooled after it leaves the compressor, it is said to pass through an *aftercooler*.

The compressors can be cooled also. Compressor cooling is accomplished with either water or air, depending on the application. Air cooling is less costly because less equipment is needed. But water cooling is more economical because it is more efficient. Air cooling is generally limited to smaller compressors (below 25 hp) and water cooling to the larger units.

Several types of intercoolers are manufactured for compressors. They may be air-cooled or water-cooled, depending on the compressor application. *Air-cooled* intercoolers are usually one or more lengths of finned tubing connecting the first and second stages of the compressor. Hot compressed air passing through the tube gives up its heat to the tubing, which in turn transfers it to the cooling fins. Because the fins spread the heat over a larger surface area, the transfer rate is much better than the small surface area of the tube.

The amount of cooling obtained depends on the amount and temperature of the compressed air, the length of tubing, the number of fins, and the temperature and speed of the cooling air moving over the tubing. To aid in transferring the heat from the finned tubes, most air-cooled compressors have a fan mounted on the compressor crankshaft. This fan is usually cast as a part of the compressor flywheel, and it forces air over the fins while the compressor is running.

The *water-cooled* intercooler is simply a shell-and-tube heat exchanger consisting of a bundle of tubes mounted in a shell. Cooling water is connected to pass through the shell, and compressed air is connected to flow through the tubes. The hot discharge air transfers its heat to the tubes and then to the water. Some heat exchangers are connected in the opposite manner, with the water running through the tubes and air passing through the shell.

Whatever arrangement is used, *the fluid and gas usually flow in opposite directions, commonly called counterflow*. Air entering the heat exchanger is cooled by liquid that is leaving. As the air moves further through the heat exchanger, it is cooled by cooler water. Finally, just before the air leaves the heat exchanger, it is being cooled by the inlet water just entering the heat exchanger. This is a more efficient way of transferring heat than *parallel flow*, where the water and air flow in the same direction.

As in air cooling, the amount of heat transferred in the heat exchanger depends on the amount of air cooled, the amount of water used, how hot the air is, how cool the water is, and the size of the heat exchanger. Caution must be used with a water-cooled intercooler. The air must not be cooled to a temperature where the water vapor in the air condenses. If it is, water forming in the intercooler can be carried into the compressor high-pressure stage and cause damage.

Compressor Lubrication

Proper compressor lubrication is very important if the compressor is to have an efficient service life. Because of the large amount of heat generated in a compressor, the working surfaces of the compressor must be protected with the proper lubricating oil. All manufacturers recommend the type and viscosity of oil to be used in their compressor. If the oil is too thin, it will not lubricate properly and much of it will be carried away in the compressed air. If it is too thick, it will not flow properly and parts of the compressor will not be lubricated sufficiently.

The two main functions of compressor lubricating oil are to reduce friction and to improve the transfer of heat in the compressor. That is why there are two kinds of lubrication in reciprocating compressors: crankcase lubrication and cylinder wall or piston ring lubrication. Each of these types performs separate functions; crankcase lubrication oils all of the rotating parts in the lower portion of the compressor; cylinder wall lubrication prevents the piston rings from wearing, and removes heat (Fig. 3.5).

Compressor lubrication is usually accomplished by splash or pressurized lubrication methods. Many compressors rely on both of these methods at the same time. The most common is splash lubrication. *Splash-lubricated* reciprocating compressors have dippers on the connecting rods which scoop lube oil

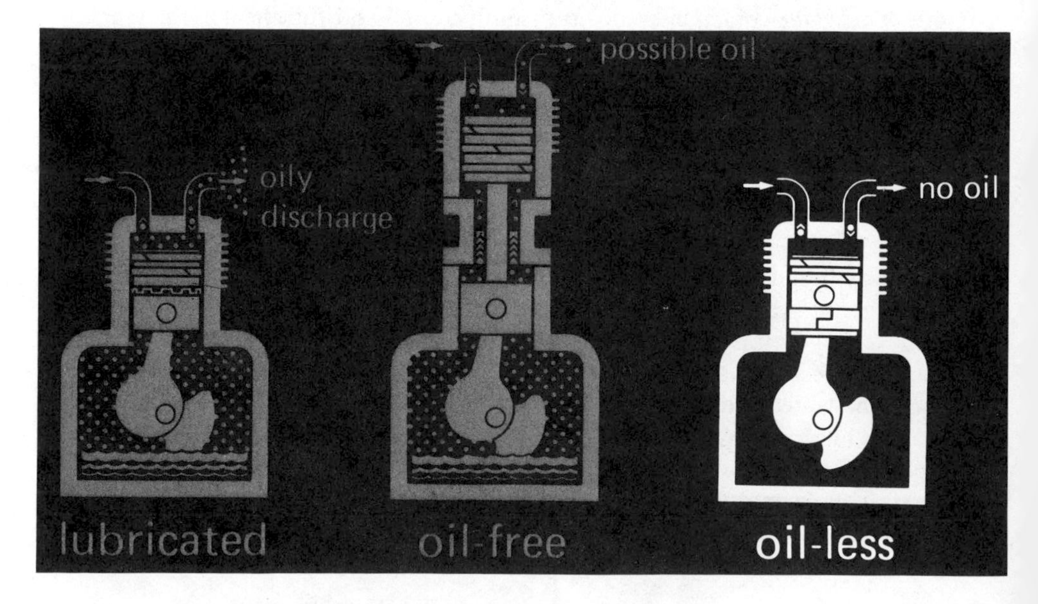

Figure 3.5 Oil-free and oil-less air compressors.

out of the crankcase. This oil is forced through drilled oil passages in the connecting rods to the compressor bearings. The spray and fog created also lubricate the cylinder walls and piston rings.

Large or heavy-duty compressors rely mainly on *pressurized lubrication*. Pressurized systems use a gear pump to force oil through drilled passages to bearings and pistons. In addition, some pressurized oil is sprayed onto the cylinder walls. Pressurized oil systems provide a more positive means of lubrication, and they can also meter the oil to ensure proper distribution. To ensure a proper oil condition many large compressors have thermometers for measuring oil temperature and also heat exchangers for cooling the oil when necessary. Many compressors are also equipped with low-oil-pressure shutoff switches.

Nonlubricated Compressors

If oil-free air is required, compressors are used that require no lubrication or have only well-isolated bearing lubrication. Nonlubricated compressors are manufactured in both reciprocating piston and rotary types. Reciprocating compressors ordinarily have graphite- or Teflon-coated piston rings to seal the cylinder. Small compressors use ringless pistons. Bearings may be provided with oil seals to prevent lubricating oil from getting into the air that is being compressed.

Compressor Controls

When the compressor starts up, various electrical and pneumatic controls must operate. The larger the compressor and pneumatic system, the more controls used. If an electric motor drives the compressor, it must be started through a suitable electrical switch without causing an overload. On large compressors, this can be done only by unloading the compressor (relieving all internal pressure) so that it does not compress air and is easier to start. When the compressor is turning at sufficient speed, the unloading controls are deenergized, allowing the compressor to load up and compress air. Small compressors usually do not require cylinder unloading and are provided only with a holding-type starting control and a high-starting-torque motor.

In small compressors up to 25 hp, system pressure is usually maintained by starting or stopping the compression. When air is needed to raise the pressure of the system, the compressor is started. When the line pressure is high enough, the compressor is stopped. Starting and stopping is accomplished with a diaphragm- or bellows-operated pressure switch that starts the compressor motor whenever the air tank pressure drops below a preset cut-in pressure. When the tank pressure reaches the preset cutout pressure on the switch, the motor stops. The switches are usually snap-acting, to protect their contacts from excessive wear.

Compressors may also be equipped with a mechanical *centrifugal pres-*

sure release, which operates independently of the pressure switch that starts and stops the compressor. When the compressor is stopped, pressure is vented from the compressor and any lines or accessories ahead of the discharge line check valve. During startup, the pressure release vents the compressor until the proper rpm rate is reached. The release then closes and the compressor delivers compressed air.

Compressor output may also be varied by changing the speed of the compressor. This is accomplished by changing the speed of its prime mover. However, it is usually more economical to use an unloading mechanism, or to simply start and stop the compressor. Usually, on-off controls are used when adequate air storage is available and the air requirement is less than 75% of the delivery capacity of the compressor.

An unloading mechanism can consist of a simple device to hold the inlet check valve open when the system is up to pressure and sufficient air is in the receiver. The air then sweeps in and out of the piston as it reciprocates at pressures which are created by the frictional factors. Little energy is used.

Compressor Accessories

Other accessory equipment supplied for compressors includes pressure gauges, safety devices, controls, thermometers, oil filters, cooling water controls, belt drives, flywheels, pulleys, guards for moving parts, and sometimes, flexible discharge pipes. An intake filter is an essential part of every compressor. An intake filter removes the dirt from the air and protects the moving parts of the compressor. Filters are discussed in more detail later.

An oil pressure safety device furnished with many compressors is connected to the lubricating oil system. Some devices unload the compressor whenever lubricating oil pressures are below a predetermined level. If very low or no oil pressure exists, the compressor will shut down. On nonpressurized compressors, a warning light may be turned on by a low-oil-level switch when the oil level in the compressor drops too low.

3.1.2 Rotary Compressors

Low-pressure applications are often referred to as *process applications,* and are used for cooling, heating, moving bulk materials, and aiding combustion. Compressors that supply air for these applications are usually of the rotary type, and are manufactured in several different styles or models. Some are designed to supply air in large quantities at intermediate pressure, while others supply only low-pressure process air. The construction and operating principles of rotary compressors are discussed below.

Compressor Classification

Rotary compressors with mechanically separated inlet and discharge ports are classified as *positive displacement.* Dynamic compressors have no method of separating the inlet and discharge ports, and permit free passage of air when they are not running. Positive-displacement rotary air compressors include sliding vane, dry screw, wet screw, liquid ring, and impeller or lobe types. *Dynamic* compressors include centrifugal and axial flow types that are similar to liquid pumps. Dynamic compressors use several stages or steps, each consisting of an independent impeller, to achieve their operating pressures. How these various compressors operate will be reviewed next.

Vane Compressors

The sliding vane compressor is a rotary positive-displacement compressor. This compressor is a compact unit that is less efficient than a piston compressor but is more efficient than any other rotary compressor. Sliding vane compressors closely resemble vane pumps, but are larger. The compressor consists of a rotor with radial slots mounted off center in a round casing. The rotor is equipped with rectangular vanes placed in the slots. As the rotor turns, the negative pressure created by the vanes causes air to be drawn through the inlet into the compressor. As the rotor continues to turn, the vanes confine the air in a space that gets progressively smaller. As the vanes approach the discharge portion of the casing, the air is compressed and then discharged to the receiver through the discharge port.

Single-stage vane compressors are capable of developing up to 50 psi discharge pressure. Although this is suitable for some applications, it may be low for others. If higher pressure is needed, a two-stage model is used. Two-stage compressors are capable of pressures up to 125 psi.

Vane compressors are smaller in physical size than piston compressors having comparable discharge pressure and flow capabilities. However, their operating efficiency is also slightly lower. At 100 psi they deliver slightly less than 4 cfm/hp, compared to approximately 5 cfm/hp for piston compressors. Their delivery in cfm/hp is higher than that of dynamic compressors. Because most vane compressors operate at motor speeds of 1200 to 1800 rpm, pressurized lubrication is required for the bearings and other rotating parts.

Rotary Screw Compressors

The *dry* rotary screw positive-displacement compressor combines the compactness of a rotary compressor with a more constant pressure than can be obtained from a piston compressor. The design of rotary screw compressors is quite different from that of other compressors (see Fig. 3.6). They are constructed with two rotors: one with a concave (female) profile and the other with a convex (male) profile. The rotor may or may not have the same number of lobes or blades, depending on the manufacturer. The rotors are made of steel, aluminum alloys, or other high-strength metals. The rotors mesh with about 0.003 (three one-thousandths of an inch) clearance between them. The rotors are driven, and prevented from touching by a set of timing gears (Fig. 3.7).

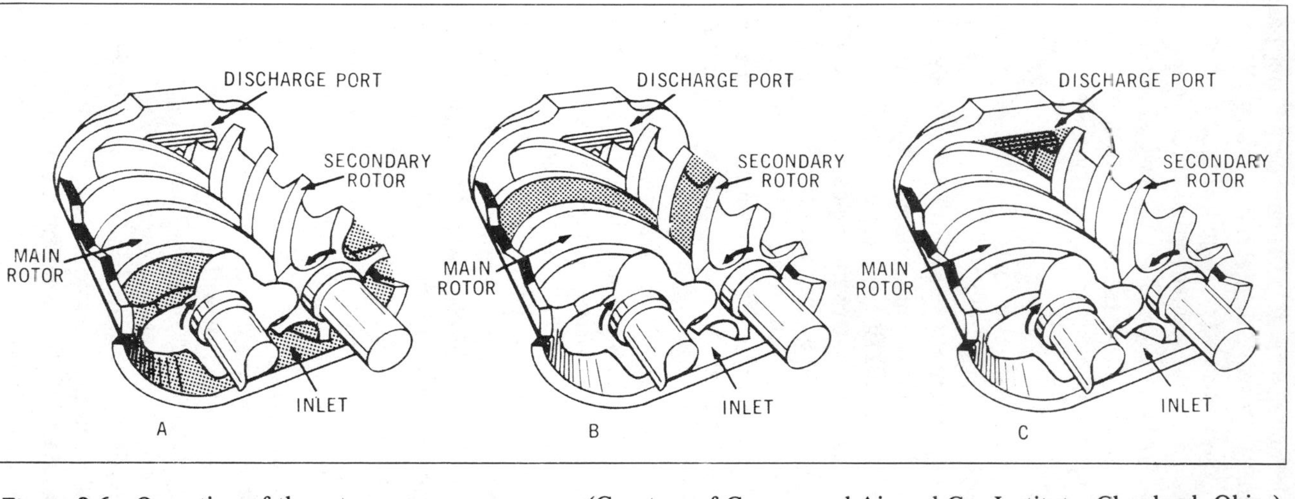

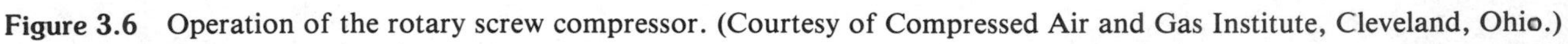

Figure 3.6 Operation of the rotary screw compressor. (Courtesy of Compressed Air and Gas Institute, Cleveland, Ohio.)

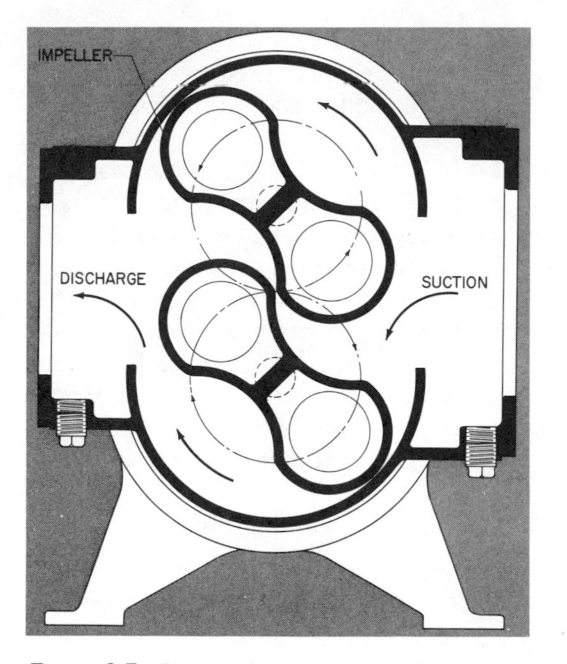

Figure 3.7 Rotary air compressor. (Courtesy of Compressed Air and Gas Institute, Cleveland, Ohio.)

As the rotor revolves, a negative pressure area is created at the inlet allowing air to be drawn into the casing. Air entering the inlet is then trapped between the rotors and the casing. As the male and female rotors mesh, the air is carried along in a progressively smaller space until it is discharged. Successive pockets of air are picked up, compressed, and discharged in this manner. Each cavity completes its discharge as the following cavity begins discharging. This provides a smooth, continuous, and shock-free flow of air.

Single-stage compressors are powered by motors as large as 75 hp, and produce approximately 300 cfm at 50 psi. Two-stage compressors have smaller second-stage screws fed by a direct flow passage from the first-stage discharge port. The pressure range is about 150 psi for two-stage compressors and 250 psi for three-stage compressors. Usually, no intercooler is used. Heat is dissipated by the compressor structure, which has fins on it to increase the cooling surface area.

The *wet* rotary screw compressor is similar in design and construction to the dry rotary screw compressor. However, the wet type does not use timing gears. One lobe powers the other. To prevent wear and reduce the air temperature, oil is sprayed into the inlet chamber and is carried along by the air. The oil remains as a liquid as it passes through the compressor, and is separated in the receiver or by an oil separator.

Both wet and dry rotary screw compressors are driven at high speeds (3000 to 12,000 rpm). Where electric motors or other lower-speed drives are used, a speed increaser is used to bring the compressor up to an efficient operating range.

The *liquid ring* compressor is a different type of rotary compressor. Although it appears similar to a vane compressor, it is quite different. The main components include a casing, an off-center drive shaft, a rotor with fixed blades, and a liquid (usually water). During operation the liquid is carried around the inside of the casing by the rotor blades. As the rotor turns, the liquid (reacting to centrifugal force) follows the contours of the casing. Because the rotor and casing are not centered, the liquid forms a flexible interior compression chamber. As the rotor blades pass the inlet port, air is drawn into the compressor by the increasing size of the fluid-formed chamber. The chamber size begins to decrease near the compressor outlet, and the air is discharged where the chamber is smallest (Fig. 3.8). Liquid ring compressors have capacities of up to 5000 cfm at 75 psi in single-stage models.

Low-Pressure High-Volume Compressors

The impeller compressor is a low-pressure compressor. These compressors are often referred to as *blowers* because they are designed for high volume rather than pressure. The units are made up of a housing and two interlocking, timed, gear-driven impellers. The impellers may have two, three, or four vanes. The vanes can be straight or have a slight twist or helix configuration. The curvature reduces turbulence and smoothes out the pressure pulsations at this discharge point. The timing gears prevent the lobes from coming in contact.

The operation of these compressors can be compared to that of a gear pump. As the impellers rotate, they create a negative pressure at the intake, which draws air into the housing. As they revolve, the impellers carry the air along between the lobe and housing, toward the discharge port. As the lobes mesh, the air is squeezed out the discharge port. The impellers are machined to very close tolerances, and have only 0.003 to 0.006 in. clearance between them. To be efficient, they must be operated at the recommended speed.

These compressors handle relatively large amounts of air at pressures up to 10 psi. If higher pressures are required, a second stage can be added to increase the discharge pressure to 30 psi.

Diaphragm Compressors

Diaphragm compressors are used for many different light-duty applications, and are classed as positive displacement compressors. The diaphragm of the compressor is usually connected to an eccentric drive by a piston and connecting rod arrangement.

The diaphragm is made of flexible, reinforced, rubber-type material that

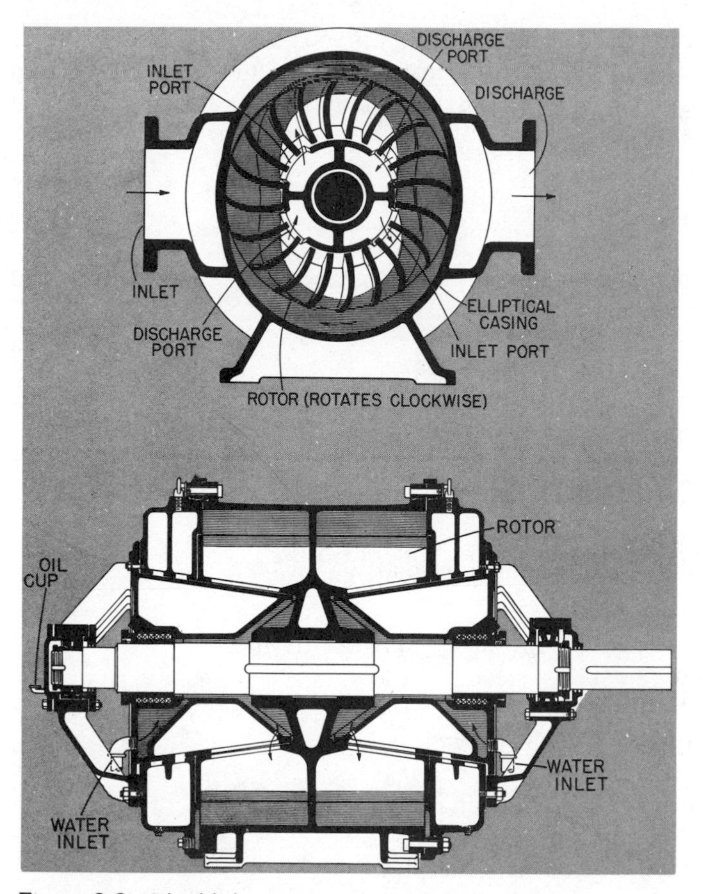

Figure 3.8 Liquid ring compressor. (Courtesy of Compressed Air and Gas Institute, Cleveland, Ohio.)

is oil and moisture resistant. The drive eccentric is fitted with a prelubricated sealed for life bearing. As the piston or diaphragm plate travels downward, the diaphragm flexes inward, drawing air into the pumping chamber. When the diaphragm plate reverses its stroke, the diaphragm flexes outward, forcing air through the discharge valve. These compressors are not as efficient as other types, but they do furnish small quantities (1 to 3 cfm) of air at 30 to 40 psi.

Dynamic Compressors

Dynamic compressors are designed to deliver large amounts of air (as high as 100,000 cfm) at pressures up to 125 psi. They are used primarily to provide process air, but may also be used for ventilation. Their minimum capacities of about 3000 cfm make them too large for most pneumatic power systems. Dynamic compressors are more compact and powerful than the blowers or fans used in heating or ventilating installations, but are less efficient and more noisy. The two main types of dynamic compressors are centrifugal and axial flow (see Fig. 3.9).

3.1.3 Troubleshooting

1. Low discharge pressure, probable cause:
   a. Air leaks
   b. Leaking valves
   c. Restricted air intake
   d. Slipping belts
   e. Unloading mechanism not operating properly
   f. Blown gaskets
   g. Worn compressor rings
   h. Defective gauge
   i. Compressor too small to load
2. Knocking
   a. Loose motor pulley or compressor flywheel
   b. Loose belts
   c. Lack of oil in crankcase
   d. Oil pump not functioning
   e. Worn connecting rod bearings
   f. Worn piston pin bearing
   g. Worn main bearings
   h. Excessive crankshaft end play
   i. Excessive motor shaft end play
   j. Loose valve assemblies
   k. Loose piston
   l. Piston hitting the head, due to foreign matter or carbon deposits
3. Overheating
   a. Poor ventilation
   b. Dirty cooling surfaces
   c. Incorrect flywheel rotation
   d. Flywheel too close to wall
   e. Restricted air intake
4. Compressor fails to attain speed
   a. Loose belts
   b. Low voltage

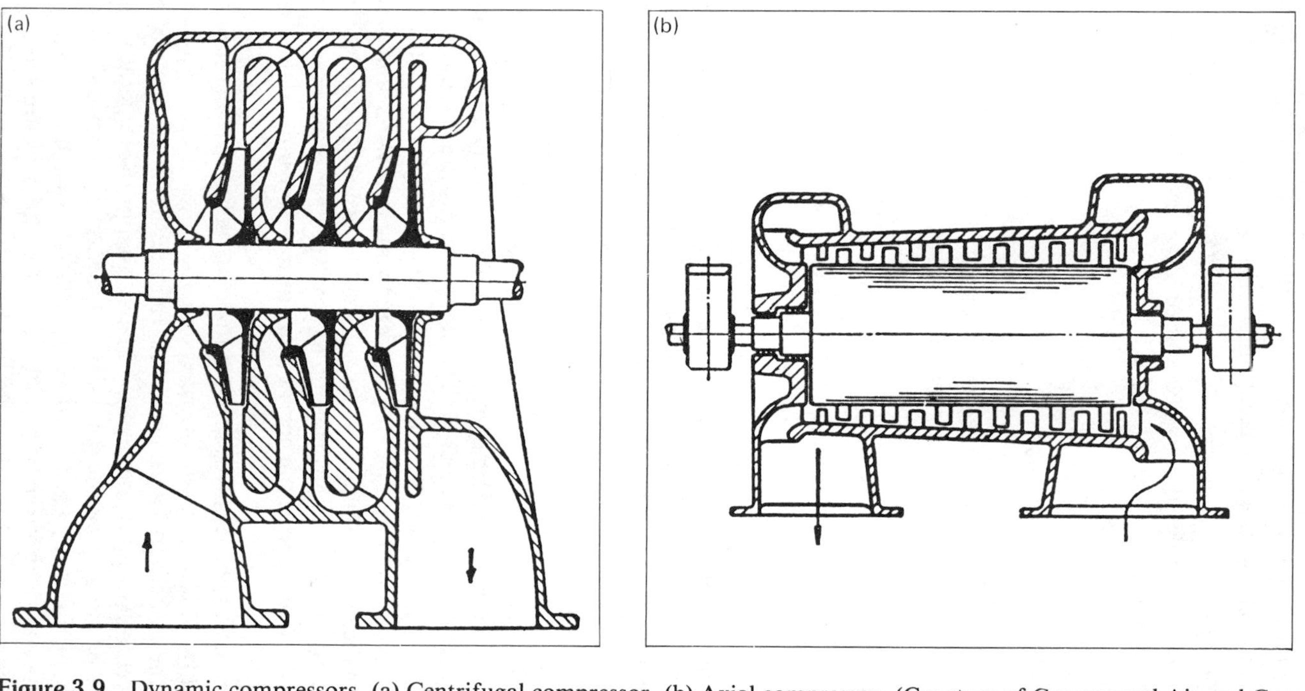

Figure 3.9 Dynamic compressors. (a) Centrifugal compressor. (b) Axial compressor. (Courtesy of Compressed Air and Gas Institute, Cleveland, Ohio.)

c. Overloaded motor
d. Check electrical installation

5. Excessive belt wear
   a. Motor pulley or flywheel out of alignment
   b. Belt too loose or too tight
   c. Belt slipping
   d. Flywheel wobble
6. Oil in the discharge air
   a. Worn piston rings
   b. Compressor air intake restricted
   c. Restricted crankcase breather
   d. Excessive oil in compressor
   e. Wrong oil viscosity

3.2 PNEUMATIC VALVES

The basic pneumatic valve is a mechanical device consisting of a body and a moving part which connects and disconnects passages within the body. The flow passages in pneumatic valves carry air. The action of the moving part may control system pressure, direction of flow, and rate of flow.

3.2.1 Valve Types

Control of Pressure

Pressure in a pneumatic system must be controlled at two points: at the compressor and after the air receiver tank. Control of pressure is required at the compressor as a safety measure for the system. Control of pressure at the point of air usage provides a means of safeguarding the system and of maximizing the efficiency—compressed air will not be wasted. Characteristically in a pneumatic system, energy delivered by a compressor is not used immediately, but is stored as potential energy in an air receiver tank in the form of compressed air.

A compressor generally is selected to be of such a size that it operates intermittently. The compressor operates until high pressure is reached in the tank; then the air delivery is curtailed. When air pressure in the tank decreases, the compressor delivery is once again initiated. Compressor operation in this manner is a power savings for the system. One way to accomplish this function is with the use of a *pressure switch*. In a pressure switch, the system pressure is sensed at the bottom of a piston through the pressure switch inlet. When pressure in the system is at its low level, a spring pushes the piston down. In this position, contact is made, causing an electrical signal to turn the compressor on. As pressure in the receiver tank rises, it forces the piston upward. With system pressure at its high level, the piston breaks the electrical contact, shutting down the compressor.

The *safety relief valve,* used to protect a system from excessively high or dangerous pressures, is a normally closed valve. The poppet of the safety relief valve is seated on the valve inlet. A spring holds the poppet firmly on its seat. Air cannot pass through the valve until the force of the spring biasing the poppet is overcome. Air pressure at the compressor outlet is sensed directly on the bottom of the poppet. When air pressure is at an undesirably high level, the force of the air on the poppet is greater than the spring force. When this happens, the spring will be compressed and the poppet will move off its seat, allowing air to exhaust through the valve ports.

In order to control or regulate the air pressure downstream of the receiver tank, a *pressure regulator* is utilized. A pressure regulator is a pressure reducing valve and consists of a valve body with inlet and outlet connections, and a moving member which controls the size of the opening between the inlet and outlet. It is a normally open valve, which means that air is normally allowed to flow freely through the unit. With the regulator connected after the receiver tank, air from the receiver flows freely through the valve to a point downstream of the outlet connection. When pressure in the outlet of the regulator increases, it is transmitted through a pilot passage to a piston or diaphragm area which is opposed by a spring. The area over which the pilot pressure signal acts is rather large, which makes the unit responsive to the outlet pressure fluctuations. When the controlled (regulated) pressure nears the preset force level, the piston or diaphragm moves upward, allowing the poppet or spool to move toward its seats, thereby controlling the flow (increasing the resistance). The poppet or spool blocks flow once it seats and does not allow pressure to continue building downstream. In this way, air at a controlled pressure is made available to an actuator.

Control of Direction

To change the direction of airflow to and from a cylinder, we use a directional control valve. The moving part in a directional control valve will connect and disconnect internal flow passages within the valve body. This action results in a control of airflow direction.

The typical directional control valve consists of a valve body with four internal flow passages and a moving part, a spool, which alternately connects a cylinder port to the supply pressure or to exhaust. With the spool in one extreme position, supply pressure is connected to port B and port A is connected to exhaust. If cylinder port A is connected to the valve port A and cylinder port B to the valve port B, then with the spool in the other extreme position, supply pressure is connected to cylinder passage A and cylinder passage B is connected to exhaust. With a directional control valve in a circuit, the cylinder's piston rod can be extended or retracted and work performed.

The valve described is called a four-way direction control valve—it has four connections or flow paths that are either connected or not. In a pneumatic system, two-way and three-way valves are also used.

A two-way directional valve consists of two ports connected to each other with passages that are connected and disconnected. In one extreme spool position, port A is open to port B; the flow path through the valve is open. In the other extreme, the path between A and B is blocked; the flow path is blocked. A two-way valve provides an on-off function. This function can be used in many systems as an interlock and to isolate and connect various system parts.

A three-way directional valve consists of three ports connected through passages within a valve body: port A, port B, and port C. If port A is connected to an actuator, port B to a source of pressure, and port C is open to exhaust, the valve will control the flow of air to (and exhaust from) port A. The function of this valve is to pressurize and exhaust one actuator port. When the spool of a three-way valve is in one extreme position, the pressure passage is connected with the actuator passage. When in the other extreme position, the spool connects the actuator passage with the exhaust passage. Three-way valves may be used singly to control single-acting cylinders or in pairs to control double-acting cylinders.

One type of movable member quite frequently used to accomplish the direction change function is a *poppet*. This type of valve is shown in Fig. 3.10. The poppet configuration is available in two-way, three-way, and four-way arrangements. However, to achieve the four-way function, two movable poppets are needed.

We have seen that a movable member can be positioned in one extreme position. The member is moved to these positions by mechanical, electrical, pneumatic, or manual means. Directional valves whose spools are moved by muscle power are known as manually operated or manually actuated valves. Various types of manual actuators include levers, pushbuttons, and pedals.

A very common type of mechanical actuator is a plunger equipped with a roller at its top. The plunger is depressed by a cam which is attached to an actuator. Manual actuators are used on directional valves whose operation must be sequenced and controlled at an operator's discretion. Mechanical actuation is used when the shifting of a directional valve must occur at the time an actuator reaches a specific position. Directional valves can also be shifted with air. In these valves, pilot pressure is applied to the spool ends or to separate pilot pistons.

One of the common ways of operating a directional valve is with a *solenoid*. A solenoid is an electrical device which consists basically of a plunger and a wire coil. The coil is wound on a bobbin, which is then installed in a magnetic frame. The plunger is free to move inside the coil. When electric current passes through the coil of wire a magnetic field is generated. This

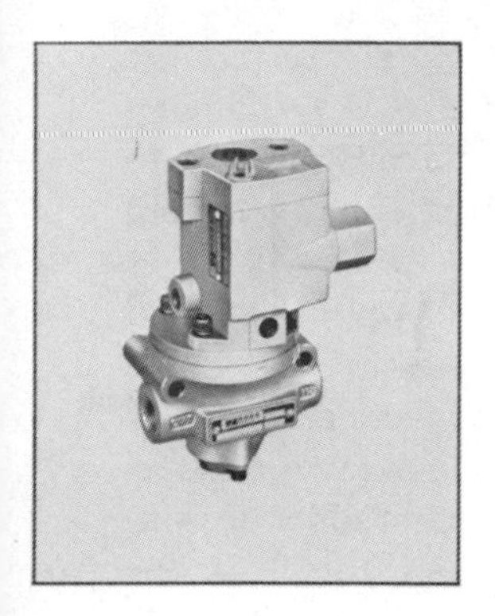

2/2 VALVES

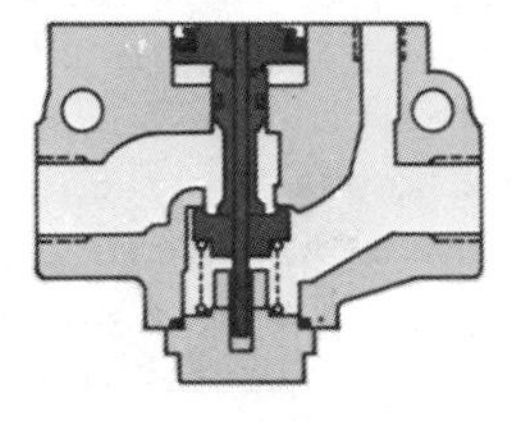

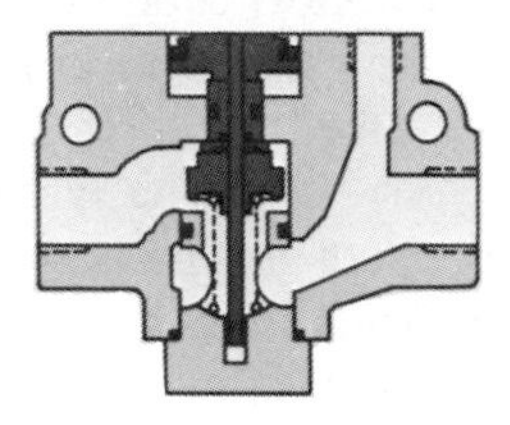

3/2 VALVES

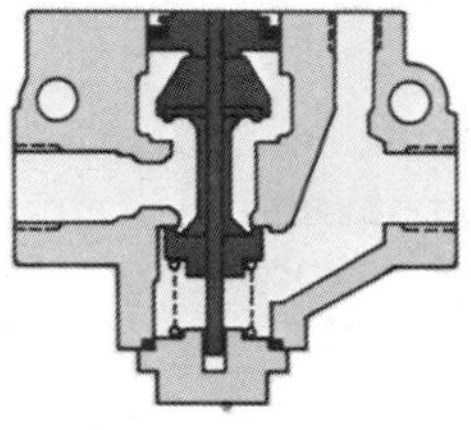

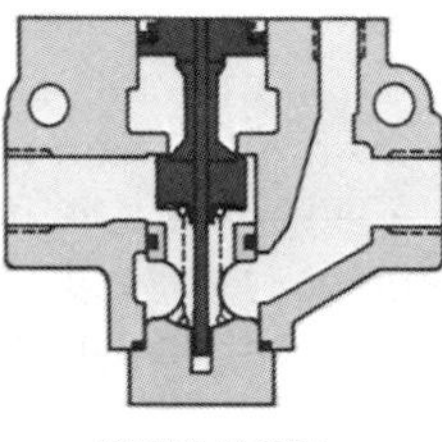

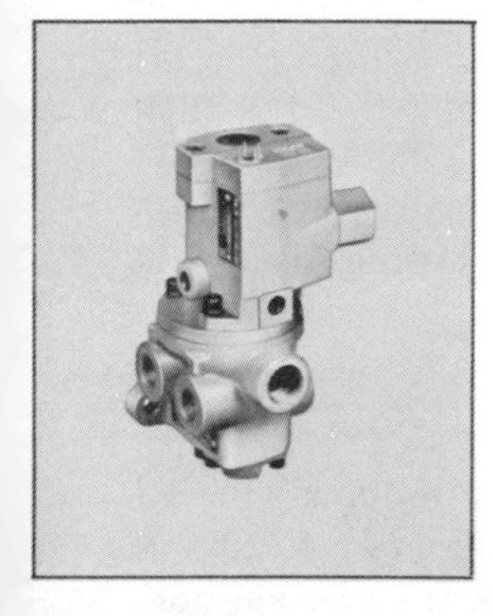

4/2 VALVES

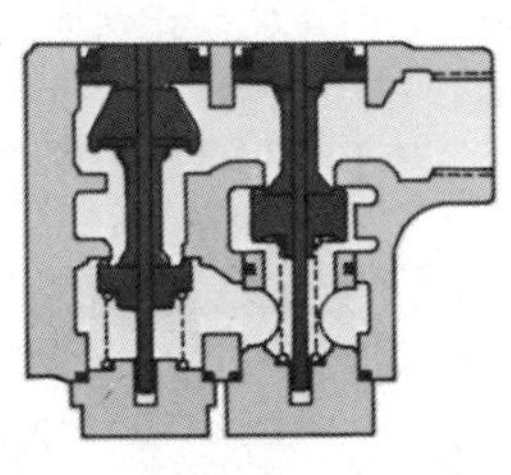

Figure 3.10 Poppet-type pneumatic direction control valves. (Courtesy of Ross Operating Valve Co., Detroit, Michigan.)

magnetic field attracts the plunger and pulls it into the coil. As the plunger moves in, it can either cause a spool to move or seal off a surface changing the flow condition. When the motion of the solenoid is directly coupled to the shifting mechanism of the valve, it is called a *direct-operated* solenoid valve.

Another type of valve actuation is the *pilot actuator*. A pilot-actuated valve uses air pressure to move the valve spool. This air pressure may come from a variety of sources. Pilot-actuated valves are used where actuation is required in remote locations. They are also useful to control low pressures and are a requirement in pressure-centered valves. Since actuation forces increase with increasing pressure to the actuating ports, high forces can be generated to shift the valve members. Pilot-actuated valves may be internally or externally piloted. A valve is considered to be *externally piloted* if the air pressure received for shifting comes from an external source. If the pilot pressure comes from a source within the valve, it is said to be *internally piloted.*

Internally piloted valves use part of the pneumatic energy delivered to the pressure port to position the movable member within the valve body. Such a valve has a definite minimum and typically a maximum pressure requirement. The application of this type of valve may impose circuit restrictions on the designer. This is due to the need to maintain the minimum specified pressure whenever the valve element requires shifting.

Conversely, externally piloted valves require a pressurized source which is derived externally to the valve. This means that the pressure can be selected in such a way that the valve will energize and deenergize in the same amount of time. If slow shifting times are required, a low pressure may be used. If rapid shift rates are necessary, we can use a high-pressure signal or the rapid exhausting of one of the pilot ports. However, when using high pressure to shift the main valve member, caution should be used. Extremely high velocities may cause high impact forces on the valve member, leading to reduced life.

Control of Flow Rate

In a pneumatic system, actuator speed is determined by how quickly the actuator can be filled and exhausted of air. In other words, speed of a pneumatic actuator depends on the force available from the pressures acting on both sides of the piston, a result of the number of cfm flowing into the inlet and out of the exhaust port.

A pressure regulator will influence actuator speed by portioning out to its leg of a circuit the pressure required to equal the load resistance at an actuator. This additional pressure is used to develop airflow. Even though this is the case, pressure regulators are not used to vary actuator speed. In a pneumatic system, actuator speed is affected by a restriction such as that obtainable with a needle valve or a needle valve with a bypass check, often called a flow control valve. This valve will meter flow at a constant rate only if the system resistance, the total load (friction and load), and the pressure at the inlet of the cylinder and at its outlet do not vary throughout the entire stroke length. This valve does not control flow; it only affects flow. For example, when the load encounters an added resistance, actuator speed would decrease or the machine will stop. Once the resistance has been overcome, there would be a very rapid increase in speed. This could be a very dangerous condition for the operator or the machine.

As indicated previously, a needle valve in a pneumatic system affects the operation by causing a restriction. The typical needle valve consists of a valve body and an adjustable part. The adjustable part can be a tapered-nose rod which is threaded into the valve body. The more the tapered-nose rod is screwed toward its seat in the valve body, the greater the restriction to free flow.

By restricting exhaust airflow in this manner, a back pressure is generated within the actuator, thus reducing the forces available to create motion. This means that a larger portion of regulator pressure is used to overcome the resistances at the actuator and less pressure energy is available to develop flow. With less air flowing into the actuator, actuator speed decreases. Again, by controlling the amount of restriction developed by a needle valve, the speed of an actuator can be controlled, but only if the total load is constant.

3.2.2 Maintenance and Troubleshooting

Pneumatic valves are rugged, reliable system components and are used extensively in various types of production equipment. This applies to metal-to-metal spool, resilient seal spool, and poppet-type valve styles. However, best performance and long service life are influenced greatly by proper and regular maintenance. When malfunctions or failures occur, it is important for the maintenance personnel to be able to determine quickly the cause and source of the problem. Then the appropriate corrective action can be taken.

Maintenance

Following is a listing of the maintenance and service requirements for proper valve operation.

1. *Supply clean air*: Experience has shown that foreign material lodging in the valve is a major cause of breakdowns. An air line filter should be used that is capable of removing solid and liquid contaminants. Accumulated liquid should be drained from the filter frequently. If the filter is located where frequent routine maintenance is difficult, use a filter with an automatic drain.

2. *Supply lubricated air*: Most valves and the mechanisms they control require light lubrication. A good lubricator should put atomized oil into the air line in direct proportion to the rate of airflow. Either excessive or inadequate lubrication can cause the valve to malfunction. For most applications an oil flow rate of 1 drop per minute is adequate. Another lubrication check can be made by holding a piece of clean white paper near the valve's exhaust port for three or four cycles. A properly lubricated valve will produce only a slight discoloration of the paper. Suitable lubricating oils must be compatible with the materials in the valve used for seals and poppets. Generally speaking, any light-bodied mineral- or petroleum-base oil with oxidation inhibitors, an aniline point between 82°C (180°F) and 104°C (220°F), and an SAE 10, or lighter, viscosity will prove suitable.
3. *Clean valve periodically*: The internal surfaces of a valve may gradually build up a deposit of varnish or dirt. This can lead to sluggish or erratic valve action, especially in spool valves. It is recommended that a schedule be established for the periodic cleaning of all valves. To clean valves, use a water-soluble detergent or a solvent such as kerosene. Do not scrape varnished surfaces. Also, avoid chlorinated solvents (trichloroethylene, for example) and abrasive materials. The former can damage seals and poppets, and abrasives can do permanent damage to metal parts.
4. *Clean electrical contacts*: In the external electrical circuits associated with the valve solenoids, keep all switches or relay contacts in good condition to avoid solenoid malfunctions.
5. *Replace worn components*: In many cases it is not necessary to remove the valve from its installation for servicing. Ordinarily, the only items that will need replacing after a long period of use are moving seals and possibly springs. These can be installed in the valve without removal from its mounting location. Before disassembling a valve or other pneumatic component, or removing it from its installation, shut off and exhaust the entire pneumatic circuit, and verify that any electrical supply circuit is not energized.

Troubleshooting

Troubleshooting is the process of observing a valve's trouble symptoms, such as a buzzing solenoid or sluggish action, and then relating these to their most likely causes. By careful analysis of the symptoms, the experienced troubleshooter can quickly determine the trouble, identify the cause, and take the appropriate correction action.

A pneumatic valve troubleshooting chart is presented next to assist in the process of identifying the cause and analysis of symptoms. The chart contains the most common trouble symptoms that valves experience and their probable causes.

Problem	Cause
Valve blows to exhaust when not actuated	Inlet poppet not sealing Faulty seals Faulty valve-to-base gasket Cylinder leaks
Valve blows to exhaust when actuated	Faulty valve-to-base gasket Faulty seals Damaged spool Cylinder leaks Air supply pressure too low Water or oil contamination
Solenoid fails to actuate valve, but manual override does actuate valve	Loose pilot cover or faulty solenoid Low voltage at solenoid
Solenoid fails to actuate valve and manual override also fails to actuate valve	Faulty seals Varnish deposits in valve Pilot pressure too low Water or oil contamination
Airflow is normal only in actuated position	Broken return spring
Solenoid buzzes	Faulty solenoid Low voltage at solenoid Varnish in direct-operated spool valve
Solenoid burns out prematurely	Varnish in direct-operated spool valve Incorrect voltage at solenoid
Pilot section blows to exhaust	Loose pilot cover Pilot poppet not sealing
Poppet chatters	Air supply pressure low Low pilot or signal pressure Faulty silencer/muffler
Valve action is sluggish	Faulty seals on spool valve Varnish in spool valve Air supply pressure low Low pilot or signal pressure Poor or no lubrication Faulty silencer/muffler Water or oil contamination

(continued)

Problem	Cause
Sequence valve gives erratic timing	Faulty piston seal Excessive lubrication Fluctuating air pressure Accumulated water Faulty gasket
Flow control valve does not respond to adjustment	Excessive lubrication Incorrect installation or dirt in valve

Before disassembling a valve to investigate a system malfunction, check other possible causes of the malfunction. Because malfunctions in other components can affect valve action, the valve is sometimes blamed for a problem which, in fact, lies elsewhere. Leaky cylinder packings, poor electrical contacts, dirty filters, and air line leaks or restrictions are just a few of the things to be considered when troubleshooting a pneumatic system. Consideration of these possibilities can sometimes save an unnecessary valve disassembly job. The following paragraphs detail the corrective actions that can be taken for various causes of pneumatic valve failures.

Main Inlet Poppet Not Sealing. Foreign particles may be holding the poppet off its seat. Cycle the valve several times to see if the flow of air through the valve will flush out the particles. If not, it will be necessary to disassemble the valve (Fig. 3.11).

To disassemble the valve, first disconnect or turn off the electrical circuit to the valve; shut off and exhaust the air supply; then disassemble the valve body assembly. The inlet poppet should be pulled off the valve stem and checked for dirt and damage. If damaged, the poppet must be replaced. If the poppet is swollen or has deteriorated, improper lubricants or solvents may be the cause.

Also, check the poppet seats for dirt and damage. If there is damage to a seat, the entire valve body assembly must be replaced (Fig. 3.12). If there is no damage to the poppet or seats, clean the parts thoroughly, lubricate lightly, and reassemble.

Faulty Seals. The materials of which seals are made can be attacked by substances such as chlorinated hydrocarbons (trichloroethylene, for example) and some lubricating oils. This can produce swelling or shrinking of the seals and result in erratic valve action or blowing to exhaust. Swollen seals may cause some in-line poppet valves to stick in a partially open position so that the valve blows to exhaust. Swollen seals on a spool valve can result in sluggish or erratic valve action, or even failure of the spool to move at all. Badly nicked or torn seals can produce blowing to exhaust in resilient seal

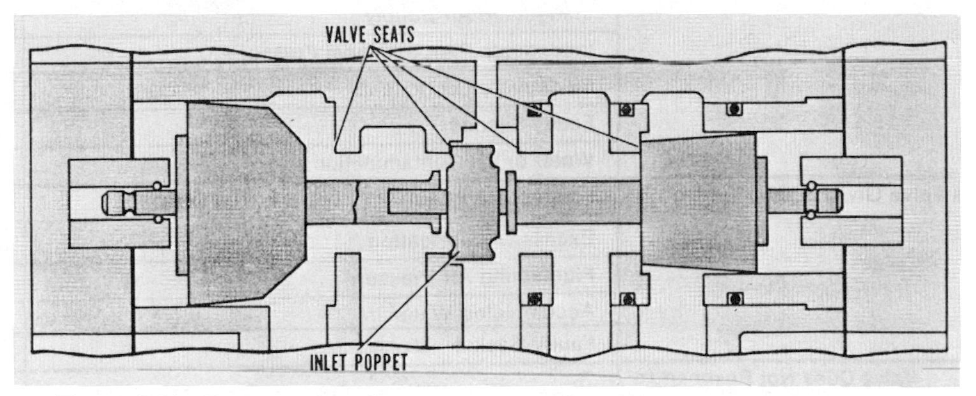

Figure 3.11 Poppet valve with poppet not sealing. (Courtesy of Ross Operating Valve Co., Detroit, Michigan.)

spool valves by allowing air to pass from one port area to another. Small leaks in piston poppet seals can affect the timing accuracy of sequence adapters on in-line valves, or even render the valve inoperable.

Varnish Deposits in Valve. Varnish deposits can cause a valve to act sluggishly or even prevent movement of the valve element altogether, especially after a period of inactivity. A spool valve frozen in position by varnish can cause a direct-acting solenoid to buzz, and eventually leads to solenoid burnout (Fig. 3.13).

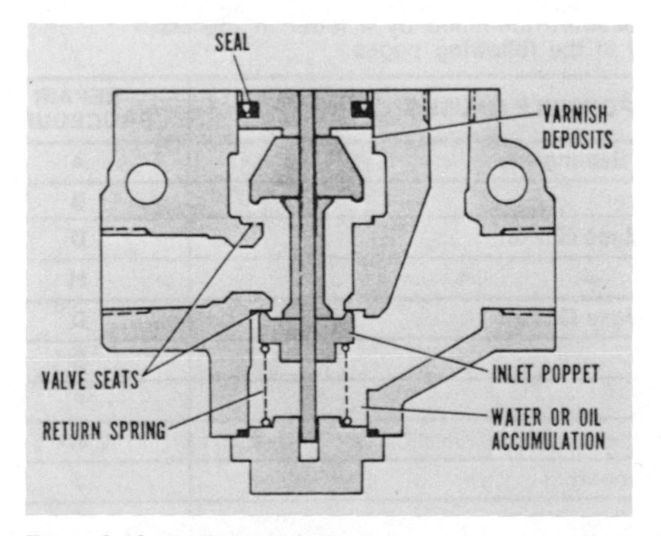

Figure 3.12 In-line-mounted poppet valve contamination. (Courtesy of Ross Operating Valve Co., Detroit, Michigan.)

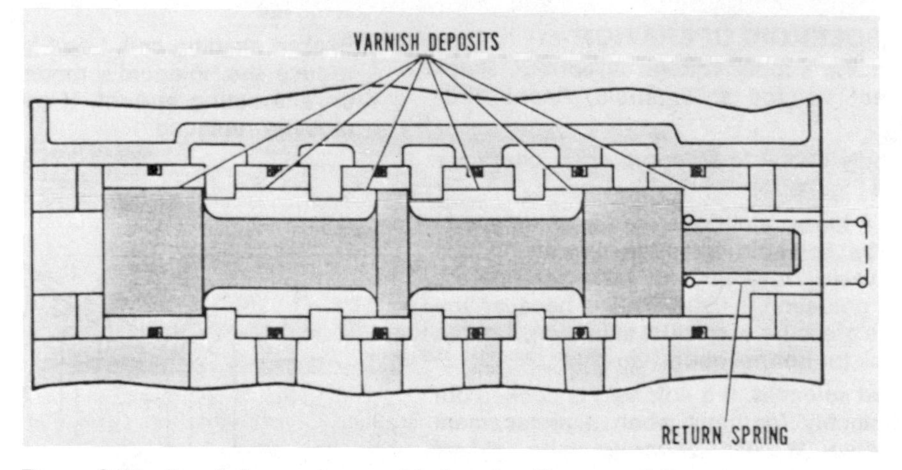

Figure 3.13 Spool sleeve valve varnish deposits. (Courtesy of Ross Operating Valve Co., Detroit, Michigan.)

Varnish results from the action of oxygen on the lubricating oil, and can be aggravated by excess heat. Varnish can also come from overheated compressor oil carried over into the air lines. Properly lubricated valves do not usually suffer from varnish deposits (Fig. 3.14).

To remove varnish, use a water-soluble detergent or solvent such as kerosene. Do not scrape off the varnish. Also, avoid chlorinated solvents (trichloroethylene, for example) and abrasive materials. The former can dam-

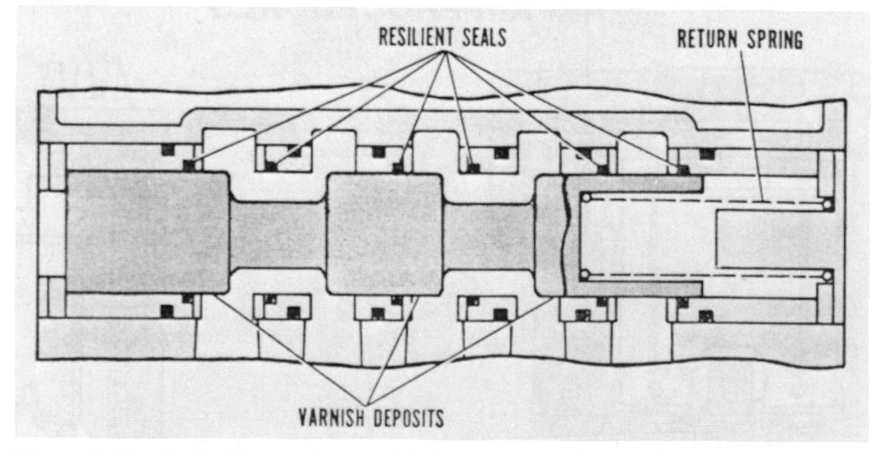

Figure 3.14 Packed bore valve varnish deposits. (Courtesy of Ross Operating Valve Co., Detroit, Michigan.)

age seals and poppets, and abrasives can do permanent damage to metal parts. After cleaning, lightly lubricate moving valve parts and reassemble the valve.

Broken Return Spring. A broken return spring on a spool valve can cause the spool to remain in an actuated position, or to be only partially returned. In the latter case, several abnormal flow patterns may result, depending on the valve configuration. If a spool valve has a normal flow pattern only in an actuated position, a broken return spring is the most likely cause of the trouble. A broken return spring on an in-line poppet valve is less likely to prevent closing of the inlet poppet, but should be considered as a possible cause of the valve's blowing to exhaust when not actuated, especially in a low-pressure application.

Damaged Spool. If a spool is badly scored or nicked, it can allow air to pass from one port area to another. This can result in unwanted pressurizing of an outlet port or blowing to exhaust. The problem can be further aggravated by the spool's cutting the resilient seals and increasing the leakage. A damaged spool cannot be repaired, but must be replaced.

Inadequate Air Supply. An inadequate air supply volume causes an excessive pressure drop during valve actuation. Pilot air pressure may be great enough to begin movement of the valve element, but the pressure drop resulting from the filling of the outlet volume depletes the pilot air supply. This may result in chattering or oscillating of the main valve, or may simply keep the main valve partially actuated so that it blows to exhaust.

Check the pressure drop shown on the gauge at the pressure regulator. If the pressure falls more than 10% during actuation of the valve, the air supply may be inadequate. Inspect the system for undersize supply lines, sharp bends in the piping, restrictive fittings, a clogged filter element, or a defective pressure regulator. Remember, too, that the air volume supplied can be insufficient if more pneumatic devices are connected to a circuit than the compressor is designed to serve.

Fluctuating Air Pressure. If a valve with a timed sequence adapter suffers from erratic timing, the cause can be a fluctuating supply pressure. Consistent timing requires a consistent supply pressure. If the supply pressure varies considerably, install a pressure regulator set at the system's lowest expected pressure.

Inadequate Pilot or Signal Pressure. Pilot or signal pressure below the minimum requirement can produce chattering, valve oscillation, or sluggish valve action. Check the valve specifications for minimum pilot or signal pressure requirements.

Undersized or Dirty Silencer. An undersized silencer, or one that is partially plugged, restricts the exhaust flow. The resulting back pressure can cause erratic motion of valve elements. Remove the silencer to see if valve performance is improved. Clean the silencer to see if valve performance is improved.

Verify that the silencer is of adequate size. Do not reinstall an undersized silencer. Install a larger silencer and check the valve performance again.

Lubrication. Some valves require lubrication to operate properly. Check the system lubricator to see that it is working as it should. Do not lubricate excessively. Excess oil can accumulate in low points of the system and restrict the flow of air. It can also form pools that will produce a dash-pot effect and slow valve action. A visible oil fog exhausting from the valve is a sure sign of excessive lubrication. A properly lubricated valve will produce only a slight discoloration on a piece of white paper when held close to the exhaust port for three or four cycles.

Air Cylinder Leaks. Four-way valves sometimes blow to exhaust because of leaking packings in the air cylinder connected to the valve. Before looking for faults in the valve, check the cylinder for leaks. To check for cylinder leakage, the following steps can be taken (see Fig. 3.15):

1. Disconnect the air line to the end of the cylinder which is not under pressure. If air comes out of the open port, the cylinder packings are leaking and must be repaired. If there is no leakage, reconnect the line.
2. Reverse the position of the valve and disconnect the other air line to the cylinder. Again, check for air coming out of the cylinder port. If there is air coming out, the cylinder packings must be repaired.

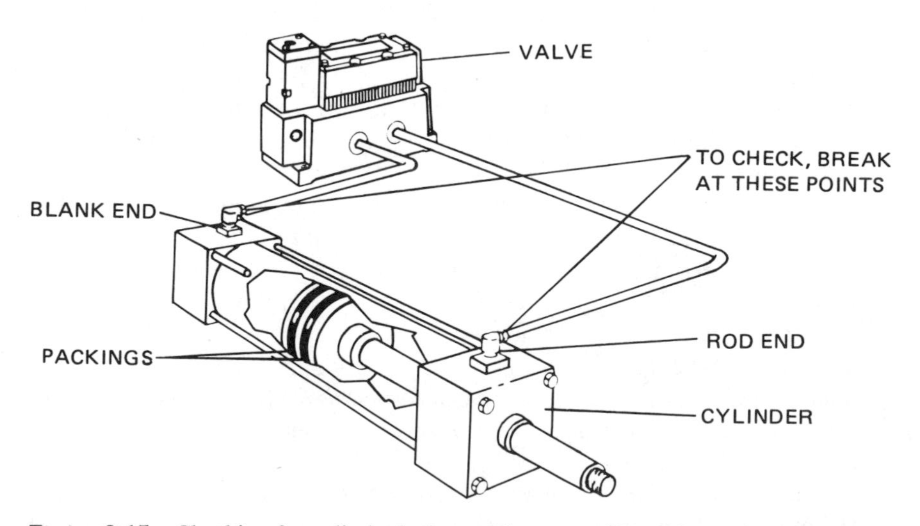

Figure 3.15 Checking for cylinder leakage. (Courtesy of Ross Operating Valve Co., Detroit, Michigan.)

3. If there is no leakage at the cylinder, reconnect the air line and proceed with troubleshooting of the valve itself.

Water or Oil Contamination. Accumulations of water or oil have an especially bad effect on devices with small orifices, such as timed sequence adaptors. Accumulations in such a device can change the effective size of the timing orifice or even block it completely. The device must be disassembled, cleaned, lightly lubricated, and reassembled. It may be necessary to install a filter in the supply line to prevent recurrence of the problem.

Accumulations of water or oil can also occur at low points in pilot supply lines. This can result in pressure fluctuations that produce erratic timing. The best cure is to reroute the pilot supply lines to eliminate low points. Water and oil can also accumulate at low points in a valve, and hinder movement of the valve element, perhaps completely preventing its motion. This is especially true of a valve operating in a subfreezing environment where accumulated water can turn to ice. It is important in such applications to ensure that the supply air is dry, and that the air line filter is drained frequently.

Internally Piloted Valve Shifts Improperly. An internally piloted valve may shift partially, then stall. Air blows steadily through the exhaust port. This is a sign that the pressure at the inlet port of the valve has fallen below the valve's minimum operating pressure. Increase the supply pressure or provide a local air accumulator to maintain pressure at the valve during periods of high flow.

If the valve is provided with an internal pilot exhaust, change it to exhaust externally if possible. Check for restrictions in the supply line. A gauge can be used to check the pressure available at the inlet port just before the valve fails to shift. Common causes of restricted supply lines are clogged filters, restrictive lubricators, and undersized hose or fittings.

On some valve designs, especially momentary contact types, full shifting may be obtained by restricting the exhaust port to allow the valve to maintain a pressure level above the minimum operating pressure. In extreme cases, provide a local accumulator for the pilot circuit or use pilot pressure from a remote source.

Valve Occasionally Malfunctions or Circuit Reaction Slow. This type of problem can be caused by icing. Rapidly exhausting air often cools entrained moisture below its freezing point, causing ice particles to restrict exhaust flow. Mufflers can also freeze. Check the exhaust ports of the valve and/or muffler for signs of icing. Dry the incoming air or provide a means for heating areas that tend to freeze.

Alternating-Current Solenoid Failures. The failure of alternating-current (ac) solenoids can be due to a number of conditions. Following is a listing of the most common causes (Figs. 3.16 and 3.17).

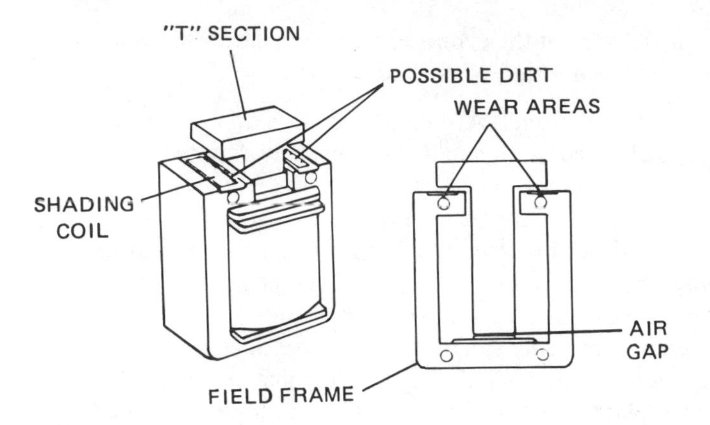

Figure 3.16 Typical pilot solenoid. (Courtesy of Ross Operating Valve Co., Detroit, Michigan.)

Low applied voltage: Not enough magnetic force will be developed to allow the armature of the solenoid to seat. The unit will continuously draw high inrush current and burn out. Voltage should be checked at the coil with the solenoid energized. Possible causes of low voltage include high-resistance connections, too much load on the electrical circuit, and low voltage on the control transformer that powers the solenoid.

Valve spool or poppet stuck: The armature may be held unseated because the spool or poppet will not shift. The solenoid will draw high inrush current for too long a time and burn out.

A metal-to-metal spool type of valve may be varnished in place, or dirt may prevent the spool from shifting. Clean, lubricate, and reassemble the valve.

A packed resilient seal spool type of valve may not shift because swollen seals hold the spool in place, or because dirt prevents the spool from completely shifting. Clean the valve and repack the seals or replace the spool.

Solenoid Corroded. Use valves with adequate protection against moisture, coolants, and so on, which may come in contact with them. Provide sealed electrical connections. Make sure that dirt and moisture covers are securely in place.

Solenoids Energized Simultaneously. On momentary-contact, direct-actuated valves, check to make sure that both solenoids are never energized at the same time. This check is easily made by wiring a small indicator lamp temporarily across the coil of each solenoid. If both lamps are lit at the same time, the last solenoid to be energized will burn out. Correct the electrical circuit to prevent this.

High Transient Voltages. Solenoid burnout may be caused by high transient voltages that break down coil insulation, causing short circuits to ground. High transient voltages are most common where solenoids are connected to lines operating above 120 V, which also control motors and other inductive load equipment. Switching of such loads can create very high voltage peaks in the circuit. The remedy is to isolate solenoid circuits from main power circuits.

Temperature Too High or Too Low. Solenoid failures can be expected when a valve is operated above its rated temperature. Insulation may fail because it is not suitable for high-temperature use. Specify solenoids designed for the ambient temperature, place the valve in a cooler location, or consider the use of a pilot-actuated valve at the hot location, controlled by a remote pilot valve.

Solenoids also can fail when valves are operated at lower-than-rated temperatures. Metal parts may shrink and lubricant viscosities may increase to a point where solenoid motion is retarded or stopped. This can be avoided by moving the valve to a warmer place or by providing a heated enclosure for the valve. If trouble persists with closely fitted valves, change to a type that can work at low temperatures.

Direct-Current Solenoid Failures. Direct-current (dc) solenoids generally do not burn out due to the valve spool or poppet sticking, low voltage, and high voltage. This is due to the fact that the dc solenoid does not have a high inrush current at the outset of the armature travel. The maximum current to the coil is the holding current, which is the same whether the solenoid armature is seated or not. With an ac solenoid, current to the coil at the outset of armature travel is three to four times greater than when it is seated (has completed its stroke). This absence of high inrush current in dc solenoids prevents coil burnout when the valve is stuck or the armature fails to move for other reasons.

Valve failures to shift can, of course, occur with dc as well as ac solenoids. Corrosion and temperature can also effect dc solenoid operation. The general guides for ac solenoids can be applied in these cases. High transient voltage problems can be handled in the same manner as for ac solenoids.

If the solenoid noise level on an ac solenoid is very high and occurs each time the solenoid is energized, check to see that the armature is seating. Most direct solenoid actuated valves are provided with a manual override. If the solenoid noise decreases when the override is operated, incomplete solenoid motion is indicated. Check to determine if rubbing parts can move properly and that the correct voltage is available at the coil. Extremely loud ac hum can be caused by a broken part within the solenoid.

In marginal noise problems, consider mounting the valve so that the armature works up and down rather than horizontally. Maximum quieting

Observable damage	Possible causes	How it happened	Remedy
Coil insulation is burned; plunger is in open position; nylon coil bobbin is melted under the plunger	1. Low line voltage	Insufficient force to close plunger; high inrush current continues and generates excessive heat[a]	Replace coil only;[b] correct low voltage
	2. Ambient temperature too high	Plunger eventually will not close because undissipated heat has reduced current flow (closing force) while electrical resistance increases and generates more heat[a]	Replace coil only[b]
	3. Cycling too fast		Install a continuous-duty model
	4. Load too high, or valve spool is blocked	Plunger blocked in open position permits electrical resistance to increase and generate excessive heat[a] (see also note below)	Replace coil only;[b] correct high load condition
Coil insulation is burned; plunger is in closed position (no melted nylon)	1. Voltage too high	Extra force of excessive holding current holds plunger in closed position while electrical resistance increases, generates excessive heat, and burns out coil	Replace coil only;[b] correct high-voltage condition
	2. Solenoid too large for light load		

Frayed and burned lead wires	External mechanical short	Water-based coolant, metal chips, etc., have created contact between wires	Replace coil only;[b] shield from coolant
Small pinhole burn in coil wrap	Transient short to ground	High-voltage surge causes spark to jump between coil winding and solenoid C stack (or other nearby ground)	Replace coil only[b]
Spongy insulation on coils and lead wires	Internal mechanical short	Fire-resistant fluids (phosphate esters) dissolve coil insulation, coil varnish, paint, etc.; cause short between coil turns	Install solenoid with proper insulation
Deep scoring at all seating surfaces	Overvoltage or reduced load (or wrong size)	Excessive closing force causes T bar to wear through copper shading coils at top of C stack; plunger also hammers base, resulting in destruction	Replace the entire solenoid, not just the coil

[a]Excessive heat burns insulation off coil wires, permitting electrical short and melting nylon bobbin.

[b]If solenoid is the encapsulated type, the entire solenoid must be replaced.

Note: In the case of double-solenoid valves, the cause of solenoid burnout may be that both solenoids were actuated simultaneously. Usually, one solenoid will burn out from inability to close properly.

Figure 3.17 Causes and cures for solenoid burnout. (Courtesy of Parker Hannifin Corp., Cleveland, Ohio.)

can be obtained by using dc solenoids. This is practical even on large valves of the solenoid-controlled, pilot-actuated type. A chart summarizing the solenoid valve troubleshooting procedure is presented in Fig. 3.18.

The following list provides a summary of the maintenance and service procedures to be followed so that trouble-free valve operation is obtained.

1. Air pressure may be too high or too low at the valve inlet.
2. Compressors should be followed by aftercoolers. Hot (450°F), wet air from a compressor may cause sticking valves (particularly with metal-to-metal spool designs), erratic cylinder operation, early seal failure, and reduced component life.

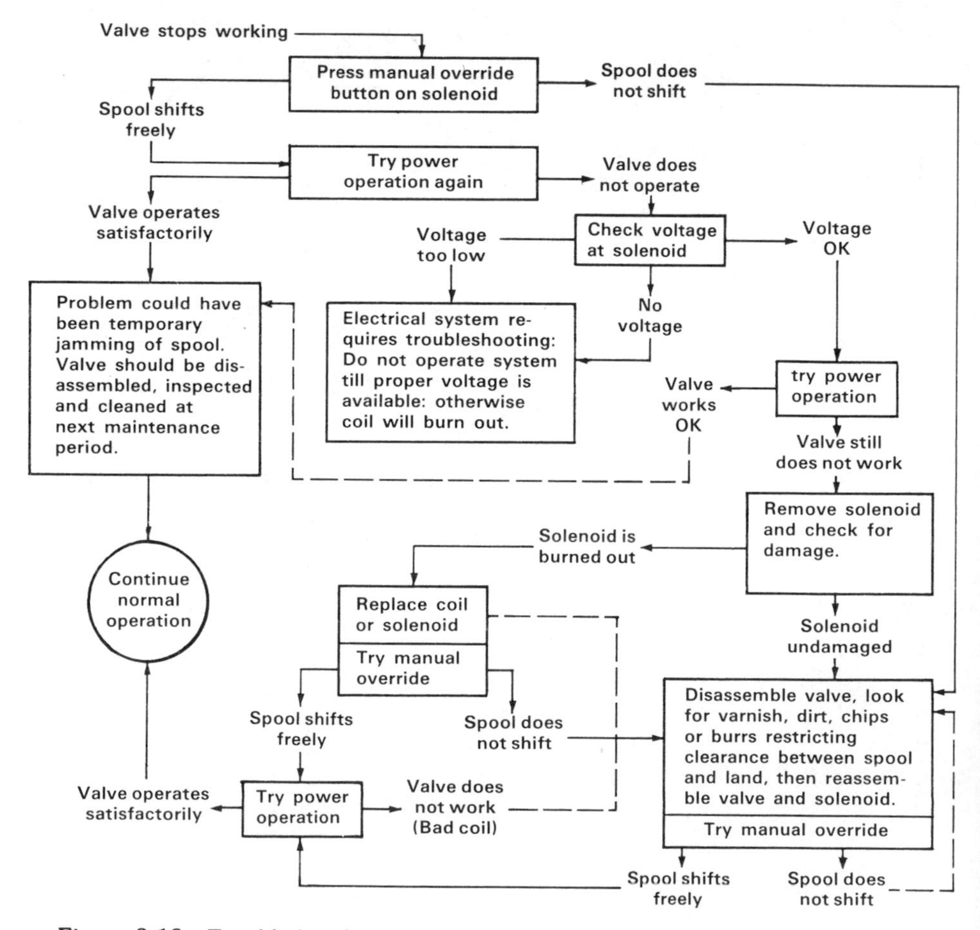

Figure 3.18 Troubleshooting guide for single-solenoid direction control valve. (Courtesy of Parker Hannifin Corp., Cleveland, Ohio.)

3. Old or poorly maintained compressors, or certain compressor lubricating oils, may produce excessive oil breakdown (varnish). Where this is a severe problem, use a coalescing filter ahead of the valve inlet port to remove all dirt and liquid.
4. Do not restrict valve inlet supply lines. Use high-flow-capacity quick disconnects. Piping inside diameter (ID) should be no less than the minimum pipe ID of the valve port. Tubing ID should also not be less than the pipe ID of the valve, with smooth 90° bends. The quantity of bends should be held to a minimum.
5. Use filters and lubricators wherever possible.

3.3 AIR MOTORS

In the pneumatic tool industry, the designation *air motor* is used with two different meanings. First, it identifies the driving force of a rotary pneumatic tool. In the second sense, it classifies a complete tool which is used in literally hundreds of applications, from winding reels to driving nuts in multiple-spindle applications.

Since it is the heart of a rotary tool, a thorough knowledge of the basic air motor is highly important. Only two air motor designs, vane and piston, are of major importance. Other types, not used widely, are lobed-rotor, V, diaphragm, and turbine motors. *Lobed-rotor* motors are similar in design to lobed-rotor compressors, and were developed for special high-temperature applications. *V motors,* which are a version of radial-piston motors, have pairs of cylinders in a V configuration, with pistons driving a crankshaft. *Diaphragm motors* obtain rotary motion through a reciprocating diaphragm and ratcheting mechanism. *Turbines* utilize air velocity for rotation, such as a windmill does. Incoming air strikes impellers or blades in a rotor. This type of motor has a very low starting torque, and is used principally in small, very high-speed die grinders.

A *piston*-type motor has more positive starting and better low-speed characteristics than a *vane*-type motor, but is slower and heavier. Motors with vanes are available in power ranges of 1/8 to 25 hp, and in speeds up to 30,000 rpm. Axial piston motors are available from ½ to approximately 4 hp, and radial piston types range from 1½ to 25 hp. Maximum speed of piston motors is about 5000 rpm.

Radial piston motors have pistons arranged radially, with all connecting rods linked to a common crankshaft. In *axial piston* motors, pistons are aligned along an axis parallel with the output shaft; force is transmitted to the shaft through an inclined "wobble" plate. The number of cylinders in a piston motor can range from three to six.

In a piston motor, torque is developed by pressure on the pistons confined within the cylinders. Power in this type motor is determined by inlet pressure, number of pistons, area of piston face, stroke, and speed. Speed in a radial piston motor is limited by inertia of the reciprocating parts. Inertia is not a limiting factor in an axial piston motor, and this type is in a more compact package. Vane motors, with their wide ranges of power and speeds and high power/weight ratio, are used as power sources for all except a very few rotary pneumatic tools.

3.3.1 Operation and Characteristics: Vane Motors

Principle of Operation

It is a common belief that air motors convert energy into mechanical motion by controlled expansion of air, and to some extent, this is correct. Basically, however, the rotor is turned from the force of compressed air on the rotor blades, or vanes, and the pressure imbalance between the blades. Air admitted to the motor is expanded only about 25% before it is exhausted. If it were expanded to anywhere near the initial compression ratio (7.12 to 1 for a line pressure of 90 psi), rapid expansion would quickly freeze any moisture in the motor, to clog exhaust ports and reduce power and speed (Fig. 3.19). By keeping compressed air behind the rotor blades during a major portion of the cycle, maximum torque is obtained at stall or near stall speed. At stall, air pressure within the motor builds up quickly to line pressure.

Longitudinal vanes fit in a radially slotted rotor, which is mounted eccentrically in a cylinder. Air enters the motor unit through ports in the cylinder close to the "seal point" of the motor, that portion of the surface of the cylinder which is closest to the rotor. This seal point prevents the compressed air from escaping; therefore, pressure builds up against the first blade the air contacts. This forces the blade away from the seal point, to start motor rotation.

Compressed air is sealed off from blade 1 as soon as the rotor turns far enough to expose the next blade to pressurized air. At this point, air which contacted the first blade at 70 psi, for example, has dropped in pressure to approximately 35 psi because of expansion. Even though direct exposure to pressurized air by the first rotor blade has been cut off by exposure of the second blade, blade 1 will provide rotating force to the rotor because air pressure against one side is greater than pressure on the other side. Blade 1 will continue to apply rotating force until it reaches the exhaust ports in the motor housing. At this point air pressure is equalized on both sides of blade 1 and it applies no more rotating force.

A vane-type motor, complete with planetary gear reducer to lower the shaft output speed, in Fig 3.20, and Fig. 3.21 depicts such a motor used in an impact tool.

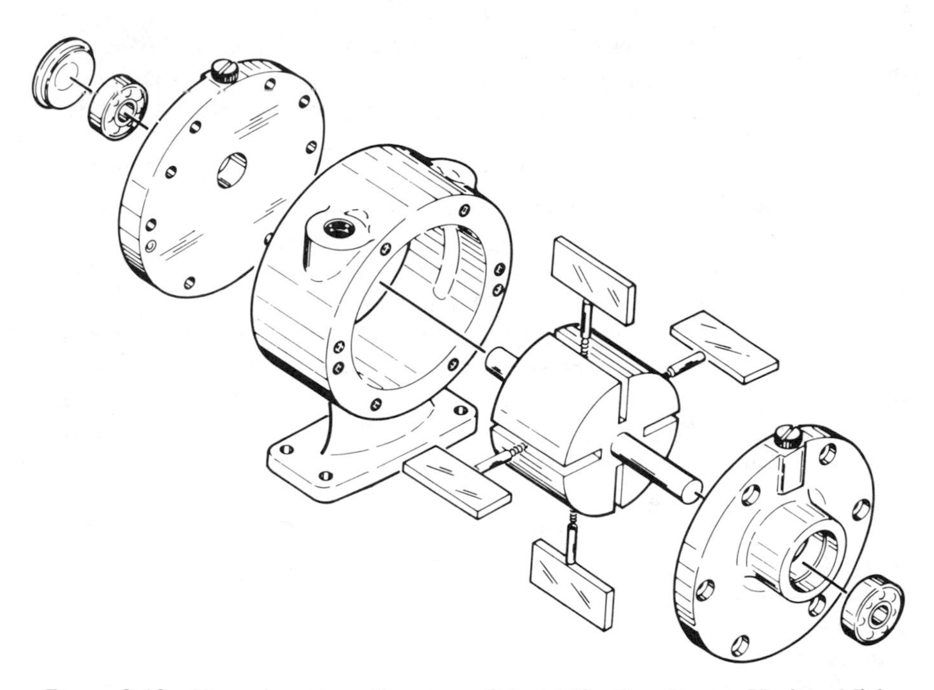

Figure 3.19 Vane air motor. (Courtesy of Gast Mfg. Co., Benton Harbor, Michigan.)

Characteristics

Torque. In air motors, starting torque is as important as maximum torque which is reached just above zero speed and decreases linearly as the motor approaches free speed. The starting torque is generally 75 to 85% of maximum torque because of internal static friction. Starting torque is the limiting factor of a motor, since the motor must overcome the applied load if it is to be useful.

In vane motors, effective torque output is proportional to the exposed vane area, pressure imbalance across the blade, and the moment arm through which the pressure is working. Torque at any given speed can be increased by:

1. Increasing the exposed vane area
2. Increasing inlet air pressure, thereby increasing pressure imbalance across the blade
3. Increasing housing ID or length and/or rotor outside diameter (OD) or length

The amount that any one of these can be increased is limited. Increasing exposed vane area increases porting problems and shortens blade life. Increasing the cylinder ID increases the speed of the blade rubbing against the cylinder wall; this also shortens blade life. Available line pressure limits the amount that inlet air pressure can be increased. Higher-than-normal pressure also shortens tool life.

Power. An air motor rating is usually expressed as the maximum horsepower produced at 90 psi. However, motors rated at different pressures can be compared by assuming a change of 14% in horsepower for each 10-psi difference in air pressure. A motor rated at 4 hp at 90 psi, for example, would be the equivalent of one rated at 3.4 hp at 80 psi.

Air motor torque reaches its maximum just above zero speed, but at the same speed, brake horsepower is near its minimum. In an ungoverned motor, power increases to maximum at approximately 50% of free speed, or maximum speed under no-load conditions. Power of a governed motor follows the same pattern, but the maximum is reached at approximately 80% of maximum governed speed.

Air Consumption. Air consumption by a motor varies with speed and motor size. At 90 psi, consumption per horsepower can be as low as 21 cfm in a large motor or as high as 45 cfm in a fractional horsepower motor. For estimating air consumption by a motor, use a range of 30 to 40 cfm/hp for motors of 1 hp or less, and 23 to 30 cfm/hp for those over 1 hp.

Variable Speed. Air motor speed is easily adjustable over a wide range using simple valving. The exact speed of an air motor is difficult to predict ahead of time because, like an air cylinder, speed is a function of balance between inlet pressure and load. If anything changes this balance, the motor speeds up or slows down to find a new balance. Also, like an air cylinder, the best way to size an air motor is to overpower it with respect to the load, then use a pressure regulator or a needle valve to bring it into balance with the load.

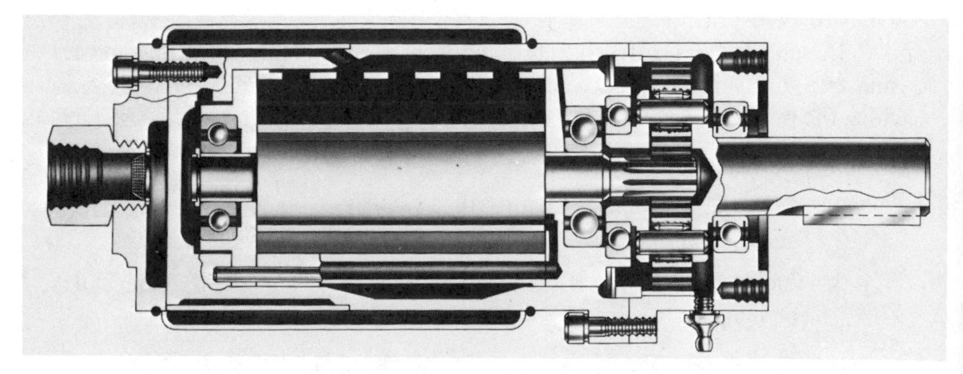

Figure 3.20 Typical multivane air motor features. (Courtesy of Ingersoll Rand Co., Washington, New Jersey.)

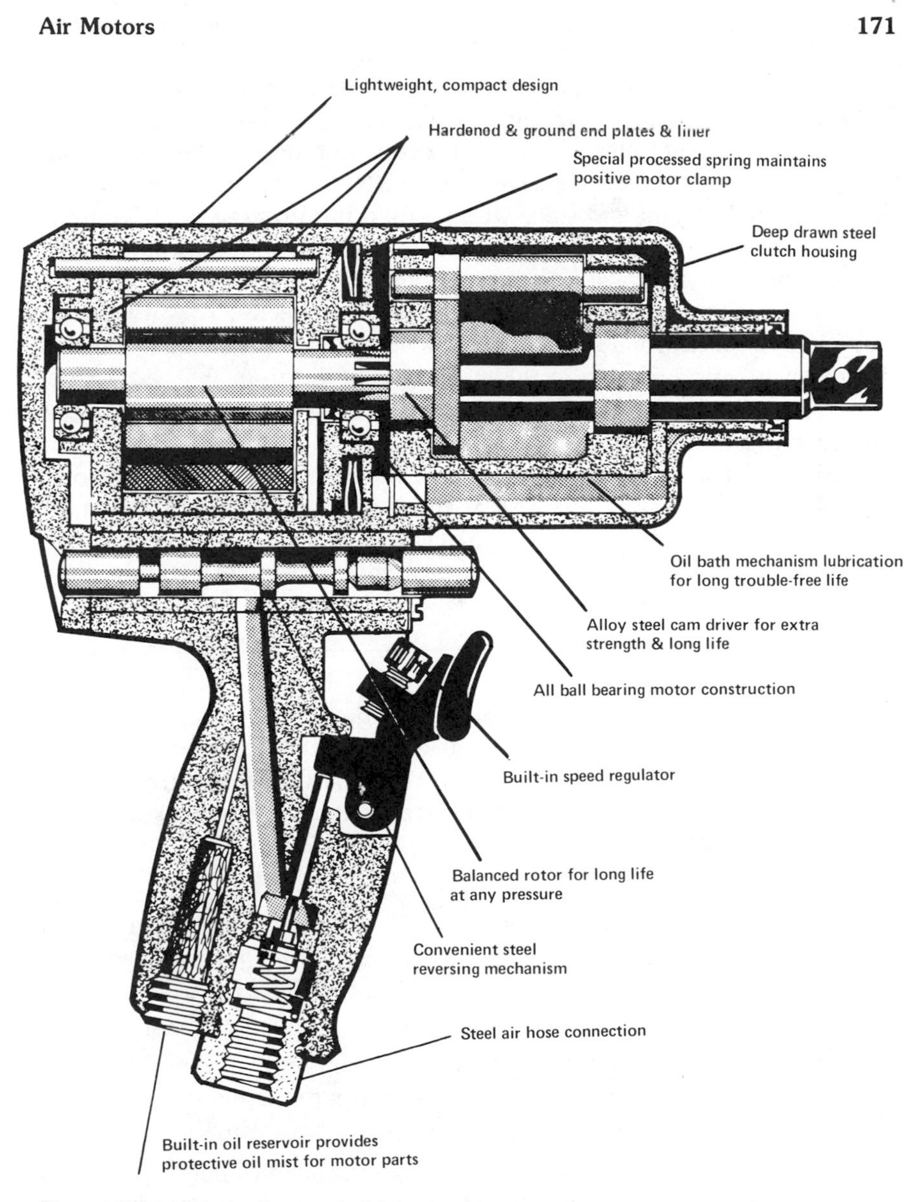

Figure 3.21 Vane-air-motor-driven impact wrench.

Motor speed varies inversely with any change in shaft loading. If the load and the inlet pressure remain constant, the speed remains constant. The ability to change speed quickly and easily is an important consideration on many applications.

Stalling. Air motors, like air cylinders, can be stalled indefinitely without damage. They can be indefinitely overloaded with no overheating.

They will maintain a constant tension on a reel, or can be used to supplement other drives, such as an electric motor, and will adjust their speed to that of the main drive; and by means of a pressure regulator, they can be made to take any proportion of the total load. They are especially useful on power conveyors that may tend to jam and stall, which would burn out an electric motor.

Environment. Air motors can operate under conditions unfavorable to other motors. They will tolerate very high or very low ambient temperatures, extreme heat, and corrosive or explosive atmospheres. They will operate completely submerged in liquid. They can be mounted to run in any position if proper attention is given to the method of lubrication.

Ease of Control. Single direction of rotation air motors can be controlled with simple two-way shutoff or needle valves placed in the inlet air line or in the discharge port of the motor. Reversible models are controlled with four-way air valves. Because of their low rotational momentum, air motors are more instantly reversible than electric motors. They can be started, stopped, and reversed as often as necessary without overheating and without damage to other equipment.

3.3.2 Maintenance

Other than low air pressure, two factors are the primary cause for loss of power by a vane-type air motor: internal leakage at the seal point and around the rotor blades, and interference of revolving parts. Before attempting to locate the source of power loss in the motor, service instructions should be studied for correct disassembly and reassembly procedure. The following will assist in assuring a correctly reassembled motor:

1. When removing the motor unit from the housing, note the position of the motor; many motors are rotationally positioned in the housing with respect to inlet hole location. This is particularly true of reversible motors.
2. Disassembly of different types of motor units will vary. When disassembling one for the first time, the parts should be placed on a clean work area in the order of their disassembly.
3. Remember that almost all nuts or governor assemblies which screw onto or into the rear end of a clockwise-rotation rotor have left-hand threads, to prevent their working loose in service.
4. Rotor bearings are usually pressed on the rotor shaft and are slip-fit into the bearing plates. Whenever possible, the force applied to remove bearings should be applied to the inner race.

After the motor unit has been disassembled, the following parts should be checked for wear, and replaced if necessary.

Rotor Bearings

Normally, it is good practice to replace all bearings when an air motor is repaired; this is particularly true if the motor has been run on air that has not been filtered properly. When removing the bearings from a motor, the exposed side on each should be noted. They may be flush ground on one side, shielded or sealed on one side only, or designed to take both thrust and radial loads. To reassemble the bearings on the rotor shaft, a driver that slips over the rotor shaft and locates against the inner race of the bearing should be used. Bearings are inexpensive in comparison to the parts they support. A worn bearing can allow the rotor to drift, resulting in poor performance and damage to more expensive parts.

Rotor Blades

Rotor blades are also inexpensive, and should be replaced when a motor is repaired. They can be reused, provided that (1) the contact edges are free of chips and have no foreign matter embedded in them, and (2) blade wear is less than the amount specified by the manufacturer as sufficient to necessitate blade replacement. If this information is not readily available, the old blades can be compared with new ones. The width at both ends of the rotor blade must be greater than the maximum distance between the OD of the rotor and ID of the cylinder; this distance is 180° from the seal point. Air leakage under the blade will result from wear, and a worn blade may fail due to poor engagement in the rotor.

Rotor

The rotor may be of solid or floating type. The *solid* rotor has an extremely long life provided that bearing failure or wear has not caused damage to the diameters accepting the bearings. Bearing wear will also cause wear between one or both ends of the rotor and bearing plates. The highest point of wear of solid rotors which have not been subjected to rub is at the rotor blade slots; rate of wear at this point is slow, and the rotor does not require replacement if wear is slight.

A *floating* rotor slips freely on a shaft which is positioned by bearings. This type of rotor usually has a shorter life than the solid type because of wear between it and the shaft; this wear allows the rotor to move away from the seal point on the cylinder, resulting in leakage and loss of power.

Bearing Plates

A bearing plate can be expected to have a long service life unless the rotor has rubbed the plate because of bearing failure. If this should happen, a bearing plate need not be replaced if galling is only slight. Sanding the plate to remove the burrs will permit it to be used; however, if rubbing has been excessive, replacement is necessary to obtain maximum power output.

Cylinder or Housing
Cylinder wear is a major source for internal leakage and consequent power loss; rate of wear depends on the quality of air received. When a cylinder wears to the point that the motor is producing insufficient power, it must be replaced; it cannot be honed out, as air leakage at the seal point will result in a great loss of pressure against the rotor blades on the power side of the cylinder, and excessive air leakage through the exhaust slots.

3.3.3 Maintenance of Air Tools

There are two basic ways to control the maintenance of air tools:

1. Operate the air tool until it ceases to run, then overhaul.
2. Follow a program for periodic inspection and overhauling of air tools.

The first method can obviously be very costly in terms of repairs and downtime. When tools are operated until critical parts are worn out, the cost of parts replacement can be higher than necessary. A tool seldom fails when it is least needed in production. Failure usually occurs in peak production periods.

The second method requires a program for inspecting each tool at regularly scheduled intervals. The length of this interval will be determined by operating conditions, and where extreme conditions prevail, tools should be examined more often.

Preventive Maintenance

Periodic Examination. The first objective of any maintenance program should be the prevention of breakdowns. Each tool should therefore be examined according to a regular schedule, so that any parts showing evidence of wear may be replaced promptly. A simple card system can be set up to show date and nature of repairs.

Some maintenance departments prefer that all air tools be subjected to inspection at fixed intervals of 45 to 90 days. Color coding can be used effectively in such a system. For example, all tools scheduled for a specific month can be spotted with a touch of black paint, and those for the following month spotted with red. In this way the supervisor can easily identify the tools that are to be sent in for maintenance. Spare tools should always be available for emergency use. This simple procedure has been used in many factories and has helped reduce the cost of repairs.

Lubrication. Correct lubrication is a major requirement in preventive maintenance. It is likely that some 80% of tool failures can be traced back to inadequate lubrication. The absolute necessity of oil in any automobile engine crankcase is accepted without question. Application of the same standard to air tools would eliminate the primary cause of faulty performance and early tool failure. Many types of precision-built air tools develop motor speeds of 17,000 rpm or more. Sustained operation at such speeds requires filtered air and correct lubrication.

Most portable air tools have built-in reservoirs, but in practice these features are almost never utilized. It is not practical to attempt to incorporate reservoirs in smaller portable tools. Therefore, air motors can be lubricated most efficiently by means of standard air line oilers. Such oilers usually have a minimum capacity of a pint or more, with a transparent bowl which allows visual inspection of the oil level. Responsibility for replenishing air line oilers should be clearly assigned. An air tool may run for many hours without lubrication but with decreased efficiency, and eventually a critical part will wear out and fail.

The recommended lubricant for air motors is a good grade of light spindle oil SAE-5. Some common types are Mobil DTE 24 Oil or Non-Fluid Oil A-88-NR. Most major oil companies offer comparable lubricants.

For planetary gear systems, right angles, flat angles, and impact tools, use a good gear grease, as recommended by the tool manufacturer. One-half ounce of grease should be added to these assemblies after 40 to 60 hr of use. This quantity is usually dispensed by one stroke of the average lever-type grease gun. Do not pump too much grease into the gearing of an air tool, since this will tend to reduce speed and power.

As a final precaution, a check should be made to ensure that air tools are actually receiving oil from the air line lubricator. Simply place a piece of white paper over the exhaust port and with the tool running for a few moments, a slight discoloration of the paper should show when oil is being supplied to the air motor.

Air Hose. The air hose supplying the tool should have an adequate inside diameter. The correct size is usually one size larger than the inlet thread on the tool. About 60% of air tools returned for repair have reducing nipples inserted at the air inlet. Such nipples cause a reduction of air power and their use should be discouraged, especially on grinders and sanders, where air volume is important. Note that hoses ½ in. in diameter or larger require the use of safety devices to prevent hose whipping.

Automatic Air Couplers. Self-sealing or automatic shutoff couplers, with mating connectors, are recommended for connecting a tool to the air line. The coupler is actually a valve that supplies air; it can be added to the branch line or to a hose leading from the branch. The connector is attached to the tool or attached to the far end of a whip hose on the tool. The coupler and connector are easily snapped together to form an airtight connection, yet the hose can still swivel freely. The coupler should have adequate capacity for the tool, without excessive pressure drop.

Air Tool Repair

Proper maintenance of air tools often depends on the ability of a repairperson to determine whether a part or assembly is worn to the point where it should be replaced. Most portable air tools are powered by a rotary vane motor.

Replacement of each of the air motor parts is recommended when a certain degree of wear (based on past experience) has been detected. Parts that have reached this point may still be usable for a long period, but they will not yield performance equal to a new tool. The following standards are therefore recommended.

Rotor. Examine the end faces for roughness and smooth with a honing stone if necessary. Normally, there should be no noticeable wear on such faces, as the rotor is a few thousandths shorter than the cylinder. Inspect the blade slots for wear or burrs. A new blade should move in and out of the slot without binding. Use a honing stone to break away any sharp edges found on the corners of the slots. Examine the spline or gear teeth at the driving end. If they have become so worn that a step can be seen next to mating surfaces, the rotor should be replaced.

Cylinder. Examine the cylinder on the inside diameter for rough circular grooves. If the grooves are from 0.005 to 0.010 in. deep, replace the cylinder. Such grooves are usually caused by foreign matter in the air line. Installation of an air filter is the best means for preventing such grooving and scoring. A badly scored cylinder cannot be restored to usefulness by honing.

End Plates. Any upper or lower end plate whose face shows wear greater than a depth of 0.005 in. should be replaced. Such wear is usually caused by incorrect rotor spacing. Light score marks can often be lapped out with a 150-grit abrasive cloth on a flat surface plate.

Bearings. Shielded or sealed bearings should never be washed in solvent because:

1. It is probable that more dirt will be introduced and trapped within the bearing than is removed.
2. It is probable that the lubricant within the bearing will be diluted or washed away completely.

Open bearings may be washed only in fresh clean solvent and should then be repacked with any good bearing grease. To check a bearing, hold the inner race and rotate the outer race by hand. If rough movement or end play are detected, replace with a new bearing.

It is also possible to compare a used bearing with a new one in order to determine the amount of wear. Sometimes a bearing shield may become dented and cause a drag. It is possible to lift off the bearing shield by use of a sharp pick. The removal of this damaged shield should free the bearing. A bearing should be reassembled into a tool with the open side in direct contact with the end plate. This method will give protection from dirt and will help retain the bearing lubricant. Preloaded bearings should always be reassembled according to instructions furnished with the air tool. These bearings are usually found on air motors whose end plates are fastened in place by a nut, screw, or governor weight assembly.

Rotor Blades. A rotor blade is probably the most expendable part of an air motor, since it is subjected to a high degree of movement and friction. Most blades are machined from a fibrous form of laminated phenolic. By means of various treatments, blades are stabilized so that they can withstand a wide latitude of temperature, humidity, water, and oil. They are almost immune to warpage. All these characteristics are necessary for the smooth cycling action of the blades as it moves in and out of the rotor slot.

It is a generally accepted rule that if a rotor blade loses 20% or more of its width, it should be replaced. Width can be checked by comparing the old blade with a new one. Blades narrowed by wear will eventually tilt at the edge of the slot, and this will create a groove mark on the side of the blade. The groove mark will then deepen enough to cause the blade to break or else to bind on the edge of the slot. Blade breakage at high speed can cause severe damage to the interior of the air motor. A blade that binds on the edge of the slot will simply freeze the motor and keep it from moving.

Rotor blades found to be within wear limits may be cleaned up by a simple lapping operation on each side. Place a piece of 400-grit waterproof sandpaper on a flat surface and lap each side of the blade. This will clean the blade but will not remove enough of the material to affect the performance of the air motor.

A good set of rotor blades might be considered the heart of an air motor. If blades are maintained in good condition and replaced when necessary, the motor will have much longer life and greater freedom from trouble.

Repair Facility

Often, the most costly repairs are those applied to tools that have been allowed to run until they stop. Air tool users are therefore encouraged to set up, in their own plant, a tool maintenance center along the following lines:

1. Assign specific maintenance responsibilities to one or more persons.
2. Set up proper repair facilities.
3. Maintain an adequate stock of repair parts.
4. Establish a systematic preventive maintenance program.
5. Train maintenance personnel in proper air tool servicing techniques.

The location of the tool maintenance department should be near the area where tools are being used, be clean and well lighted, and have the appearance of an efficient repair area. There should be at least one workbench with a 4-in. soft-jaw vise. The majority of repairs can be made with ordinary hand tools. The department should be equipped with proper service parts sheets, plus simple records for logging all repairs.

There should also be facilities for testing repaired air tools. Adequate testing can often be done with a flowmeter (to indicate the consumption of cubic feet of air per minute), and a tachometer. It can be shown that if a tool has proper airflow and is running at proper speed, it will almost surely be developing its rated power.

The question of whether to repair or replace an air tool is a problem common to all users, but there are wide differences in prevailing practices. Most companies, for example, have an amortization program in which they write off all air tools in a 5-year period. Yet many users are reluctant to scrap an air tool after 5 years, and it is not uncommon to find tools in use for periods up to 10 and 12 years. Retention of old air tools in a plant can be uneconomical, particularly if the tools are being used constantly. Cost studies have shown that a power tool performing at less than 90% efficiency is actually costing money. An actual survey in a typical plant showed that about 25% of the air tools were performing well below their rated efficiency. Where systematic maintenance is not practiced, it is not uncommon to find the average air tool running at about 60% of maximum efficiency.

A first step toward the solution of this problem is the use of a *controlled service record.* These records should be reviewed periodically by someone with responsibility and authority for weeding out those tools which, because of overage or extremely hard usage and/or abuse, are giving a substandard performance or generating expensive repair costs. A helpful rule for the maintenance department would be that a tool should be scrapped if its repair cost exceeds 50% of the cost of replacement.

4

Contaminants in a Hydraulic System

Nearly all hydraulic component manufacturers and users agree that *dirt* is responsible for a majority of malfunctions, unsatisfactory component performance, and degradation. Dirt comes from many sources and in various forms, such as chips, dust, sand, moisture, pipe sealant, weld splatter, paints, and cleaning solutions. The aggregate of all these forms is termed *contamination.*

4.1 SOURCES

The sources of contamination are numerous (see Fig. 4.1) and include:

Being introduced with new oil

Built in at the time of hydraulic system construction

Introduced with the air from the environment

Generated by wear within the hydraulic components

Introduced through leaking or faulty seals

Introduced by shop maintenance activities

The removal and control of contamination necessitates the use of a filter. The selection of the correct filter and its proper location in the system requires as much care and expertise as the selection of other critical components, such as pumps, valves, and actuators.

Dirt is introduced into hydraulic systems at the time of fabrication of the components and during the manufacture of the system. Some of the con-

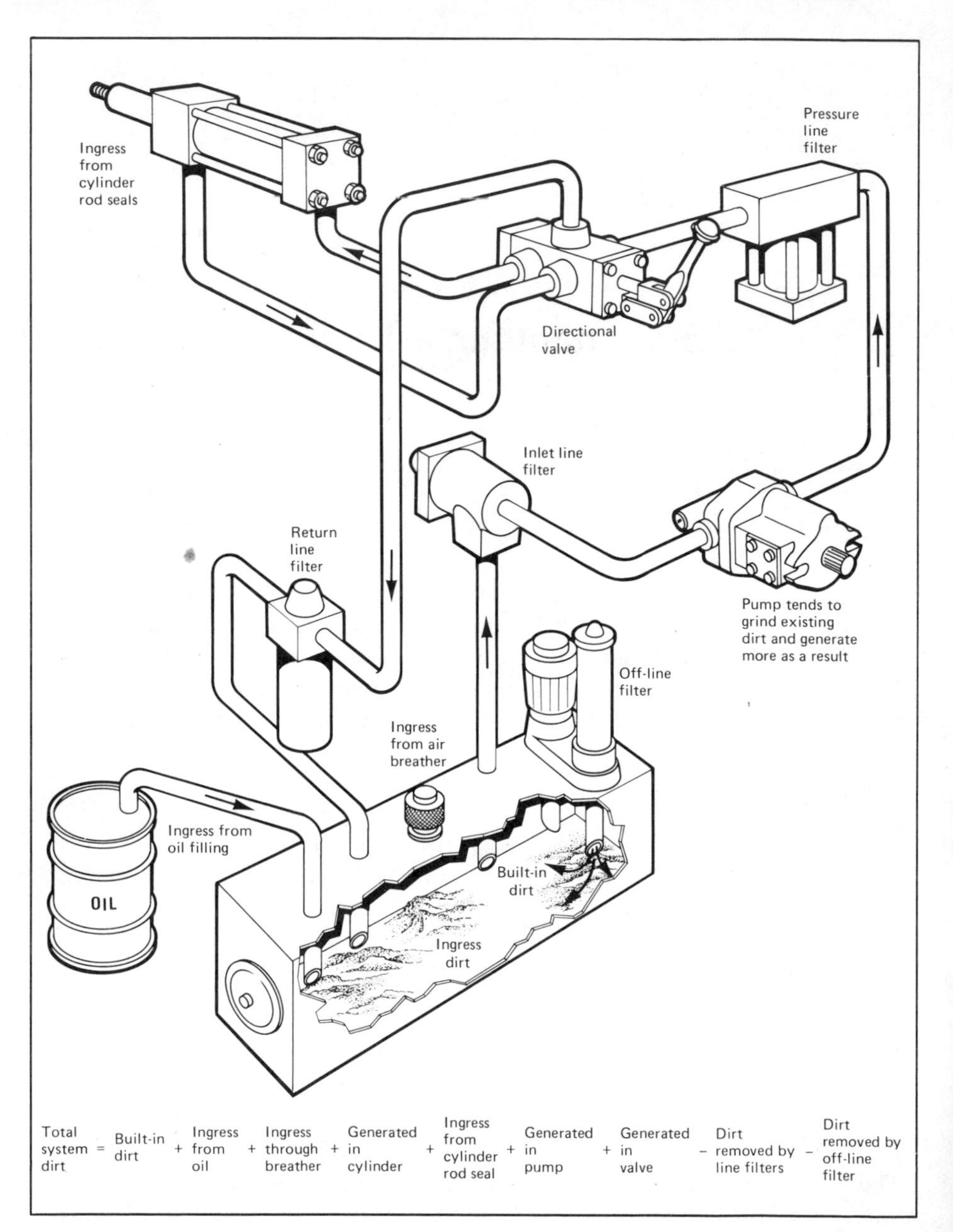

Figure 4.1 Pictorial representation of filtration equation. (Courtesy of Sperry Vickers, Troy, Michigan.)

tamination found in oil samples taken from a system after a short run-in period of a new machine includes:

Metal chips from:
- Tubing burrs
- Pipe threads
- Component manufacture particles
- Tank fabrication

Pipe dope

Teflon tape shreds

Lint from wiping cloths

Welding scale and beads

Pipe scale

Factory dust

Rubber particles

Debris from hose fabrication

Although oil is refined and blended under relatively clean conditions, it is usually stored in drums or in a bulk tank at the user's factory. At this point, it is no longer clean because the filling lines contribute metal and rubber particles, and the drum always adds flakes of metal or scale. Storage tanks can be a real problem because water condenses in them to cause rusting, and contamination from the atmosphere finds its way in unless satisfactory air breather filters are fitted.

If the oil is being stored under reasonable conditions, the principal contaminants on delivery to a machine will be metal, silica, and fibers. Using a portable transfer unit or some other filtration arrangement, it is possible to remove much of the contamination present in new oil before it enters the system and is ground down into finer particles.

Contamination that enters the oil from the environment surrounding the hydraulic system can do so by following several paths. These entry points include air breathers, access plates, and seals.

Air breathers: Very little information appears to be available on what these filters will actually achieve, and purely nominal ratings are usually specified. The amount of air passing through the filter will depend on draw-off, which means that single-acting cylinders operating in dirty atmospheric conditions will result in a greater introduction of dirt.

Access plate: In some plants it cannot be assumed that access plates will always be replaced. In power unit design, good sealing is vital, and in bad environments such items as strainers should not be positioned inside the reservoir if access requires the refitting of removable plates. Other removable items will allow dirt to enter during maintenance, and good design practice should minimize this.

Seals: Wiper seals in cylinders cannot be 100% effective in removing very fine contaminants from the cylinder rod. If they were, they would remove the oil film from the piston rod. A completely dry rod would quickly wear out the seals. Where cylinders remain extended in a heavily contaminated atmosphere, considerable quantities of fine particles can get into the system unless protection such as a bellows is provided.

Dirt is continually introduced into operating hydraulic systems because of wear and degradation of the working components. The wearing action of working parts in components such as pumps, fluid motors, valves, and cylinders generates contamination. Rust scale from the reservoir caused by condensation above the oil level is also a source of dirt. Burrs on tubing and piping break loose during service; and flexing of components continuously releases particles which were not removed during the initial cleaning of the system (Fig. 4.2).

Contamination is added to the system by shop maintenance activities. This is done in various ways, including the following:

1. Leaks are caught in a bucket that may be dirty and which is open to the atmosphere when returned to the tank.
2. Dirty buckets are used to catch oil when components are repaired or lines broken; then the oil is returned to the tank.
3. Oil is stored in drums that are not sealed.
4. Occasionally, supposedly clean oil, which is new or reprocessed, will contain lint particles from the refining filters.
5. When equipment is being repaired:
   a. Components pick up contamination from dirty workbenches.
   b. Dirty rags are used to clean components.
   c. Tubing and hose is left on the floor or exposed to dirt.
6. Suction line strainers are often left off and lying in the tank or perforated with a long screwdriver by maintenance people who "don't want to be bothered" cleaning strainers under oil or are in a hurry to get the machine going.
7. Strainers are taken off for cleaning. The procedure stirs up the dirt in the bottom of the tank. Pumps are started up because "It won't hurt to run the pump a few minutes while I am cleaning the strainer."
8. Fill caps are left off the tank.

	Oxide scale	Plastics and elastomers	Oil additives	Metal particles	Airborne dirt	Silica sand	Lapping compound	Process residues	Fibers
Oil	X		X		X	X			X
Tank	X	X		X	XX	X			X
Pressure relief valve		X		XX	X	X	X		X
Accumulator (bladder and piston types)	X	X		X	X			X	X
Filter	X	X		X	XX				X
Piping, fittings and rubber tubing	X	X		X	XX			X	X
Control valves		X		X	X		X		X
Actuators		X		X	X				X
Pump		X		XXX	X	X			X

Note: X, noticeable contamination contribution potential.
XX, medium contamination contribution potential.
XXX, strong contamination contribution potential.

Figure 4.2 Contamination from components. (Courtesy of Sperry Vickers, Troy, Michigan.)

The situations described above occur in most shops. It is economically impractical to control these activities. Repair work is done on the machine, on portable workbenches, or in maintenance shops. All of these tend to be dirty by nature. Management cannot hope to provide working areas and tools that are "clean" enough to prevent contamination from being added during repair and maintenance activities. The solution to this apparent dilemma is to make the importance of system cleanliness known to the maintenance personnel through training activities. The awareness of the sources, the effect on the system, and control of contamination will develop habits within the maintenance people that are conducive to minimizing the introduction of dirt by shop activities.

4.2 EFFECTS

It is well known that contaminant particles are of all shapes and sizes, and that the finer they are, the more difficult it is to count them and to determine the material of which they are composed. However, we can say that the majority are abrasive and that when interacting with surfaces, they plough and cut little pieces from the surface. This wear accounts for about 90% of the failures due to contamination or dirt. The effect of these contaminant particles on various system components reflects itself differently depending on the mechanism of operation. Following is a list of what dirt does to hydraulic components (Fig. 4.3).

1. *Pumps*
   Erodes wear plates.
   Causes sticking of vanes creating erratic action.
   Causes the vanes to wear out the cam ring.
   Wears out rotor slots.
   Increases shaft journal and bearing wear.
   Increases gear wear, with resultant inefficiency.
   Increases piston and sleeve wear, with resultant inefficiency.
   In compensator controls on *variable-volume pumps,* it causes sticking, slow response, and erratic delivery.
   Creates excessive heat and inefficient use of horsepower.
2. *Relief valves*
   Dirt causes chatter.
   Accumulated dirt causes relief valve to "fail-safe;" pressure becomes erratic.
   Dirt causes seat wear.
   Dirt causes plugged orifices on balanced piston-type valves.

Symptom	Possible component malfunction	Possible cause
Slow movement of hydraulic cylinder	Excessive leakage in control valve Excessive internal leakage in pump	Parts worn by particles circulating in fluid; this type of abrasion or erosion is often caused by silt-size contaminants
Stuck manually operated control valve Burned-out solenoid on electrically operated control valve	Jammed, stuck, silted spool-sleeve-type control valve	1. One or more large, hard particles physically preventing movement of spool; or 2. Silting of valve caused by large number of particles; or 3. Gums or lacquers from deteriorated fluid plating out on valve surfaces
Poor positioning of actuator Overheating of system	Worn control valve	Increased valve clearances often result in poor positioning accuracy and an increased internal leakage, which can result in overheating
Repeated failures of hydraulic motors, pumps, or hydrostatic transmissions	Same	A chain of failures often indicates that the debris from the first failure was not adequately removed and resulted in a quick second failure

Figure 4.3 Problems caused by particulate contaminants. (Courtesy of Sperry Vickers, Troy, Michigan.)

3. *Directional valves*
   Dirt causes wear to spool and housing lands creating excess leakage.
   Dirt deposits cause spools to stick, which can cause solenoid failure.
   Sticking valves can cause excessive shock loads, damaging hose, piping, fittings, and other components.
4. *Check valves*
   Dirt permits fluid to bypass check.
   Dirt causes wear on the ball and seats, creating leakage.

5. *Flow control valves*
   Dirt causes erosion of orifices, changing the flow setting characteristics.
6. *Servo valves*
   Dirt causes erosion of sharp edges, which affects metering.
   Dirt plugs nozzles because they contain very small orifices.
   Dirt leads to buildup of varnish deposits.
7. *Cylinders*
   Dirt causes excessive wear of the cylinder rod, piston seals, rod seals, and the tube bore.
   Dirt causes cushions to malfunction.
8. *Fluid motors*
   Dirt causes wear similar to pump wear
9. *Fluid*
   Dirt acts as a catalyst, thereby breaking down the molecular structure of the oil, causing gummy residue (varnish) to form.
   Dirt in the tank tends to attract additives, which changes the composition of the fluid.

From the foregoing, it can be seen that failures arising from dirt or contamination can be classified into three categories:

1. *Catastrophic failure*: occurs when a large particle enters a pump or valve. For instance, if a particle causes a vane to jam in a rotor slot, the result may well be complete seizure of the pump or motor. In a spool valve, a large particle trapped at the right place can completely stop a spool from closing.
2. *Intermittent failure*: caused by contaminant on the seat of a poppet valve, which prevents it from reseating properly. If the seat is too hard to allow the particle to be embedded into it, the particle may be washed away when the valve is reopened. Later, another particle may prevent complete closure only to be washed away when the valve opens. Thus a very annoying type of intermittent failure occurs.
3. *Degradation failure*: follows wear, corrosion, and cavitation erosion. They cause increased internal leakage in the system components, but this condition is often difficult to detect.

There are several philosophies regarding the sizes of particles that cause degradation failures. Two schools of thought are prevalent:

1. Degradation wear is related to the dirt level in the system. Compo-

nent wear is accelerated as dirt levels increase above the level a component can tolerate. Littlc, if any, increase in component life is obtained by reducing dirt levels below the level of component tolerance.

2. Wear is related to ultrafine particles of less than 5 μm, known as *silt particles*. (Larger particles tend to be associated with catastrophic failures.) Abrasion and erosion often create hard, metallic-wear products which enter the hydraulic fluid and cause additional wear. The total number of these particles also causes oxidation and other breakdown, hastening the deterioration of the hydraulic fluid. Very small amounts of free water in hydraulic fluid will cause corrosion, oil breakdown, acid formation, and other problems.

4.3 CONTROL

Nearly any discussion regarding dirt in a hydraulic system reduces to a consideration of particle sizes. For the most part, these are smaller than a grain of salt, which is 100 μm in size; 40 μm is about the smallest particle that can be seen with the naked eye. One micrometer is equal to 39 millionths of an inch; 25 μm equals one-thousandth of an inch. Thus when a 60-mesh filter (238 μm) is considered for use in a system, the openings are twice as large as a grain of table salt (100 μm). Thus the filter cannot be effective in removing these small, harmful particles (see Fig. 4.4).

Sizes of Familiar Objects

	Size	
Substance	μm	in.
Grain of table salt	100	0.004
Human hair	70	0.0027
White blood cell	25	0.001
Talcum powder	10	0.0004
Red blood cell	8	0.0003
Bacterium (average)	2	0.00008
Lower limit of visibility (naked eye)	40	0.00156

The performance requirements that a filter must meet are as follows:

1. It must be capable of reducing the initial contamination to the desired level within an acceptable period of time.
2. It must be capable of achieving and maintaining the desired level, and allow a suitable safety factor.

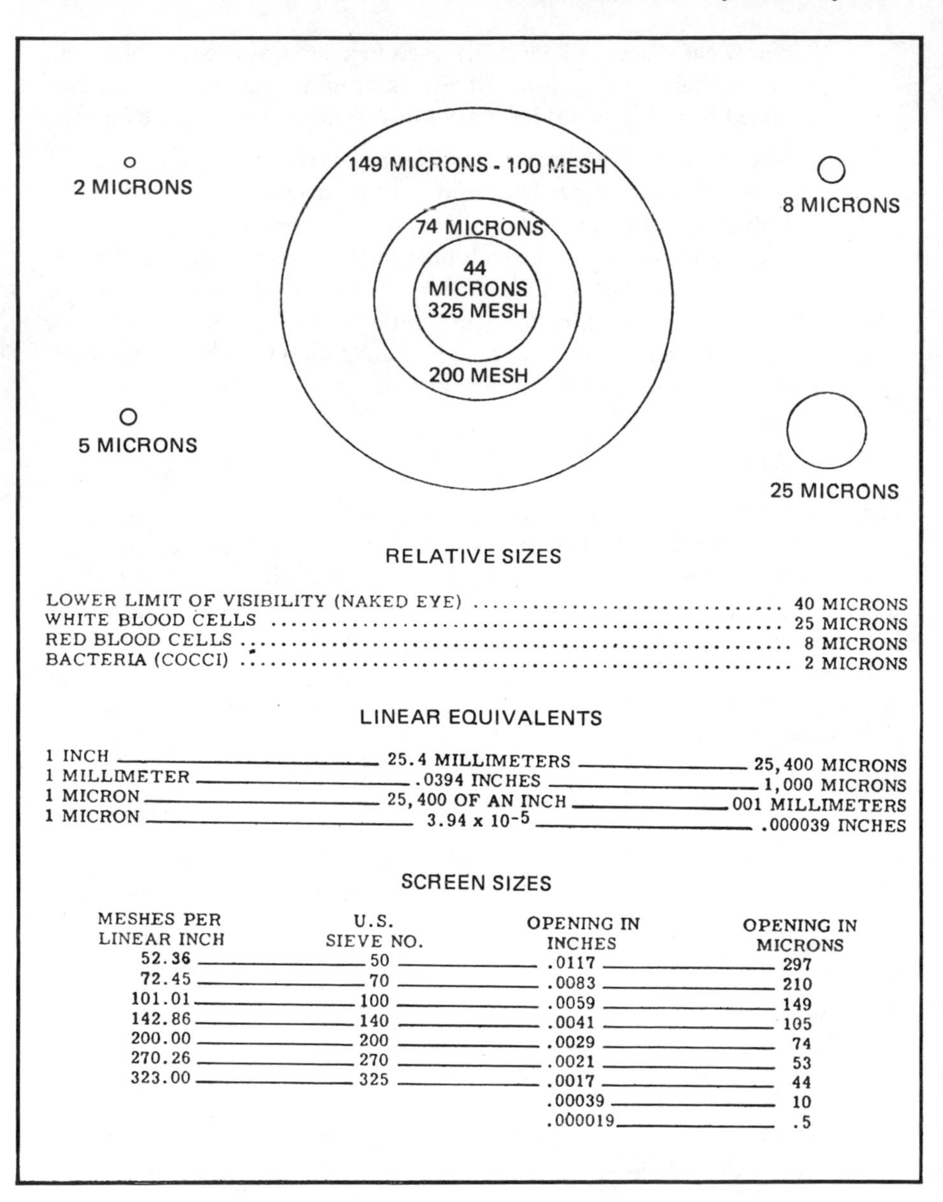

RELATIVE SIZES

LOWER LIMIT OF VISIBILITY (NAKED EYE)	40 MICRONS
WHITE BLOOD CELLS	25 MICRONS
RED BLOOD CELLS	8 MICRONS
BACTERIA (COCCI)	2 MICRONS

LINEAR EQUIVALENTS

1 INCH	25.4 MILLIMETERS	25,400 MICRONS
1 MILLIMETER	.0394 INCHES	1,000 MICRONS
1 MICRON	25,400 OF AN INCH	001 MILLIMETERS
1 MICRON	3.94×10^{-5}	.000039 INCHES

SCREEN SIZES

MESHES PER LINEAR INCH	U.S. SIEVE NO.	OPENING IN INCHES	OPENING IN MICRONS
52.36	50	.0117	297
72.45	70	.0083	210
101.01	100	.0059	149
142.86	140	.0041	105
200.00	200	.0029	74
270.26	270	.0021	53
323.00	325	.0017	44
		.00039	10
		.000019	.5

Figure 4.4 Relative size of micronic particles.

3. The quality of maintenance available at the user location should be considered.
4. Filters must be easily accessible for maintenance.
5. An indication of filter condition to suit the user's requirements must be provided.
6. In continuous process plants, facilities must be provided to allow elements to be changed without interfering with system operation.
7. The filters must provide sufficient dirt-holding capacity for an acceptable interval between element changes.
8. The use of a filter in the system must not produce undesirable effects on the operation of components, such as high back pressures.

To satisfy the performance requirements, a number of factors must be considered in selecting hydraulic filters. These include:

Degree of filtration

Flow rate

Pressure drop

Dirt capacity

Compatibility

Element cleanability or replacement

System pressure

Temperature

It is generally recommended that filtration to at least a 25 μm range be provided for a hydraulic system. This recommendation is made with respect to particle removal and is subject to the following qualifications:

Filters located in a pump suction line must be correctly sized to prevent cavitation.

There are systems where better filtration is required because of very small clearances. In such systems, recommendations should be sought individually from component manufacturers.

When filtering fire-resistant fluids, specific recommendations of the component and fluid manufacturers should be followed.

4.4 FILTER TYPES

Filter elements are divided into two general classifications: depth-type filter elements and surface-type filter elements. *Depth-type elements* force the fluid to pass through multiple layers of material and the dirt is caught because of the intertwining path that the fluid takes. These are generally absorbent-type elements since the dirt particles are trapped by the walls. An absorbent element causes the dirt to stick to the surface. Because of its construction, a depth-type element has many pores of various sizes. Since there is no one consistent pore size, the element is generally given a nominal rating which is based on its average pore size. For example, an element with a nominal rating of 25 μm means that the average pore size is at least 25 μm; and, initially, it will remove most contaminants 25 μm and larger in size.

Surface-type elements consist of a single layer of material through which the fluid must pass. The material layer consists of perforated metal or woven-wire mesh screening. The manufacturing process of these material layers can be very precisely controlled so that these elements can have very consistent pore or opening sizes. Because of this fact, surface-type elements can be identified by their *absolute rating*. The absolute rating signifies the largest opening in the filter element. This rating thus provides the largest, hard, spherically shaped particle that can pass through it.

A decision as to which type of filter element to select—depth or surface type—requires consideration of the advantages and disadvantages of each. Following is a list of these for each type of element.

Advantages of Depth Elements

High efficiency on a one-time-fluid-pass basis

Large dirt-holding capacity

Usually inexpensive

Disadvantages of Depth Elements

Impractical to clean.

Limited compatibility with fluids.

High initial clean pressure differential.

Paper and synthetic depth elements are adversely affected by temperatures above 260°F.

Paper elements have limited shelf life.

Paper elements are not recommended for use with water-base fluids.

Paper elements are impractical to use with fluid viscosities above 1000 SSU.

Advantages of Surface Elements

Strength and resistance to fatigue, temperature, and corrosion.

Surface elements can be cleaned and reused.

Maximum pore size can be controlled.

Low initial clean pressure differential.

Disadvantages of Surface Elements

Usually expensive.

Not as efficient as a depth element on a one-time-fluid-pass basis.

Some dirt particles in a system are magnetic. They are built into the system while the machinery is being fabricated. They are also generated within the system from the action of moving parts and fluid erosion. They can also enter the system through the reservoir openings and air breather.

These particles are normally abrasive and can react chemically with hydraulic oil to decompose the oil. They should be removed from the system. However, most of this type of dirt consists of very small particles. A fine filter would have to be used, which means an increase in cost and probably an increase in maintenance. Both of these factors can be avoided by using a magnet. If many magnetic particles are present in a system, a relatively coarse element used in conjunction with a magnet can be as effective as a finer filter. The magnet will catch the small metal particles. The element will catch the larger dirt and will not become clogged as quickly as will a finer element. Cost and maintenance are reduced.

It is especially recommended that magnets be used with fire-resistant fluids. Petroleum oil allows many of the metal particles to settle to the bottom of the reservoir. Fire-resistant fluids are more detergent and tend to keep these particles in suspension. Consequently, there can be more magnetic particles in a stream of fire-resistant fluid than in a stream of petroleum oil.

The degree to which the hydraulic fluid should be filtered is another important consideration in the process of providing a means for controlling dirt. Often, the component manufacturer will provide information in the catalog relative to particle size sensitivity. If not, the following rules of thumb may be applied:

1. With standard vane and gear pumps operating at pressures of 1000 psi or less, 74 μm filtration can be selected.

2. Circuits operating up to 2000 psi should be equipped with 40 μm filtration.
3. For circuits operating above 2000 psi, 25 μm filtration should be selected.

The filtration levels listed are based on typical loads imposed on the components by forces related to pressure. These loads cause the component parts within a pump or valve that move relative to each other to rub against each other. The moving parts, however, are separated by a thin layer of oil that fills the clearance space between them. If this oil film is loaded with dirt particles, rapid degradation, wear, and eventual seizure or binding of the parts will occur. With low-pressure units, relatively large clearances are tolerated in the components and contamination has less effect; also, there is less force available to drive the particles into the clearances. Thus increasing the pressure is of major significance in determining the effect of dirt on hydraulic components such as pumps and valves (Fig. 4.5).

Filter elements are available in a wide variety of materials, ranging from paper to woven-wire mesh. Figure 4.6 lists some of more common materials together with the rating available with each medium.

In addition to the degrees of filtration, the location of the filters within a system is an important consideration. Hydraulic filters can be installed in the intake, pressure, or return lines of the system or in the reservoir (Fig. 4.7).

Intake line filters prevent catastrophic failure of pumps by collecting "large" contaminant particles. The filters are usually mounted either partly or completely outside the system reservoir, near the pump intake. They are usually relatively coarse, which makes them ineffective in controlling dirt levels (see Fig. 4.8). Most designs provide for servicing of the filter element through the filter housing without opening or draining the system reservoir. Some designs even allow servicing without shutting down the system. Intake filters must be sized carefully to assure that the pump can tolerate pressure drop at maximum flow and minimum operating temperature, assuming a plugged filter element. Otherwise, the pump may "starve" and cavitate.

Pressure filters are used on the outlet side of the pump or just ahead of valves and other highly dirt-sensitive components, such as servo valves and proportional valves. Pressure filters remove critical contaminants passing through or generated by the pump before they get into the remainder of the system. Any contamination generated in the system downstream of the pressure filter will increase dirt levels in the reservoir unless controlled by a return line filter. The same is true for dirt that enters the system through cylinder rod seals (see Figs. 4.9 and 4.10).

Item	Micrometers	Typical clearance (in.)
Gear pump (pressure loaded)		
Gear to side plate	½–5	0.000,02–0.000,2
Gear tip to case	½–5	0.000,02–0.000,2
Vane pump		
Tip of vane	½–1[a]	0.000,02–0.000,04
Sides of vane	5–13	0.000,2 –0.000,5
Piston pump		
Piston to bore (R)[b]	5–40	0.000,2 –0.001,6
Valve plate to cylinder	½–5	0.000,02–0.000,2
Servo valve		
Orifice	130–450	0.005 –0.018
Flapper wall	18–63	0.000,7 –0.002,5
Spool sleeve (R)[b]	1–4	0.000,04–0.000,16
Control valve		
Orifice	130–10,000	0.005 –0.40
Spool sleeve (R)[b]	1–23	0.000,04–0.000,90
Disk type	½–1[a]	0.000,02–0.000,04
Poppet type	13–40	0.000,5 –0.001,5
Actuators	50–250	0.002 –0.010
Hydrostatic bearings	1–25	0.000,04–0.001
Antifriction bearings	½[a]–	0.000,02–
Sleeve bearings	½[a]–	0.000,02–

[a]Estimate for thin lubricant film.

[b]Radial clearance.

Source: Machine Design, May 25, 1967.

Figure 4.5 Typical clearances for fluid system components. (Courtesy of Pall Corp., Glen Cove, New York.)

Return line filters are normally placed in the system return line to clean the hydraulic fluid before it enters the reservoir. Return line filtration is especially recommended for startups of new systems because most initial contamination comes from within the system itself: contaminants inadvertently left in piping and components, metal from manufacturing and assembly, moving surfaces and threaded fittings, and particles from thread-sealing materials, hoses, elastomer seals, and so on.

Each filter location offers advantages over any other. For example:

Only a return line filter can reduce the amount of dirt entering the system through a cylinder rod seal.

Only an intake filter can prevent a large particle that may be in the reservoir from entering the pump and possibly causing a catastrophic failure.

Only a pressure line filter at the pump outlet can prevent particles from a wearing or failing pump to be spread throughout the system. These particles could cause valves, cylinders, and motors to fail.

All locations are important; however, economics generally dictates that all three filters cannot be placed in the system. The use of only one or two filters in a system requires the careful analysis of the dirt-level tolerances of the components and the consequence of a failure. The analysis should also include safety, potential machine damage, maintenance, and repair costs. Figures 4.11 and 4.12 depict various filtration arrangements commonly used in hydraulic systems. For example, if a designer is permitted only one filter in the hydraulics system of a heavy-duty dump truck operating in a quarry, where should it be put? That is, which component should be protected? The answer: a pressure filter may be recommended to protect the truck's telescopic cylinder, because its failure might involve the operator's safety, and it is the most expensive hydraulic component to repair or replace.

Filter Bypass Valve

If filter maintenance is not performed, the pressure differential across a filter element will increase. An unlimited increase in pressure differential across a filter on the pressure side means that the filter element will eventually collapse or dirt may be pushed through. To avoid this situation, a simple relief

		Rating (μm)		
Medium[a]	Type	Nominal	Mean	Absolute
100-mesh screen	Cleanable	135	140	220
200-mesh screen	Cleanable	70	74	105
Sintered Dutchtwill woven wire mesh	Cleanable	10	17	25
Resin-impregnated paper	Disposable	10	18	30
Ultrafine	Disposable	0.45	0.9	3

[a]In addition to those tabulated, other types of filters and ratings are: paper, cellulose, felt, and glass, nominal ratings from 0.5 to 100 μm, and absolute ratings from 2 to 50 μm × nominal; sintered powders of metal, ceramics, or plastics, with nominal ratings of 2 to 65 μm and absolute ratings of 13 to 100 μm.

Figure 4.6 Filter materials. (Courtesy of Pall Corp., Glen Cove, New York.)

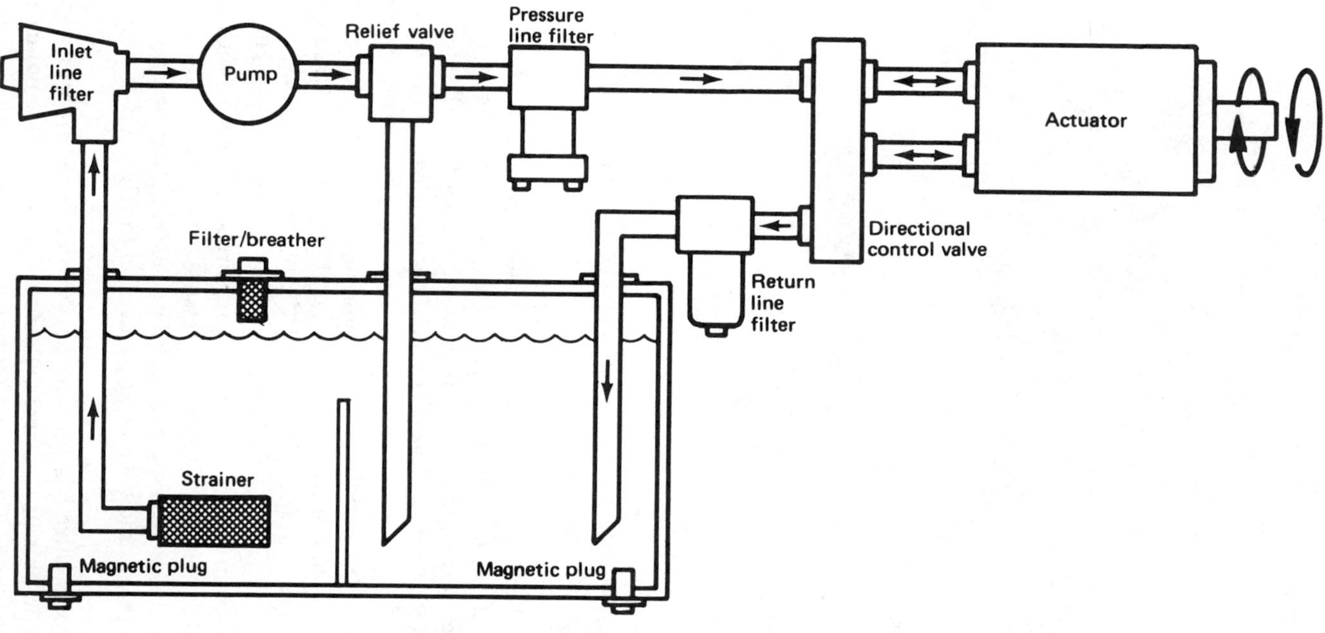

Figure 4.7 Typical locations of filters in a hydraulic system. (Courtesy of Sperry Vickers, Troy, Michigan.)

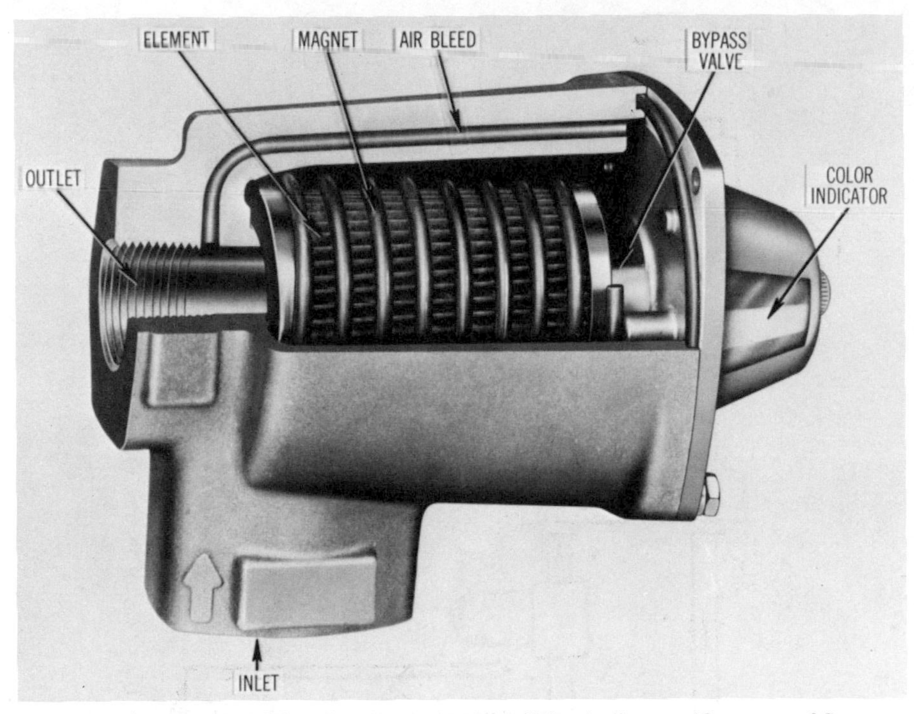

Figure 4.8 Inlet line filter with magnet and condition indicator. (Courtesy of Sperry Vickers, Troy, Michigan.)

valve is used to limit the pressure differential across a full-flow filter. This type of relief valve is generally called a *bypass valve* (Fig. 4.13).

There are several types of bypass valves, but they all operate by sensing the difference in pressure between dirty and clean fluid. A bypass valve consists basically of a movable piston, housing, and a spring that biases the piston. Pressure from dirty fluid coming into the filter is sensed at the bottom of the piston. Pressure from the fluid after it has passed through the filter element is sensed at the other side of the piston, on which the spring acts. As the filter element collects dirt, the pressure required to push the dirty fluid through the element increases. Fluid pressure after it passes through the element remains the same. When the pressure differential across the filter element, as well as across the piston, is large enough to overcome the force of the spring, the piston will move up and offer the fluid a path around the element.

A bypass valve is a fail-safe device. In an inlet filter, a bypass limits the maximum pressure differential across the filter if it is not cleaned. This protects the pump. If a pressure or return line filter is not cleaned, a bypass

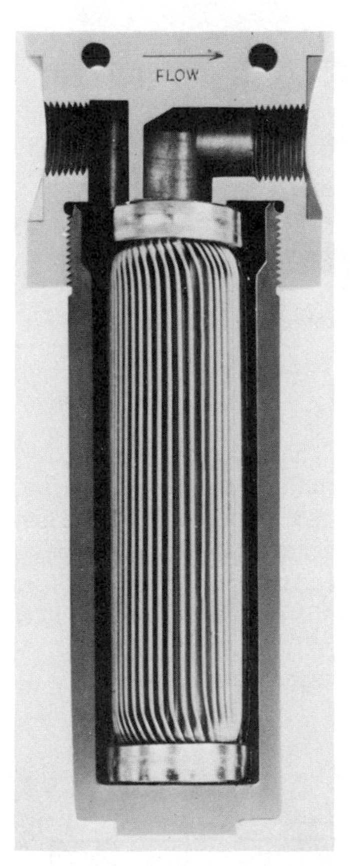

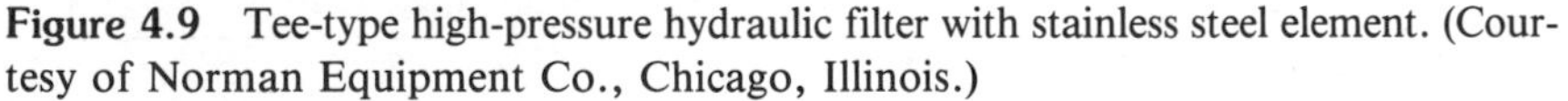
Figure 4.9 Tee-type high-pressure hydraulic filter with stainless steel element. (Courtesy of Norman Equipment Co., Chicago, Illinois.)

will limit the maximum pressure differential so that dirt is not pushed through the element or the element is not collapsed. In this way the bypass protects the filter.

***Filter Condition Indicator* (see Fig. 4.14)**

The bypass valve protects the filter by offering the fluid an alternative path when the pressure difference across the filter element exceeds a safe value. When the bypass valve is open, the filter is in effect out of the circuit and does not effect any dirt control. This condition is not tolerable for any length

Figure 4.10 In-line high-pressure hydraulic filter. (Courtesy of Norman Equipment Co., Chicago, Illinois.)

of time since system degradation will take place. The remedy is to clean, replace, and maintain the filter whenever such a condition is encountered. To help in this regard, a filter can be equipped with a condition indicator. This item indicates when the element is clean, needs cleaning, or is in the bypass condition. Indicators are available which are dependent on the motion of the bypass piston, as are units that sense the pressure differential across the filter element.

If the filter element is not changed when the indicator shows that it is

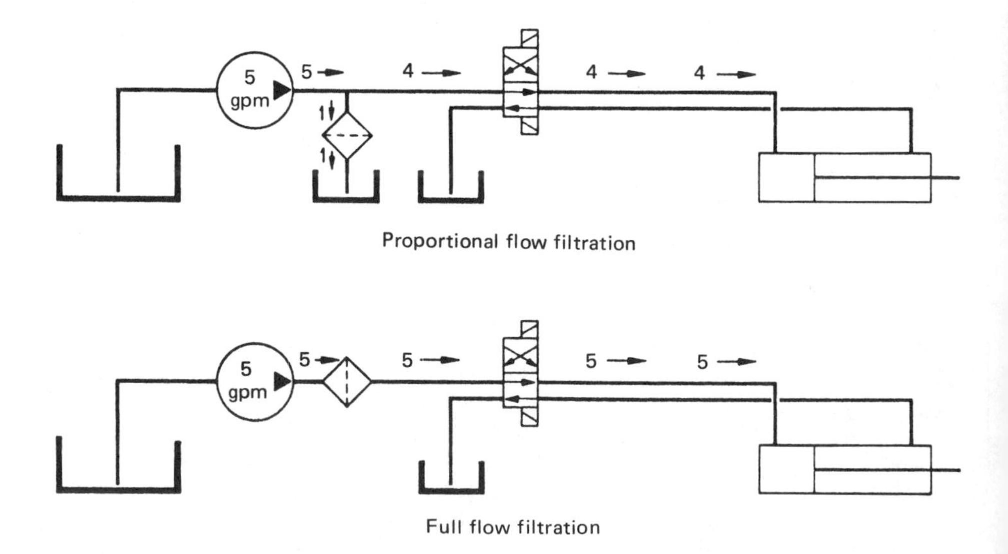

Figure 4.11 Proportional and full-flow filtration arrangements.

Pressure line filtration with bypass filter.

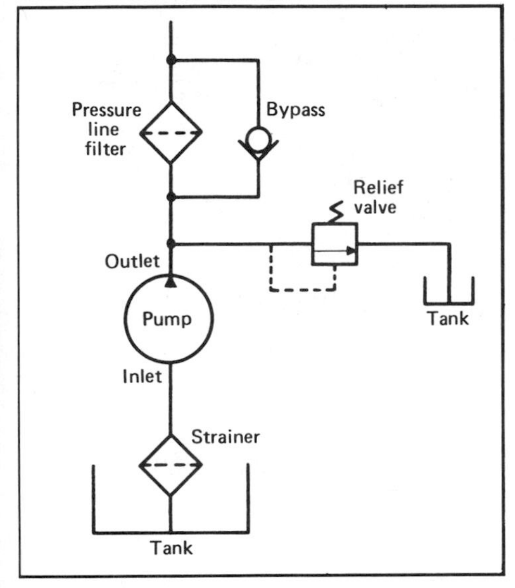

Pressure line filtration with non-bypass filter.

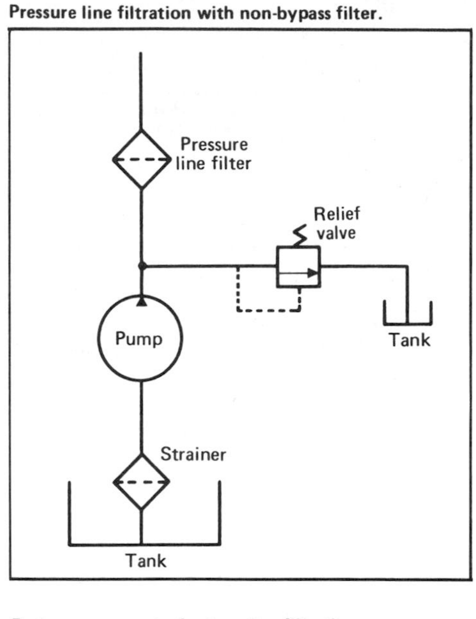

Locating pressure filter before relief valve gives constant flow through filter.

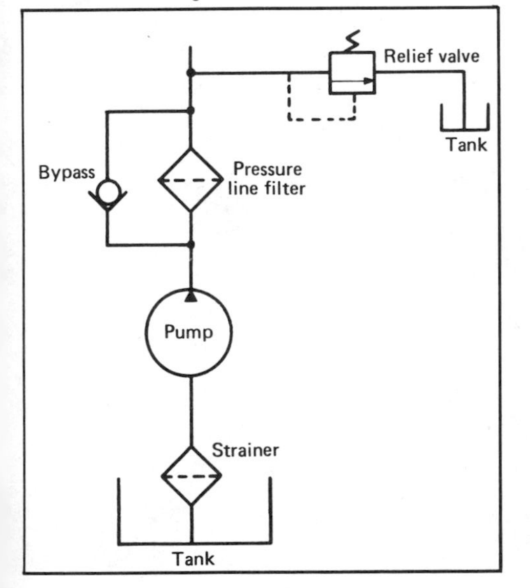

Basic arrangement of return line filtration.

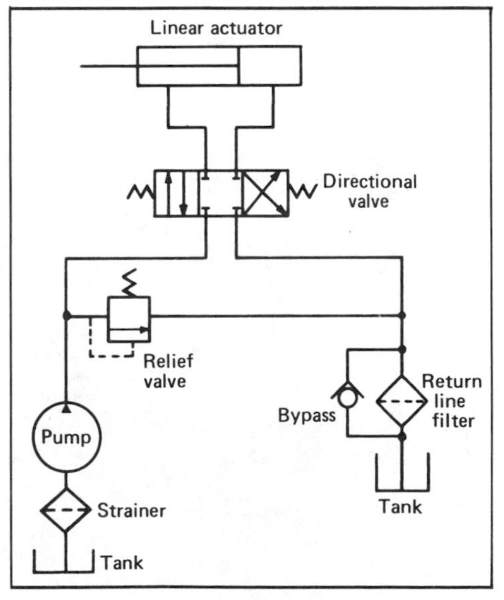

Figure 4.12 Hydraulic system filtration arrangements. (Courtesy of Sperry Vickers, Troy, Michigan.)

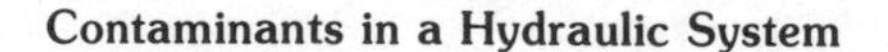

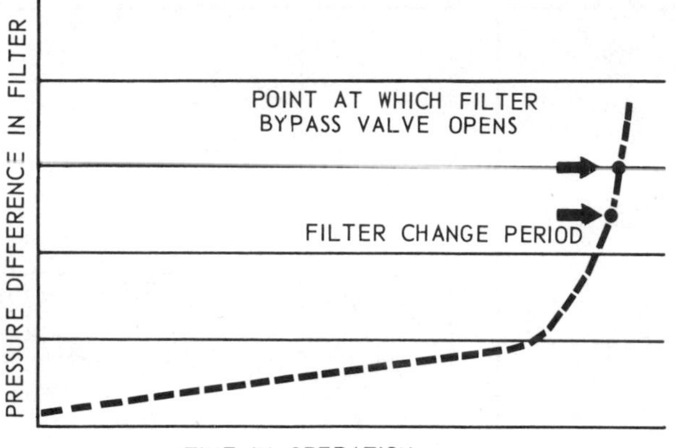

Figure 4.13 Life of a filter element. (Courtesy of John Deere and Co., Waterloo, Iowa.)

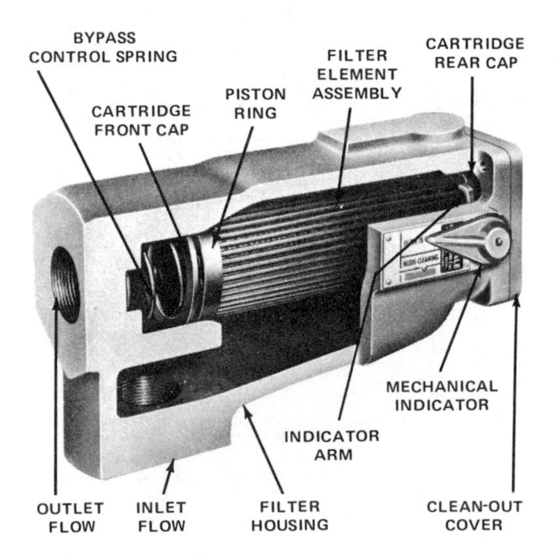

Figure 4.14 Hydraulic filter with mechanical condition indicator. (Courtesy of Parker Hannifin Corp., Cleveland, Ohio.)

dirty, the effectiveness of the filter is lost because the large pressure differential needed to force fluid through the filter may collapse the element. If a bypass valve is used, it will open and again, the filter will not function. The result in the hydraulic system is excessive wear, as shown in Fig. 4.15.

4.5 INSTALLATION, MAINTENANCE, AND TROUBLESHOOTING OF HYDRAULIC FILTERS

Filter location depends on the intended function of the filter. A filter used to protect hydraulic components such as pumps, servo valves, and hydraulic motors should be located close to the components being protected. Typically, a suction filter provides pump protection and a pressure filter protects other sensitive components. A return line filter can be located anywhere downstream of the application, and is usually the last restrictive device the fluid sees before entering the reservoir. One primary use of a return line filter is to limit the contamination level in a hydraulic reservoir to a specified value.

4.5.1 Installation

Suction Filters

Several precautions should be taken when using a suction filter. A low spring rate should be selected for the bypass control spring (2 or 3 psi) to minimize the maximum pressure drop across the filter. When sizing a suction filter,

Figure 4.15 Abnormal wear created by sludge. (Courtesy of John Deere and Co., Waterloo, Iowa.)

the clean flow pressure drop should be approximately one-third that of the bypass spring setting. This allows a relatively long element replacement life.

All external piping should be kept to a minimum to prevent any added pressure loss. The pipe leading to the filter inlet should not be smaller than the inlet port size; do not reduce the inlet port with a reducer bushing.

Return Line Filters

A return line filter is positioned in the circuit just before the reservoir. Being one of the last restrictive devices in a system, pressures are usually below 300 psi. Caution should be taken when sizing a return line filter. Due to discharging components (cylinders, actuators, etc.) in a system, flow rates through a return line filter can vary by 500%. When sizing a return line filter, consider the maximum flow through the filter and size for a clean flow pressure drop of approximately one-third of the bypass spring rate. This allows a relatively long element replacement life.

Filter Element Materials

Paper. Paper media consist of cellulose fibers treated with a resin coating. This medium is not recommended for water-base fluids, as the fibers may swell and shorten element life. All disposable media elements are affected by heat, time, and humidity. To avoid fatigue failures, temperature should not exceed 250°F and shelf time should not exceed 2 years.

Synthetic. Synthetic media consist of spun-bonded polyester material. This medium is recommended for petroleum-base hydraulic fluids where temperatures do not exceed 300°F, or for high-water-base fluids where temperatures do not exceed 120°F. It has a shelf life of 2 years.

Composite. Composite media are multilayered, consisting of wire mesh backing, special paper layers, and a bonded fiberglass inner layer. This medium is recommended for all common hydraulic fluids where temperatures do not exceed 400°F.

Wire Mesh. Wire mesh media are composed of stainless steel or Monel metal mesh suitable for application in all common fluids up to a temperature of 400°F. This medium is cleanable and usually has an absolute micrometer rating.

4.5.2 Maintenance

A hydraulic system may be equipped with the best filters available and they may be positioned in the system where they do the most good; however, if the filters are not taken care of and cleaned when dirty, the money spent for the filters and their installation has been wasted. Thus the whole key to good filtration is filter maintenance. Following are some suggestions that help to provide proper filter maintenance.

1. A filter maintenance schedule should be set up and followed diligently. This service action should be part of the preventive maintenance program for the system.
2. Inspect filter elements that have been removed from the system for signs of failure and of impending system problems.
3. Any fluid that has leaked out should not be returned to the system.
4. The supply of fresh fluid should be kept tightly covered.
5. Use clean containers, hoses, and funnels when filling the reservoir.
6. Use commonsense precautions to prevent entry of dirt into components that have been temporarily removed from the circuit.
7. All cleanout holes, filler caps, and breather cap filters on the reservoir should be properly fastened.
8. Do not run the system unless all normally provided filtration devices are in place.
9. Make certain that the fluid used in the system is of a type recommended by the manufacturers of the system or components.
10. Before changing from one type of fluid to another (e.g., from petroleum-base oil to a fire-resistant fluid), consult component and filter manufacturers in selection of the fluid and the filters that should be used.

Regular maintenance of filter elements is the key to clean fluid and long component life. Filters do not function when elements are clogged with contaminant to the point of bypass. Therefore, it is important to do three things:

1. Check the filter indicator frequently to determine when the element needs servicing.
2. Make sure that the filter is fitted with the correct replacement element. The correct replacement element should be listed on the filter nameplate.
3. Periodically, analyze a fluid sample of the system.

Servicing Elements

When replacing or cleaning a filter element, use the following procedure:

1. Shut down the system and relieve all pressure in the filter line.
2. Loosen the screws on the cover. Turn slightly to clear the cover screws and remove the cover. Allow excess fluid to drain.
3. Remove the element from the assembly.

a. *Disposable type*: Replace the element if a disposable paper or synthetic medium is used.
b. *Cleanable*: Soak the element in an ultrasonic cleaner for 15 min. If an ultrasonic cleaner is not available, soak in hot soapy water and ammonia solution for 15 min. Swish around and blow dry from the outside in.

4. Replace the element and install in the filter housing. Make sure that the O ring seats properly in the head.
5. Inspect the cover O ring seal and replace it if defective.
6. Replace the cover and tighten the screws until they are snug; do not overtorque.

4.5.3 Troubleshooting

Occasionally, system problems arise and it is difficult to locate the source of trouble. To help determine if the problem might be associated with the filter, the following troubleshooting chart lists problems and solutions to hydraulic system problems.

Problem	Cause	Solution
Suction application		
Noisy pump	1. Entrained air	Tighten and seal all fittings; add oil to reservoir; check O-ring seal on filter
	2. Cavitation	Clean clogged inlet line Replace or clean filter element Check pump inlet with vacuum gauge
Always repriming pump	1. Entrained air (aeration)	Check O-ring cover seal on filter Tighten and seal all fittings
Indicator reads "bypass" (mechanical) Engages switch (electrical)	1. Oil viscosity too high	Allow system to run for a short time Indicator should return to normal
	2. Bypass spring too weak (soft)	Change poppet assembly and indicator (see instructions)
	3. Element dirty	Clean or replace element

(continued)

Problem	Cause	Solution
Return, low-pressure application		
Indicator reads "bypass"	1. Element dirty	Replace or clean element
	2. Oil viscosity too high	Allow system to run for a short time Indicator should return
Dirty oil	1. Plugged cartridge	Replace cartridge
	2. Partial bypass—continuous	Correct filter size and oil viscosity
	3. Improper micrometer rating	Check particle size rating and switch to proper size rating
	4. Improper changes	Correct maintenance procedure or add bypass indicator
	5. Faulty (leaking) or ruptured filter	Replace filter
Broken housing	1. Too much pressure	Remove downstream restriction or change to corresponding filter of higher pressure or flow rating
	2. Too much mechanical shock	Add shock-absorbing material

A periodic fluid analysis is very important in maintaining an acceptable contamination level in a hydraulic system. The fluid analysis is also a good tool for monitoring possible changes or problems in a hydraulic system. To take a sample, follow this procedure:

1. Select a fitting on the pressure side of the system.
2. Clean the area with a solvent.
3. Allow the system to reach operating temperature, crack the fitting, and allow the fluid to flush the sampling point.
4. Fill a clean bottle with the fluid (at least 16 oz).

Using an automatic particle counter, a laboratory analysis can determine the concentration and distribution of contaminant in the sample. Other fluid characteristics can also be measured, such as:

1. Specific gravity
2. Viscosity
3. Water content
4. Gravimetric
5. Neutralization number
6. Visual, microscopic inspection
7. Particle shape

Any unusual characteristics detected are listed in the laboratory report comparing the particle count and fluid analysis with contamination standards and fluid specifications. This method will inform the maintenance staff of the fluid condition as well as the hydraulic system status, and of the possibility that component degradation may be occurring.

5

Compressed Air Filtration, Lubrication, and Moisture Control

5.1 CONTAMINANTS IN THE AIR SYSTEM

Dirt contaminates a pneumatic system. It is very similar to industrial smokestacks spewing soot. Dirt is only one form of contamination in a pneumatic system. To determine what other elements harm a pneumatic system and how they are dealt with, let us follow the step-by-step passage of air through a pneumatic system. A logical place to start our journey is with the industrial air made available to the compressor (Fig. 5.1).

In an industrial environment, air carries several undesirable elements as far as a pneumatic system is concerned. Two of these are dust and water vapor. Depending on the type of system and the degree of sophistication, dust and water vapor must be removed either partially or totally if a degree of system dependability is to be realized. A compressor is sometimes considered the heart of a pneumatic system. If the compressor fails, the system cannot function.

Air is often described as a mixture of gases made up of 21% oxygen, 78% nitrogen, and 1% other ingredients. Air is colorless, odorless, and tasteless. Identifying air by its physical properties is slightly misleading because in a pneumatic system air is not dealt with in this pure form. Besides being all the things described above, air is also a carrier.

Pneumatic system pollution is measured using the micrometer scale. One micrometer is equal to one millionth of a meter or 39 millionths of an inch. A single micrometer is invisible to the naked eye and is so small that even to imagine it is extremely difficult. To bring the size more down to earth, two everyday objects measured using the micrometer scale have the following values (see Fig. 5.2): An ordinary grain of table salt measures 100 μm, and the average diameter of human hair measures 70 μm. Twenty-five micrometers is approximately one thousandth of an inch. The lower limit of visibility for an unaided human eye is 40 μm. In other words, the average person can see dirt that measures 40 μm and larger. Some of the harmful dirt particles of a pneumatic system are below 40 μm in size.

A compressor is protected by an intake filter which is engineered into the inlet by the compressor manufacturer. It is designed to remove dust particles commonly found in an industrial environment. Industrial dust particles are not necessarily the same type of dust you might find on a piece of furniture in your living room. Depending on the type of manufacturing being done, industrial dust can be iron, carbon, grinding wheel particles, silicates, fiberglass, and abrasive materials of all kinds. This pollution must be prevented from entering a compressor or it will harm compressor seals as well as interfere with the operation of downstream system components such as regulators, directional control valves, flow control valves, and cylinders.

An intake filter is a very important part of a pneumatic system. It is the first line of defense against industrial air. For this reason, intake filters should be serviced at regular intervals to achieve optimum dependability.

Being filtered to a large extent by the compressor intake filter, air moves to the piston chambers. If it is a two-stage compressor, air is compressed, with a consequent temperature rise in the first stage; passed through an intercooler, where its temperature is decreased; and then on to the second stage. In the second stage, the air is compressed more and has another increase in temperature. It is then pushed into the system.

The air at this point has increased in potential energy because it is compressed. This is desirable, but the air is now hot and contains water vapor. It may contain some dirt which was not removed by the intake filter, and it may carry some oil vapor which was picked up while passing through the compressor. This is not desirable. Instead of discharging the air from the compressor outlet directly into an air receiver for storage, the air is often passed through an aftercooler. As its name implies, an aftercooler cools compressed air. This is accomplished by passing cooling water or air over the aftercooler chamber. Besides being the point where air cools, an aftercooler is also the place where some dirt and oil vapor fall out of suspension, and a good portion of entrained water vapor condenses.

Moisture can affect the operation of a pneumatic system in several ways:

1. Moisture can wash away lubricant from pneumatic components, resulting in faulty operation, corrosion, and excessive wear.
2. Moisture that collects in pneumatic cylinders could cause cushions to become ineffective.

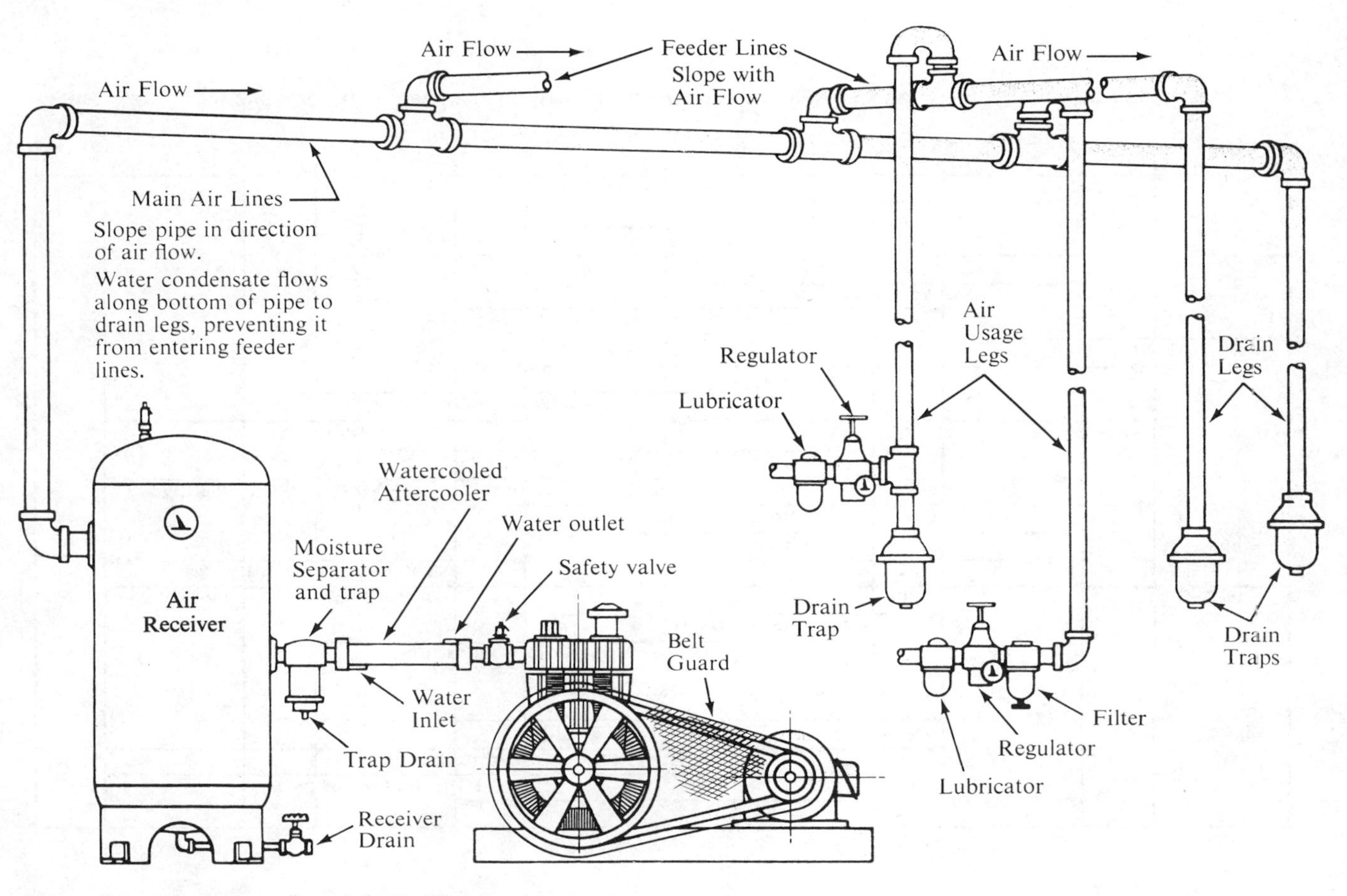

Figure 5.1 Air compression and conditioning system.

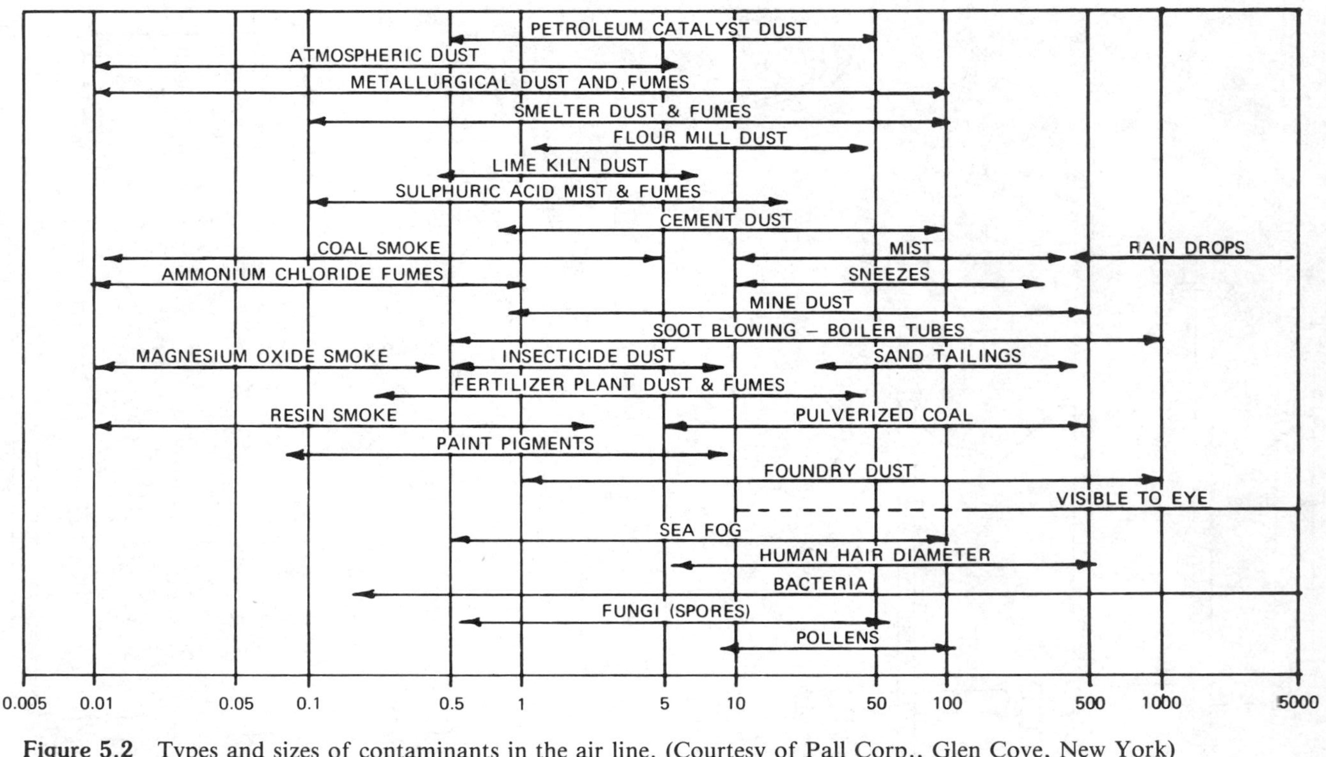

Figure 5.2 Types and sizes of contaminants in the air line. (Courtesy of Pall Corp., Glen Cove, New York)

3. Moisture that condenses at work outlets could cause an operation or final product to bc ruined.
4. Moisture condensing at work outlets is also an annoyance and inconvenience to the operator.

As air travels through piping, it picks up pieces of dirt such as rust and pipe scale (Fig. 5.3). It will also contain any moisture that was not removed in the aftercooler. Before air reaches a directional valve and actuator downstream, it must be filtered again. This is the function of an air line filter.

5.2 AIR LINE FILTERS, LUBRICATORS, AND PRESSURE REGULATORS

5.2.1 Compressed Air Filters

Filtration is the first, and often the only, line of defense against contamination in a pneumatic system. Adequate filtration is seldom expensive and will return the investment many times over in improved component life and reduced downtime.

Contaminants

Compressed air contains all the dust, pollen, and other solid impurities that were in the free air that entered the compressor inlet—but with one major difference: At 100 psi, the concentration of contaminants is eight times greater than it was in free air. In addition, the compressed air can also pick up rust, scale, and other dirt particles from the steel piping, which carries it to the point of use. Excess pipe sealant (from careless plumbing) is another contaminant. Other construction debris are bound to cause problems long after installation is completed (Fig. 5.4).

In many cases, even the compressor contributes metal fragments or fine solid particles of carbon or Teflon from sliding seals. If there is a desiccant dryer in the system, desiccant particles may be carried downstream.

Air leaving the compressor is usually not only saturated with water vapor, but has probably picked up oil vapor from the compressor. As air moves through the cooler areas of the whole pneumatic system, some of these vapors condense as liquid particles, which are carried along in the airstream (Fig. 5.5). For predictable operation and reliable service life, both solid and liquid particles must be removed by the pneumatic filters.

Filter Operation

A typical industrial air filter should technically be described as a combination of a dynamic separator and a mechanical or static filter. An air line filter consists basically of a housing with inlet and outlet ports, deflector plate, filter

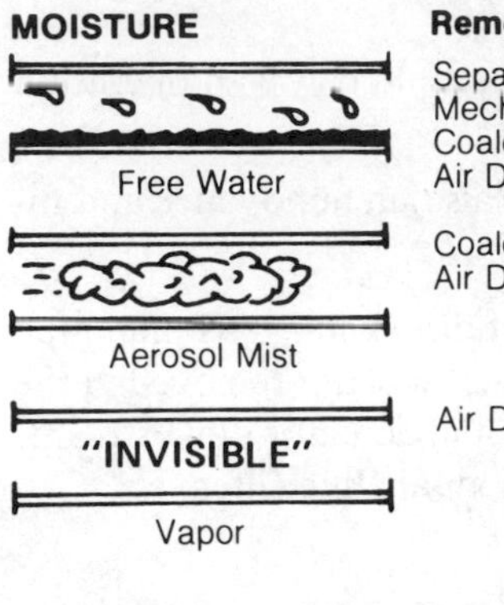

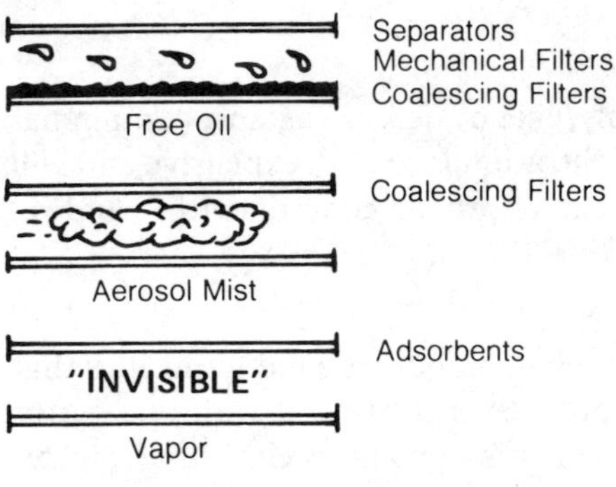

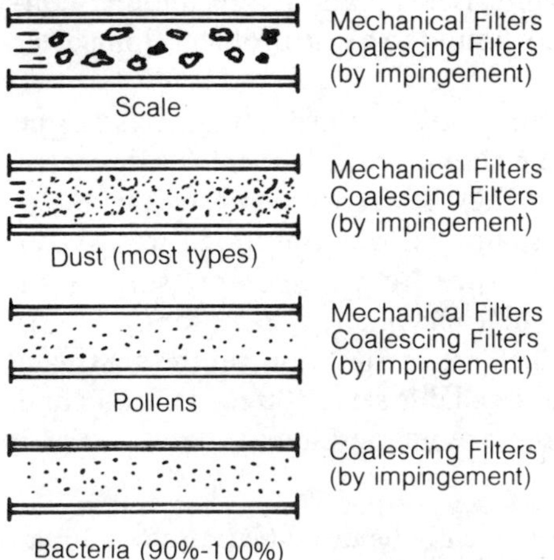

Figure 5.3 Compressed air contaminants. (Courtesy of Zurn, Inc., Erie, Pennsylvania.)

Application	Problem
Pneumatic instruments and controls	Corrosion of internal parts and deterioration of seals, resulting in malfunction
Outside air lines	Freezing, line blockage, rupture
Chemical blanketing	Corrosion of containers, piping, and valves
Pneumatic tools	Rust and pitting of internal parts
Sand or grit blasting	Clogging of feed lines and spray nozzles
Material handling	Wet cakes, sticking, and plugged lines
Spraying—paints and coating	Surface blemishes
Air bearings	Corrosion

Figure 5.4 Typical problems caused by wet, dirty air. (Courtesy of Pall Corp., Glen Cove, New York.)

element, baffle plate, filter bowl, and drain (Fig. 5.6). Air entering the inlet is deflected by the deflector plate. This causes a swirling action which throws out large pieces of dirt and liquid droplets against the wall of the bowl. Dirt and liquid fall along the side of the bowl to the area beneath the baffle. When the swirling air hits the baffle, its direction is again changed. It then passes through a filter element, where fine pieces of dirt are removed. It then exits the outlet port.

The area below the filter baffle is known as the *quiet zone*. This is the

	Contaminants				
	Water		Oil		
Compressor type	Liquid	Vapor	Liquid	Vapor	Dirt
Oil lube reciprocating	●	●	●	●	●
Oil lube screw	●	●	●	●	●
Oil-free reciprocating	●	●			●
Oil-free screw	●	●			●
Centrifugal	●	●			●

Figure 5.5 Air quality from compressors. (Courtesy of Zurn, Inc., Erie, Pennsylvania.)

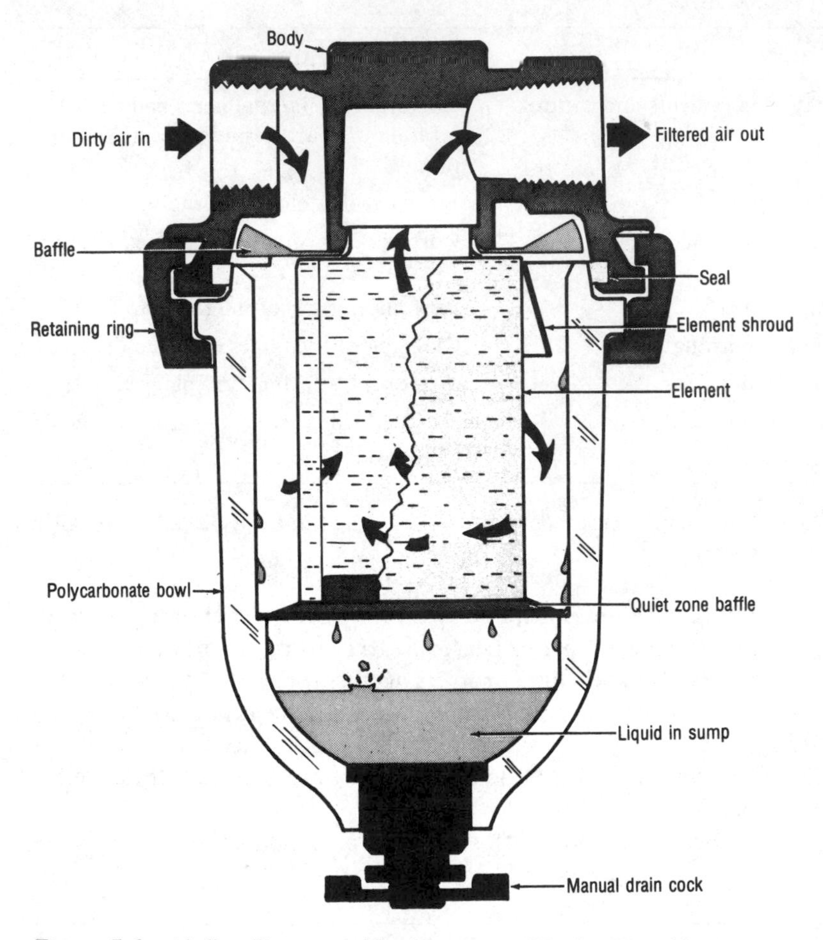

Figure 5.6 Air line filter assembly. (Courtesy of Parker Hannifin Corp., Cleveland, Ohio.)

place where large contaminants and liquid collect. Often, filter bowls are made of a transparent plastic material so that buildup in the bowl can be observed. When the collection of contaminant and liquid approaches the baffle, the drain cock in the bottom of the bowl should be opened. Some air line filters are equipped with an automatic drain.

After separation of large solid and liquid particles, the compressed air is directed through a filter element which strains out smaller particles, subject to the size limitations of the element's design. Air flows from the outside to the inside of the element.

Filter elements used in air line filters are divided into depth and edge types. *Depth*-type elements force air to pass through an appreciable material thickness. Dirt is trapped because of the intertwining path the air must take. Depth-type elements in pneumatic systems are frequently made of porous bronze. *Edge*-type elements offer an airstream with a relatively straight flow path. Dirt is caught on the surface or edge of the element that faces the air flow. Edge-type elements in pneumatic systems are often made of resin-impregnated paper ribbon.

Because of its construction, an air line filter element has many pores of various sizes. Many of the pores are small. A few pores are relatively large. Since it has no one consistent hole or pore size, an air line filter element is given a *nominal rating* (Fig. 5.7). A nominal rating is an element rating given by the filter manufacturer. It is designed to indicate the expected average hole size of the element. For example, a depth element with a nominal rating of 40 μm indicates that on the average, the pores in the element are 40 μm in size.

Air line filter elements found in industrial pneumatic systems generally have nominal ratings ranging from 50 to 5 μm. These ratings usually indicate the elements average pore size. Since dirt in a pneumatic system comes

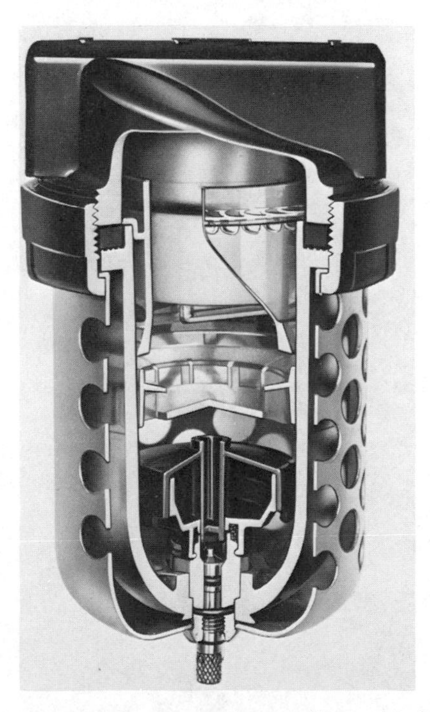

Figure 5.7 Porous bronze filter with float-type drain. (Courtesy of Norgren Corp., Littleton, Colorado.)

in all sizes, shapes, and materials, no guarantee is made by a nominal rating as to what size particles will be removed from a compressed airstream.

Drains

There are many manual and automatic drains to remove collected liquids from filter bowls. Miniature filters, where size is a consideration, use a tire valve type of drain with an external pushpin to act as an actuator. Larger filters are designed with metal or plastic drain cocks that are actuated by wing nuts or knobs. Some manufacturers equip their manual drains with a flexure tube which opens the drain whenever the tube is pushed away from its normal vertical position (Fig. 5.8).

The automatic filter drain (Fig. 5.9) is a valuable option when filter location makes servicing difficult or dangerous, or when filters are hidden and may be inadvertently overlooked. The two basic types of automatic drains are

Figure 5.8 Manual drain on air line filter. (Courtesy of Parker Hannifin Corp., Cleveland, Ohio.)

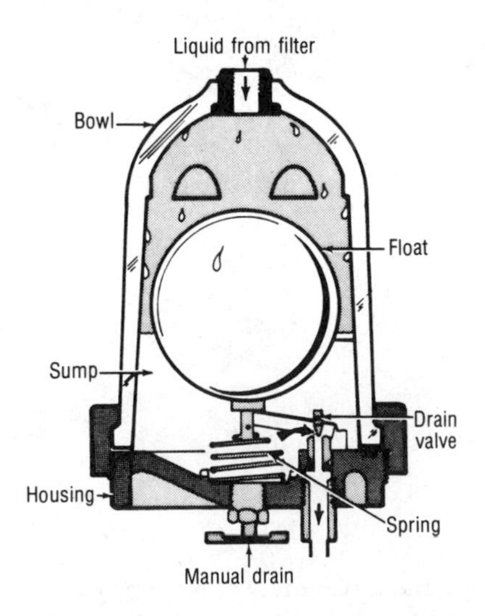

Figure 5.9 Float-type automatic drain for air filter. (Courtesy of Scovill-Schrader Corp., Akron, Ohio.)

pulse-operated for intermittent air flow and float-operated for continuous flow. In the *pulse-operated* drain, the drain valve is linked mechanically to a piston or diaphragm which fits across the bowl above the surface of the sump. When fluid pressure is equal on both sides of the piston, the larger top area holds the drain valve seated. As flow starts through the filter, the pressure drops temporarily above the piston; the piston unseats the drain valve, emptying the sump. Pressure above the piston then reseats the drain valve. Flow through the filter must be intermittent for the pulse-operated drain to work.

In *float-operated* drains, accumulated liquids in the sump raise a float that either opens the drain mechanically, or pilots a valve that opens the drain. Bowl pressure forces the liquid out rapidly, the float lowers, and the drain closes.

Types of Elements

Industrial air filter elements are available in a variety of materials, including felt, paper, cellulose, metal and plastic screening, metal ribbon, sintered bronze, sintered plastic, glass fiber, and cloth. Elements are rated according to the minimum-size particle they will remove from the airstreams. Elements are also rated for flow through the filter at specific pressure. Some manufacturers classify elements as edge or depth type. Theoretically, an edge-type element traps particles only on its surface; a depth type picks up additional contaminants internally, as the air follows a random path through the element. Elements are also listed as cleanable and disposable. Edge-type elements are easy to clean: depth-type elements are more difficult to clean.

Coalescing describes an act of growing together. Coalescing elements for oil aerosol removal starts with a random bed of borosilicate microfibers. To make the fibers self-supporting, they can be bonded with resin, or another, stiffer fiber can be mixed with the borosilicate. Another construction method supports the borosilicate fibers mechanically with external layers of strong material. A sleeve over the outer surface of a coalescing element helps prevent the airstream from reentraining liquid oil from the wet surface of the element. Adsorption is described as adhesion of the molecules of a gas, liquid, or dissolved substance to a surface. Adsorption of oil vapors takes place in either packed columns of activated carbon (the adsorptive medium) or in elements that have activated carbon particles mixed with filtering fibers.

Filter Application

Certain general considerations apply to all air line filter units. Most important, units should be sized according to airflow demands, not according to pipe size. For instance, an oversized filter may not impart enough swirling action to separate all the moisture droplets. Conversely, an undersized filter creates too much turbulence in the air being processed. The turbulence prevents the liquids from settling into the quiet zone. Undersized units may also add excess pressure loss to the system. Flow capacities for a given port size also vary among manufacturers because of variations in internal construction.

Air drop lines to conditioning units should be attached to the top of the main header to decrease the amount of liquid that might run down to equipment. A separate drain should be provided for those liquids that accumulate at low points in the mainheader line.

Conditioning units should be installed as close as possible to the point of use, consistent with accessibility for inspection and possible servicing. Bends in the air lines should be avoided since they cause additional pressure drop in the system. In addition, momentum forces tend to deposit lubricant at bends. Necessary turns should have as large a radius as possible.

Filters should be installed upstream from, and as close as possible to, the devices to be protected. Filters remove only droplets and solid contaminants; water vapor is not affected. Thus vapor can condense between the filter and the components being served. To minimize this possibility, filters should be located as close as possible to the point of use. The filter bowl must hang vertically so that free moisture and oil droplets cling to the inner surface of the bowl as they work their way down to the quiet zone at the bottom of the bowl. Filters should be located so that manually drained bowls can be observed and drained when necessary. Drain lines should be provided for filters that incorporate automatic drain devices.

Normally, filters have no moving parts to service or adjust. However, the filter element should be replaced when the pressure differential across the filter unit exceeds 10 psi. Although a manual drain requires regular attention, it is necessary to drain the collected liquids and solids only when their level approaches the lower baffle.

5.2.2 Lubricators

After compressed air passes through an air line filter, it is clean and relatively free of entrained water. But for some applications, the air must be conditioned in still one more way—it must be lubricated. Lubrication of compressed air may be necessary to provide seal lubrication, prevent sticking of moving parts, and control wear. Air line lubricators work on the basis of developing a controlled mist of lubricant in an airstream (see Fig. 5.10).

Air entering the inlet runs into a constriction. This results in a higher pressure at the inlet to the constriction with respect to the small opening of the mist chamber. This pressure differential is sensed across the pickup tube, pushing fluid up the tube and into the mist chamber, where it is mixed with air. The air then exits the outlet carrying suspended oil particles. The amount of oil mist that exits from this type of lubricator depends on the adjustment of the needle valve.

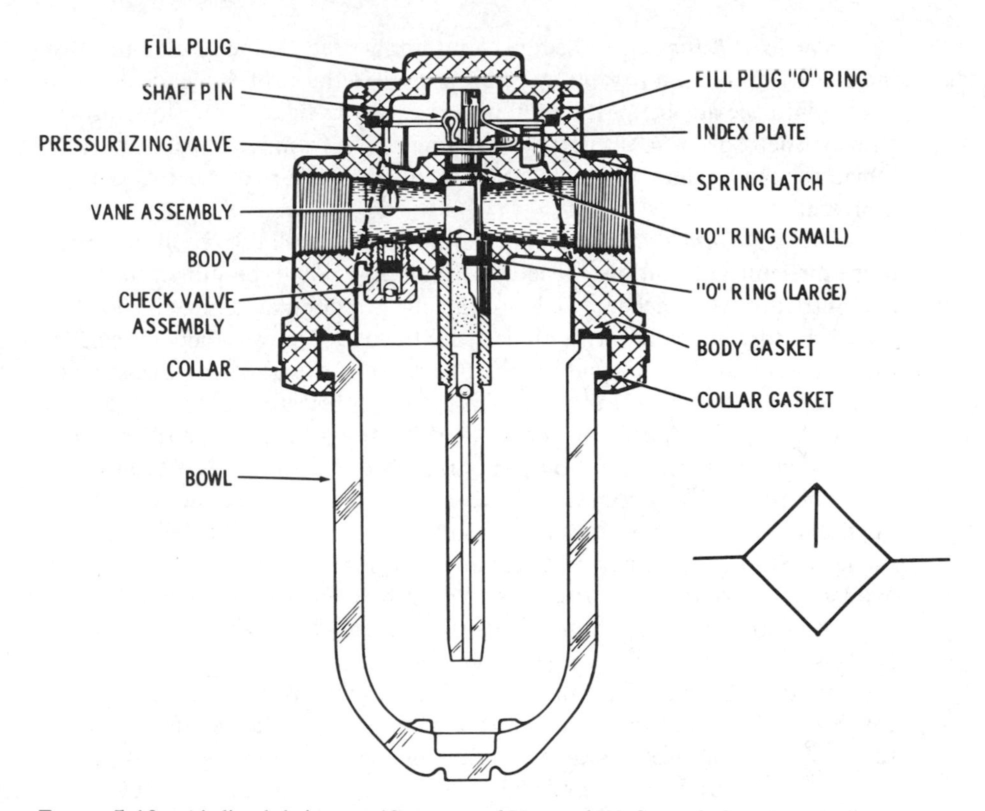

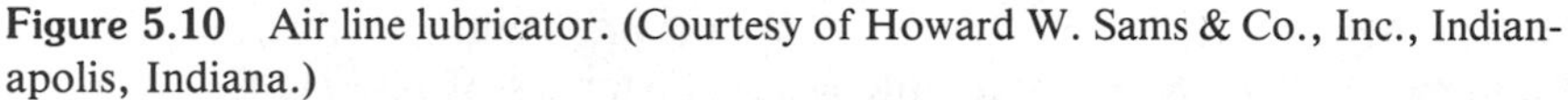

Figure 5.10 Air line lubricator. (Courtesy of Howard W. Sams & Co., Inc., Indianapolis, Indiana.)

Refer to the manufacturer's instruction manual as to the method of adjusting oil flow and safety precautions while servicing the unit.

Remember that oil injected into an air stream is eventually blown out as an exhaust. If too much is blown into a confined space, the oil content of the air may become greater than allowed by regulations.

A pneumatic system may be equipped with the best filters and lubricators available, and they may be positioned in the system where they do the most good; but if they are not serviced when required, the money spent for their use has been wasted. Lack of proper lubrication in pneumatic power components creates excessive maintenance costs, production inefficiency, and premature failure. Proper lubrication allows components to operate with a minimum of friction and corrosion, and it minimizes wear on the component parts. The most economical way to lubricate pneumatic components is with an air line lubricator. In most cases it is also the most efficient.

The *Mist Lubricator,* the most commonly used in-line lubricator, is of the direct-flow type. It permits every drop of oil in the sight dome to go downstream into the air line. All the oil observed in the sight dome flows downstream in aerosol form, the particles ranging in size from 0.01 to 500 μm. This lubricator can be filled while the air line is under pressure because it contains a pressurization check valve.

The *micro-mist lubricator* was developed to fulfill the requirements of more difficult applications. It is a recirculating-flow-type lubricator. This means that the oil, once injected into the air, is recirculated into the bowl of the lubricator, where the larger oil particles fall out of the air. Only the smaller oil particles that float around in the upper part of the bowl go downstream. The particles going downstream in this type of lubricator range in size from 0.01 to 2 μm. This lubricator, because of its design, cannot be filled under pressure unless a remote fill button is incorporated into the lubricator.

Oil mist flowing from the lubricators will remain suspended over varying distances. Direct-flow mist lubricators can achieve oil particle suspension in the line over a distance of 20 ft, utilizing the greater portion of the oil supply available. By comparison, the recirculating-flow micro-mist lubricator can keep the majority of the oil suspended in the piping for approximately 100 ft. This can be achieved because of the absence of the larger oil particles in the output of the micro-mist lubricator. The larger particles in the mist lubricator tend to "eat up" or "coalesce" the smaller particles as that aerosol mixture flows downstream. Because of the absence of these larger particles in the micro-mist lubricator, this "coalescing" or "eating up" affect is not prevalent. Consequently, the oil particles that do come from the lubricator can remain suspended over considerable longer distances of piping.

Lubricator Types

Mist Lubricator. The Mist Lubricator injects oil into the air. The resulting air-oil mixture is then carried directly downstream. Air passes through the vortex generator, where the oil drips into the airstream and is atomized. Then the air-oil mixture flows directly into the secondary port cavity and downstream. During very low airflows the air passes through the center of the vortex generator. As the airflow increases, the flexible bypass disk starts to bend, permitting this increasing airflow to bypass the center of the vortex generator. The bypass disk allows the pressure drop to be controlled in the lubricator as the flow changes, thereby providing linear oil delivery. Mist lubricators start delivering oil when a minimum flow of 1 scfm is reached. This permits the lubricator to be used for low-flow as well as high-flow applications. It can be filled while under pressure simply by hand-removing the fill plug. The air in the bowl will automatically exhaust as soon as the O-ring seal is broken on the fill plug. The lubricator can then be filled with oil and the fill plug replaced and hand-tightened. No wrenches or screwdrivers are needed. When the pressure in thc bowl is exhausted, the pressurization check valve reduces the flow of air to the bowl to a minute induced bleed. It is this bleed that repressurizes the bowl once the fill plug has been replaced.

Micro-Mist Lubricator. The micro-mist lubricator injects oil into the airstream and then recirculates the resulting air-oil mixture into the bowl of the lubricator. It is here that the larger oil particles fall out of the air. Only the smaller oil particles, ranging in size from 0.01 to 2 μm, go into the airstream.

This represents approximately 3% of the oil seen dripping in the sight dome. (To get one drop downstream, adjust the lubricator to a drip rate of 33 drops per minute.) All of the oil particles above 2 μm in size (which represents approximately 97% of the oil dripping in the sight dome) fall out of the air into the bowl. The vortex generator and bypass disk assembly is the same in both lubricators. Consequently, the micro-mist lubricator will also deliver oil with a minimum of 1 scfm of airflow. The ability to support large reverse flow with no oil delivery in reverse-flow conditions is another feature of the micro-mist lubricator. This ability makes the micro-mist lubricator ideal for piping between a valve and a cylinder.

Micro-mist lubricators are particularly well suited for the following applications: (1) where instant oil delivery and lubrication are required, whether the air line is wet or dry; (2) in systems where multipoint lubrication is required, such as lines that contain several valves and cylinders; (3) where large oil drops from the air exhaust would contaminate either the environment or the material being processed (the micro-mist lubricator actually permits less oil to be injected into the air line to perform the lubrication job); (4) where the air line includes (after the lubricator) valves, bends, and similar baffling points that would tend to separate out heavier particles; (5) where ease of drip-rate adjustment is required; (6) for high-cycle pneumatic applications; (7) where very low flow rates are necessary; (8) where the oil particles stay in suspension in the air lines for 100 ft or more; (9) where the oil particles must stay in suspension for more than 10 min; and (10) where the oil mist becomes a vaporous part of the air and travels as part of the air until it is baffled out by bearings, vanes, and turbulence.

5.2.3 Pressure Regulators

The amount of pressure needed in a system depends on the force required to do a job. Because some applications require less force than others, it follows that the required pressure varies accordingly. It is also essential that once a system pressure has been selected, air be supplied at constant pressure to each tool or actuator regardless of variations in flow and upstream pressure.

Thus it is important to add to the system a pressure regulator which performs the following functions:

Supplies air at constant pressure regardless of flow variation or upstream pressure.

Helps operate the system more economically by minimizing the amount of pressurized air wasted. This happens when the system operates at pressures higher than those needed for the job.

Helps promote safety by operating the tool or actuator at a reduced pressure.

Extends component life because operating at higher-than-recommended pressures increases the wear rate and reduces equipment life.

Increases operating efficiency.

Type of Regulators

Unbalanced Poppet, Non-Pilot-Operated Type. Figure 5.11 shows an unbalanced poppet regulator, the simplest pressure regulator. Supply pressure enters the inlet port and flows around the poppet. However, as shown, the

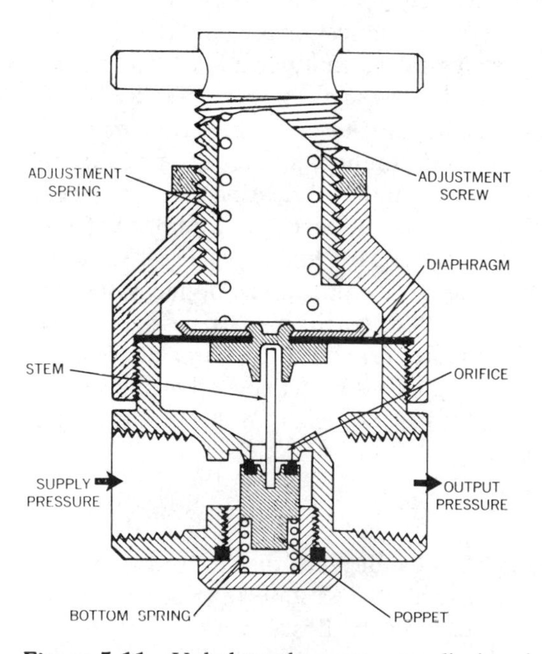

Figure 5.11 Unbalanced poppet, nonrelieving air line pressure regulator. (Courtesy of Watts Regulator Co., Lawrence, Massachusetts.)

poppet is seated on the orifice, not permitting flow. Turning the adjustment screw down, compresses the adjustment spring, forcing the diaphragm downward. The diaphragm pushes the stem down and the poppet moves away from the orifice. As downstream pressure builds, it acts on the underside of the diaphragm, balancing the force of the spring, and the poppet throttles the orifice to restrict flow and produces the desired pressure. As the demand for downstream flow varies, the regulator compensates by automatically repositioning the poppet in relation to the orifice. The bottom spring of the poppet assures primarily that the regulator will close at no-flow.

Unbalanced Poppet, Non-Pilot-Operated, with Diaphragm Chamber. The next more complex regulator is shown in Fig. 5.12. It is also an unbalanced poppet non-pilot-operated type. However, it has a diaphragm chamber which isolates the diaphragm from the main airflow. This reduces the abrasive effects of the air on the diaphragm.

The diaphragm chamber is connected to the outlet chamber through an aspirator tube. This tube causes a slightly lower than outlet pressure in the diaphragm chamber as flow increases through the regulator. This reduces the pressure droop experienced when the regulator opens.

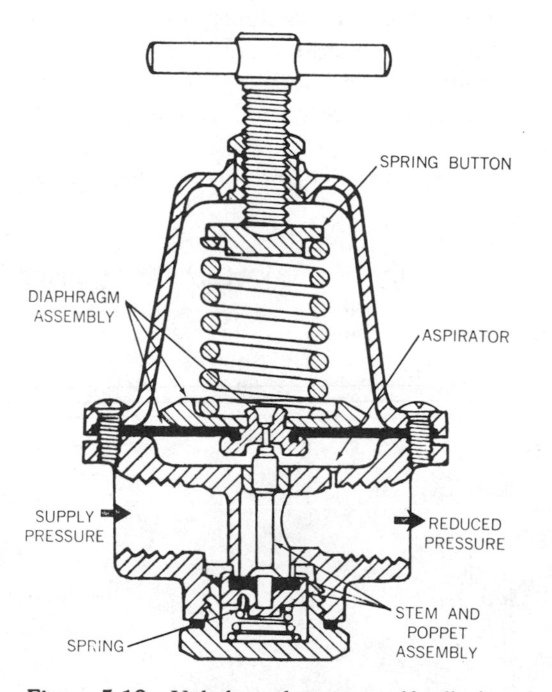

Figure 5.12 Unbalanced poppet, self-relieving air line pressure regulator. (Courtesy of Scovill-Schrader Corp., Akron, Ohio.)

Remote-Controlled, Pilot-Operated, Balanced Poppet. With some applications, the regulator must be installed where it cannot be easily adjusted. In such cases, the regulation and pressure-setting mechanisms are separated. A small air pilot line connects the regulator (in the air line at the point of use) to the remote setting mechanism, where it is convenient to make the adjustment (such as a control panel) (Fig. 5.13). The remote setting mechanism is a small regulator used to produce a control air signal.

Piston-Type Regulators. Although diaphragm-type regulators have been used to illustrate these basic types of regulators, almost every example is available as a piston regulator, where a piston replaces the diaphragms.

Air Line Pressure Regulator Application

A regulator should be located for convenient servicing. The pressure setting should be adjusted to the value required by typical operating conditions. The no-flow pressure setting differs from the secondary pressure under flow conditions. A slight decrease in secondary pressure with increasing flow through the unit is normal.

Regulator flow characteristics are affected by the spring rate used to cover the desired pressure range. Generally, a standard unit is rated for 250 psig primary supply pressure, and adjustable secondary pressure is rated from 3 to 125 psig. Low-pressure and high-pressure units are also available. If an application requires 30 psig, the standard unit will probably be satisfactory. If optimum performance is required, the low-pressure unit can be used, since its lower rate spring gives better performance characteristics. The standard, adjustable secondary pressure range is satisfactory for most industrial pneumatic applications.

All secondary pressure requirements are covered by one pilot-controlled regulator, since the secondary pressure range of a pilot-controlled regulator

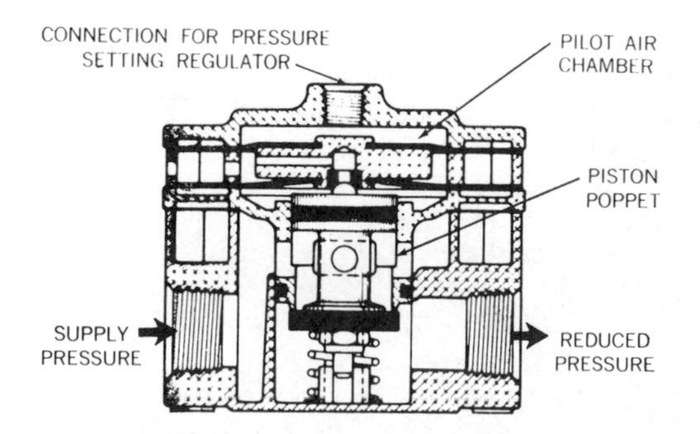

Figure 5.13 Remote-controlled, pilot-operated pressure regulator. (Courtesy of Parker Hannifin Corp., Cleveland, Ohio.)

is determined by the range of its pilot regulator. If the pressure and flow requirements of a system vary widely, a pilot-controlled regulator with a high-pressure pilot regulator provides the best performance. If the decrease in pressure under flow conditions seems excessively high for a particular application, the supply lines, the regulator unit, or other system components may be undersized for the application. A clogged filter element can also cause an excessive decrease in pressure under flow conditions.

To lower a setting, the regulator should be reset from a pressure below the final secondary pressure desired. For example, to lower the secondary pressure from 80 psig to 60 psig, decrease the secondary pressure to 55 psig or less, then adjust it upward to 60 psig.

In a nonventing regulator some secondary air must be exhausted to lower the setting. Many regulators include a constant bleed of secondary air to obtain good pressure cracking characteristics. A small fixed bleed in the secondary line provides the same effect. This modification is especially important in no-flow conditions, and permits the regulator to provide immediate response.

Pilot-controlled regulators are applied in the same way as standard spring-controlled regulators. The distance between the pilot-controlled regulators and its pilot regulator is not critical. Regulators do not require routine maintenance in normal service when properly protected by filters. The secondary pressure setting should be checked, however, whenever system requirements change. The combination of an air line filter, pressure regulator, and lubricator is commonly called an *F-R-L* or *trio unit.* A typical such unit is shown on Fig. 5.14.

5.2.4 Choosing an F-R-L of Adequate Size

A rule-of-thumb sizing method can be used for applications where only one cylinder is being operated. The trio unit may be selected with the same pipe size as the cylinder. Where several cylinders are drawing air from the F-R-L at the same time, the unit should be at least one pipe size larger than the port of the largest cylinder. An alternative sizing method, for multiple cylinders operating simultaneously, is to add piston areas of all cylinders; then refer to a cylinder catalog to find the port size of one large cylinder having about the same piston area as the combined area of the smaller cylinders, and use this for the trio size.

On existing installations where cylinder speed is too slow, the F-R-L should be checked for excessive pressure drop. First, replace the element in the filter. If this does not improve performance, check catalog specifications for the F-R-L to see if it is overloaded. Loss through a F-R-L should be less than 10 psi with the cylinder traveling under full load. If the cylinder speed

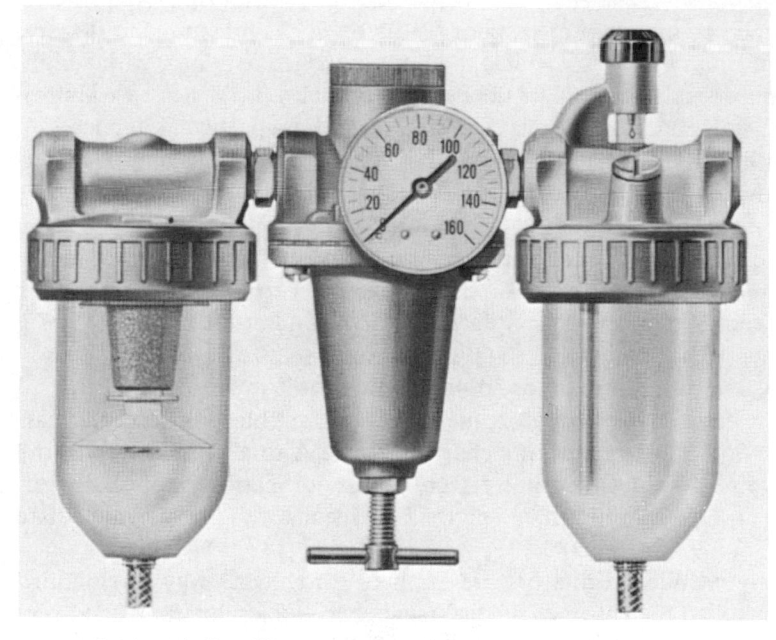

Figure 5.14 Air line filter, pressure regulator and lubricator. (Courtesy of Parker Hannifin Corp., Cleveland, Ohio.)

is still too slow after checking the F-R-L, see if the piping is large enough to carry the airflow. Increasing pipe diameter, especially on long runs, will quite often increase the speed of an air cylinder to a worthwhile degree.

Safety: Metal Bowls

Normally, clear plastic bowls perform with no difficulty, and they are highly desirable from the standpoint of visibility of condensate level or oil level (see Fig. 5.15). Metal bowls are available for most brands of trio units. They are intended to provide a greater measure of safety when one of these unusual operating or environmental conditions is encountered:

1. *High temperatures*: If internal or external temperature will exceed about 150°F, most plastic bowls lose their mechanical strength and should be replaced with metal bowls.
2. *Solvents*: If solvents are present either in the airstream or in the atmosphere surrounding the trio, metal bowls should be used. The composition of plastic bowls varies some between manufacturers, and the literature should be checked to determine if a given solvent is harmful to the particular plastic bowl in use. Harmful conditions

CAUTION: These products are specifically designed for compressed air service. Use with or injection of certain hazardous fluids or gases in the system (for example, alcohol or liquified petroleum gases) could be harmful to the unit or result in a combustible condition or hazardous external leakage.

Beware of materials that will attack polycarbonate plastic bowls. When dirty replace bowl or clean with dry, clean cloth. A bowl guard should always be used with all polycarbonate plastic bowls.

THE FOLLOWING MATERIALS WILL HARM A POLYCARBONATE BOWL:

Acetaldehyde
Acetic acid (conc.)
Acetone
Acrylonitrile
Ammonium fluoride
Ammonium sulfide
Benzene
Benzoic acid
Benzyl alcohol
Bromobenzene
Butyric acid
Carbolic acid
Carbon disulfide
Carbon tetrachloride
Caustic potash solution (5%)
Caustic soda solution (5%)
Chlorobenzene
Chloroform
Cresol
Cyclohexanol
Cyclohexanone
Cyclohexene
Dimethyl formamide
Dioxane
Ethane tetrachloride
Ethyl ether
Ethylamine
Ethylene chlorohydrin
Ethylene dichloride
Formic acid (conc.)
Freon (refrigerant & propellant)
Gasoline (high aromatic)
Hilgard Co.'s hil-phene
Hydrochloric acid (conc.)
Methyl alcohol
Methylene chloride
Milk of lime (CaOH)
Nitric acid (conc.)
Nitrobenzene
Nitrocellulose lacquer
Phenol
Phosphorous hydroxy chloride
Phosphorous trichloride
Propionic acid
Pyridine
Sodium sulfide
Styrene
Sulfuric acid (conc.)
Sulphural chloride
Tannergas
Tetrahydronaphthalene
Thiophene
Toluene
Xylene
Perchlorethylene and others

TRADE NAMES OF SOME SYNTHETIC COMPRESSOR OILS & RUBBER COMPOUNDS THAT HARM POLYCARBONATE BOWLS.

■ Cellulube #150 and #220 ■ Houghton & Co. oil #1120, #1130, and #1055 ■ Keystone penetrating oil #2 ■ Marvel Mystery Oil ■ Some Loctite Compounds ■ Sinclair oil "Lily White" ■ Haskel #568-023 ■ Parco # 3106 Neoprene ■ Sears Regular Motor Oil ■ Garlock #98403 (Polyurethane) ■ Kano Kroil ■ Pydraul AC ■ Stillman #SR 269-75 (Polyurethane) ■ Stillman #SR 513-70 (Neoprene) ■ Tenneco Anderol #495 and #500 oils

Figure 5.15 Caution on the use of polycarbonate plastic bowls. (Courtesy of Scovill-Schrader Corp., Akron, Ohio.)

might include the use of a lubricator as a chemical injector in a gas line, a fire-resistant liquid mist in the airstream, or an antifreeze chemical in the lubricator to prevent moisture freeze-up in the lines.

3. *High pressure*: Plastic bowls usually carry a lower pressure rating than metal bowls, and this is primarily because of the possibility of accidental exposure to solvents or high temperature. Under the usual atmospheric temperature conditions, most plastic bowls are as safe as metal bowls.

The pressure gauging port on the side of a pressure regulator is usually 1/8- or 1/4-in. National Pipe Thread (NPT) and is connected into the regulator secondary. This is a convenient place to make another air takeoff for a small quantity of air. On some brands the gauge port has a full 1/4-in. opening into the airstream and can be used for full capacity of a 1/4-in. line. On other brands, while the gauge connection may be 1/4-in. NPT, the actual passage inside the port may be only 1/16-in. in diameter. A visual inspection will determine its size.

5.3 ROUTINE MAINTENANCE OF F-R-L UNITS

What Oil to Use

Air line lubricators are designed to work on petroleum oil of low viscosity. The precise grade, viscosity, or brand is not extremely critical. For general use, a nondetergent motor oil of not greater than SAE 10W weight is acceptable. Avoid the use of oils with additives because of the possible effect of these additives on components that may be downstream (see Fig. 5.16).

How Much Lubrication

While oil feed rate is not extremely critical, an effort should be made to get near the optimum amount. Too much oil can be detrimental. Experience has shown that a feed rate of 1 drop of oil to about every 20 standard cubic feet of air (free air) used is about right for most applications.

Excessive Oil Feed

Oil feed is excessive if a surplus drips out the exhausts of directional control valves. Reduce the lubricator feed rate and check again in a few days after the system has stabilized.

Cleaning Plastic Bowls

Every once in a while the plastic filter bowl should be removed and cleaned. Use only a damp cloth, with a little soap if necessary. Do not use lacquer thinner, gasoline, or other solvents because of their possible harmful effect on the plastic. A bowl weakened by solvent may blow later when under pressure.

A good lubricant must be compatible with the materials used in the system for seals and valve poppets. In general this requires a light bodied mineral or petroleum base oil with oxidation inhibitors, an aniline point between 82°C (180°F) and 104°C (220°F) and an SAE No. 10 or lighter viscosity. A partial list of such compatible lubricants is shown below.

MAKER	BRAND NAME
Amoco	American Industrial Oil #15
Citgo	Pacemaker #32 Pacemaker #46
Gulf	Harmony 46 Paramount 46 Security 46
Mobil	DTE (light) Vectra (light)
Shell	Tellus #11
Texaco	Rando 32
Union	Unax RX 150 Unax AW 150 Turbine Oil XD 150

Figure 5.16 Compatible lubricants. (Courtesy of Scovill-Schrader Corp., Akron, Ohio.)

While cleaning the bowl, inspect and clean the filtering element. Replace it if necessary.

5.3.1 Troubleshooting

Air Leak at Regulator

A continuous air leak from the small vent hole in the regulator bonnet indicates a leaky main poppet or diaphragm. In either case repair parts should be ordered at once and the regulator scheduled for repair. Overhaul kits with diaphragm and seals are available for most standard regulators.

Important! Replace diaphragm or seals as soon as possible after leak is discovered. A complete failure of the diaphragm might permit full inlet pressure downstream. Raising the pressure on some solenoid valves may cause them to shift by themselves, creating a safety hazard to personnel.

Pressure Regulator Problems

If adjusting the T handle fails to reduce the pressure downstream, check these possible causes:

1. The regulator may be connected backward, with inlet pressure connected to the regulator outlet. It positively will not work if connected in reverse.
2. The regulator may be connected improperly. A gauge port may have been accidentally used as one of the main ports. This is not a full-flow port on some regulators.
3. There may be broken parts inside the regulator. If the diaphragm is cracked or broken, a high-velocity air leak at the vent hole will pinpoint this trouble.
4. If the regulated pressure is too low and cannot be raised by screwing the T handle all the way down, the range must be changed by replacing the main spring with a heavier one.

Filter Problems

Very few problems are encountered with air line filters provided that the element is cleaned or replaced regularly. Here are a few suggestions for best filtering action:

1. Make sure that the filter is correctly plumbed, with incoming air entering at the inlet. It will filter in the reverse direction, but the centrifugal throw-out baffles will be useless, and large particles of contamination will collect on the inside of the element and cause severe air restriction, making the element difficult to clean.
2. If the filter manufacturer offers a choice, install a fine filtering element, a 5 μm rating being preferred. With present-day microfinishes on cylinder barrels, a finer degree of filtration becomes increasingly desirable.
3. Service the filter regularly, cleaning or replacing the element. If allowed to become overly contaminated, the pressure drop through the unit may increase to the extent that speed in the air circuit will be significantly reduced.
4. Drain the condensate before it rises above the baffle. If allowed to rise above the baffle, air turbulence will pick up condensed water and carry it downstream. If regular draining is a problem, install an automatic drain in the petcock line, or replace the inner assembly with a type that has automatic drain feature.

5.4 MAINTENANCE OF COMPRESSED AIR FILTERS

A filter cannot perform its intended function when liquids carry over into the outlet air because the sump has risen too high, or when a clogged filter causes excessive pressure drop, interfering with the operation of pneumatic equipment. A filter maintenance program should be aimed at avoiding these two situations.

Manufacturers choose transparent polycarbonate for the bowl material so that sump conditions can be easily and readily inspected while the filter is on line and operating. The sight gauge on a metal bowl performs the same function. Visual inspection determines when a manual drain should be actuated or whether an automatic drain requires servicing.

A filter maintenance program should include physical inspection of the element. A clogged element may be subjected to pressure differential forces which exceed the element's structural limitations, causing the element to distort, collapse, or even break apart. Contaminants then bypass the element; in some cases, broken element pieces migrate downstream. An element that collapses, but maintains its seal, can cause a severe pressure loss in the downstream system.

The usual signal for indicating a dirty element is pressure drop across the element. A variety of accessories are available to indicate the degree of pressure drop.

> A pressure gauge upstream with a second gauge connected downstream shows pressures on either side of the element.
>
> A differential pressure gauge connected across the filter eliminates the arithmetic required by two gauges. The gauge displays the *pressure difference* between the inlet and the outlet air pressure.
>
> A pneumatic maintenance alarm, adjustable for pressure drop, is connected across the filter and actually sounds a whistle when the element is clogged. The same pneumatic signal can also be used to shift a valve, operate a buzzer, or turn on a light.
>
> A pop-up indicator in a transparent housing is an integral part of one manufacturer's filter body.
>
> A colored ball indicator shows its green zone at zero pressure differential. As differential pressure increases, the ball rotates and starts to show its red zone. At 15-psi pressure drop, the visual signal is all red.

It is impossible to predict accurately the service life of an air filter element because of varying conditions of the incoming air and diverse operating cycles. However, most manufacturers mention weeks or even months of operation before cleaning or replacement should be necessary. Even if more frequent service should be required, the user still gets the degree of filtration specified, and the equipment is protected.

Cleanable elements are blown clean with compressed air or washed in water or solvent according to the manufacturer's directions. Good safety practice calls for the user to replace the dirty element with a spare, then clean the dirty element elsewhere, preferably in an enclosed area designed for that purpose.

Most filter designs make it possible to remove and replace a filter element without tools. Be sure to check element orientation and end cap seating before reassembling. While removed, the bowl can be cleaned for better visibility.

Oil-absorbing elements act as a wick when wet, and feed oil back into the air stream. They should be changed before this condition occurs. Oil-absorbing elements function properly while wet, and need to be changed only when solid particle accumulation generates an excessive pressure drop.

With few, if any, moving parts, no adjustments, and moderately priced replacement elements, a regular maintenance program on a properly specified and installed industrial air line filter assures that the filter will protect downstream pneumatic components satisfactorily over years of service life.

5.5 AFTERCOOLERS AND AIR DRIERS

5.5.1 Aftercoolers

As its name implies, an aftercooler cools compressed air after compression has been completed. This is accomplished by passing cooling water or air over the aftercooler chamber. Besides being the point where air cools, an aftercooler is also the place where some dirt and oil vapor fall out of suspension, and a good portion of entrained water vapor coalesces out. The aftercooler must have a moisture separator, preferably having an automatic drain.

We know that when a gas cools, its specific volume decreases. This change usually results in a change (decrease) in the pressure of the gas. Also, when air cools, its ability to carry water vapor decreases. In an aftercooler, compressed air and the water vapor it carries are cooled and its water content is made to condense into "rain." As air passes to the air receiver, it has much less potential energy than when it entered; it is cooler, cleaner, and holds less water. The air leaving the compressor is typically very humid. This high humidity requires an aftercooler to remove some of the water vapor.

In a typical water-type aftercooler, the direction of water flow is opposite that of the airflow. A good aftercooler will cool the air flowing through it to within 15°F of the cooling water temperature. It will also condense up to 90% of the water vapor originally contained in the air as it enters the receiver tank.

Water removal is as vital as heat removal; 1000 ft^3 (28.3 m^3) of air after compression can release as much as 1.4 quarts (1.3 liters) of water. Thus a modest-sized 100-scfm (472-dm^3/s) system could produce over 50 gal of condensed water in a single 24-hr day. If demands of a system require drier air, the compressed air can then be put through the following processes:

1. Overcompression
2. Refrigeration
3. Absorption
4. Adsorption
5. Combination of the methods above

Overcompression

In the overcompression process, the air is compressed so that the partial pressure of the water vapor exceeds the saturation pressure. Then the air is allowed to expand, thereby becoming drier. This is the simplest method, but the power consumption is high. It is usually used for very small systems and therefore is not as common in industry as are the other methods.

Refrigeration

As was mentioned before, when the temperature of air is lowered, its ability to hold gaseous water is reduced. This is what takes place in an aftercooler. However, the typical minimum air temperature attainable in an aftercooler is limited by the temperature of the cooling water or air. If extremely dry air is needed, a refrigerant-type cooler is employed. In these devices, hot incoming air is allowed to exchange heat with the cold outgoing air in a heat exchanger. The lowest temperature to which the air is cooled is 32.4°F (0.6°C), to prevent frost from forming. This type of air-drying equipment has relatively low initial and operating costs.

Absorption

Water vapor in compressed air can be removed from compressed air by methods such as absorption. There are typically two basic absorption methods. In the first, water vapor is absorbed in a solid block of chemicals without liquefying the solid. The chemicals used in the solid insoluble type are typically dehydrated chalk and magnesium perchlorate. Another type uses deliquescent drying agents such as lithium chloride and calcium chloride, which react chemically with water vapor and liquefy as the absorption proceeds. These agents must be replenished periodically.

Some problems tend to exist with the deliquescent drying process because most of these drying agents are highly corrosive. Also, the desiccant pellets can soften and bake at temperatures exceeding 90°F (32°C). This may cause an increased pressure drop. In addition, a fine corrosive mist may be carried downstream with the air and corrode system components. However, this type of air dryer has the lowest initial and operating cost of the more common air dryers. Maintenance is simple, requiring periodic replacement of the deliquescent drying agent.

Adsorption

Adsorption (desiccant) drying is another industrial method for drying air. Adsorption chemicals hold water vapor in small pores in the desiccant chemicals. Processes of this sort typically seek to make use of chemicals such as silica gel (SiO_2) or activated alumina (Al_2O_3). This type of air drying is the most costly of the drying methods discussed here. This is because of moderate to high initial and high operating costs. However, maintenance costs may be lower than the absorption type, because there are no moving parts. Also, the replacement of the desiccant is eliminated.

5.5.2 Air Driers

When there is wet air in compressed air lines, money can be lost because of production losses from damage to machinery, product spoilage, air lost through corroded and worn components, and instrument and tool repair and replacement costs. If you total the cost of replacement parts, labor, standby inventory, and downtime, the figure can be staggering. Eliminating even one of the above by drying the compressed air can offset the cost of installing and operating equipment to do the job.

Types of Dryers

In the broadest terms, there are three basic types of air dryers: deliquescent, regenerative desiccant, and refrigeration. *Deliquescent* dryers contain a chemical desiccant that reacts with the moisture in the air and absorbs it, regardless of whether it is already condensed or is still water vapor. The chemical is consumed in the drying (water-removing) process, and must be replenished periodically. The drain solution from these dryers contains both condensed water and some of the chemical. The dryer must be drained daily. Deliquescent dryers reduce the dew point of the air 15 to 25°F below the inlet air temperature, so if the incoming air has a temperature of 90°F, it will leave a deliquescent dryer at a dew point of about 65°F. Depending on operating conditions, some deliquescent dryers can produce dew points as low as 40°F.

Regenerative desiccant dryers remove water from air by adsorbing it on the surface of a solid desiccant, usually silica gel, activated alumina, or molecular sieve. The desiccant does not react chemically with the water, so it need not be replenished, but it must be dried, or regenerated, periodically (see Figs. 5.17 and 5.18).

Heatless regenerative dryers use two identical chambers filled with desiccant. As air moves up through one chamber and is dried, a portion of the

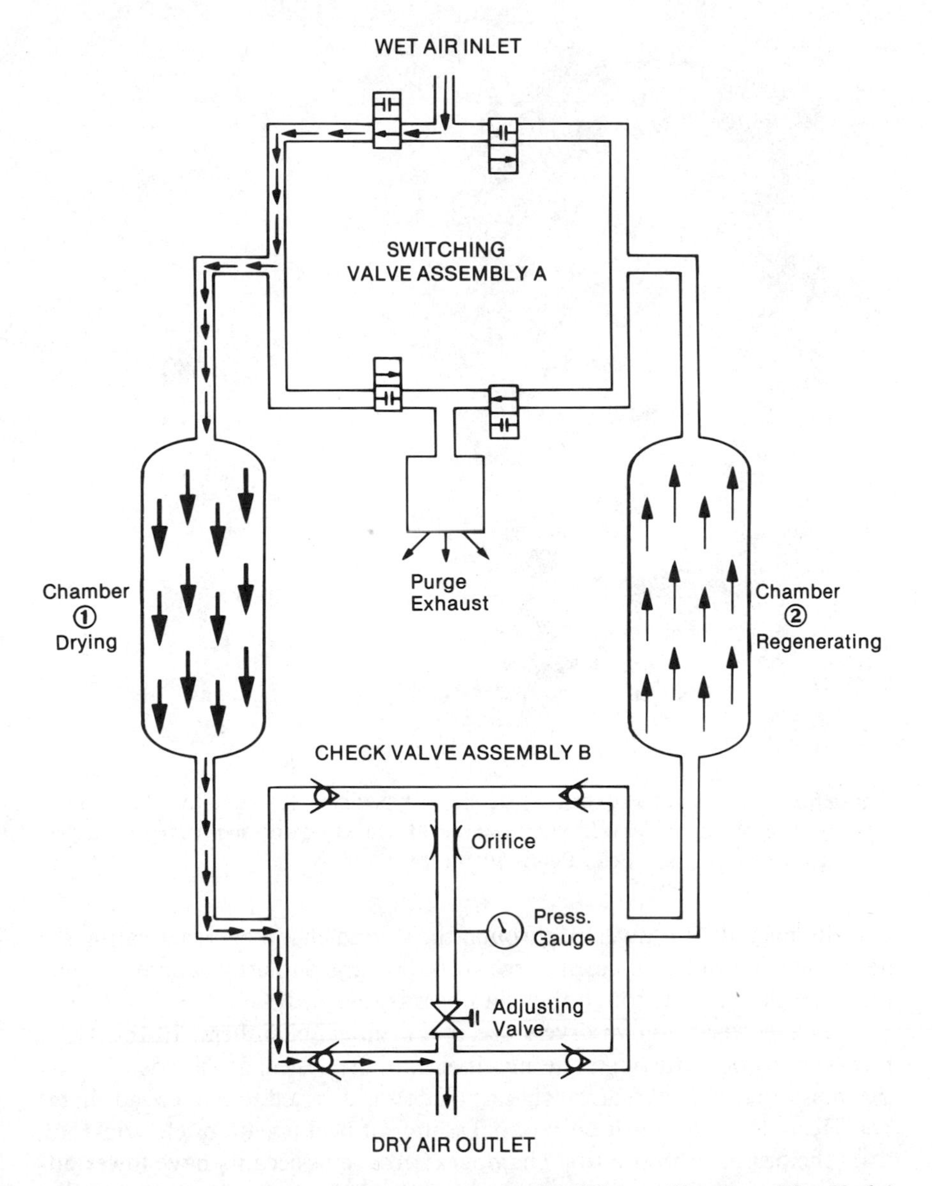

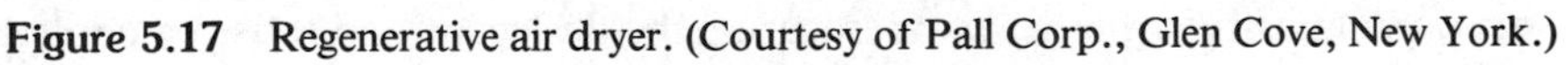
Figure 5.17 Regenerative air dryer. (Courtesy of Pall Corp., Glen Cove, New York.)

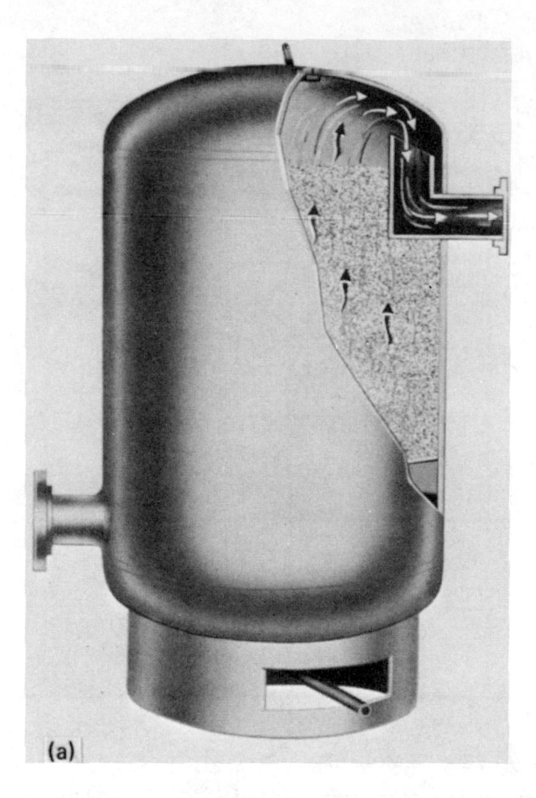

Figure 5.18 Types of air dryers. (a) Dessicant-type dryer. (b) Tube-to-tube refrigerator dryer. (c) Water chiller refrigerator dryer. (d) Direct expansion refrigerator dryer. (Courtesy of Zurn, Inc., Erie, Pennsylvania.)

dry, discharged air is diverted through the second chamber, reactivating the desiccant. The moisture-laden purge air is discharged to atmosphere. A short time later, air flow through the chambers is reversed.

Heated regenerative dryers use two identical chambers. In this type, however, air flows through one chamber until the desiccant has adsorbed all the moisture it can hold, at which time airflow is diverted to the second chamber. Heated outside air or an external source of heat (steam or electric) then dries the desiccant in the first chamber. Because desiccants have lower adsorption capacity at higher temperatures, the desiccant bed must be cooled from the temperature it reached during regeneration. The regeneration cycle in these dryers usually lasts several hours, divided between heating (75%) and cooling (25%) time.

Refrigeration dryers condense moisture from compressed air by cooling the air in heat exchangers chilled by refrigerants such as Freon gas. These

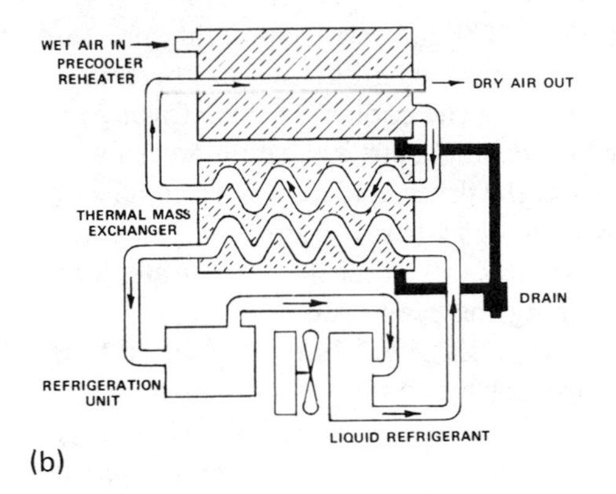

(b)

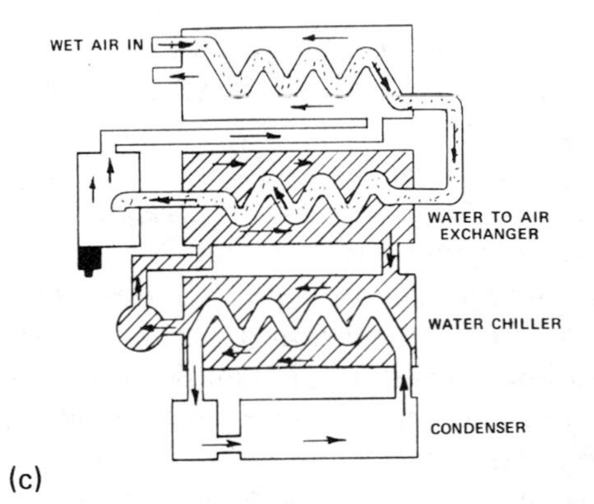

(c)

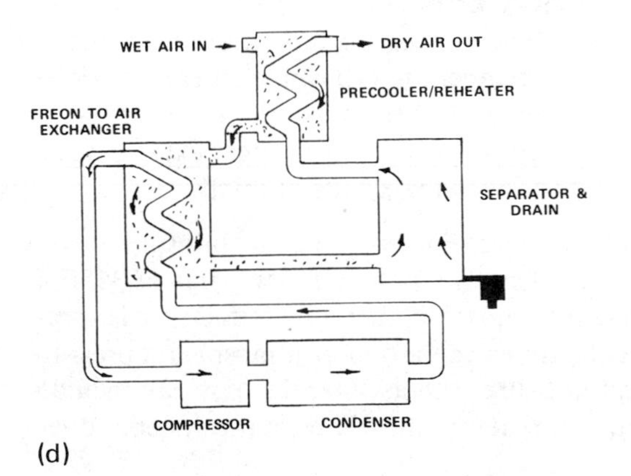

(d)

Figure 5.18 (continued)

dryers will produce dew points of 33 to 50°F at system operating pressure. Many refrigeration dryers reheat the compressed air after it has been dried, either with a heating element or by passing the cooled air back through the heat exchangers in contact with the hot incoming air. Reheating prevents condensation on air lines downstream from the dryer, and also helps cool incoming air. Refrigeration dryers must not be used where the ambient temperature is less than 35°F because lower temperatures will freeze the condensate, blocking air passages, and possibly damaging the evaporator.

Refrigeration dryers can be further classified into three basic types. *Tube-to-tube* refrigeration dryers operate by cooling a mass of aluminum granules or bronze ribbon that in turn cools the compressed air. These dryers can produce dew points of 35 to 50°F. *Water-chiller* refrigeration dryers use a mass of water for cooling. An extra heat exchanger is necessary to maintain chilled water flow through the condenser, as in a water pump. The dew point is 40 to 50°F. *Direct expansion* refrigeration dryers use a Freon-to-air cooling process and achieve a dew point of 35°F under maximum operating conditions. No recovery period is necessary.

The tube-to-tube refrigeration dryer is a cycling dryer. A thermometer in the mass senses its temperature. As the temperature rises, a switch turns on the refrigeration unit; when the temperature drops to a cutoff point, refrigeration stops. Dryers of this type provide a dew point that varies within a few degrees; check with the manufacturer for further information. Water-chiller and direct expansion refrigeration dryers run continuously. This continuous operation may add slightly to power requirements, and may result in freezing during light-load conditions if adequate controls for bypassing hot gas and/or reducing compressor capacity are not included. Figure 5.19 presents a chart comparing the various types of compressed air dryers.

How Dry Must the Air Be? The most important criterion in choosing an air dryer is the dew point, the temperature to which air can be cooled before water begins to condense from it. The required dew point of the air system determines how dry the air must be, and, to a great extent, which type of dryer you choose. Keep the following points in mind.

1. Dew point varies with pressure. For example, an atmospheric dew point of −12°F is equivalent to a pressure dew point of 35°F at 100 psig. Be sure that you know whether a manufacturer is specifying the dew point the dryer can attain at atmospheric pressure or at a typical system pressure such as 100 psig. You can then determine what the minimum dew point will be at the system's operating pressure.
2. Required dew point varies with the application. If you are concerned primarily with preventing condensation in compressed air

Type of dryer	Initial cost 1000-scfm unit	Pressure dew point (100 psig) achieved w/100°F inlet air (°F)	Operational cost[a] per 1,000,000 ft³ (including depreciation) at maximum flow: 100°F, 100 psig inlet	Yearly costs per 100-hp compressor capacity[b]
Refrigeration	$6000	40	$3.91	$185
Twin-tower desiccant				
Heated	8000	0	6.64[c]	315
Heatless	7000	−70	23.00[c]	1091
Deliquescent desiccant				
Salt or urea	3000	80–85	20.00[d]	949

[a]Energy costs based on $0.035/kWh.

[b]Operating 80% of time, 8 hr/day, 260 days/year

[c]Includes cost of regenerating heat or air plus maintenance of prefilters/after-filters.

[d]Includes cost of replacement desiccant, freight, handling, storage, downtime, and maintenance of prefilters/afterfilters.

Figure 5.19 Dryer comparison chart. (Courtesy of Zurn, Inc., Erie, Pennsylvania.)

lines, the lowest ambient temperature to which the air lines will be subjected will be the controlling factor (there may be fluctuations of ±35°F or more from summer to winter). However, for some applications, dew-point requirements will be more severe, possibly as low as −40°F at atmospheric pressure. Do not inject too great a safety factor by starting a dew-point level that is not really needed. A safety margin of 20°F is about the maximum recommended.

3. You may require extremely low dew points at only a few isolated points. Consider using individual dryers at each point of use to attain these low dew points, in tandem with a less expensive dryer that will dry the air to less stringent requirements for use in the rest of the air system.

What Flow Capacity Is Needed? An air dryer must not only be able to dry the compressed air to the required dew point, it must also be able to handle the airflow required without causing excessive pressure drop. The flow capacity of a dryer depends on the operating pressure, inlet air temperature, ambient air or cooling water temperature, and required dew point. When any of these conditions change, the flow capacity of the dryer also changes.

Most manufacturers can supply performance curves that show the relationship of their dryers' flow capacities to these four factors. They merit careful consideration.

Installation and Maintenance of Driers

Where you install an air dryer can affect how well it performs. The amount of maintenance a dryer requires can add to its total yearly cost in terms of both labor and materials. If all the compressed air will be used inside a building where temperature is maintained at a stable level, the required dew point can be fixed within a range of a few degrees. If, however, some or all of the compressed air is subjected to outdoor temperature variations, the required dew point can change from day to day, or even from hour to hour. There is an upper limit to ambient air temperature for refrigeration air dryers of about 100 to 110°F. Above this level, there is no efficient heat sink, and the dryer will not operate properly. Water-cooled condensers can tolerate higher ambients. (*Caution*: Refrigerant air dryers should not be exposed to ambient temperatures much below 40 to 55°F without the addition of low-ambient-temperature controls, available from some manufacturers.)

When using deliquescent dryers, if the dryer is used in a central system, add bypass piping around the dryer to maintain the air supply while adding desiccant to the dryer. Also make certain there is no set of conditions (such as a valve that can be opened) that will reduce system pressure and cause high airflow velocity through the dryer, possibly carrying the chemical into the air lines (Fig. 5.20).

Refrigeration and deliquescent dryers should be drained regularly, depending on the amount of moisture accumulation. Most refrigeration dryers have automatic drains. These are available as options on deliquescent dryers. Some refrigeration dryers require a prefilter to remove oil and dirt that can coat the inside of the dryer, lowering heat transfer. With regenerative desiccant dryers, oil from the compressor can coat the desiccant, rendering it useless; install equipment designed for oil removal ahead of the dryer.

Troubleshooting the Compressed Air Dryer

Compressed-air-dryer problems generally fall into one of four types:

1. Water carry-over
2. Oil carry-over
3. Excessive desiccant usage
4. Excessive pressure drop

Often, the trouble encountered is the result of more than one of the problems. Thus the entire list should be considered.

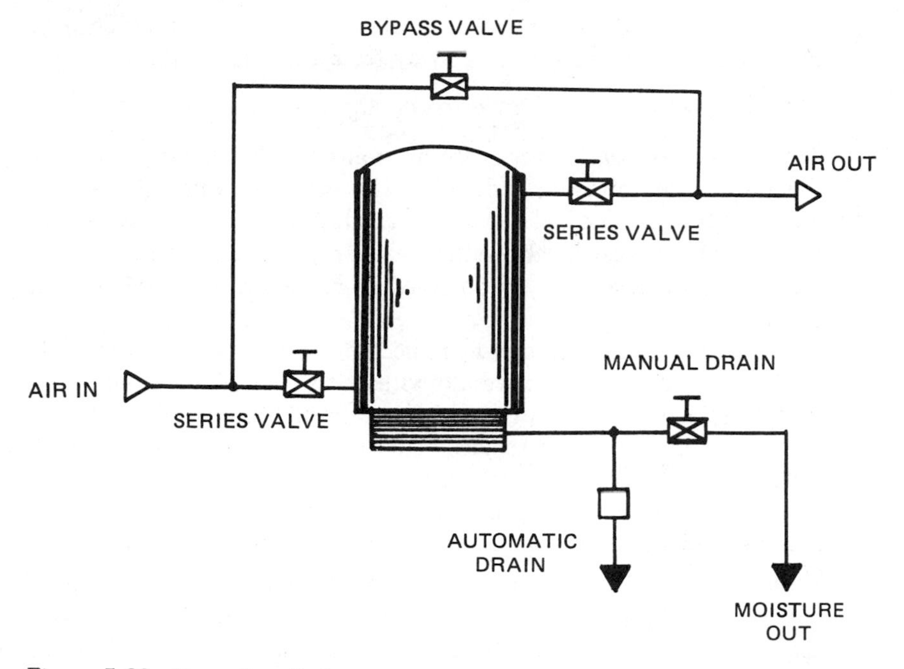

Figure 5.20 Dryer installation.

Water Carry-Over into the Downstream Air Line.

Dryer undersize for the application: Perhaps the original estimate of airflow in the circuit was in error, or perhaps additional equipment has been added since the dryer was installed resulting in an undersized unit. The existing dryer need not be discarded. Drying capacity can be increased by adding a second dryer, of either the same or different size, in parallel with the first.

High air temperature: The inlet air may be too warm for proper drying action. This may occur only at certain times of the day, so a 24-hr check should be kept by recording temperature every hour, or perhaps oftener. First, feel the inlet pipe. If it is above body temperature by even the slightest amount, this is a fairly certain indication of potential trouble, such as excessive desiccant use and water carry-over.

Radiant heat: There may be radiant heat soures such as hot air ducts, steam or hot water pipes, ovens, cupola, or furnaces near the upstream piping or near the dryer shell. Radiant heat picked up by the airstream causes caking of the desiccant near the bottom of the bed, even though the top of the bed looks normal. Radiant heat picked up by the shell may cause the desiccant to pull away from the sides of the vessel.

Oil Carry-Over into the Downstream Air Line.

Oil source: In nearly every case, unwanted oil in the air line comes from the air compressor. Desiccant dryers will usually precipitate only a very small amount of oil—that which will emulsify with the water that is removed. Any excess above this amount will collect on the surface of the pellets and can easily be observed by inspection of the bed through the access cover.

A compressor that is badly in need of an overhaul (pumping oil) will not provide satisfactory results with an air dryer. If the compressor uses the splash system of lubrication, and the oil carry-over is excessive, the remedy is overhaul or replacement of the compressor. Oil extracting filters installed ahead of the dryer may help to reduce the severity of the problem. Rotary compressors that inject oil as a lubricant into the compressor inlet are generally not satisfactory for use with air dryers. Oil-less rotary compressors give very satisfactory results.

Excessive desiccant consumption: The specific amount of desiccant that should be used per day or per month cannot be firmly established in rule form. However, a good procedure is to keep a record of the desiccant quantity added over a period of time, and from this a weekly rate or monthly rate of consumption can be established. This should be done when the dryer is installed. Subsequently, if consumption increases, the various factors that cause excessive consumption can be investigated. These include the following: drain stoppage, high inlet air temperature, and intermittent or continuous overloading with excessive airflow.

Excessive pressure drop: If the dryer has been sized correctly, the pressure drop through it should generally not exceed 2 psi. Excessive pressure drop is nearly always caused by overloading of the dryer. The increased velocity of the air causes increased pressure drop. The overloading may occur either because of excessive air consumption downstream, or from low inlet pressure to the dryer.

Miscellaneous Information on Compressed Air Dryers

A rule of thumb states that "humidity increase is in proportion to absolute pressure increase, provided that the temperature remains constant." This rule illustrates clearly why we have so much trouble with water in compressed air. If the air is compressed eight times, the humidity increases to eight times the original humidity. Thus if compressor inlet air has a humidity greater than 12½% and is compressed eight times or more, the humidity will exceed 100%

and the excess water will drop out. But this dropout does not all occur immediately after compression because the air is still hot from the heat of compression and will hold more water. But as the compressed air gradually cools, the excess water drops out, and this may occur all the way along the line from compressor to air tool. The only remedy is to dehumidify the air after compression so that any further drop in temperature will not precipitate water.

6
Hydrostatic Transmissions

6.1 PRINCIPLES OF OPERATION

Generally, hydrostatic transmissions consist of an axial piston pump connected in a closed loop to an axial piston motor. The pump or the motor or both can be variable-displacement units. The most commonly used configuration has a variable-displacement pump and a fixed-displacement motor. The variable-displacement pump is driven by a prime mover (e.g., internal combustion engine, electric motor) and the fixed-displacement motor, which is driven by the fluid from the pump, drives the machine (vehicle, hoist, conveyor, etc.). The direction of rotation and speed of the motor output shaft depends on the flow from the pump. The system pressure is determined by the machine load. A relief valve is installed on a hydrostatic system main loop to limit the maximum pressure and protect all components from catastrophic failure. Figure 6.1 depicts a closed-loop hydrostatic transmission.

The speed of the pump input shaft can be varied by changing the prime mover speed. The speed of the input shaft controls the flow output of the pump because the unit is a positive-displacement device. The flow output can also be changed by mechanically or hydraulically moving the piston stroke control device (swashplate or yoke assembly). This motion varies the pump displacement from zero to maximum. The oil flow from the pump controls the motor speed. Thus, by varying the pump output flow, the motor speed is correspondingly varied.

The piston-type pump and motor have a small amount of internal leakage which is removed from the pump/motor closed-loop circuit. This results in fluid loss from that circuit which must be replenished to prevent cavita-

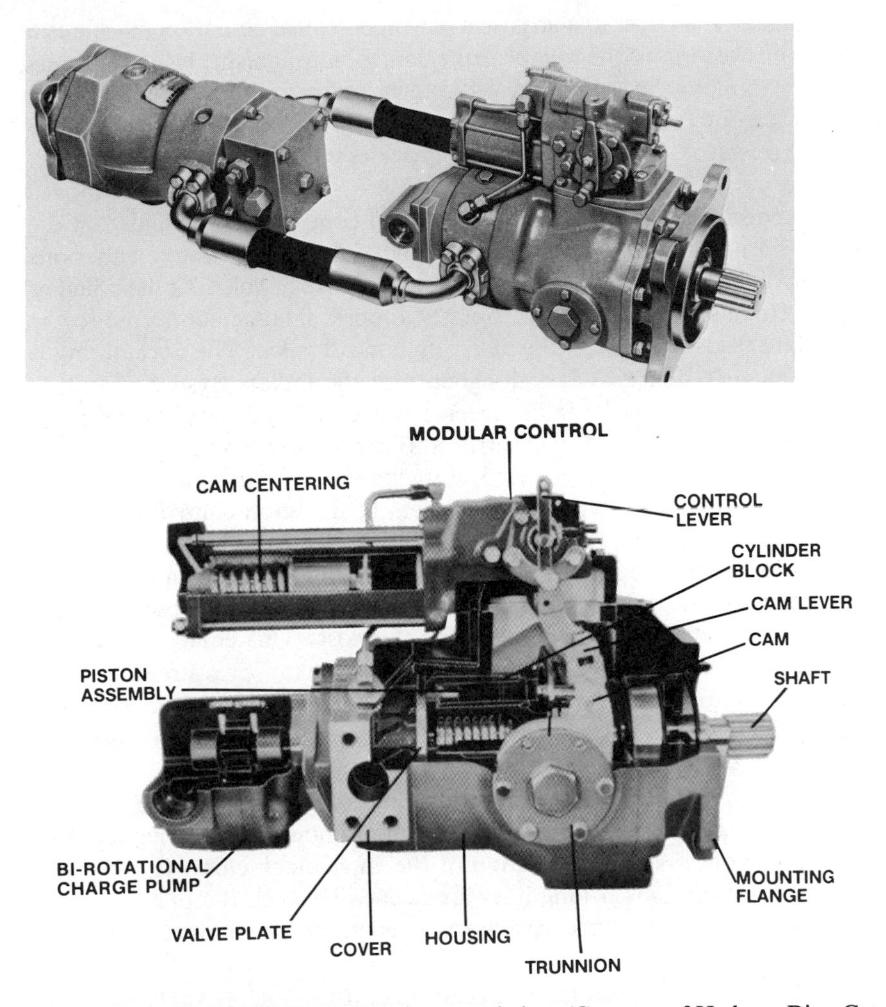

Figure 6.1 Closed-circuit hydrostatic transmission. (Courtesy of Hydreco Div., General Signal, Kalamazoo, Michigan.)

tion. To accomplish this task, a fixed-displacement pump is added to the circuit (generally a gear-type pump). It is driven directly by the prime mover through the piston pump shaft (the two pumps are coupled in tandem). The pump is called a *charge* or *replenishing* pump and it makes available a sufficient amount of fluid required to replenish the leakage losses in the pump/motor circuit. The pressure in the charge pump circuit is limited by a relief valve so that any fluid not needed by the circuit is discharged through this valve back to the system reservoir.

Some of the larger systems use a servo-controlled actuation mechanism to vary the position of the pump stroke control mechanism. Usually, forces necessary to move the swashplate or yoke assembly on high-capacity piston pumps are quite large and a mechanical or hydraulic advantage is needed to assist the operator. The servo-controller provides this advantage hydraulically.

The servo system works as follows. A speed control lever is moved which is connected directly to the directional control valve, usually on the pump. Charge pump flow is diverted through this control valve to either one of two hydraulic cylinders which rotate the swashplate or yoke. As the cylinders develop force and move the swashplate, another linkage connected to the swashplate tends to recenter the direction control valve. This recentering is a mechanical feedback system designed into the linkage system. In other words, a given control lever displacement is compared with the resulting swashplate rotation, and when the error (difference) between these two is zero, the direction control valve is centered, stopping further swashplate movement. Oil is trapped in both servo cylinders when the direction control valve is in neutral, thus preventing it from moving.

In closed-loop systems, an occasional sudden torque increase (pulse) to the motor can cause increases in hydraulic pressure. When long hoses are used, this pressure rise will store in expanding the hose. This change in volume could absorb the entire charge pump flow fed to the low-pressure side intake of the main pump, thus losing charge pressure. This results in erratic pump operation. To eliminate such a problem, control either the rate of pressure rise or the rigidity (sponginess) of long hose lines.

As the extra fluid is forced into the motor when the load (pressure) decreases, extra fluid can be forced into the motor housing, affecting the shaft seal or the cooler. Also, the location of the valve block closer to the pump might be advantageous if long lines are needed between the pump and the motor, which cause large pressure drops affecting charge pump system pressure.

Figure 6.2 depicts one type of hydrostatic transmission where the pump and motor are separated physically and connected only hydraulically. Figure 6.3 shows a close-coupled or back-to-back hydrostatic drive. The motor and pump shafts are not mechanically connected in the transmission. The ribs have been added to increase the heat-dissipation ability of the unit.

6.1.1 Circuit Development

To provide a more detailed and clearer understanding of how closed-loop hydrostatic transmissions function, the following step-by-step description is presented. For illustration purposes, the description uses a variable-displacement axial piston pump with a swashplate stroke control mechanism and a fixed-displacement axial piston motor.

Figure 6.2 Variable-displacement pump-fixed-displacement motor hydrostatic transmission. (Courtesy of Hydreco Div., General Signal, Kalamazoo, Michigan.)

A. Method of controlling speed and direction of rotation

1. As the swashplate of the variable pump is tilted in one direction, fluid is pumped through one line to the motor, resulting in rotation of the output shaft in one direction. The fluid is returned to the pump through the other line and the cycle repeated. The amount of flow (output speed) varies with the swashplate angle.
2. When the swashplate is tilted in the opposite direction, fluid is pumped to the motor through the opposite line, causing output shaft rotation in the oppostie direction.
3. When the swashplate is in neutral (zero angle) the axial pistons do not stroke; therefore, there is no fluid flow and no output shaft rotation.
4. The hydrostatic transmission has three major conditions: neutral, forward, and reverse. The output shaft speed is infinitely variable between neutral and forward or reverse.

B. Start with a basic closed-loop circuit

1. The first component is a variable-displacement axial piston pump, which has a control. The pump is unidirectional and requires internal part changes to reverse its direction of rotation.
2. The second component is a fixed-displacement axial piston fluid motor, which is bidirectional.

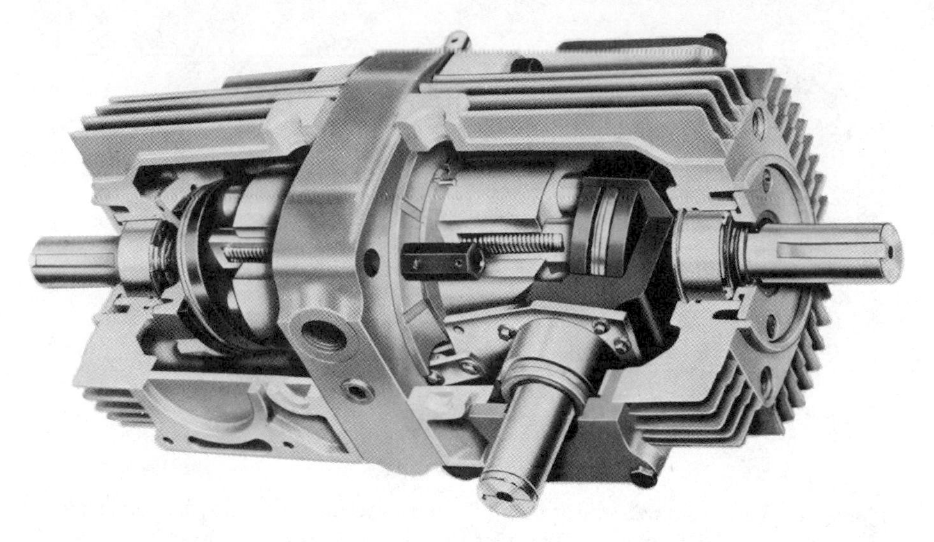

Figure 6.3 Back-to-back hydrostatic transmission. (Courtesy of Sperry Vickers, Troy, Michigan.)

3. These are connected in such a manner that oil flows from the pump through the motor, back to the pump, then repeats the cycle. This is termed a closed-loop circuit.
4. The pump swashplate can be tilted to either side of neutral so that the output (motor) shaft can be rotated in either direction.

C. Add case drain lines

1. A small amount of oil is used internally for the hydrostatic lubrication of critical wear surfaces. There is also a small amount of leakage past pistons and other parts.
2. This is the leakage flow and is removed from the circuit to prevent excessive heat. It is routed from the motor housing through the pump housing and then back to the reservoir.
3. When a heat exchanger is used, this flow is passed through it and cooled down before it enters the reservoir.
4. There is a limit to the case pressure, which is 40 psi. Exceeding this limit will result in gasket and seal leakage.
5. It is important to know what the case pressure is on a given machine as it is used to determine the working (differential) charge pressure (see below). For a given machine model, the normal case pressure will remain the same machine to machine

unless the circuit is altered, and should only have to be taken oncc for that design. In others words, the case pressure under various speed and load conditions can be determined at the factory (or in the field) on one machine and then be recorded and used for all machines of that design.

D. Add a charge pump circuit

1. In the circuit as described to this point we would soon starve the system, as we are continuously taking the leakage fluid out of the closed loop and back to the reservoir. It is necessary to replenish the circuit with fluid to keep it fully charged.
2. To accomplish this, a gear-type charge pump is added to the variable displacement pump and driven by the input shaft. The primary purpose of the charge pump is to replenish the oil taken from the system or to keep the system "charged"; hence the term charge pump.
3. It is necessary to maintain a pressure at the inlet to the variable displacement pump to ensure that if fills properly. This pressure is maintained by a charge relief valve, set at the factory for the correct differential pressure. It may range from 75 psi for some transmissions to over 200 psi for others.
   a. Differential pressure (denoted as Δpsi) is merely the difference in pressure from one side of the valve to the other. Differential pressure in this instance is obtained by subtracting the case pressure from the charge pressure gauge reading. For example:

Charge pressure gauge reads:	230 psi
Case pressure:	20 psi
Differential charge pressure is:	210 Δpsi

   b. The charge relief valve is a direct-acting poppet type that is factory set by shimming the relief valve spring. It is actually located in the charge pump housing.
4. Two charge check valves are added to direct the charge flow into the low-pressure side of the main closed loop. The high pressure in one side of the circuit acts on one of these check valves and holds it closed while charge pressure opens the opposite check valve, allowing charge flow to enter the low-pressure leg of the circuit.
5. The fluid for the charge pump comes directly from the reservoir. Since this fluid is the only fluid that enters the system, it

must be filtered to remove contamination and prevent abrasive particles from entering the system. This filter must have a 10-μm nominal rating and must not have a by pass. The transmission control system is designed to protect itself if the filter becomes plugged.

6. There is a limit to the amount of vacuum allowed on the inlet port of the charge pump. This limit is 10 in. Hg at operating temperatures.
   a. It is acceptable for this vacuum to exceed 10 in. Hg during cold startup of the system; however, the variable-displacement pump should not be stroked until the vacuum is within acceptable limits.
   b. As with case pressure, once normal inlet vacuums for various speeds have been determined for a given design, those figures can be used on all machines of that design.
7. The charge pump is sized so that it continues to replenish the system even after the leakage has increased due to wear. There is an excess amount of charge flow, which is passed over the charge relief valve and back to the reservoir through the pump case.

E. Control circuit

1. The internal forces acting on the swashplate in larger size pumps may make it desirable to use a hydraulic control on the variable-displacement pump. This hydraulic control makes it possible to move the swashplate with a minimum of operator effort. This is a servo or slave type of control and is called the *displacement control.*
2. The oil supply to operate this control is from the charge pump, thus giving a second function to that pump. This oil is fed into the displacement control through a small orifice. This orifice provides acceleration/deacceleration control so that no matter how fast the control lever is stroked, the control system can respond only at a predetermined rate.
   a. This orifice can vary in diameter, depending on the application. When dealing with a variety of pump models, it is necessary to ensure that proper orifice sizes are maintained.
   b. Because of its small diameter, this orifice can become plugged, resulting in sluggish transmission response to operator commands.

3. The internal pumping forces that are acting on the swashplate are such that charge pressure must be maintained to ensure proper control. When the charge pressure drops, the swashplate will automatically return to neutral (due to the internal forces) and the machine will stop. This feature provides protection if for any reason charge pressure is lost, including a plugged inlet filter. The swashplate centers because the charge pressure decreases and the machine will not operate. When the swashplate returns to neutral, the motor output shaft also stops.

F. High-pressure relief valves

1. It is necessary to protect the hydraulic system from overloads. Two pilot-operated high-pressure relief valves are added at the fixed motor to provide this protection. These are nonadjustable, factory-set, cartridge-type relief valves.
2. The system relief valves open when their pressure setting is reached and pass just enough flow from the high-pressure side of the closed loop to the low-pressure side to maintain that pressure setting. These valves are generally factory set.
3. These high-pressure relief valves will handle the flows normally required to limit system pressure; however, they are not large enough to handle full pump flow. If full pump flow is spilled across one of these valves, even for a short period of time (less than 10 sec), the transmission will heat up excessively and can be severely damaged.

G. Addition of cooling circuit

To maintain proper operating temperatures within the hydrostatic transmission further components must be added. These consist of a shuttle valve and a second charge relief valve, located at the fixed-displacement motor (which is the hottest part of the circuit). In neutral (0° swashplate angle) the shuttle valve is blocked and this cooling circuit does not function; however, in forward or reverse the shuttle valve is shifted by the high system pressure and opens the low-pressure side of the closed loop to this second charge relief valve. The charge relief valve at the motor is factory set so that it offers the path of least resistance for the excess charge oil. The fluid from the charge pump will now flow out of this valve instead of the charge relief valve at the pump. When in forward or reverse the entire charge pump flow enters the main system (closed loop) through the charge check valve, into the low-pressure side, passes through the variable-displacement pump, then to the fixed-displacement motor, out of the motor, through the shuttle valve, over the charge relief valve at the motor, through the case drain system and the heat exchanger, and finally back to the reservoir.

H. Critical circuit data

1. Charge pressure: as recommended by the manufacturer.
2. Case drain pressure: 40 psi maximum
3. Charge pump inlet vacuum: 10 in. Hg maximum at operating temperature
4. Inlet filter
   a. 10 μm (nominal)
   b. No bypass

Figure 6.4 contains the entire circuit in diagram form. All of the components along with their interconnections are shown to illustrate their function in the hydrostatic transmission circuit.

6.2 TRANSMISSION STARTUP PROCEDURE

1. After the transmission has been installed, remove the threaded plug from the side of the main pump housing. To read charge pressure at this port, install a pressure gauge. Also, install a vacuum gauge at the charge pump inlet for reading the inlet vacuum.
2. Check all fittings to make sure that they are tight.
3. Fill the pump and motor cases through the upper case drain openings with recommended fluid. It is recommended that all fluid be passed through a 10-μm filter. Reinstall and tighten the case drain lines.
4. Loosen the charge pump line, coming from the filter-reservoir, at the inlet to the charge pump.
5. Fill the reservoir with fluid. When fluid appears at the loosened hose of the charge pump inlet, install and tighten the hose and continue filling the reservoir. Leave the reservoir cap loose so that air will escape.
6. It is recommended that the control linkage to the pump control valve be left disconnected until after initial startup. This will allow the pump to remain in positive neutral.

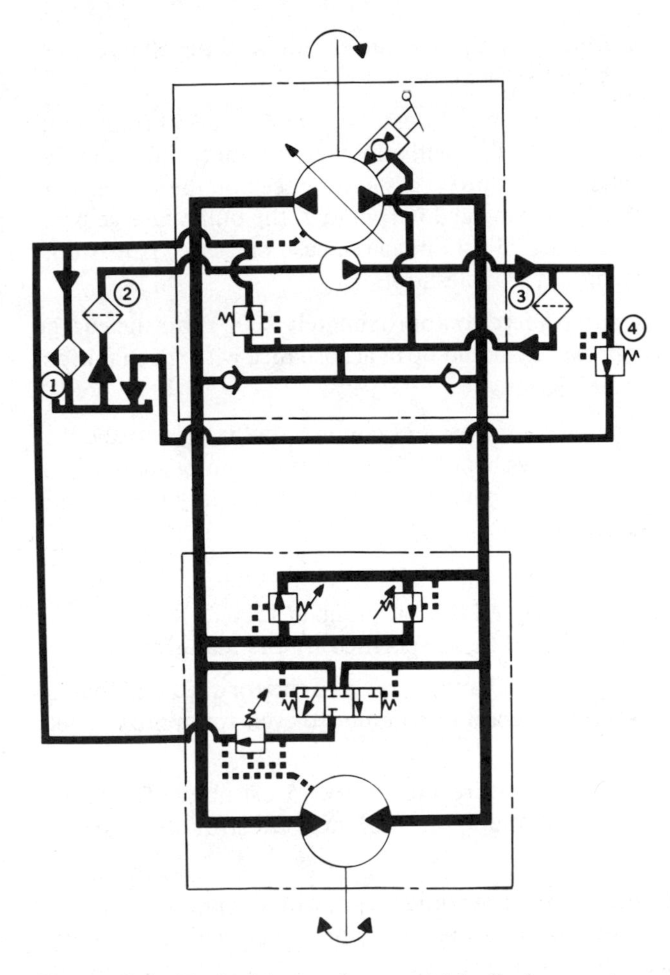

Figure 6.4 Typical heavy-duty variable-displacement pump–fixed-displacement motor transmission schematic. (Courtesy of Dynapower, a Unit of General Signal, Kalamazoo, Michigan.)

7. If the prime mover is:

   a. *Engine (diesel, gasoline, or liquefied petroleum)*: Remove the coil wire, close the injector rack or leave the gas turned off, and turn the engine over until the charge pressure reaches 30 psi or more.

b. *Electric motor*: Jog the starting circuit until the charge pressure reaches 30 psi or more.

8. Start the prime mover and, if possible, maintain a 750-rpm pump shaft speed for 5 min. This will allow the system to fill properly. During this phase, pressure surges may be seen on the gauge. *This is normal.* While running at 750 rpm idle, the pump charge pressure must be at least 75 psi above the case pressure. If it is not, stop the system and troubleshoot.
9. Increase the pump speed to approximately 1000 rpm; the charge pressure on the gauge should be in accordance with the manufacturers recommendations.
10. Shut down the prime mover and connect the linkage to the displacement control valve handle. *Caution*: If the motor shaft is connected to the drive mechanism, the necessary safety precautions must be considered.
11. Check the fluid level in reservoir and add fluid if necessary.
12. Start the prime mover and run the pump at 1500 to 1800 rpm; the charge pressure should be as specified by the manufacturer.
13. Move the pump control handle slowly to the forward and then to the reverse position. Repeat or continue to cycle for approximately 5 min.
14. Should the charge pressure fall below 75 psi above the motor case pressure, discontinue startup until the trouble has been found.
15. Run the prime mover at maximum rpm with the pump in neutral. Observe the reading at the vacuum gauge connected to the charge pump inlet. This reading should not exceed 10 in. Hg at normal operating conditions.
16. Shut down the prime mover, remove all gauges and replace all plugs or lines. Check the reservoir fluid level and tighten the oil fill cap. The machine is now ready for operation.

6.3 SYSTEM MAINTENANCE

Fluid: Generally, a fluid change interval of 2000 hr is adequate with a sealed reservoir system. A more frequent fluid change is required if the fluid has become contaminated by water or other foreign material or has been subjected to abnormal operating conditions. An open reservoir system with an air-breathing filler cap requires the fluid to be changed every 500 hr.

Filter: As a general recommendation, with a sealed reservoir system, the 10-μm inlet filter should be changed every 1500 hr. With an open reservoir system utilizing an air-breathing filler cap, the filter should be changed every 500 hr.

Reservoir: The reservoir should be checked daily for proper fluid level and the presence of water in the fluid. If fluid must be added to the reservoir, use only filtered or strained fluid. Drain any water as required.

Hydraulic lines and fittings: Visually check daily for any fluid leakage. Tighten and repair or replace as required.

Heat exchanger: The heat exchanger core and cooling fins should be kept clean at all times for maximum cooling and system efficiency. Inspect daily for any external blockage and clean as required.

6.4 TROUBLESHOOTING

I. The troubleshooting guides that will be used are of the fault-logic diagram type. They have been arranged so that the simplest operations are first, with the most complex (removing the transmission) last. They are further arranged by the frequency of occurrence of problems based on experience. The primary goal of these troubleshooting guides is to get the machine back in operation. The steps outlined can actually be taken in any order that makes troubleshooting easier. For example, some transmissions are so surrounded by the structure of the machine that it is easier to remove the transmission than perform the troubleshooting tasks.

The problems likely to be encountered have been condensed into five basic statements and set up these guides accordingly. The first step is to fit the problem into one of these basic statements, then follow the guide. The problem statements, such as "system operating hot," refer to what is normal for the equipment involved. To troubleshoot properly one needs a set of baseline data, either written or by experience, that define how the equipment should operate.

II. *Neutral difficult or impossible to find* (Fig. 6.5)

A. Check control linkage. This statement is referring to the machine control linkage. Disconnect that linkage from the

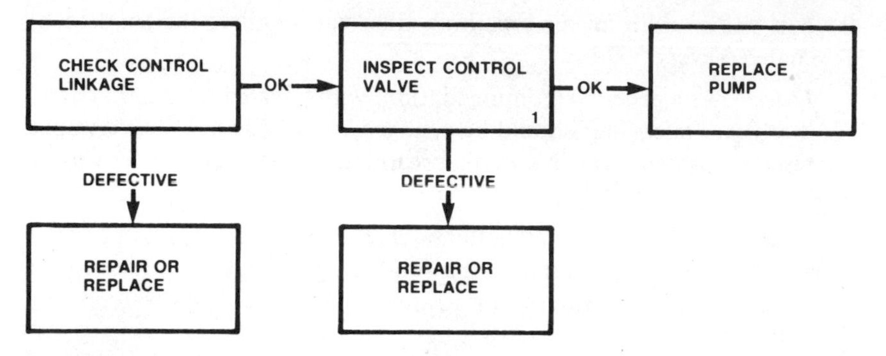

Figure 6.5 Neutral difficult or impossible to find. (Courtesy of Sundstrand Corp., Ames, Iowa).

displacement control and see if neutral is still hard to find by using the lever at the pump. Inspect the linkage for missing or damaged pins, cables, or rods.

B. Inspect control valve. With the machine control linkage disconnected, check the lever on the displacement control to be certain that it is still spring centered. The problem may be that the control spool is out of adjustment. It is recommended that the control be replaced; however, this adjustment can be made. The problem may also be internal, such as a sticking control spool or damaged centering spring. Disassembly of the control is *not* recommended and the control should be replaced and the unit checked again.

C. Replace the pump. This is the only instance when it is recommended to replace the pump only. In all other cases replacing the transmission (pump and motor) is recommended. This is because any problem (such as contamination) that has resulted in the malfunction of one unit will cause damage to some degree to the other unit. Replacing the pump may get the machine running today, but in 2 weeks the motor may also have to be replaced because of a repetition of the same problem.

III. *System operating hot* (Fig. 6.6)

A. Check oil level in reservoir. This is a common problem. Over a period of time the unit is worked on or a hose blown or a fitting leaks and the operator neglects to replenish the reservoir with lost oil and the level has become too low.

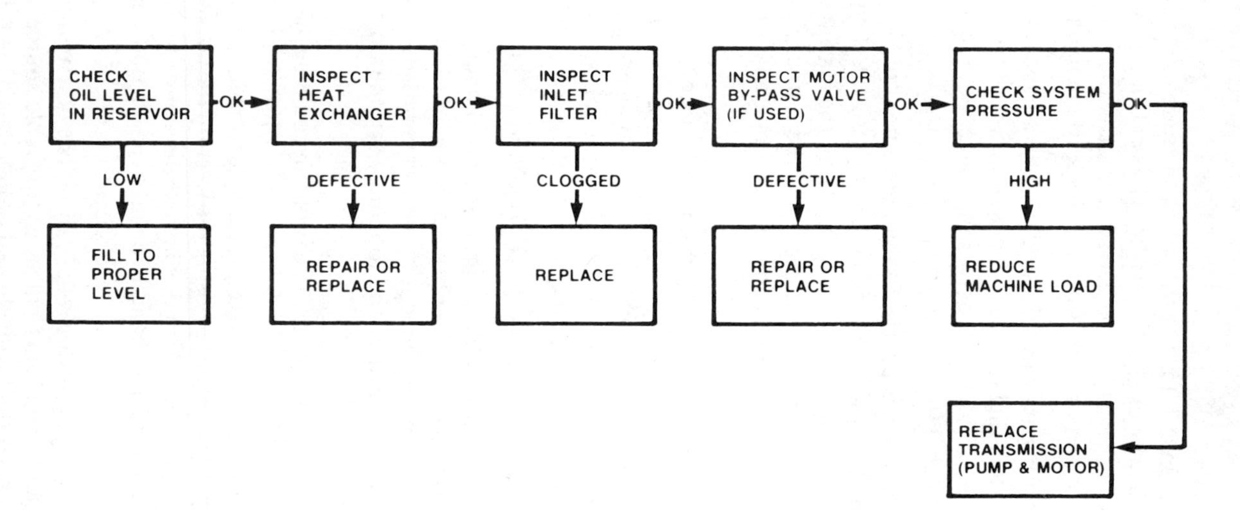

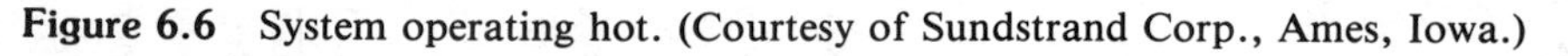
Figure 6.6 System operating hot. (Courtesy of Sundstrand Corp., Ames, Iowa.)

B. Inspect inlet filter. At this point in troubleshooting the only reliable way to check the filter is to put in a new element. It is also necessary to be certain that it is the correct element (proper manufacturer's number for a 10-μm nominal element).

C. Inspect bypass valve (if used). When a bypass valve is used, if it is left slightly open, the system can operate hotter than normal, as fluid spills across this valve. Be sure that the valve is fully shut.

D. Check system pressure. System pressure should be checked at this point. If it is too high (i.e., higher than it should be for the conditions at which it is checked), the machine load is too high. The problem could be that the output drive is in the wrong gear or the machine is being operated on too steep a grade.

E. Replace transmission (pump and motor). If the system pressure is satisfactory, the transmission should be replaced. This should be the last resort, as it is time consuming and exchange units must be available. The pump and motor should both be replaced for the reasons discussed previously.

IV. *Transmission operates in one direction only* (Fig. 6.7)

The nature of this problem automatically eliminates certain components as a source of the problem. Since the system *does* operate satisfactorily in one direction, we can eliminate pump, motor, charge pump, and filter and look only at components related to one side.

A. Check control linkage. The problem could be that the control linkage of the machine is out of adjustment or damaged. Disconnect the linkage from the displacement control and see if the system operates by using the lever on the displacement control (if possible). With this linkage disconnected, check the displacement control handle to make certain that it is still spring centered.

B. Inspect high-pressure relief valves. The best method for checking the system relief valves as a trouble source is to remove them from the manifold and switch them side for side. If the problem changes sides, one of the system relief valves is at fault and should be replaced. Remember that this is a pilot-operated relief valve and the light spring load that can be felt on the main spool does *not* hold the spool shut against system pressure.

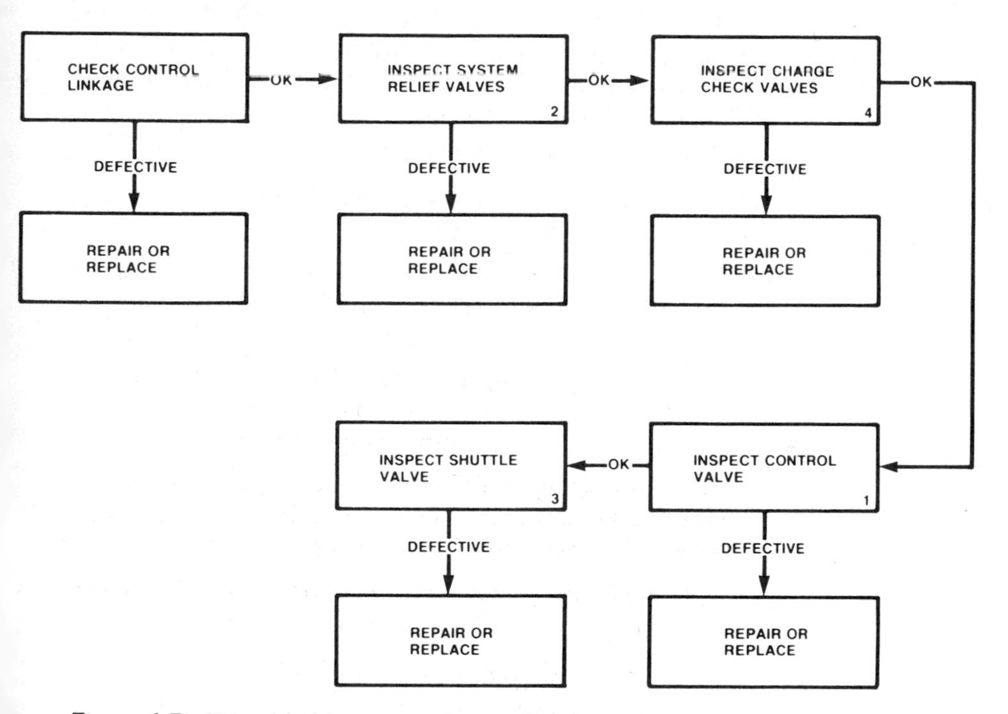

Figure 6.7 Transmission operates in one direction only. (Courtesy of Sundstrand Corp., Ames, Iowa.)

C. Inspect charge check valves. The charge check valves are located under the charge pump, which must be removed first. Like the system relief valves, a simple procedure is to switch these two valves side for side and see if the problem changes sides. If it does, the problem is in one of the valves and it should be replaced. These valves can be checked further by pushing against the internal ball to make sure that it is still spring loaded.

D. Inspect control valve. As stated previously, check the control lever for proper spring loading with the machine linkage disconnected. The control spool can be out of adjustment, causing this type of problem. If so, replacing the valve is suggested; however, this adjustment can be made, as previously discussed, provided that caution and patience are used. The problem could also be internal, such as a sticking spool, in which case the valve should be replaced as disassembly of this component is not recommended.

E. Inspect the shuttle valve. The shuttle valve spool can be sticking of one of its centering springs broken. Remove the two plugs, check the parts, and make certain that the spool moves freely in the bore and is not binding.

V. *System response sluggish* (Fig. 6.8)

A. Check the charge pressure (the location of the various pressure check points are shown on Fig. 6.9).

1. Remember that there are two charge pressures (neutral and forward or reverse) that must be read.
2. Inspect the charge relief valve at the pump. If the charge pressure is *low in neutral,* inspect the charge relief valve at the charge pump. Remove the plug, spring, spacers, and poppet and inspect for dirt, galling, and broken parts.
3. Inspect the charge relief valve at the motor. If the charge pressure is okay in neutral but *low in forward* (or reverse), inspect the charge relief valve at the motor.
4. If, however, the charge pressure is *low in both neutral and forward* (or reverse), take the following steps.

a. Inspect the charge relief valve at the pump.

b. Inspect the inlet filter. If that is not the problem, inspect the inlet filter. The most reliable way to accomplish this is to replace the element with a new one, making certain that it is the correct element.

c. Inspect the charge pump. Inspect the charge pump next. This is done by removing the pump and replacing it with a new one. Internal repair of the charge pump is not recommended. The charge pump is made up of several sections, and once taken apart these are difficult to reseal.

d. Replace the transmission (pump and motor). If the problem is not the charge pump, replace the transmission.

B. Inspect the control valve. If the charge pressure is okay, inspect the control valve. The small orifice could be partially plugged or the spool might be sticking.

C. Inspect the bypass valve (if used). Next, check the bypass valve (if used) to be certain that it is not partially open or defective.

D. Replace the transmission (pump and motor). The last step is to replace the transmission if the trouble cannot be isolated in any of the others areas.

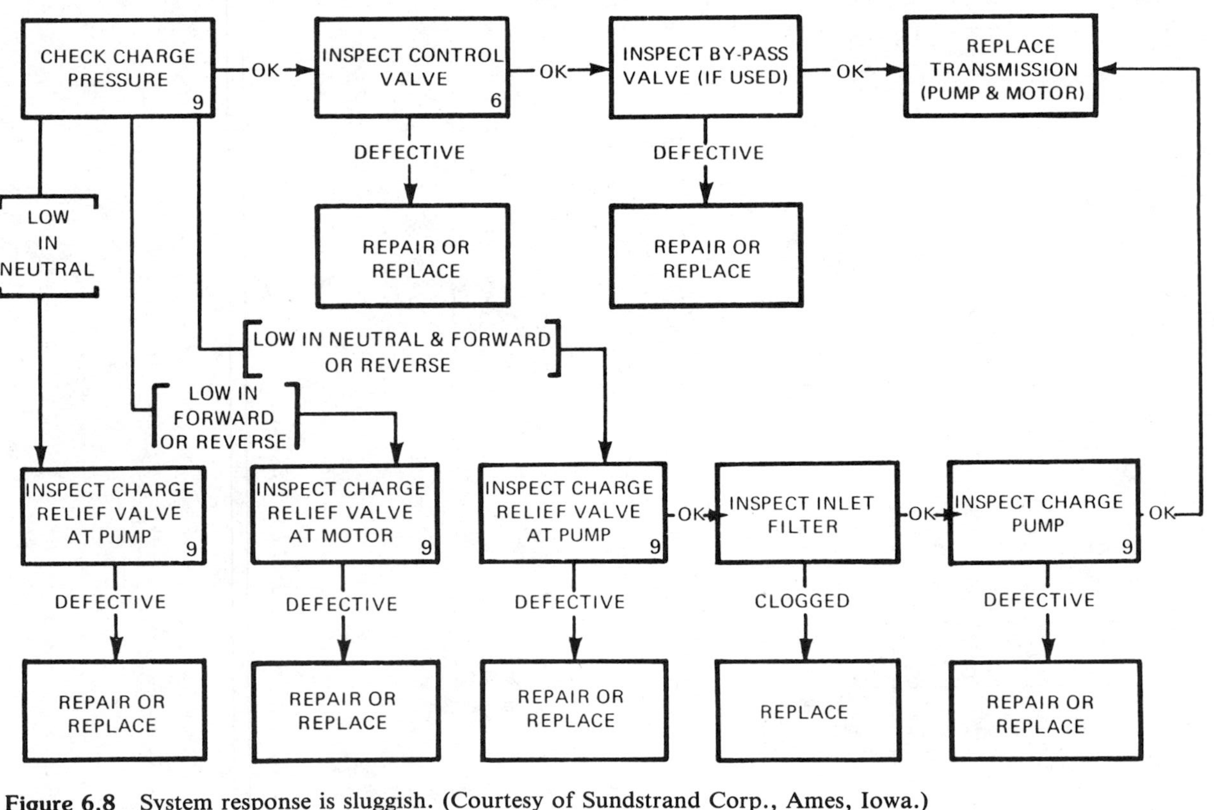

Figure 6.8 System response is sluggish. (Courtesy of Sundstrand Corp., Ames, Iowa.)

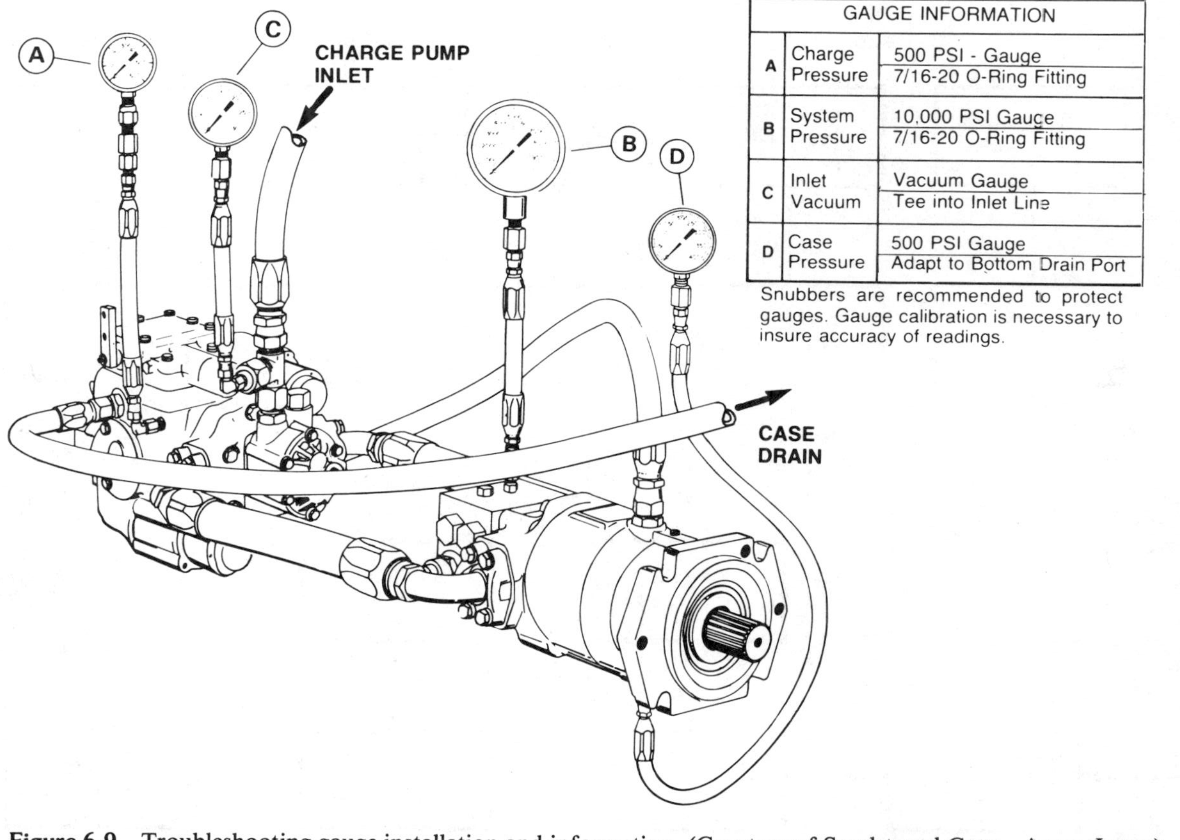

Figure 6.9 Troubleshooting gauge installation and information. (Courtesy of Sundstrand Corp., Ames, Iowa.)

VI. *System will not operate in either direction* (Fig. 6.10)

A. Check the oil in the reservoir.

B. Check the control linkage (on the machine).

C. Inspect the bypass valve (if used).

D. Check the charge pressure.

1. Inspect the charge relief valve at the pump if *low in neutral.*
2. Inspect the charge relief valve at the motor if okay in neutral but *low in forward* (or reverse).
3. If the charge pressure is *low in neutral and forward* (or reverse):

   a. Check the charge relief valve at the pump.

   b. Inspect the inlet filter.

   c. Inspect the charge pump.

   d. Replace the transmission.

E. If the charge pressure is satisfactory, perform the following steps:

1. Check the system pressure.
2. Inspect the pressure override (if used). If the system pressure is low, check the pressure override control to be certain that it is properly adjusted and not malfunctioning.
3. Inspect the control valve. If the system pressure is low (lower than it should be for the test conditions), the fault could be in the displacement control valve.
4. Replace the transmission (pump and motor).

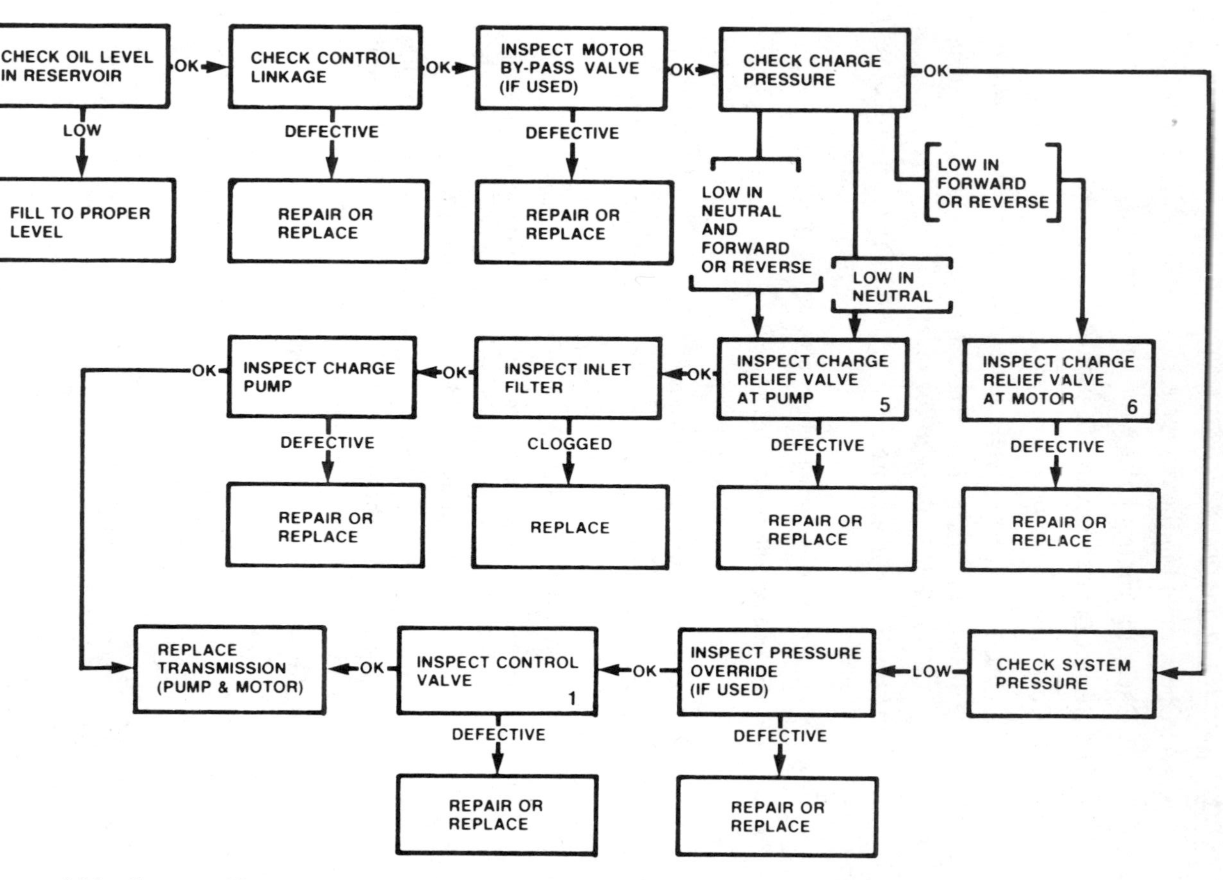

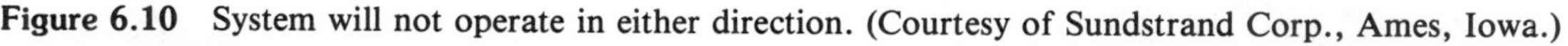

Figure 6.10 System will not operate in either direction. (Courtesy of Sundstrand Corp., Ames, Iowa.)

7
Temperature Considerations

A hydraulic system that is allowed to overheat can cause costly seal deterioration and fluid oxidation or breakdown. This results in corrosion and formation of sludge and varnish, which may in turn clog orifices and accelerate valve wear. In some cases, extreme temperatures will cause seizure of valves, pumps, and other components. High temperatures can also create a safety hazard should high-temperature fluid leak, and operating personnel could be severely burned. It is generally acknowledged that the ideal operating temperature of industrial oil hydraulic systems should not exceed 150°F. However, the preferred maximum operating temperature is below 140°F because above that level the life of most fluids is shortened (see Fig. 7.1). To design a system that will maintain thermal stability, it is necessary to understand how hydraulic systems generate and dissipate heat.

7.1 HEAT GENERATION

Heat is generated in a hydraulic system whenever oil flows from a higher to a lower pressure without doing mechanical work. This means that if a relief valve is allowing the oil to flow back to the tank, and the system pressure is being maintained, the difference in pressure or loss is the difference between the system pressure and the tank line pressure. Pressure losses can occur from flowing through inadequately sized valving or piping, and kinked or sharp bends in hose or tubing. The sun can cause excessive heat. High oil temperature can also be caused from metal parts being worn, or from the seals being worn, allowing the oil to bypass and cause heat. By checking the reser-

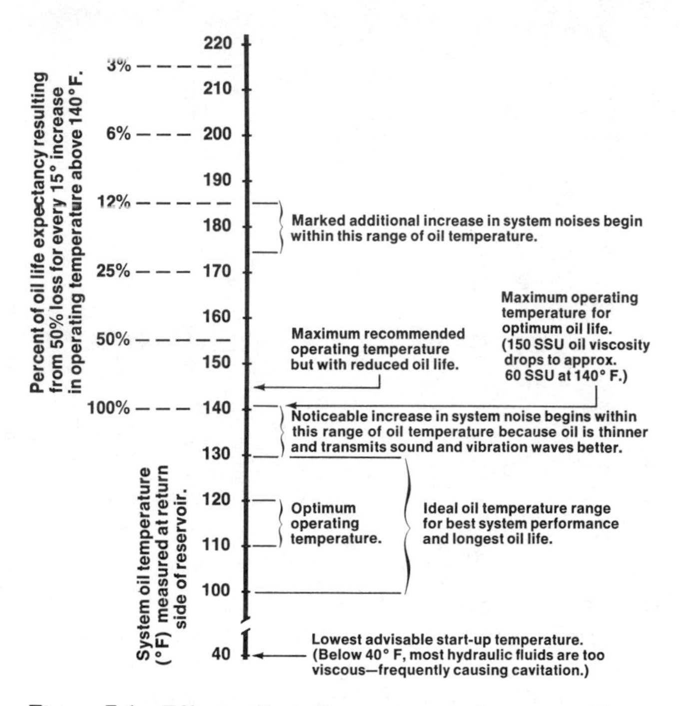

Figure 7.1 Effects of hot oil on system performance. (Courtesy of Parker Hannifin Corp., Cleveland, Ohio.)

voir temperature, a system problem can sometimes be corrected before serious damage can be done.

In most systems, the main heat-generating component is the relief valve. Generally, the relief valve is used only in a short period of the cycle. It is, therefore, necessary to find the maximum rate of heat generation by using the horsepower formula, then calculate an average for an entire hour.

Pressure reducing valves and pressure-compensated flow controls are another source of heat generation. Some pressure reducing valves should be checked for heat generation if they are passing high flows or reducing the pressure from extremely high pressure. Pressure-compensated flow controls, when used to reduce the flow to or from a cylinder or to and from a motor when using a fixed-delivery pump, will cause the excess oil not used to move the actuator to flow over the relief valve. Sometimes it is best to use a pressure-compensated pump in a system like this or use what is called a *bleed-off circuit*. The bleed-off circuit allows the excess oil to flow back to the tank unrestricted. This type of circuit will help reduce heat generation.

Sometimes it is helpful to figure that 15 to 20% of the electric motor horsepower will go into heat. This will help calculate this difficult-to-estimate point in a system, such as: miscellaneous valves, fluid friction in pipes, tubing or hose, mechanical friction, and slippage of pumps, fluid motors, and cylinders. In cases of marginal to high heat generation, it is a good practice to install heat exchanger connections in the main oil return line and relief valve return line. This will allow a heat exchanger to be added at any time.

Pressure drop across controls, or pressure that is the result of pump slippage or line loss, will create heat. The amount of heat generated can be calculated using the formula

$$q = 1.5(P_1 - P_2)Q$$

where

q = heat generated, Btu/hr
Q = flow rate, gpm
P_1 = higher pressure, psi
P_2 = lower pressure, psi

Some helpful conversion factors are:

1 hp = 2545 Btu/hr
1 hp = 746 W
1 hp = 33,000 ft-lb/min

Pressure drop across controls, or pressure drop that results from pump slippage or line loss, will generate heat, as has been stated previously. The following examples use the formula given above to compute the amount of heat generated (Fig. 7.2).

Pump Slippage. The volumetric efficiency of a pump is indicative of fluid leaking to the drain or back to the pump inlet, resulting in a pressure drop. A typical 11-gpm pump will deliver 11.4 gpm at 100 psi and 10.6 gpm at 1000 psi. This difference of 0.8 gpm is pump slippage; pressure drop is 1000 psi. The heat generated is

$$1.5 \times 0.8 \times 1000 = 1200 \text{ Btu/hr}$$

Line Loss. As oil flows through lines and fittings, there is a pressure drop, or line loss, that generates heat. A fluid with a viscosity of 150 SUS and a flow of 10 gpm through a ½-in. standard pipe has a pressure drop of 0.55 psi/ft. A 90° elbow has an equivalent length of 1.7, or a pressure drop of 0.94 psi. Ten feet of pipe and two elbows would have a line loss of 7.38 psi, and the heat generated would be

$$1.5 \times 7.38 \times 10 = 110 \text{ Btu/hr}$$

The heat generated by line loss is usually not a serious problem when tubing or pipe is used because lines dissipate heat almost as readily as it is generated.

System component	Cause of heat	Pressure drop	Heat loss[a] (Btu/hr)
11-gpm pump at 1000 psi	Pressure drop and surface tension during slippage back to pump inlet or to reservoir through drain	1000 psi (0.8 gpm)	1200
½-in. pipe or hose at 10 gpm	Normal line pressure drop and surface tension at walls	0.55 psi/ft	7.25 per foot
½-in. 90° elbow pipe at 10 gpm	Pressure drop plus friction at sharp turn in flow path	0.94 psi	14.1 per elbow
3/8-in. four-way spool-type directional control valve	Leakage friction across both lands	0.86 psi	13
at 10 gpm and 1000 psi	Pressure drop plus friction at turns and orifices	41 psi	615
¾-in. angle check valve at 20 gpm	Pressure drop plus restricted passage of fluid through valve while providing back pressure for shifting piloted valves	65 psi (spring tension)	190
Relief valve (on press) at 10 gpm	Pressure drop plus restricted passage of relief fluid through valve during feed, dwell, and reload cycles	1000 psi	10,500

[a]For oil of 150 SSU viscosity.

Figure 7.2 Examples of heat sources in a typical hydraulic system. (Courtesy of Parker Hannifin Corp., Cleveland, Ohio.)

An exception is when the lines are in a hot ambient atmosphere, in which case they actually absorb heat and transfer it to the system.

Valve Leakage. Spool valves that permit leakage across the sealing lands will generate heat. A typical directional control valve has a nominal leakage of 1 in.^{3}/min/land/1000 psi. There are two lands, so the heat generated is

$$1.5 \times (2/231) \times 1000 = 13 \text{ Btu/hr}$$

This form of heat generation can normally be neglected.

Valve Pressure Drop. Valve pressure drop is best divided into two types: The first type is pressure drop that is the result of fluid flowing through valves (inlet to outlet) which are teed into lines (similar to line loss). For example, consider a typical 3/8-in. four-way directional control valve with a loop pressure drop of 41 psi at 10 gpm. The heat generated is

$$1.5 \times 10 \times 41 = 615 \text{ Btu/hr}$$

Depending on the number of valves, their pressure drops, flow rates, and the degree to which the designer wants to analyze thermal stability, this form of heat generation may be neglected.

In a second example, the system has a ¾-in. angle check valve with a 65-psi spring, used in a 20-gpm system to provide back pressure for shifting pilot-operated valves. The heat generated is

$$1.5 \times 20 \times 65 = 195 \text{ Btu/hr}$$

In systems with a considerable number of valves, the designer should consider this source of heat generation.

The second type of pressure drop across valves is that resulting from controlling pressure, as in the case of pressure relief or pressure reducing valves. Typical of this type of valve pressure drop, relief valves in fixed-displacement pumping systems may be the single greatest source of heat generation. For example, consider a press that has a 1-min work cycle, during 30 sec of which the pump is dead-headed or holding (8 gpm at 1000 psi). The remaining time the cylinder is either advancing or retracting (useful work). The heat generated is

$$1.5 \times 8 \times 1000 \times (30/60) \times 1 = 6000 \text{ Btu/hr}$$

As another example, consider a machine with a 30-sec work cycle: 5-sec rapid advance, 10-sec feed (a flow control is used in a meter-in circuit to limit flow to 2.5 gpm), 6-sec dwell, 3-sec return, and 6-sec reload. The pump discharges 10 gpm at 1000 psi and passes oil through the relief valve when not doing useful work. The heat generated is

During rapid advance: negligible

During feed: $1.5 \times 7.5 \times 1000 \times (10/60) \times 2 = 3750$ Btu/hr

During dwell: $1.5 \times 10 \times 1000 \times (6/60) \times 2 = 3000$ Btu/hr

During return: negligible

During reload: $1.5 \times 10 \times 1000 \times (6/60) \times 2 = 3000$ Btu/hr

The total heat generated is 9750 Btu/hr.

As a shortcut, many designers frequently compute the quantity of heat generated by the major contributors, and multiply this by a factor of 1.25 to take into account those items not considered and to provide a margin of safety.

Suggestions for Heat Reduction

1. Where heat generation may be a problem in some systems, use a generous-size reservoir.

If reservoir size must be limited because of space limitations, facilities for adding a heat exchanger should be provided.

Additional baffles in the reservoir may improve the heat dissipation value by forcing the return oil along the outside walls.

2. Unload the pump when system pressure is not required.
3. Set the main relief valve for the amount of pressure that is required to do the work—not higher.
4. If possible, when using pressure-compensated flow controls with fixed-delivery pumps, connect them in a bleed-off arrangement.
5. Reservoirs should be located in the open to ensure good air circulation. Reservoirs located outside should be shaded from the rays of the sun. They should not be set directly on the floor but rather should be slightly raised to allow air to circulate past the reservoir bottom.
6. Accumulators may sometimes be used in holding or clamping circuits where high static pressures are needed.
7. On some systems, compressed air to oil intensifiers may be used to maintain system pressure with a minimum amount of heat generation.
8. Use intensifiers to develop higher pressures when long holding cycles are involved.
9. Use load-sensing circuits.

7.2 HEAT DISSIPATION

Unless auxiliary heat-removal devices such as heat exchangers are used, the major method of heat dissipation is by convection. It is therefore important that the reservoir be located so that all surfaces are exposed to freely circulating air. Because heat is also transferred by conduction and radiation, it is important to shield the reservoir from external heat sources such as the sun or furnaces. Reservoir heat loss may be approximated by

$$q = 2.5A(T_1 - T_2)$$

where

q = heat dissipated, Btu/hr
A = surface area of reservoir exposed to circulating air, ft^2
T_1 = temperature of oil, °F
T_2 = temperature of air, °F

You can calculate the temperature at which thermal stability occurs from the relationship

$$T_{max} = T_{amb} + \frac{q}{2.5A}$$

where

T_{max} = temperature at which thermal stability occurs, °F
T_{amb} = ambient temperature, °F
q = heat input, Btu/hr
A = surface area of reservoir exposed to circulating air, ft^2

For average calculations, assume heat dissipation from the sides, top, and bottom of the reservoir. The surface area of external plumbing may also be used or counted as a dissipating surface. Do not include the bottom of the reservoir unless it is exposed to free air circulation. The cooling capacity of the reservoir will increase in proportion to the square footage of dissipating surface, and also in proportion to the difference between the oil temperature and the ambient air temperature.

For steel reservoirs, this formula will give approximate results:

$$hp = (\text{heat dissipation}) = 0.001 \times TD \times A$$

where

A = square footage of dissipating surface
TD = temperature difference between oil and surrounding air
hp = cooling capacity expressed in horsepower

There should be a reasonable amount of free air circulation around the reservoir. A forced blast of air directed on the side of the reservoir is a consideration, as it can increase the heat dissipating capacity as much as 50%.

When it is found that the reservoir cannot dissipate all of the heat generated by the hydraulic system, it is necessary to use a cooler. There are two principal types of coolers: air coolers and water coolers.

7.3 HEAT EXCHANGERS: COOLERS

Hydraulic systems perform tasks and receive energy from an outside source to generate fluid pressure to do work. As the actuator operates, it uses energy generated by the pump to move the load. But when fluid flows from a high- to a low-pressure condition without doing work, the energy there in turns into heat. Although the amount of heat generated is variable, it can begin to approximate the nameplate horsepower of the driver. Even well-designed hydraulic systems convert as much as 20% input horsepower to heat.

What does heat do to a hydraulic system? Some heat has a desirable effect. When oil is cold, system operation is sluggish and unsatisfactory. Thus heating the oil will have a beneficial effect. Overheated oil, on the other hand, begins to decompose, forms varnish on component surfaces, and can burn seals; viscosity drops and system operation again becomes erratic and poor. Continued system operation often requires an oil change, seal replacement, and components can require repair or overhaul. Ideal system temperatures vary, but the modern trend is toward hotter-running systems. Therefore, heat exchangers are more of a necessity than ever.

But what is heat? *Heat* is a form of energy that transfers from one region to another because of a temperature difference or temperature gradient between the regions. It is a transient phenomenon which always and naturally flows from the hotter to the cooler region.

Heat dissipates from a hydraulic system in two ways: through natural convection and through forced convection. *Natural convection* takes place as heat moves from the various system components, the reservoir, and the conductors to the surrounding air because of the temperature gradient. It is wonderful when natural convection can dissipate system heat. If, on the other hand, natural convection cannot remove generated heat, system temperature will continue to rise. Then, heat exchange becomes necessary, and the second heat dissipation mode, *forced convection,* occurs. *Radiation,* another mode of heat transmission, occurs too, but its effect is essentially negligible.

7.3.1 Heat Exchanger Types

Water-Cooled

Shell-and-tube heat exchangers pour cooling water into the tubes of the exchanger while oil, the more viscous fluid, flows around the tubes in the shell side. Heat exchangers are made of materials such as red brass, copper, cast iron, admiralty brass, stainless steel, or other special metals (see Fig. 7.3). They have an outer flanged shell or barrel with end bonnets and appropriate gasketing to close and seal the ends. Sealed inside the shell is a precise pattern of tubing which runs the length of the shell and terminates in tube sheets or end plates. The tube ends are mechanically fastened to the tube sheets and seal each end of the shell. In another design, the tubes are all bent 180° so their ends are in one tube sheet.

The tubes of the tube bundle run through a variable number of baffle plates, which provide support for the tubes and also cause the oil to flow at right angles to the tubes as it travels from one end of the shell to the other. Most tube bundles used for hydraulic applications are permanently sealed in the shell; models with removable tube bundles are more expensive and have a different set of sealing conditions at the shell ends.

Heat exchangers are available in one-, two-, and four-pass configurations. The multiple passes result from arranging the tubes in such a fashion as to let the water flow past the oil one, two, or four times. Selection of the number of passes desirable is based on available coolant or water temperature.

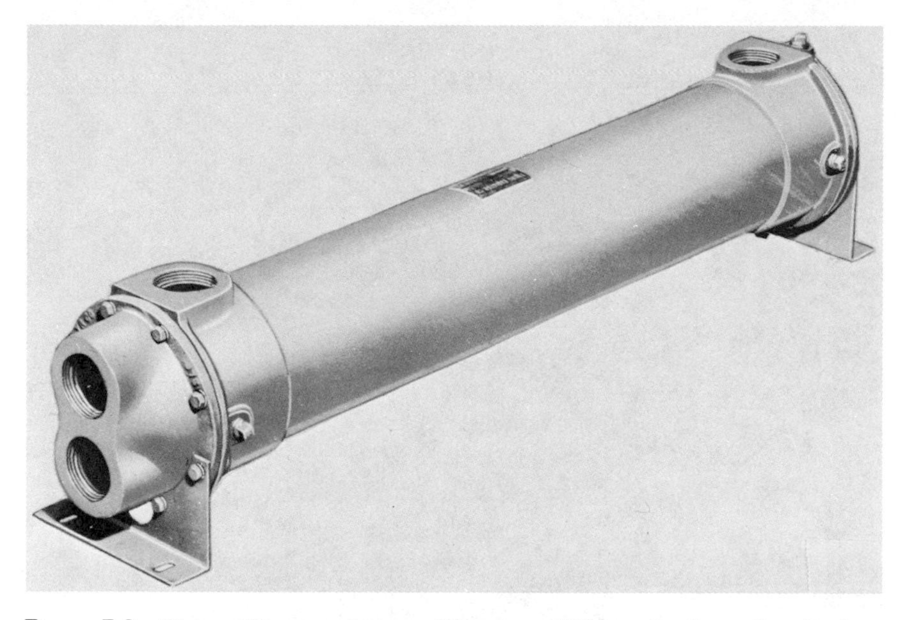

Figure 7.3 Water-oil heat exchanger. (Courtesy of Young Radiator Co., Racine, Wisconsin.)

Air-Cooled

When the air is used as the cooling media or is the choice to receive the waste heat from the hydraulic system, a heat exchanger commonly known as a *radiator* is used. Hot oil passes through the tubes of these heat exchangers. Turbulators help destroy laminar flow therein to promote efficient heat transfer from the fluid to the tube wall. The tube metals also have a high thermal conductivity factor (see Fig. 7.4).

As with shell-and-tube heat exchangers, an increase in area of the heat transfer surface helps increase heat transfer capability. Fins, physically fastened to the tubes, increase the surface area, and as an added benefit, help destroy laminar airflow.

The following are some considerations that help determine radiator core configurations:

Oval tubes provide more turbulent flow at lower liquid flow rates than do round tubes.

Round tubes give higher flow rates and lower pressure drops.

Materials generally are admiralty brass, red brass, aluminum, or steel. Choice can depend on structural needs or service environment.

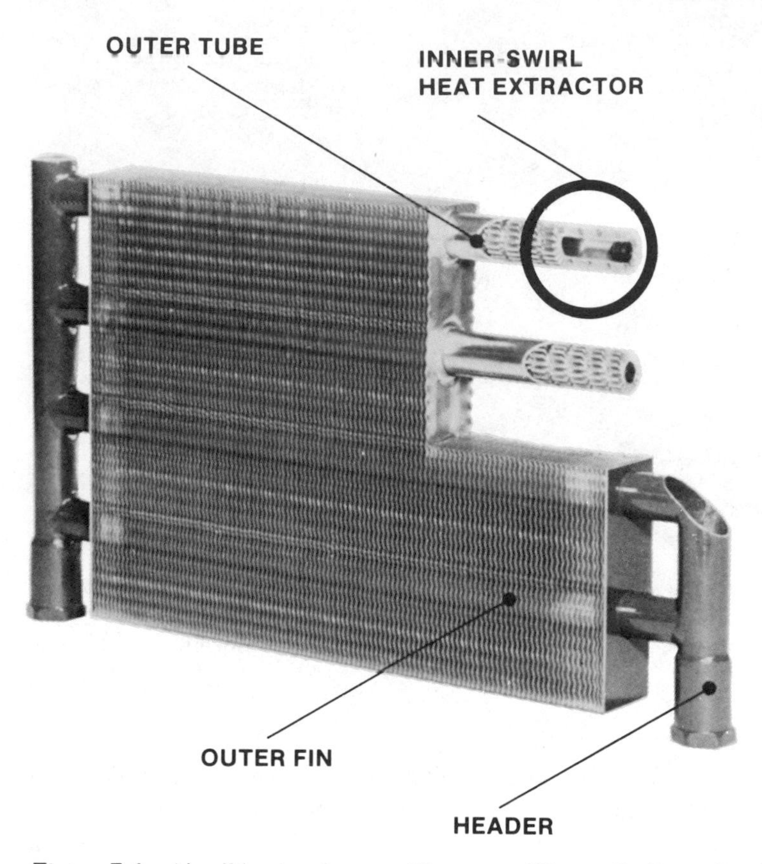

Figure 7.4 Air-oil heat exchanger. (Courtesy of Young Radiator Co., Racine, Wisconsin.)

> Fins vary in heat transfer capability and cleanability. Fin types are flat plate, humped, and louvered, ranked in order of increasing ability to generate turbulent flow and decreasing cleanability.

The common terms used with heat exchangers and their definitions are presented in Fig. 7.5.

Coolers are usually rated at a relatively low operating pressure (150 psi). This requires that they be positioned in a low-pressure part of a system. If this is not possible, the cooler may be installed in its own separate circulating system.

To ensure that a pressure surge in a line does not damage a shell-and-tube cooler, they are generally piped into a system in parallel with a 65-psi check valve. The check valve will permit fluid to bypass the cooler when the pressure in the cooler inlet line exceeds the check valve operating pressure (65 psi). Coolers can be located in a system's return line, after a relief valve, or in a case drain line of a variable-volume, pressure-compensated pump. Sizing the cooler to have the capacity for dissipating the necessary heat is a complex problem. It is recommended that the heat exchanger manufacturer be contacted or that the selection procedures outlined in their catalogs be followed.

As with all hydraulic equipment, mount heat exchangers so that they are shielded from sources of heat radiation, including the sun. Protect water-cooled heat exchangers from freezing. Air-blast radiators should also be protected from accidental physical damage caused by swinging booms, lift trucks, and other moving equipment.

Installation of Air-Oil Type

If the air is furnished by an electric-motor-driven fan, the air should be forced under pressure through the radiator core rather than being pulled by a partial vacuum. This gives slightly better heat transfer and provides a safety precaution against fan damage if the fan should become loosened on its shaft.

Core: That section of an oil cooler assembly which is comprised of the heat transfer surfaces.

Face area: Area defined by the core width times core height (oil-to-air coolers).

Face velocity: The velocity of air approaching the core (Volume per unit time divided by face area).

Fin: Extended heat transfer surface. Shell-and-tube oil coolers may have fins or other extended surface.

Header: This term has a dual meaning. It is sometimes used synonymously with tube sheet or tank.

Heat dissipation: The quantity of heat, usually expressed in British thermal units per minute, that an oil cooler can dissipate under specified conditions.

Inlet temperature differential: The difference in temperature between the fluid being cooled and the cooling medium at the point each enters the heat exchanger.

Multipass oil cooler: An oil cooler that is so circuited that either fluid passes across or through the core more than once.

Operating pressure: That fluid pressure to which the oil cooler is normally exposed during operation.

Peak pressure: The highest pressure to which the oil cooler is intermittently subjected.

Pressure drop: The pressure differential between inlet and outlet at a specified fluid flow rate and viscosity.

Notes: (1) Air side is measured in inches (millimeters) of water. (2) Oil side is measured in psi (kPa), (3) Water side is measured in psi (kPa).

Figure 7.5 Definitions of heat exchanger terms. (Courtesy of Perfex Corp., Milwaukee, Wisconsin.)

Additional Cooling Capacity

Additional cooling capacity can be obtained by a favorable location of the air-oil cooler. If practical, it should be placed so that the high-velocity air discharge through the radiator core will strike the hydraulic reservoir and keep air moving past the pump and other components. This can appreciably improve the heat-radiating ability of the tank.

On moving vehicles driven with an engine, a radiator core, without fan, may be mounted in front of the regular engine radiator, to make use of the engine fan blast. The additional restriction to airflow does not appreciably affect the engine cooling system.

7.3.2 Adding a Second Heat Exchanger

Another heat exchanger may have to be added to a system already containing one if a further increase in cooling capacity is found to be necessary. For greatest compatibility with the existing system, the second heat exchanger should be an identical model to the first. The question then arises as to whether it should be plumbed in parallel or in series with the original heat exchanger (Fig. 7.6).

Oil Circuit

Assuming that the original exchanger was correctly selected to suit the gpm flow volume of both water and oil circuits, the new heat exchanger should have its oil circuit plumbed in series with the first. The baffle spacing inside the shell has been selected to obtain an oil flow velocity between the limits of 1 to 6 ft/sec, with optimum velocity at 3 ft/sec. If the hydraulic oil were divided between the two units in a parallel connection, the flow velocity might be below the permissible range. In a series connection, the original velocity is maintained through both units.

Water Circuit

For the same reason, the water circuit should normally be connected in series through the two units to maintain adequate velocity. Again, optimum velocity is about 3 ft/sec, with maximum at 6 ft/sec. Beyond this, the water erodes the insides of the tubes, causing early failure. The modulating-type water control valve normally keeps the water velocity (and volume) throttled as low as possible to still give the amount of heat removal required. For this reason the modulating valve makes more efficient use of the water than does an "on-off" valve such as a solenoid type operated with a temperature switch.

Important! Inlet hydraulic oil should always be connected to the oil port that is closer to the end with the water connections. Inlet water should first be plumbed to the heat exchanger that is downstream in respect to the oil circuit. This gives the nearest approach to true counterflow that is possible with multipass heat exchangers, and the maximum heat transfer.

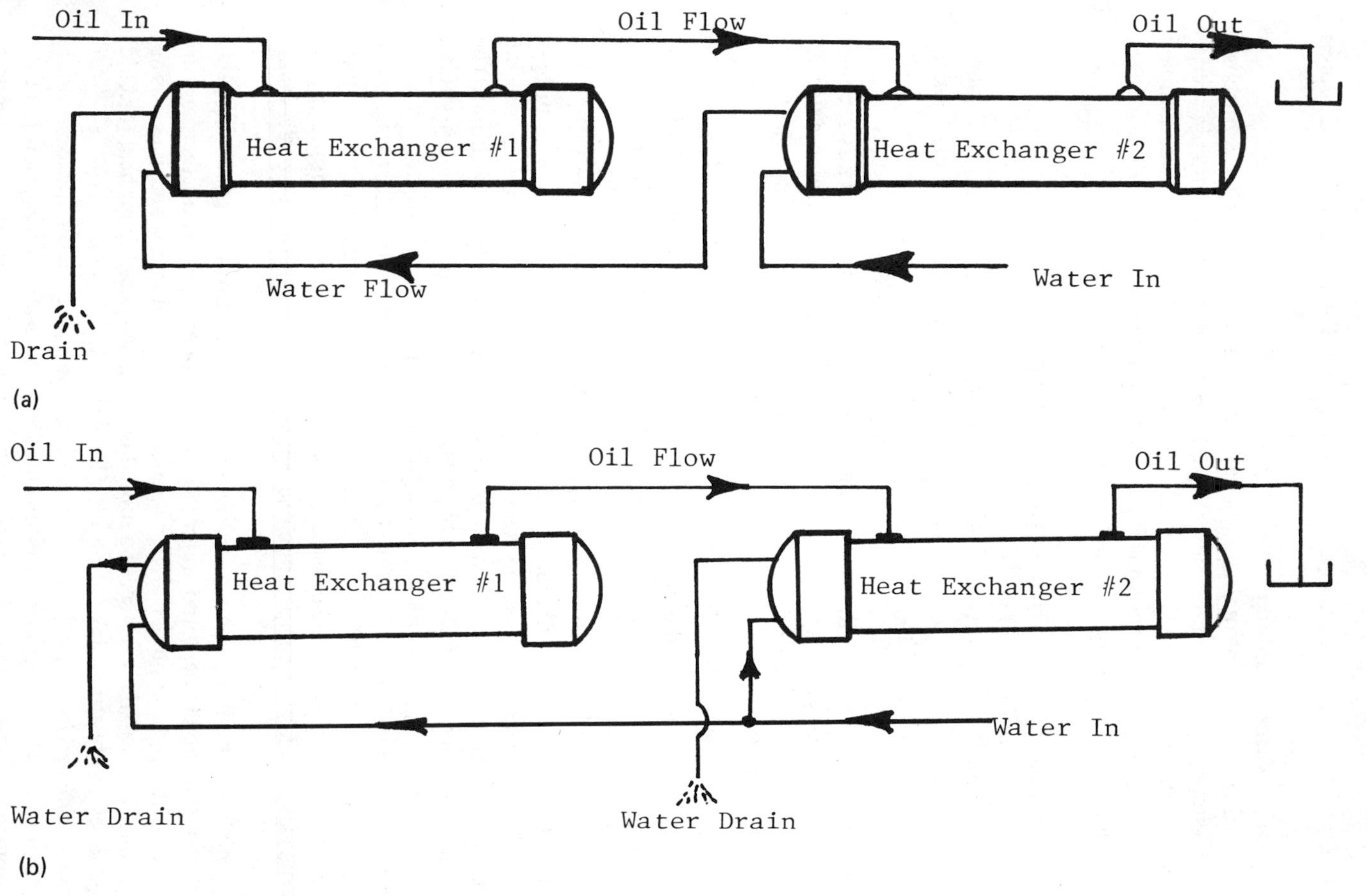

Figure 7.6 Adding a second heat exchanger.

Parallel Water Plumbing

A further increase in total cooling capacity can be obtained by connecting the water supply for parallel flow through both units, while maintaining series flow in the oil circuits. However, the water flow volume should be doubled to maintain the same internal flow velocity in the tubes for which the exchanger was originally designed.

7.3.3 Miscellaneous Tips on Heat Exchangers

Source of Cooling Water

In those locations where cooling water is scarce or expensive, a free source is sometimes available by intercepting water intended for other purposes, using it for heat exchanger cooling, then piping it on for its normal use. Water used for toilets, batch mixing, lawn sprinkling, washing, and many similar purposes can be utilized, and the slightly increased temperature as it comes out of the heat exchanger would make no difference in its ultimate use. In the usual heat exchanger the cooling water would have a temperature rise of 10 to 30°F.

Water Flow Control

To conserve water and avoid overcooling the oil, water flow to the heat exchanger should be regulated automatically. Two methods of doing this will be discussed here.

Modulating Valve. The usual, and best method of controlling the water is through the use of a modulating-type water valve. The poppet gradually opens and closes in response to a signal picked up by the sensing element, which is immersed in the hydraulic oil at a representative point. The sensing element is connected to the main valve by means of a capillary tube. When ordering a modulating valve for the control of water into a heat exchanger, the user should specify the valve to open on temperature rise. Some manufacturers term this a *reverse-acting* type. If ordering the valve for shunt control of water or of oil to bypass a radiator cooler, the user should specify the valve to close on temperature rise. This is sometimes termed a *direct-acting* type.

The modulating control is limited to a remote distance of 6 to 10 ft between sensing element and water valve. On longer runs the air temperature surrounding the capillary tube influences the control and it becomes less accurate. The capillary connecting tube should be protected from heat (or cold) sources by means of insulation or a heat shield.

Solenoid Valve. Where for some reason there is a separation of more than 10 ft between temperature-sensing point and water control point, a solenoid valve may be used to turn the water on and off in response to a temperature switch. Although this method is not as conservative with the water supply and is not as accurate in holding temperature, it must sometimes be used.

The solenoid valve must be suitable for water use and will usually be of the pilot-operated type. The installation recommended is to install a throttling valve such as a needle type between the solenoid valve and the heat exchanger, and to adjust the valve to limit water flow to a reasonable value when the solenoid valve is open. The best setting of the needle valve can be determined experimentally while operating with a full heat load. It should be throttled as much as possible and still allow enough water flow to hold oil temperature to the desired value. The needle valve should be downstream from the solenoid valve.

7.3.4 Tips on Heat Dissipation by Radiation

The most economical heat transfer system will carry away heat gained in functional operation through the tank, conductors, and components. Assuming that the hydraulic system has been carefully engineered and that a certain amount of heat will be necessary for proper functioning of the equipment, some installation procedures can aid in heat dissipation. The following list of suggestions is useful in making maximum use of heat dissipation by radiation:

1. Provide a light sheet-metal hood over the electric motor so that the air passing through the motor also passes by the pump to pick up heat radiated from the pump.
2. Keep the tank and plumbing away from steam lines, cleaning lines using hot water, cleaning tanks, and other sources of heat.
3. Insulate the platens of steam presses and other equipment from the ram or other hydraulic parts to minimize objectionable heat transfer.
4. Do not operate at fluid pressures higher than necessary. Keep fluid pressures as low as is consistent with good machine functioning.
5. Keep line sizes large enough to ensure a minimum of frictional losses and subsequent heating.
6. Baffle the reservoir so that fluid will be forced to flow along the outside walls of the tank.
7. Keep the bottom of the tank off the floor so that air can flow around all tank surfaces.
8. Do not pile objects against the tank. They can stop airflow to the tank surfaces.
9. Remove all debris from the top of the tank that may be acting as a natural insulator.

10. Consider light metal lines for return flows to increase radiation capabilities.

7.4 MAINTENANCE AND SERVICE

7.4.1 Replacement of Zinc Anodes

Most brands of shell-and-tube heat exchangers come equipped with replaceable zinc anodes. These are screwed into one or both end bonnets in contact with the fluid that circulates through the inside of the tubes. This should be the water circuit when the exchanger is used for cooling hydraulic oil. The purpose of these anodes is to protect the tubes from being eaten away by galvanic corrosion. Two-metal galvanic corrosion is a rapid corrosion of a metal due to its being electrolytically connected to another metal in the presence of an electrolyte. The electrolyte could be any fluild that would conduct an electric current. Water circulated through a heat exchanger is nearly always a conductor of electricity because of impurities in it. The cause of galvanic corrosion is the difference between the individual tendencies of metals to be acted on, or to combine with, electrolytic materials.

By placing zinc anodes in the water, the galvanic corrosion takes place on the zinc, which is gradually eaten away. These anodes should be removed and inspected every few months, or oftener if found necessary. Spare anodes should be ordered from the heat exchanger manufacturer and always kept on hand as spare parts. Anodes should be replaced when they have been consumed to less than half their original volume.

7.4.2 Cleaning Shell-and-Tube Heat Exchangers

Fixed Tube Bundle

The inside of the tubes can be cleaned by removing the end bonnets, then using a suitable wire brush or cleaning rod running through each individual tube. The construction does not permit removing the bundle from the outer shell, so the outside of the tubes cannot be mechanically cleaned. The best that can be done is to use a solvent inside the shell if cleaning is deemed necessary. Should steam be available, it can be used for cleaning the outside and the inside of the tubes. It is not necessary to clean the inside surface of the shell since no heat transfer takes place from this surface. The fact that the inside of the tubes can be cleaned makes it normally desirable to flow the fluid through them which is most apt to be fouling. This will generally be water on oil-cooling applications.

When making the original heat exchanger selection, use a model with small-diameter tubes if the water supply is clean. This gives the best heat transfer for a given-size exchanger. If the water is apt to foul the tubes, select a model with larger-diameter tubes so that it can more easily be cleaned from time to time. Water with a high mineral content will usually deposit a coating over the tube surface in the course of time, reducing heat transfer capability.

Removable Tube Bundle

The construction of these units is such that the entire tube bundle can be removed from the shell for easier cleaning. These types of heat exchangers are more costly than the fixed-tube-bundle models, and the ability to clean outside the tubes is not of particular advantage on hydraulic oil applications since the circulated oil is very clean and does not dirty the shell. They are of great advantage when both fluids may cause deposit formations.

7.4.3 Cleaning Air-Oil Heat Exchangers

Dirt that may lodge in the radiator core or fins of these units may be removed with a brush or with an air blow gun with the airstream directed opposite that of airflow from the fan. The important thing is to be aware of dirt buildup and to remove it promptly. Check to see that objects which would restrict the airflow are not allowed near the radiator core.

7.4.4 Checking for Leaks

Shell-and-tube heat exchangers may occasionally develop a leak between the shell and tubes. If water leaks into the oil, some of it may emulsify with the oil, giving the fluid a milky appearance. The excess water will settle out.

To test for water in the oil, draw a small fluid sample and allow it to stand overnight. This test will detect relatively large amounts of water. Small amounts of water can form a relatively permanent emulsion that does not readily settle out.

8
Hydraulic System Fluids: Selection and Care

Leading hydraulics system designers regard hydraulic fluids as the single most important group of materials in hydraulic systems. The fluid is literally the lifeblood of the system, the one element that ties everything together. With proper selection and handling of the fluid, most of the potential problems with the system can be prevented and the system can function more nearly as it was designed.

Mistakes and oversights in selecting, storing, and installing the fluid or in maintaining the hydraulic system can cause no end of trouble for operators and maintenance people. According to fluid power industry spokesmen, between, 70 and 85% of all hydraulic system problems are directly related to improper choice or handling of hydraulic fluids.

Heat and contamination degrade fluids while they are attacking other system components: pumps, seals, cylinder walls, and so on. The result may take some time to surface, but when it does, you can face costly downtime while the cause is determined and a remedy applied.

People in the industry believe that fully 85% of all hydraulic fluid ever installed leaks out: either slowly or in major line breaks or failures of fittings, seals, and the like. Besides the obvious wastefulness, fluid jets, sprays, and gushers can be very dangerous. Some are capable of penetrating the skin or damaging the eyes and other organs. One expert estimates that 7 million barrels of hydraulic fluids are lost each year through leaks and line breaks. Besides causing human hazards and loss of production, fluid leaks are a major cause of fires when ignition sources are nearby. Growing recognition of the serious fire-risk problem has led to the widespread use of fire-resistant fluids in many high-hazard areas.

There is no such thing as a universal or ideal hydraulic fluid. One major reason is the imposing list of fluid characteristics considered important by users, system designers, and manufacturers. For that reason, selection of the proper fluid for a given application is virtually always a compromise. To do its job well, a hydraulic fluid must do at least six things:

1. Transfer fluid power efficiently
2. Lubricate the moving parts
3. Provide bearings in the clearances between parts
4. Absorb, carry, and transfer heat generated within the system
5. Be compatible with hydraulic components and fluid requirements
6. Remain stable against a wide range of possible physical and chemical changes, both in storage and in use

Resistance to oxidation is particularly significant. Burning, of course, is an oxidation process. Slower oxidation reactions give rise to fluid degradation with resultant formulation of such troublesome reaction products as sludge, varnish, and gum, or with the formation of corrosive fluid that can attack metallic components. Other changes that need to be resisted include physical wear and pitting on pipe, tubing, and component surfaces; excessive swelling or shrinking of seals, gaskets, and other materials; significant viscosity variation; foaming; and evaporation.

Hydraulic fluid may travel through a system at velocities of 15 or 20 ft/sec or more. In compact mobile systems with small reservoirs, very high turnover may pass the fluid completely through the circuit two or more times a minute. On the other hand, where the reservoir is large, the fluid may get to "rest" a bit between cycles, allowing it to transfer more of its heat and to release entrained foam-causing air from the system.

8.1 FLUID TYPES AND PROPERTIES

The widespread use of the term "hydraulic oil" reflects the dominance of petroleum-base hydraulic fluids. Hydrocarbon oils protect well against rust, have excellent lubricity, seal well, dissipate heat readily, and are easy to keep clean by filtration or gravity separation of contaminants. Figure 8.1 tabulates the important fluid characteristics and Figure 8.2 details the characteristics desirable for the various fluid types.

Petroleum Oils

Petroleum oil is quite a serviceable industrial hydraulic fluid when specifically refined and formulated with various additives to prevent rust, oxidation foaming, wear, and other problems, as long as heat and fire hazards are not

Characteristic	Comment/definition
Favorable viscosity	Thickness or resistance to flow; needs to be suited to the requirements of the system
Viscosity-temperature relationship	Variation of viscosity with temperature, often described as viscosity index; minimal change of viscosity with change in temperature is desired
Chemical and environmental stability	Minimal change in storage or use
Good lubricity	Film formation and friction reduction
Compatibility with materials	Minimal effect, physically or chemically, on seals, gaskets, hoses, etc., as well as on metallic components
Heat transfer capability	High specific heat and thermal conductivity to carry and dissipate heat
High bulk modules	Stiffness or low compressibility of the fluid; stiff fluids permit stable handling of heavy loads without "sponginess"
Low volatility	Minimal evaporation and bubble formation reduces fluid loss and dangerous cavitation
Low foaming tendencies	Foam or entrained air reduces fluid stiffness, another cause of "sponginess"
Fire resistance	Resistance to ignition and flame propagation (no single test defines this complex property); flash point, fire point, and ignition temperature provide additional helpful information
Nontoxic and nonallergenic properties	To meet various regulatory and safety standards affecting workers, as well as acceptable odor

Figure 8.1 Characteristics of a good hydraulic fluid. (Courtesy of Sun Oil Co., Philadelphia, Pennsylvania.)

critical. But with all their useful properties, hydrocarbon oils do have one important drawback—they burn, and at temperatures a good bit lower than a number of other hydraulic fluid types. For that reason, several types of fire-resistant hydraulic fluids are available, most of them more costly than petroleum-base fluids, and several of which give away something in performance for the added resistance to burning. It should be noted that the term "fire-resistant" is far from meaning "fireproof." Under certain conditions, almost any fluid can burn. However, fire-resistant fluid resist ignition, whereas petroleum-base oils ignite quickly and propagate a blazing flame.

Fire-Resistant Fluids

Apart from the so-called water-additive hydraulic fluids, there are four principal types of fire-resistant hydraulic fluids. Following are some of their principal advantages and limitations.

Phosphate Esters. Sometimes called the straight synthetic fluids, the triaryl phosphate esters are excellent lubricants. In fact, they are the best among fire-resistant types. They have good fire resistance and are better at higher temperature ranges and at higher pressures than many other fire-resistent fluids. However, they are less useful at lower temperatures, their high specific gravity requires care in selecting pumps, and they are the most costly of all the industrial fire-resistant hydraulic fluids.

Water Glycols. These fluids are true solutions, not emulsions, containing a three-component mixture of water (35 to 40%), a glycol, and a high-molecular-weight water-soluble polyglycol. They have excellent fire resistance, good lubricating properties, and are available in a range of viscosities. However, they should not be used at temperatures higher than 120°F, and require periodic checks on water content and additive levels because of evaporation.

Water-in-Oil Emulsions. These fluids, sometimes termed *invert emulsions*, are intended for moderate-duty fire-resistent applications. They consist of 35 to 40% water dispersed in petroleum oil by means of an emulsifying additive package. They have adequate viscosity for hydraulic service; however, although superior to petroleum oil, water-in-oil emulsions do not have the inherent fire-resistance characteristics of either phosphate esters or water-glycol fluids. They require greater care to avoid contamination and should not be repeatedly frozen and thawed, which will cause the two fluid phases to separate.

Oil/Synthetic Blends. Where fire hazards are moderate, blends of phosphate esters and refined petroleum stocks are increasingly used, together with a coupling agent to stabilize the solution. They have good lubricating properties. Their fire-resistance characteristics reflect their composition, and depend largely on the ratio of phosphate ester to petroleum oil.

	Antiwear petroleum oil	Water glycol	Phosphate ester	Oil-synthetic blend	Water-in-oil emulsion (invert)	Oil-in-water emulsion (soluble oil)	Water-additive fluid
Specific gravity	0.85–0.89	1.1	1.15	1.0	0.96	1.0	1.0
Viscosity index	Good	Excellent	Poor to fair	Fair to good	Good to excellent	N/A	N/A
Lubricating quality	Excellent	Good	Excellent	Good	Fair to good	Limited	Limited
Corrosion protection	Excellent	Fair to good	Good	Good	Fair	Fair	Fair
Oxidation stability	Excellent	Good	Good	Good	Good	Good	Good
Low-temperature properties	Good	Excellent	Fair	Fair to good	Poor	Poor	Poor
Temperature range (F)	20–150	0–120	20–150	20–150	40–120	40–120	40–120
Vapor pressure	Low	High	Low	Low	High	High	High
Fire resistance	Poor	Excellent	Good	Fair	Fair	Excellent	Excellent
Spontaneous ignition temperature (F)	580	N/A	1100	840	N/A	N/A	N/A

Heat transfer	Good	Excellent	Good	Good	Excellent	Excellent	Excellent
Effect on conventional synthetic rubbers[a]	Minimal	Minimal	Severe	Severe	Minimal	Minimal	Minimal
Effect on conventional paints	None	Minimal to severe	Severe	Severe	Minimal to moderate	Minimal	Minimal
Metals attacked	None	Zinc and cadmium	None	None	None	None	None
Monitoring requirements	Viscosity, neutralization number	Viscosity, water content, pH	Viscosity, neutralization number	Viscosity, neutralization number	Viscosity, water content emulsion stability	pH, emulsion stability, oil content	pH, additive content
Relative cost ($)	1.0	4.0	7.0	4.0	1.5	0.05–0.10[b]	0.10–0.15[b]

Note: N/A, not applicable.

[a]Does not include fluorinated elastomers, which can be used with all types of fluid above.

[b]Cost assumes fluid is diluted with water and in the machine.

Figure 8.2 Comparison of hydraulic fluid characteristics. (Courtesy of Penton/IPC, Cleveland, Ohio.)

Fire-resistant hydraulic fluids need careful selection for compatibility with system components and even more care to keep them contamination-free than do petroleum-base oils. A careful evaluation of the fire hazard is important before using these fluids, both for safety and for economy.

For most users of hydraulic systems, the best rule of fluid selection is to follow the system manufacturer's recommendations. The fluid type that has been chosen meets the system's various requirements and demands, and the manufacturer has to stand behind the choice. Just as it is folly to mix two or more different hydraulic fluids in use, so is it simply asking for trouble to install a type of fluid other than the one suggested by the hydraulic equipment manufacturer.

Water

Water in hydraulic systems has had an odd history. Used in some of the very first hydraulic systems, it was soon replaced by petroleum oils because of the limitations of water: freezing, evaporation, corrosiveness, poor lubricity, and low viscosity. Now, with rising consciousness of fire risks, plus concern over petroleum supplies and the high cost of synthetic chemical fluids, water hydraulics is again attracting widespread interest.

The so-called water-additive fluids are of two basic types: oil-in-water emulsions, "soluble oils" (water is the continuous phase); and water plus chemical additives. The oil-in-water emulsions contain 2 to 10% oil dispersed in water, while the chemicals-in-water fluids contain only 2 to 5% chemicals and 98 to 95% water.

Making good use of the positive properties of water, such as its low cost, ease of disposal, and nonflammability, these substances are looked upon as the next-generation hydraulic fluids for quite a few industrial and mining hydraulic systems. Industry personnel report that manufacturers of pumps and other hydraulic system components are working hard to develop appropriate hardware.

8.2 MAINTAINING FLUID PERFORMANCE

Fluid filtration is very important to keep contamination levels down. Many of the newer fire-resistant fluids require even more attention to filtration than do conventional hydraulic oils. Selection of appropriate filters, placement in the proper parts of the system, and careful maintenance are all vital.

Checking the quantity of fluid in the system is important. Insufficient fluid can limit complete extension of the cylinders. Low fluid levels can also draw air into the system, creating "spongy" cylinder action and possibly setting up the conditions for one of the more costly hydraulic problems—pump cavitation and the resultant high or catastrophic wear.

Cavitation is a compression/expansion process in which tiny gas or vapor bubbles expand explosively at the pump outlet, causing metal erosion, leading ultimately to pump destruction. Some experts regard the action as implosion rather than explosion. Be that as it may, what the user will hear is a buzzing, rattling sound. This is an early warning signal of cavitation. The sound is that of a handful of stones tossed into the pump.

When it comes to checking fluid quality, operators should periodically inspect the cleanliness, color, thickness or viscosity, and perhaps the odor of the fluid. Beyond these checks, there are numerous standard laboratory tests that can be used to determine everything from foaming tendencies and load-carrying ability to oxidation and thermal stability. The key to checking fluids in service is to look for changes in fluid properties. Such changes represent warning signals that can indicate a need for corrective action.

Two other points bear repetition in any discussion of the proper maintenance of fluid in a hydraulic system. First, with the problems and costs associated with fluid leakage, good practice must include thorough attention to system integrity or leaktightness to minimize this difficulty. In some instances, operators have gone so far as to weld many or most connections, although this is often impractical.

Second, and bearing on the destructive effects of excessive heat on fluids and other system components, good operating practice limits the temperature range within which a hydraulic system operates. Proper design and maintenance of the fluid reservoir is important in this respect. Heat exchangers are becoming more commonplace in hydraulic systems to aid heat dissipating. Following is a list of pointers on handling hydraulic fluid.

1. Use only the type and grade of hydraulic fluid recommended by the hydraulic equipment maker or a fluid manufacturer; do not mix fluids.
2. Store fluid containers inside or under a roof and on their sides to minimize entry of water and dirt.
3. Clean the cap and the drum top thoroughly before opening.
4. Use only clean hoses and containers to transfer fluid from cans or drums to the hydraulic reservoir; use a fluid-transfer pump equipped with a 25-μm filter.
5. Use a 200-mesh screen on the reservoir filler pipe.
6. Replace fluid at the recommended interval; drain the system when it is warm to remove the maximum amount of contaminants.
7. Flush and refill the system exactly as recommended; be certain to fill to the proper level; overfilling is as troublesome as underfilling.

8. Service hydraulic filters and air breathers at recommended intervals.
9. Train operators in the proper use of quick-connect fittings to eliminate contaminant entry; fittings should be wiped clean before use and should be covered or cupped when not in use. The same care should be exercised in replacing hydraulic system components or in general maintenance.
10. Check the hydraulic system thoroughly to eliminate leaks or contaminant entry points; keeping hydraulic fluid out of plant effluent makes more sense than having to separate it later for proper disposal.
11. Never return leaked fluid to the system.

Figure 8.3 lists the viscosity for common hydraulic fluids and Fig. 8.4 tabulates the fluid viscosity range recommended for various hydraulic pumps and motors. A graphic presentation of the viscosity/fluid temperature relationship for a number of different fluids is presented in Fig. 8.5.

Fluid	Viscosity (SSU 100°F)
Plain water	30
High-water content synthetics	30
Water glycol	200
Phosphate ester	230
Oil-synthetic blend	300
Water-in-oil emulsion	450
Petroleum oil	215

Figure 8.3 Viscosities of common hydraulic fluids.

Component	Viscosity range (SSU)	
	Permissible	Optimum
Vane pump, 1200 rpm	80–1000	125–250
Vane pump, 1800 rpm	100–1000	120–250
Vane motor	80–1000	120–250
Radial piston pump	60–300	80–220
Axial piston pump or motor	40–350	80–200
Gear pump (industrial applications)	40–1000	120–250

Figure 8.4 Typical recommended viscosities.

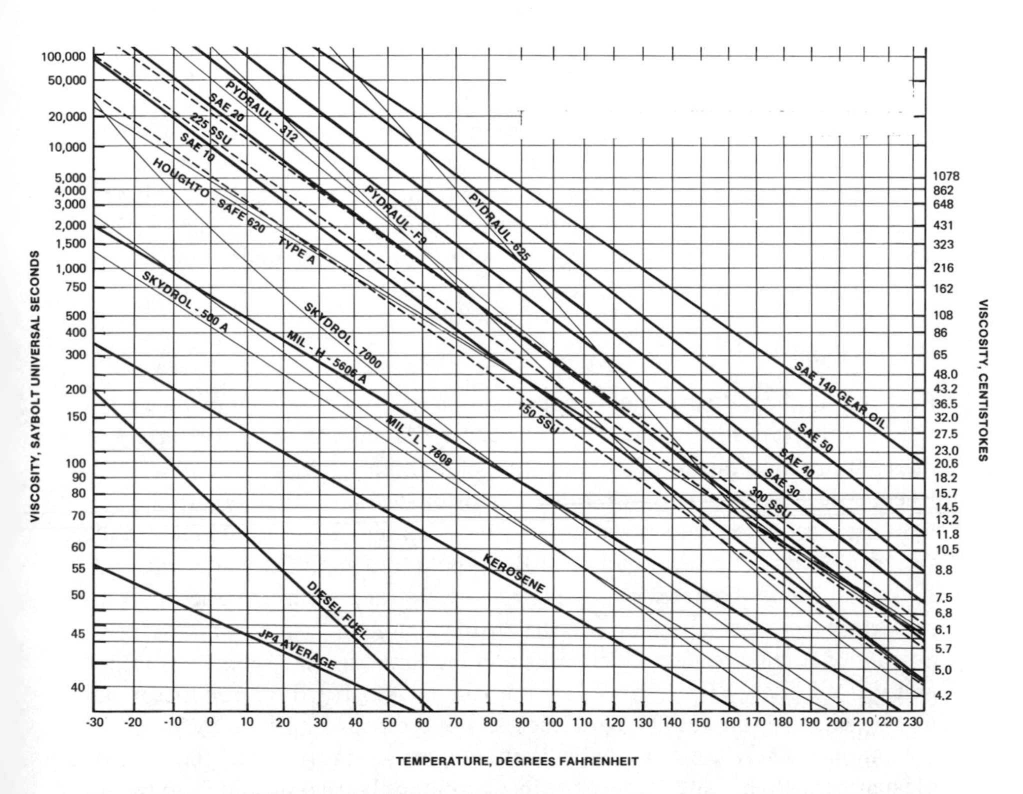

Figure 8.5 ASTM standard viscosity-temperature chart for petroleum products. (Courtesy of Sun Oil Co., Philadelphia, Pennsylvania.)

9

Fluid Conductors and Connectors

9.1 PIPE, TUBING, AND HOSE

Often, the method of marrying interacting components of a system is a minor consideration. We know that there must be some sort of fluid-carrying link between components, but the way it is done is sometimes considered arbitrary. This attitude may result in a system that is inefficient, unsafe, unattractive, and difficult to service. Conductors of a fluid power system are basically of three types: pipe, tubing, and hose.

9.1.1 Pipe

Pipe is a rigid conductor that is not intended to be bent or shaped into a desired configuration. Pipe can be manufactured and purchased in a variety of materials, such as cast iron, steel, copper, aluminum, brass, and stainless steel. Pneumatic systems generally require corrosion-resistant pipe. Hydraulic systems use steel pipe. Galvanized pipe is not recommended for use in hydraulic systems because the zinc coating of the pipe interacts unfavorably with oil.

The inside diameter of a pipe, or any fluid-carrying conductor, is an important consideration. For if the inside diameter is too small, a large amount of friction results, which translates into undesirable system inefficiency and wasted energy (see Fig. 9.1). Wall thickness of a fluid conductor determines its pressure rating. Wall thickness of a pipe is identified by its schedule number. There are 10 schedule numbers, ranging from 10 to 160. Schedule numbers 40, 80, and 160 are the most commonly used pipe in fluid power systems. Schedule 40 pipe has a wall thickness rated for low pressure, schedule 80 is for high pressure, and schedule 160 is for very high pressure (see Fig. 9.2).

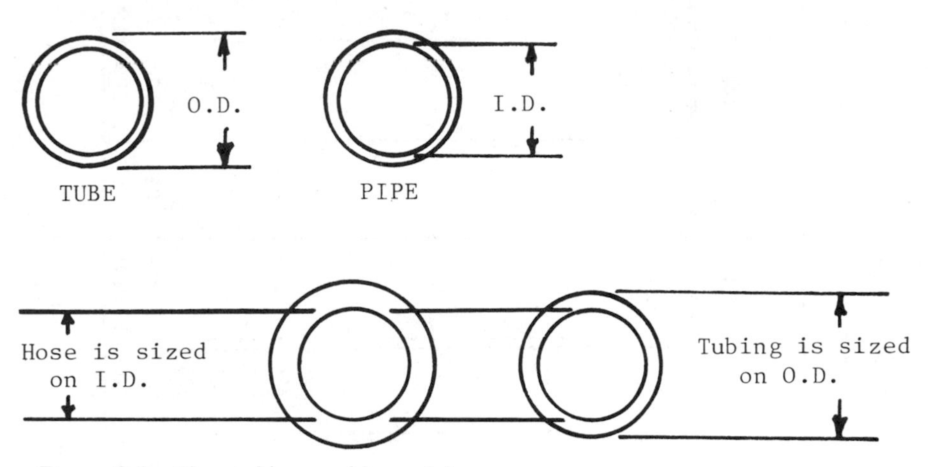

Figure 9.1 Pipe, tubing, and hose sizing.

Pipe connections are made by means of threaded joints. To join a pipe to a component or pipe fitting, a threaded end of a pipe (male end) is screwed into a female thread of a component or fitting. Pipe threads have another function besides joining, and that is sealing. To form a seal, pipe threads are tapered on the diameter 1/16 in. per inch of length. As a male pipe thread tightens into a female pipe thread, the metal-to-metal interference that occurs is intended to form a seal. In actual practice, however, temperature changes, vibration, system shock, or an imperfect thread may tend to destroy the sealing capabilities of the metal-to-metal joint. For this reason, thread sealants of various types are commonly used to help make and maintain the seal.

9.1.2 Tubing

Tubing is a semirigid fluid conductor which is customarily bent into a desired shape. The use of tubing gives a neat-appearing system, a system less susceptible to leaks and vibration, and a system whose conductors can easily be removed and replaced for maintenance purposes.

Tubing is made from a variety of materials, including steel, copper, brass, aluminum, stainless steel, and plastic. Hydraulic fluid power systems generally use steel tubing. Just as other tubular materials, tubing is measured by its outside diameter, inside diameter, and wall thickness. The inside diameter determines how much fluid flow the tubing can efficiently pass. Wall thickness determines the maximum pressure at which the material can be used for any given inside diameter. Tubing size is indicated by its actual outside diameter. For example, 1/8-in. tubing has an actual outside diameter of 1/8 in. Tubing inside diameter depends on wall thickness.

		Listed as Std., X or XX			Listed by schedule numbers								
		Inside Diameter											
Nominal Size	Pipe O.D.	Standard	Extra Heavy	Double Extra Heavy	Sched. 20	Sched. 30	Sched. 40	Sched. 60	Sched. 80	Sched. 100	Sched. 120	Sched. 140	Sched. 160
1/8	.405	.269	.215				.269		.215				
1/4	.540	.364	.302				.364		.302				
3/8	.675	.493	.423				.493		.423				
1/2	.840	.622	.546	.252			.622		.546				.466
3/4	1.050	.824	.742	.434			.824		.742				.614
1	1.315	1.049	.957	.599			1.049		.957				.815
1-1/4	1.660	1.380	1.278	.896			1.380		1.278				1.160
1-1/2	1.900	1.610	1.500	1.100			1.610		1.500				1.338
2	2.375	2.067	1.939	1.503			2.067		1.939				1.689
2-1/2	2.875	2.469	2.323	1.771			2.469		2.323				2.125
3	3.500	3.068	2.900				3.068		2.900				2.624
3-1/2	4.000	3.548	3.364				3.548		3.364				
4	4.500	4.026	3.826				4.026		3.826		3.624		3.438
5	5.563	5.047	4.813	4.063			5.047		4.813		4.563		4.313
6	6.625	6.065	5.761				6.065		5.761		5.501		5.189
8	8.625	8.071	7.625		8.125	8.071	7.981	7.813	7.625	7.439	7.189	7.001	6.813
10	10.750	10.192	9.750		10.250	10.136	10.020	9.750	9.564	9.314	9.064	8.750	8.500
12	12.750	12.080	11.750		12.250	12.090	11.934	11.626	11.376	11.064	10.750	10.500	10.126

Figure 9.2 Dimensions of welded and seamless steel pipe. (Courtesy of Parker Hannifin Corp., Cleveland, Ohio.)

Tubing is connected to system components and to other conductors by means of tube fittings. Basically, there are two types of tube fittings used: flared fittings and flareless fittings. A *flared fitting* consists of a body, sleeve, and nut. When using a flared fitting, the nut and sleeve are slipped over the tubing end. The tubing is usually flared to 37 or 45°, depending on the fitting used. When the nut is screwed onto the body, it draws the sleeve and tubing flare against the body, forming a seal.

A *flareless fitting* consists of a body, sleeve, and nut. To use a flareless fitting, the nut and sleeve are slipped over the tubing. Then the tubing is inserted into the fitting body, where it butts up against a shoulder. As the nut is screwed onto the body, the sleeve bites into the tubing; and as the nut is turned more, a ledge of tube material forms ahead of the sleeve, causing it to bow. The bowed sleeve acts like a spring, ensuring that the nut is held in place against vibration.

9.1.3 Hose

Hose is a flexible fluid conductor which can adapt to machine members that move. Hose is made up of three basic elements: inner tube, reinforcement, and cover (see Fig. 9.3). The inner tube is the lining of a hose, which contacts a fluid. Inner-tube materials are designed to be compatible with the fluid being conducted. Hose reinforcement is the fabric, cord, or metal layers that surround an inner tube. These elements give strength to the hose to withstand internal pressures and external forces. A hose cover is the outer hose layer. It is designed to protect the inner tube and reinforcing layers from chemical attack, mechanical damage, sunlight, and abrasions.

Hose size is customarily given by a *dash number,* which offers some identification as to its inside diameter. Dash numbers indicate sixteenths of an inch. A −8 is equivalent to 8/16 in. or 1/2 in. A −8 hose means that the

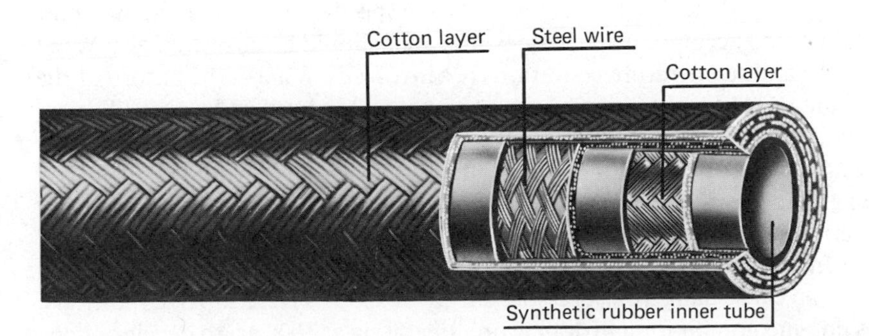

Figure 9.3 Hose construction. (Courtesy of Aeroquip Corp., Jackson, Michigan.)

inside hose diameter is 1/2 in. or a little less. The type and number of reinforcing layers of a hose determine under what system conditions it may be used. Hose pressure classifications are suction, medium pressure, high pressure, and very high pressure.

Hose is connected to system components and to pipe or tubing by means of hose fittings. Hose fittings are classified as permanent or reusable. With a *permanent* hose fitting, the hose is inserted into the fitting between nipple and socket. The socket is then crimped or swaged to hold the hose. Barbs on the outside diameter of the nipple ensure that the hose is securely held in place. *Reusable* hose fittings are screwed or clamped to a hose end. They can be removed from a worn hose and reassembled onto a replacement hose.

A *skive*-type fitting is a screw-on design for use on hoses with a thick outer cover. This cover is removed (skived) from the ends prior to fitting installation. A *no-skive*-type fitting is a screw-on design for use on hose with a thin outer cover. This cover does not require removal prior to fitting installation (Fig. 9.4). A *clamp*-type fitting is designed with a barbed nipple which is inserted into a hose. Two clamp halves are then bolted together to provide a leakproof grip.

9.2 CONNECTORS

9.2.1 Tube Fittings

Tubing is the logical choice for flow lines in many hydraulic and pneumatic systems. The reasons for this preference vary, but more often than not, the determining factor is the ease of assembly and disassembly associated with tubing lines.

The key assembly element for tubing connections is the tube fitting. Today, these fittings are produced in many types, styles, forms, and materials, offering a wide range of choice in solving specific tubing connection problems. To determine which fitting is best for a specific job, several factors must be taken into account. Typical questions that should be answered are: What pressure and temperature conditions are involved? What is the nature of the working fluid? In what kind of environment will the system be used? For example, consider a hydraulic system that is to be operated in a highly corrosive atmosphere. In a normal atmosphere, such a system would call for steel fittings and tubing, but under the circumstances, a change in fitting and tubing material is dictated.

In certain types of hydraulic systems, the operation of various components can cause dangerous pressure surges or set up severe shock conditions. It is not uncommon for a hydraulic unit operating at 1000 psi to develop shock loads of 5000 to 8000 psi when solenoid valves are used. In a similar situation, a 5000-psi system was actually found to peak at 30,000 psi, causing fail-

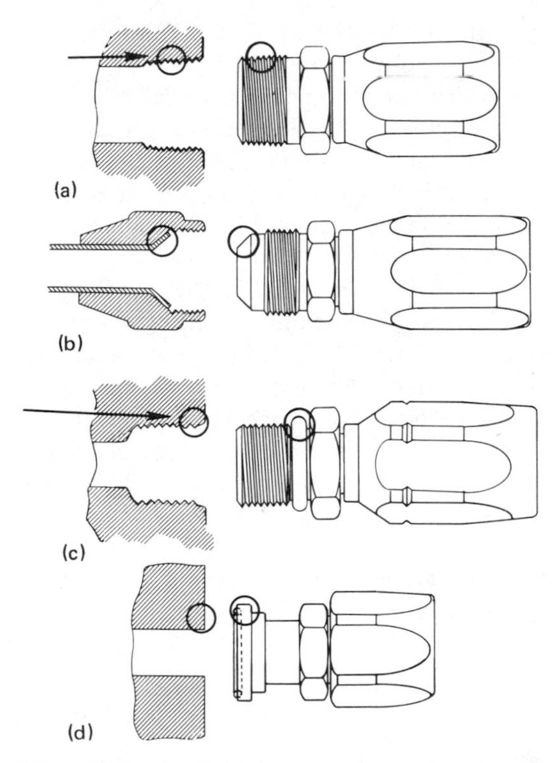

Figure 9.4 Available hose connector thread configurations. (a) Thread seal, metal to metal. (b) Flare seal to cone seat. (c) O ring seal. (d) Split-flange O-ring seal. (Courtesy of Aeroquip Corp., Jackson, Michigan.)

ure of a backwelded pipe system by stress cracking of the pipe. This type of failure can be both dangerous and costly. It could have been averted by a thorough examination of system operating conditions. Many more examples of this type could be cited.

Three basic types of fittings are in common use: (1) flared; (2) compression, or flareless; and (3) permanent (brazed, sweated, welded). Some of these fittings have been standardized. Many have not. In certain instances, operating characteristics and assembly features overlap, offering optional choices in fitting selection.

Flared Fittings

Fittings of the flared type can be further classified into two categories: the 45° style, which is commonly designated the *SAE type,* and the 37° style. The angle in each case refers to the slope of the flared tube surface with the centerline of the tubing. Both of these fitting styles have been standardized by the SAE (Society of Automotive Engineers).

The 37° flared fitting is a refinement of the 45° design. Two versions of this fitting are available. One is a two-piece unit. The other 37° fitting style is a three-piece arrangement, which is made up of an externally threaded body similar to that of the two-piece version, a shouldered sleeve insert with the 37° mating tapered seat on one end, and an internally threaded sleeve nut that is assembled on the body thread to lock the sleeve in proper seated position. When the sleeve is seated, the untapered end extends through the nut for added tube support (see Fig. 9.5).

The three-piece design is a more rugged construction that is suited for high-pressure service and severe operating conditions, including vibration, mechanical strain, and elevated temperatures. These fittings produce a fluid-tight joint that will hold beyond the burst pressure of conventional tubing lines. This fitting style is used almost exclusively for hydraulic and pneumatic systems in aircraft (see Fig. 9.5). The effectiveness and reliability of the fittings depends primarily on the quality of the tubing flare. The 37° fittings are available in steel, stainless steel, and brass.

In addition to the conventional flared fittings, a fitting classified as a *self-flaring* type is available. This fitting generally consists of four parts: body, 37° flare cone insert, compression sleeve, and nut. The aim of this design is

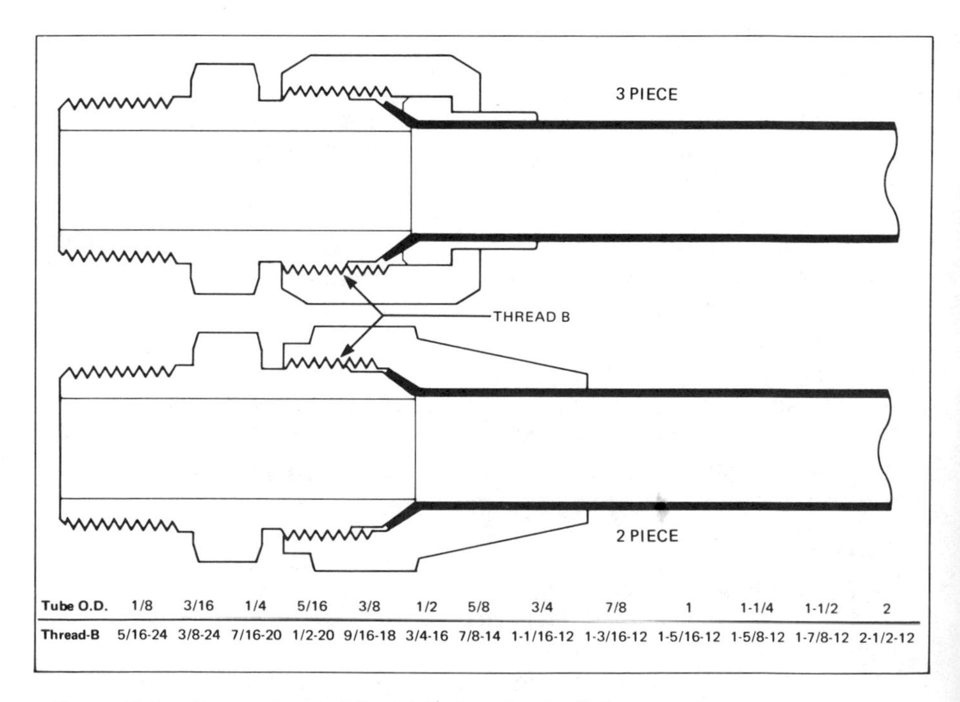

Tube O.D.	1/8	3/16	1/4	5/16	3/8	1/2	5/8	3/4	7/8	1	1-1/4	1-1/2	2
Thread-B	5/16-24	3/8-24	7/16-20	1/2-20	9/16-18	3/4-16	7/8-14	1-1/16-12	1-3/16-12	1-5/16-12	1-5/8-12	1-7/8-12	2-1/2-12

Figure 9.5 Conventional 45° or 37° flared tube fitting.

to have the fitting form a flare on the tube end during assembly, eliminating the need for a separate flaring operation. For assembly, the unflared tube end is inserted into the fitting. As the nut is tightened, the compression sleeve grips the tubing and forces it to flare out over the flare cone insert to form the required seal. A series of cone insert sizes have been developed for use with the fitting to accommodate tubes of varying wall thickness (see Fig. 9.6).

Compression Fittings (Fig. 9.7)

The need for a tube fitting that would eliminate the separate flaring operation was recognized some 60 years ago when the first compression (flareless) fitting was introduced. Although construction details vary considerably, operation of all of these fittings is based on the same general idea: As the nut is tightened on the body of the fitting, a compression grip is developed by a mechanical locking device to hold the end of the tubing in place.

From a production view, the compression fitting offers several attractive features. Tube preparation is kept to a minimum and usually can be handled with

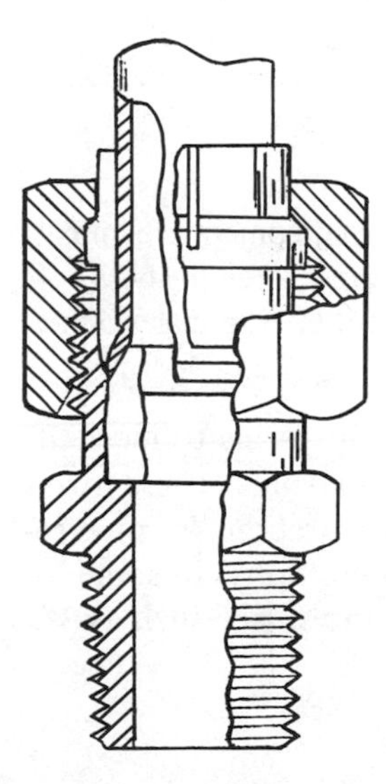

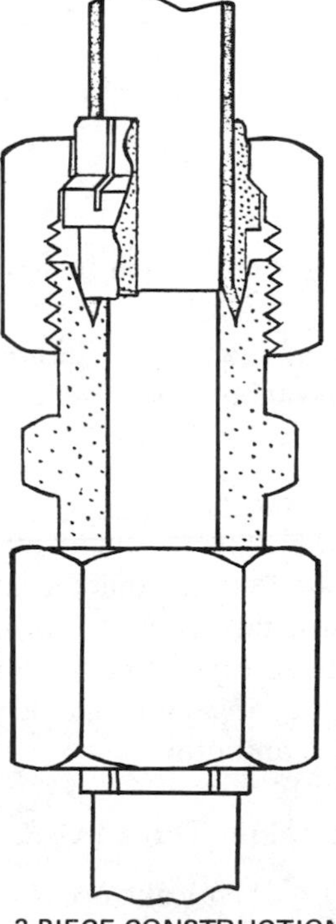

Figure 9.6 Self-flare fittings.

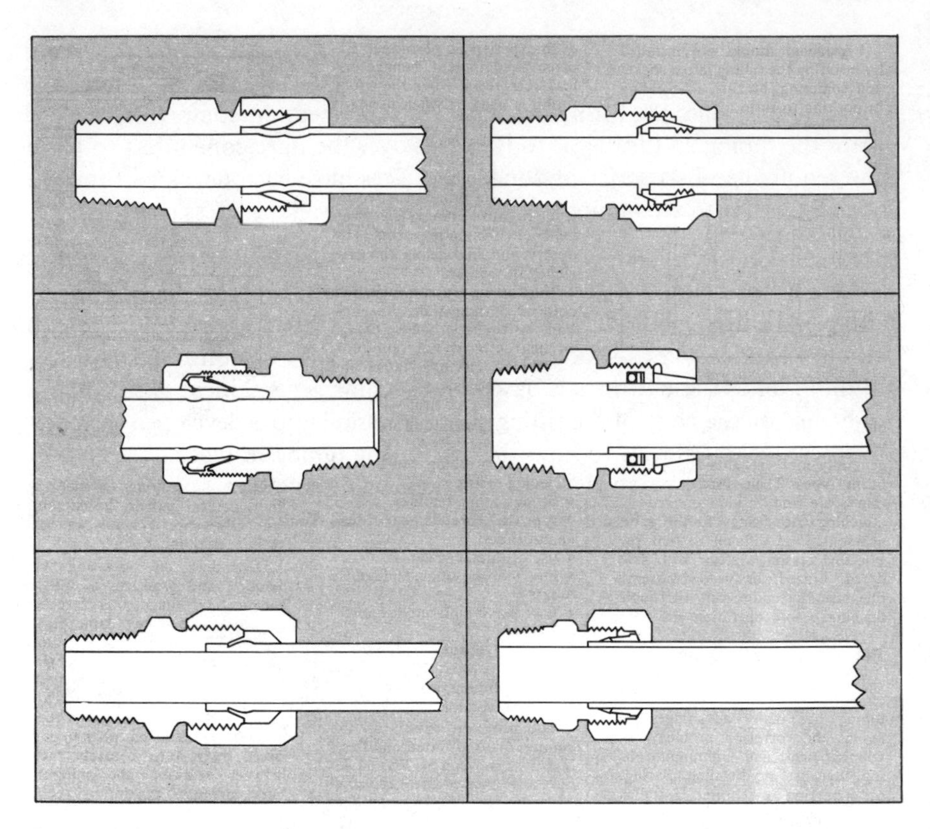

Figure 9.7 Compression fittings.

conventional equipment. The need for special tube flaring equipment is eliminated. The assembly procedure for tubing connections becomes a relatively simple operation that can be handled by semiskilled operators with little special training.

Fitting Forms

Each fitting type is usually available in many different forms to meet the various connection requirements encountered in tubing layouts. These forms include straights, elbows, tees, crosses, and other configurations for pipe-to-tube, tube-to-tube, tube-to-component, and port connections. In some instances, the range of available parts encompasses "jump sizes," such as 3/8 to 1/2 in., and others.

9.2.2 Straight-Thread O-Ring Fittings (Fig. 9.8)

In the past when leaks occurred in a hydraulic system, it was pretty much standard practice to grab a wrench and a can of pipe dope and set to work

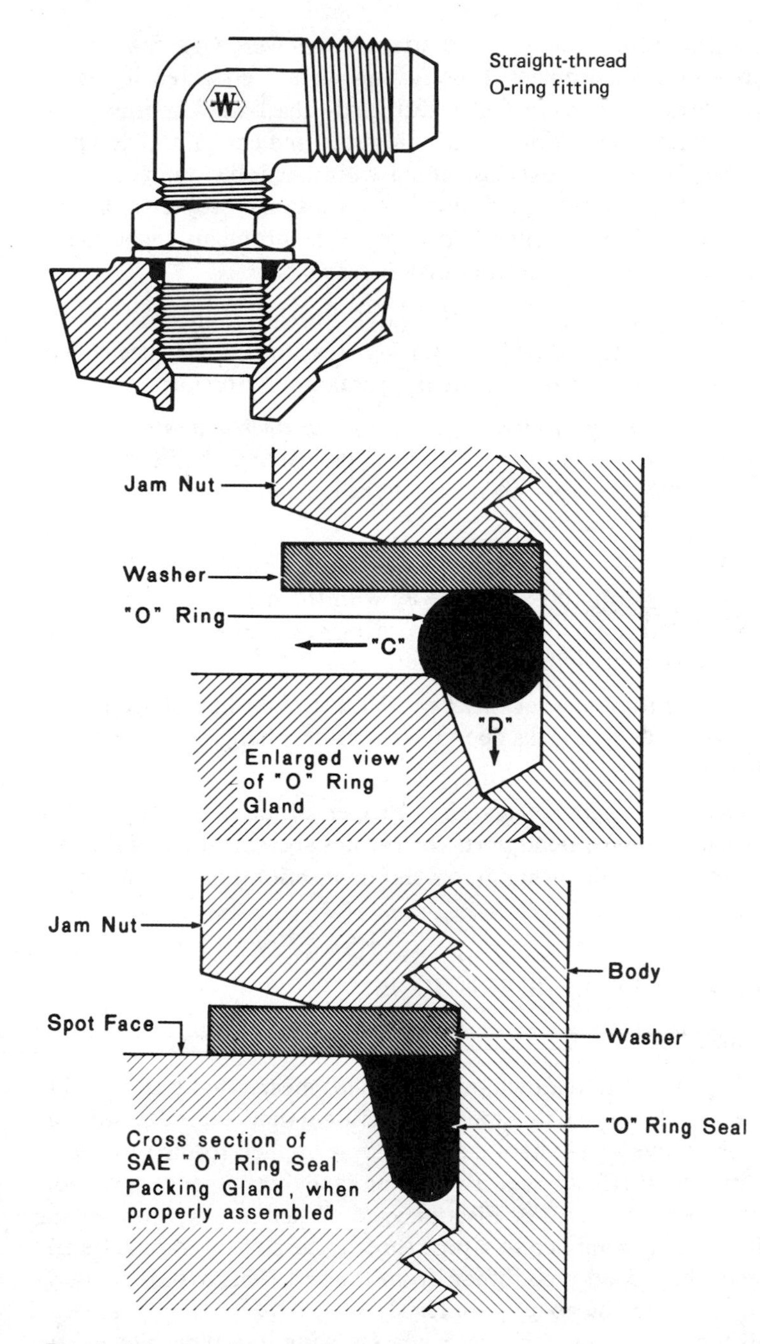

Figure 9.8 SAE O-ring boss design.

fixing the culprit. Unfortunately, these attempts to overcome leakage frequently compounded the problem. Pipe fittings were overtightened, resulting in damaged threads and in cracked or distorted valve bodies or other components in the circuit. In addition, positioning requirements of tapered pipe elbows and tees often forced overtightening with inevitable damage.

In an effort to solve these hydraulic circuit leakage problems, O-ring fittings were developed. These fittings overcome the problems encountered with pipe fittings, for at least five important reasons:

1. *No casting distortion*: SAE straight-thread connector fittings shoulder securely, providing a positive seal and a tight, mechanical joint without the hazard of leak-producing distortion.
2. *Exact positioning without "backing off" or overtightening*: With straight-thread elbow or tee fittings, you can turn the fitting to the exact position required, then tighten the locknut securely and a leak-proof seal is assured.
3. *"Pipe dope" eliminated*: Positive seal of straight thread fittings in an SAE O-ring boss does away with the need for messy "pipe dope" (or other auxiliary sealing devices) and the danger that it may contaminate the hydraulic system.
4. *No temperature or shock leaks*: With straight threads you eliminate the danger of leaks due to temperature changes or high-shock conditions.
5. *Mechanically rigid connection*: Proper torque can be applied because in straight-thread ports the fittings are working on full, perfect threads. This results in a solid connection that is completely rigid.

9.3 INSTALLATION CONSIDERATIONS

9.3.1 Pipe Installation

Threading of pipe requires removal of metal. This means that bare metal is exposed at the thread. When a threaded connection is made, it is advised that some sort of protective and sealant compound be applied to the threads to avoid corrosion and to effect a seal. Protecting pipe threads from corrosion will aid in case the joint has to be disconnected. When any type of sealant and protective coating is applied to a pipe thread, the coating should be applied only up to the second pipe thread from the end. This avoids contaminating the system with the coating material.

There is a precaution to observe when screwing in a pipe. Since pipe threads are tapered, the more a pipe is screwed into a component housing or fitting, the more likelihood there is of rupturing the housing or fitting from the wedging action of the joint.

9.3.2 Tubing Installation

In general, conductors of a system should be kept short and as straight as possible. This is one of the factors in an efficient system. However, tubing runs should not be assembled in a straight line. Bends in tubing absorb vibrations and compensate for the strain of thermal expansion and contraction. Some tightening, routing, and bending precautions that should be taken with tubing systems are presented on Figs. 9.9 to 9.11.

Line size (outside diameter, in.)	Flare nut size (across flats, in.)	Tightness (ft-lb)	Recommended turns of tightness (after finger tightening)	
			Original assembly	Reassembly
3/16	7/16	10	1/3	1/6
1/4	9/16	10	1/4	1/12
5/16	5/8	10–15	1/4	1/6
3/8	11/16	20	1/4	1/6
1/2	7/8	30–40	1/6 to 1/4	1/12
5/8	1	80–110	1/4	1/6
3/4	1 1/4	100–120	1/4	1/6

SAE 37° flare installation

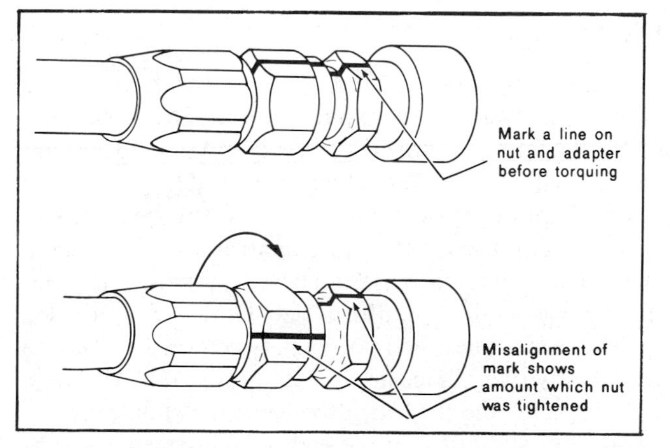

Figure 9.9 How to tighten flare-type tube fittings. (Courtesy of John Deere and Co., Waterloo, Iowa.)

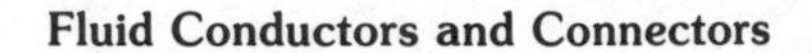

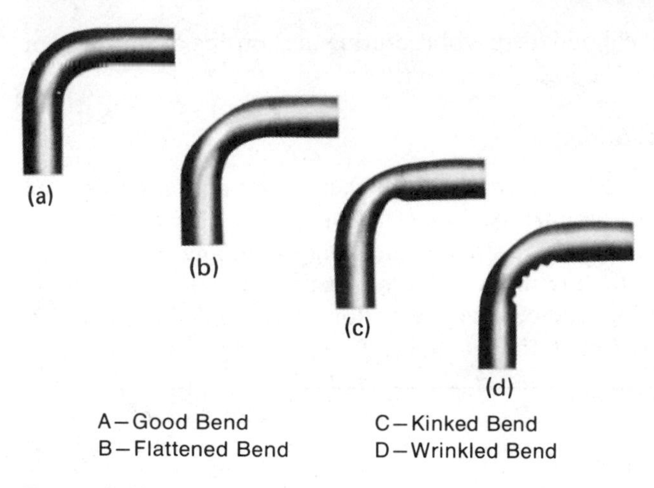

Figure 9.10 Tube bending problems. (Courtesy of Aeroquip Corp., Jackson, Michigan.)

9.3.3 Hose Installation

During operation, hose should not be scraped or chafed. This eventually weakens the hose. Hose should not be twisted during installation or system operation. Twisting a hose reduces life considerably and may help in loosening hose fittings. A twisted hose can easily be detected by examining the line along the hose cover. It should not appear twisted. Hose should not be stretched tightly between two fittings but should be allowed to sag a little. When hose is subjected to system pressure, it expands in diameter and shrinks in length.

9.4 PLANT AIR LINE PIPING SYSTEMS

Once the air has been compressed, cooled, and dried by the aftercooler and stored in the receiver tank, it is ready for the piping system. The three types of piping systems used most commonly are (1) the dead-end or grid system; (2) the unit or decentralized system; and (3) the loop system. The *grid* system is the simplest of the plant air piping systems. It is often called the *dead-end* system. It is simple in construction, consisting of a central main with small feeder lines and headers. The mains decrease in size away from the compressor, while feeder lines are generally of uniform size. Outlets are provided at convenient points on the feeder lines, and points between feeders may be served by cross-connecting any two adjacent legs (see Fig. 9.12). Although this is perhaps the simplest system and the least expensive to install, only one flow path is available and work stations near the ends of the system are subject to insufficient air supply (air starvation) when upstream demand is heavy.

The *unit* or *decentralized* system may consist of two or more grids, each with its own compressor. The individual units may be interconnected if desired. Compressors are closer to the system using the air, allowing shorter supply lines. This means lower pressure drops, resulting in a more uniform air supply and system pressure. Decentralized systems are more versatile than single grids and more easily adapted to changing requirements.

The preferred piping system is the *loop* system. This type allows the optimum conductor size and assures more equal distribution through the plant. This arrangement provides a parallel path to all work points. At points of heavy momentary demands for air, a receiver can be used to store the energy for peak demands, preventing serious pressure loss and air starvation (Fig. 9.13).

When laying out the circuit or distribution loop, give careful attention to reducing the number of fittings to a minimum and keeping bend radii as large as possible. Tubing, because it is smoother than pipe, provides better flow and reduces pressure drop. Its real advantages are the major reduction in the number of fittings required with fewer possible leak points and a smaller pressure drop. Pipe fittings generally have five times the pressure drop of an equivalent size 90° bend.

Distribution mains, feeders, and headers should have a small slope (perhaps as much as ¼ in./ft) to ensure that condensed water will be swept by airflow to drains. Water legs fitted with automatic drains should be provided at low spots on the main and on drop legs. Drop legs should be taken from the top of the main header, with the bend made of the same-size conductor as the leader line, or with a bend of 180° around a radius large enough to ensure low-pressure drops. By taking air from the top of the main, most of the dirt and water is excluded. Drains, preferably automatic, should be provided at the bottom of each drop line at the end of a water leg (Fig. 9.14).

Although pneumatic leaks do not usually promote a housekeeping problem such as that caused by hydraulic leaks, fixing them is still important. Leaks waste air and reduce system performance. In terms of air cost alone, leaks equivalent to a ½-in. hole can waste about 12 million cubic feet of air per month. At an average cost of 10 cents per 1000 ft^3 of compressed air, this leakage costs about $1200 per month. The major cost of leakage is the cost of electricity to keep the compressor running to make up the losses. Of course, the longer the compressor runs, the more maintenance is required, and finally the life of the compressor is reduced (see Fig. 9.15).

Frequently, serious leaks are audible. It is a good idea to "listen" for leaks when most of the plant is shut down for a shift change or at the end of a workweek. As another test, soapy water solutions or commercial leak-detecting fluid can be applied to suspected runs or connections. One very effective method for automatically detecting the presence of leaks and their

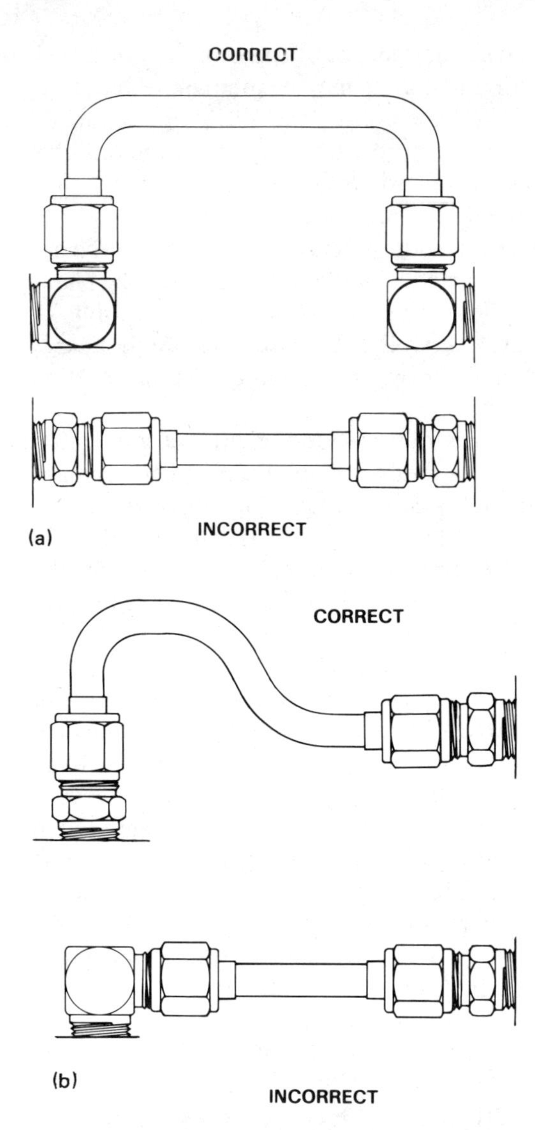

Figure 9.11 Tubing installations. (Courtesy of Aeroquip Corp., Jackson, Michigan.)

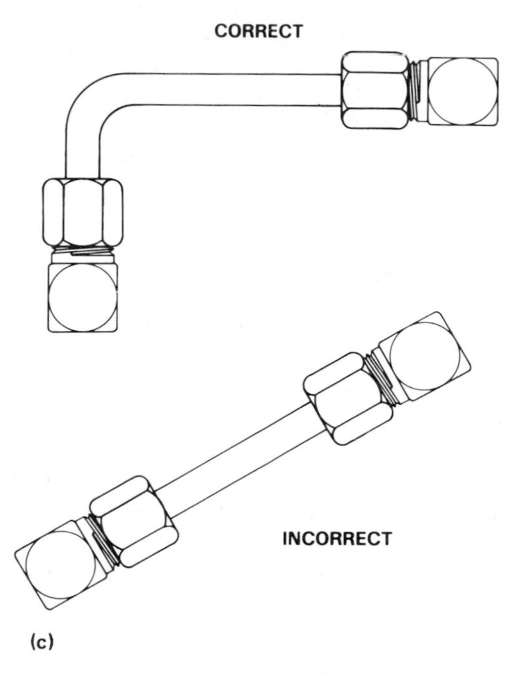

Figure 9.11 (continued)

order of magnitude is the two-clock system. The clocks are connected to the compressor system by automatic controls. One clock shows total elapsed time. The other is circuited through the motor control and runs only when the compressor is under load. Then at predetermined times, such as on weekends or at night when the air system is not in use, the clocks are started and run for a set period, usually 1 hr, and then stopped. Comparing the clocks makes it possible to determine the amount of air lost through leaks. Automatic equipment can make a permanent record.

Still another and perhaps faster procedure is to shut the system down and time how long it takes for system pressure to drop to a minimum value. This could be done at night when the plant is down. It is during this period that there is time to repair any leaks and fix other problems.

9.5 SELECTION, USE, AND CARE OF HOSE AND FITTINGS

Hydraulic hose assemblies are the lifelines of modern hydraulic systems. They carry fluids under tremendous pressure to deliver power where it is needed: power to move earth, drill through rock, and lift heavy loads. Hose and couplings have replaced bulky and expensive tubing systems in machinery, ena-

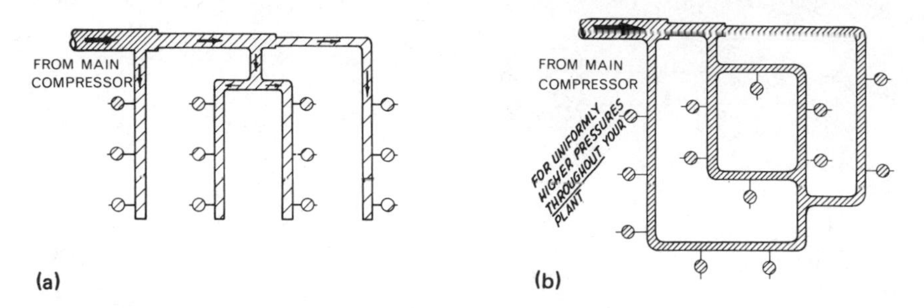

Figure 9.12 Compressed air distribution systems. (Courtesy of Parker Hannifin Corp., Cleveland, Ohio.)

bling complex production operations once thought to be impossible. Hose assemblies must perform under all types of adverse conditions. Some modern hydraulic systems reach pressures of over 10,000 psi and temperatures can be as high as 400°F. Constant shock and vibration adds to the punishment that hose assemblies must absorb. The outside environment also has an effect. Extreme heat or cold, caustic solutions, sunlight, and many other factors also affect hose life. Add to this the constant flexing, bending, abrasion,

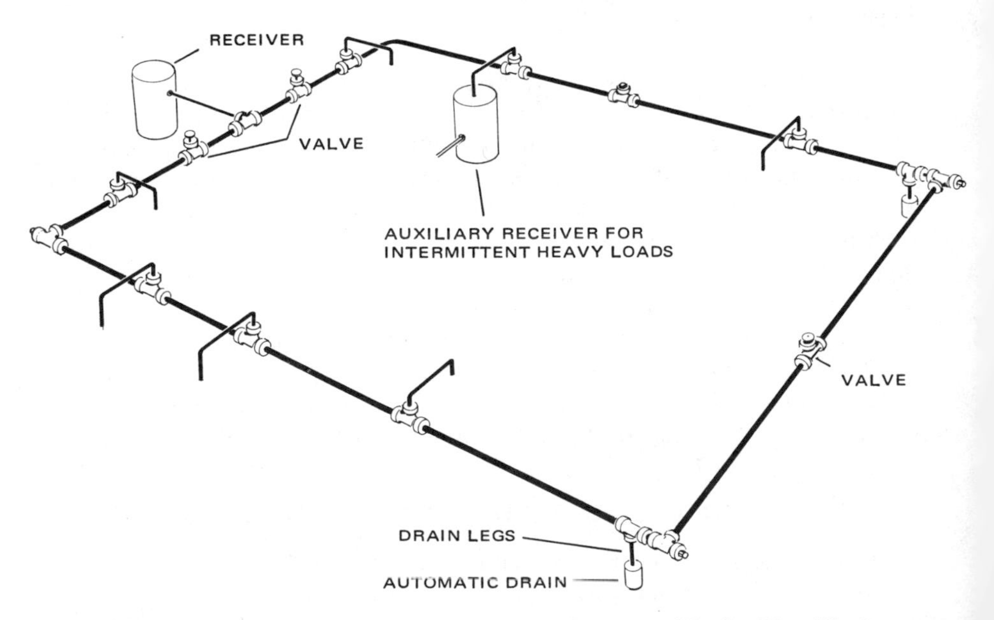

Figure 9.13 Loop-style air distribution system. (Courtesy of Parker Hannifin Corp., Cleveland, Ohio.)

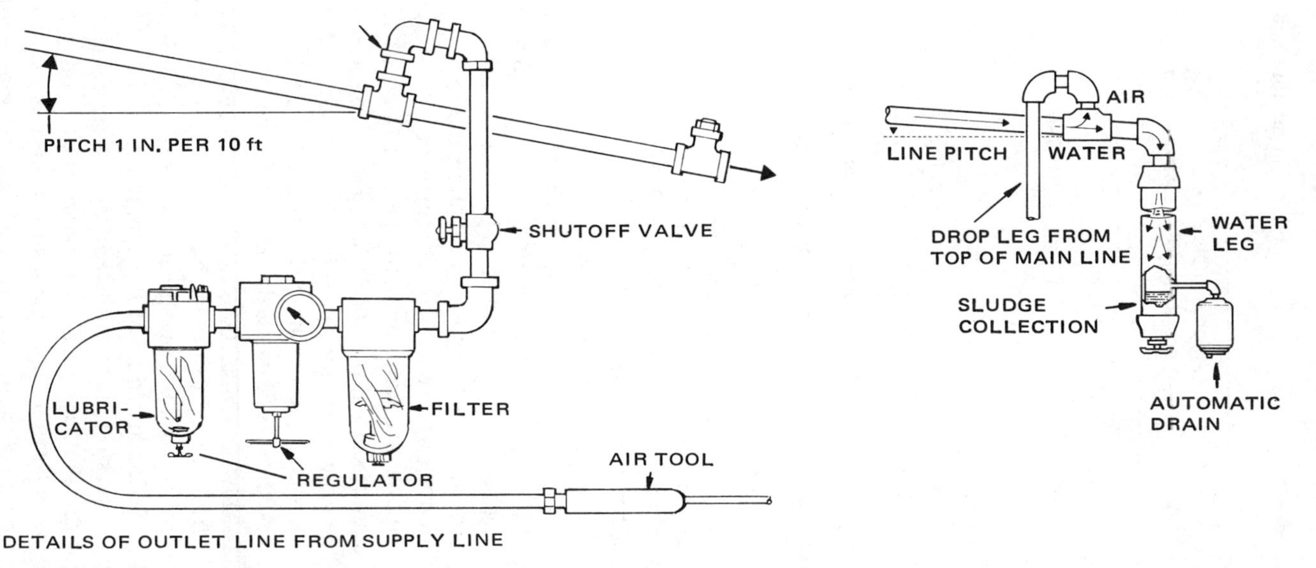

Figure 9.14 Recommended air drop line arrangement tee off top to supply all equipment. (Courtesy of Parker Hannifin Corp., Cleveland, Ohio.)

Diameter of leak (in.)	Air loss (cfm)	Approximate power loss (hp)	Air loss per month (1000 ft³)	Cost per month
1/32	1	1/5	43	6.5
1/16	4 1/4	7/8	180	27.0
1/8	17	3 1/2	730	109.5
1/4	70	14	3,000	450.0
3/8	150	30	6,500	975.0
1/2	270	54	11,700	1,755.0

Note: Based on cost of $0.15 per 1000 ft³ and orifice coefficient of 0.65.

Figure 9.15 Cost of air leaks. (Courtesy of Parker Hannifin Corp., Cleveland, Ohio.)

and all types of physical abuse a hose is subjected to and you can see why hose failures are the most common cause of hydraulic system problems. Leaks can also be real headaches. A steady drip can add up to thousands of dollars in oil over a short period of time.

While hose assemblies, like many other components, have a certain life span and are bound to wear out eventually, there are several things you can do to ensure maximum performance and longer life. We will discuss the importance of selecting the right hose and couplings for the job, following the correct assembly procedures, using good installation and routing techniques, and diagnosing failures to prevent their reoccurrence.

Hose and couplings are manufactured in a wide range of designs, each meeting certain requirements. If the hose selected does not meet the pressure requirements of the system, cannot withstand the heat, or is not compatible with the fluid used, you are in for an early failure (Fig. 9.16).

Hoses consist of three basic parts: the inner tube, which carries the fluid; the reinforcement, which provides the resistance to pressure; and the cover, which protects the hose from outside elements. The inner tube must be flexible and compatible with the fluid being carried. Synthetic rubber, thermoplastic, nylon, and Teflon are some of the compounds used. The reinforcement layer provides the strength to withstand the system's pressure. One or more layers of cotton, synthetic yarn, or wire are braided or spirally wound or wrapped over the inner tube. The spiral wire reinforcement provides better flexibility, more resistance to failure, and a greater maximum working pressure than does braid reinforcement, it is also more expensive to manufacture. The hose cover protects the reinforcement from abrasion and corrosion. The cover carries a name, part number, the hose size, SAE number or rating, and date of manufacture. This information on the cover is referred to in the industry, as the *lay line.*

The Society of Automotive Engineers (SAE) has recommended standards for the hydraulic hose industry. 100R numbers from R1 to R11 are used to indicate hose performance capabilities and construction. For example, 100R1 is rubber hose with one layer of wire-braid reinforcement. It is suitable for the most medium-pressure applications. 100R11 has a heavy wire-wrap reinforcement and is used in extremely high pressure systems.

Hose and coupling sizes are usually expressed in *dash numbers.* The inside diameter of the hose is converted to sixteenths of an inch, and the number of sixteenths becomes the dash number. A ½-in. ID would equal 8/16 in. or −8. A 1-in. ID would equal 16/16 or −16. Couplings or fittings are manufactured to very close tolerances. Each is designed to be used with a specific type and size of hose. There are both permanent and reusable couplings.

Permanent couplings are generally of one-piece design. These couplings grip the hose when the outer shell or ferrule is crimped or swaged. Die marks are visible on crimped couplings. Permanent couplings are less expensive and more compact than reusables, but when the hose fails, the entire assembly, including the couplings, must be thrown away. Reusable couplings come in three basic styles: push-on, screw-together, and clamp type. One advantage of these couplings is that they can be reused if the hose fails.

Hose size (in.)	1. Use single wire braid hose if system working pressure equals (psi):	2. Use double wire braid hose if system working pressure equals (psi):	3. Use spiral wire hose if system working pressure equals (psi):
1/4	3000	5000	
3/8	2250	4000	5000
1/2	2000	3500	4000
5/8	1750	2750	
3/4	1500	2250	3000
1	800	1875	3000
1 1/4	600	1625	3000
1 1/2	500	1250	3000
2	350	1125	2500

Figure 9.16 Selecting hose for various pressures. (Courtesy of John Deere and Co., Waterloo, Iowa.)

The connecting or threaded end of a coupling enables one to attach the hose assembly to a component in a hydraulic system. The socket accepts nipples with many different types of connecting ends. You can select a connecting end to match the port on the piece of equipment on which you are working. Be careful not to confuse 37° with 45° flares, or National Pipe Thread with SAE O-ring boss ports. Couplings must be compatible with the fluid being carried through them. Carbon steel couplings are acceptable for most applications, but when corrosive chemicals are present, stainless steel or brass couplings should be used.

While we are dealing with the subject of couplings, let us look at the basic causes of fluid system leakage. The basic causes of fluid system leakage with connections are:

1. Human error
2. Lack of quality control
3. Poor protection of fittings in handling
4. Difficulty in reaching fitting connections
5. Lack of training
6. Poor selection of materials
7. Improper design of system

A common cause of leakage in hydraulic hoses and hose assemblies is excessive twisting. This occurs most often on equipment where the hose connects two parts and one or both move. To find leaks, first pinpoint the location. To make sure that the leak is not up higher and draining down, wash down the leakage area and watch for the leak to reappear. You can place a paper towel above the connection and it will catch any fluid dropping from above. Next, you must determine if the leak is from a valve, motor, pump, shaft, a cracked casting, or if indeed it is from the fitting connection on the hose assembly. Remember that seepers or weepers can be hard to locate, and do not assume that it is the hose/fitting without first checking the rest of the system.

Leak problem areas on Dryseal Pipe Thread and SAE 45° and 37° fitting seats are due to the lack of tightening, human error, or just the lack of training. You cannot tell if the nut has been tightened by just looking at the connection. If it is more than finger-tight, you cannot tell from observation how much it has been tightened. A torque wrench is a good solution for this problem—but only when it is used. Furthermore, you must rely on the user of the wrench to be sure that the torque wrench becomes a common practice in that shop. Also, the employee must depend on his or her memory—has he or she tightened all the connections?

One foolproof method of tightening a fitting is to tighten the nut until it bottoms the seat, by hand, then mark a horizontal line on the nut and extend it on to the adapter or other connection. Now, using a wrench, rotate the nut to tighten an amount equal to one-third or more of a revolution, depending on the size of the fitting. The misalignment of the marks previously made will show how much the nut was tightened, and best of all, that it was tightened.

Now, what must be done if the joints leak after they have been tightened? First, disconnect the line and check for the following:

1. Are any foreign particles in the fittings? If so, wash them off.
2. Does the fitting have a cracked seat caused by overtightening? If it does, you will have to replace the fitting and start over.
3. Are the threads compatible? If not, select fittings that are.
4. Check for deep nicks in the seats; if present, replace fittings.
5. Look for excessive seat impression; this indicates too soft a material for the pressure involved. Threads will stretch under high pressure, so you may want to use a fitting made out of different and/or stronger material.
6. Check for chatter or tool marks on the seats. If present, replace the fitting.

Remember, on many of the leakage problems on these types of connections the indications will not show up for a number of service hours.

Leak problem areas on SAE straight-thread O-ring seals may be caused by the following:

1. Elbows loosen up after a short service
2. O-ring leakage after short service or after long service
3. Leakage right after startup

All of these problems may be caused by either human error or faulty parts. The cures for these are: (1) jam nut and washer must be to the back side of the smooth portion of the elbow adaptors; (2) lubricate the O ring—this is very important; (3) thread into port until washer bottoms onto the spot face; and (4) position elbows by backing up the adapter and tighten the jam nut.

O-ring lubrication is important because the fitting engages to a point where the O ring touches the face of the boss into which it is being fastened. Lubrication on the O ring will permit it to move in the downward direction into the chamfer or recess. When the O ring and the boss are dry, rotary motion of the assembly can cause friction and the O ring will tend to turn and twist. Also, the jam nut and washer cannot bottom fully if the O ring is between the washer and the face of the boss.

The SAE four-bolt split flange is a face seal. The shoulder, which contains the seal, must fit squarely against the mating surface and be held there with even tension on all bolts. The shoulder protrudes past the flange halves; this is to ensure that the shoulder will make contact with the mating accessory surface before the flange does. The flange halves overhang the shoulder on the ends so that the bolts will clear the shoulder.

This connection is very sensitive to human error and bolt torquing. Because of the shoulder protrusion on the flange overhang, the flanges tend to tip up when the bolts are tightened, first on one end, then on the other, in a seesaw manner. This pulls the opposite end of the flange away from the shoulder and when hydraulic pressure is applied to the line, it pushes the shoulder back into a cocked position.

To cure this, all bolts must be installed and torqued evenly. Finger tightening with the use of a feeler gauge will help to get the flanges and shoulder started squarely. When the full torque is applied to the bolts, the flanges often bend down until they bottom on the accessory or the O-ring seal. This also causes the bolts to bend outward. Bending of the flanges and bolts tends to lift the flanges off the shoulder in the center area. In order to prevent this from occurring, one should lubricate the O ring before assembly. All mating surfaces must be clean and all bolts must be torqued evenly.

To determine what hose and coupling you need, look in the appropriate parts or service manual for the equipment that is being worked on. Here you should be able to locate the part number and/or more important, the performance requirements, to help you make the proper selection (see Fig. 9.17). If a hose fails simply because of old age, check the lay line for the SAE 100R rating number and hose size. Your job becomes easy here; just replace it with the same type of hose.

For best performance, a hose should be selected to meet the service conditions under which it is to be used. Before deciding on the size and type, you must consider the pressure, temperature, fluid type, and possible contaminants. Catalogs supply information on operating and burst pressures, temperature ranges, bend radius, and fluid compatibility which can make selecting a replacement hose easier. If you are not sure which hose to pick, select a higher-pressure hose than you feel the job calls for. It is better to spend a few extra dollars than to risk an early failure.

Once you have selected the correct hose and couplings, they must be assembled properly. First, hoses must be cut to the proper length. The hose length equals the overall length of the assembly minus the distance the couplings extend beyond the ends of the hose. The cut must be clean and square

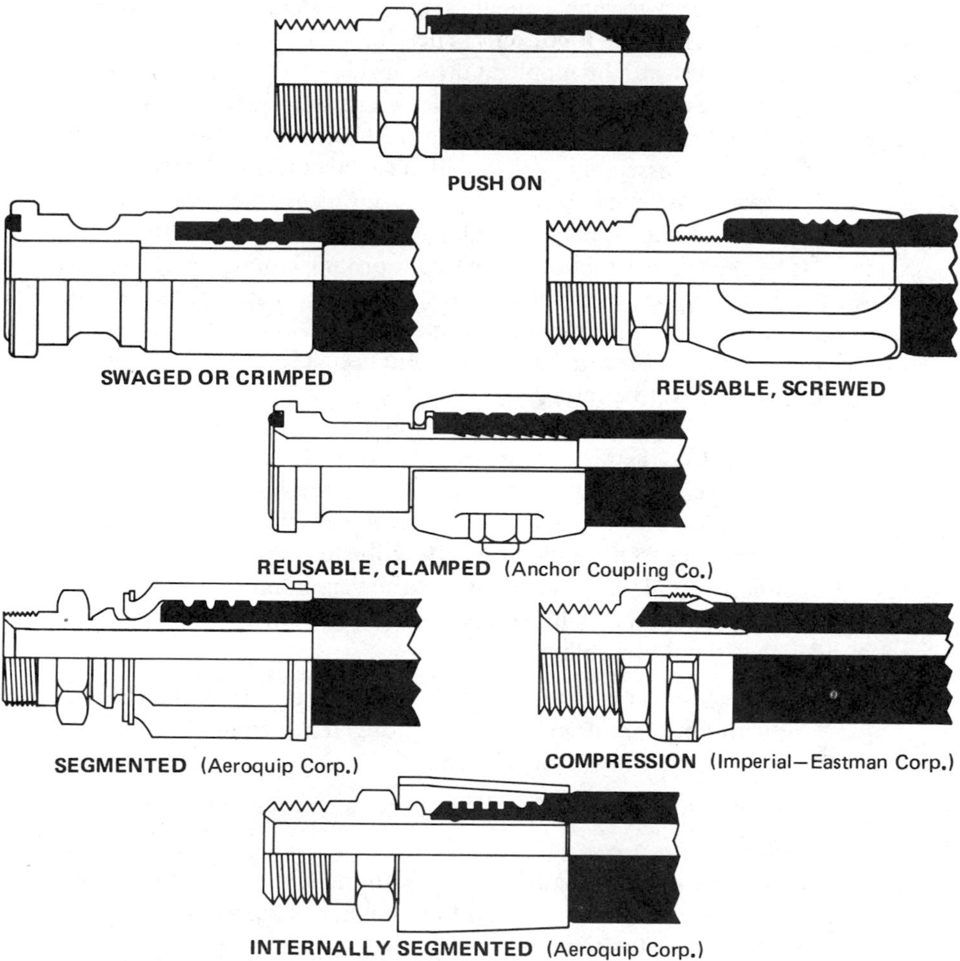

Figure 9.17 Typical hose fitting styles.

in order to seat properly. Be careful of the hose ends; those exposed wires are sharp! Under pressure, hoses can shorten up to 4% of their total length or lengthen by as much as 2%. A 50-in. assembly, for example, can shorten as much as 2 in. Provide enough slack to account for this change in dimension.

To assemble push-on couplings, place the coupling in a vise and lightly oil the nipple and the inside of the hose. Then grip the hose a few inches from the end and push it on all the way. To remove the coupling, slit the end of the hose, being careful not to nick the shank. Grasp the hose near the coupling and snap it off.

To assemble screw-together couplings, place the socket in a vise and turn the hose into the socket until it bottoms. Then back it off one-fourth to one-half turn. Next, lubricate the nipple and the inside of the hose with heavy oil. Then thread the nipple into the socket, leaving a small gap. Always make certain that the lubricating oil is compatible with the tube. Some couplings require a mandrel for assembly. It is threaded to the nipple and prevents the end of the nipple from damaging the inner tube during assembly. Some couplings are designed to bite directly into the wire reinforcement. For these, the cover must be skived or peeled off. The notch or knurl on the coupling can be used as a guide for the skiving length. To skive a hose, first mark the length, then place it in a vise and cut through the cover with a knife or with a hacksaw, using backstrokes. Make a diagonal cut and peel the cover off with pliers. Be sure not to damage the reinforcement.

To assemble clamp-type couplings, apply heavy oil to the nipple and the inside of the hose, then insert the nipple into the hose and bolt the segments together over the hose. Be sure to tighten the bolts evenly. While crimped assemblies are often assembled by the manufacturer, field crimpers allow hose assemblies to be made up on the job site.

Proper hose installation and routing is also very important. Careful attention to a few details here will prevent many early failures. Following are a few rules that should be followed.

1. Route for good appearance. This may not seem important, but a neat routing job often means less hose, fewer connections, more accessibility, and fewer problems.
2. Route hoses together if they have the same pressure rating in the system.
3. Do not twist the hose during installation. A slight twist in a hose can reduce its life by 70% and the lay line is a good guide to see if the hose is twisted.
4. Keep hose assemblies clear of sharp edges and moving parts. Use clamps or spring guards to keep the cover from being damaged.
5. Where hoses must flex, they should be long enough to allow for the required movement.
6. Route hoses so that they bend freely, without twisting.
7. If hoses crisscross, a sawing action can destroy their covers. Use clamps or guards to prevent this.
8. Always route hoses away from hot manifolds and other sources of heat. Use sleeves or shields to protect hose covers from extreme heat.

9. The use of adapters is essential in making connections more accessible and neat. Proper use of adapters eliminates unnecessary bends and reduces the amount of hose needed for the assembly.

The final point we would like to discuss here concerns diagnosing hose failures and methods to prevent their reoccurrence. When a hose assembly fails, do not just replace it—study it. Determine what caused the failure. This will help you when selecting a replacement or when installing and routing the new assembly. For example, a hose may have burst and an examination shows that the wire braid is rusty and the cover is worn. Chances are that abrasion of the cover allowed air and moisture to rust and weaken the reinforcement. Reroute or protect the replacement hose to keep this from reoccurring. An inspection of another burst hose by stripping back the cover reveals random wires broken along the entire length of the hose. High-frequency pressure impulses have caused friction and wear at the crossover points in the braid. A spiral wrap reinforcement eliminates the crossover points and should be used where high-pressure impulses are a problem.

A clean burst usually means that a hose has been subjected to pressure that exceeds its rating. A higher-pressure hose may be the answer; however, first check the system to see if it is operating at the rated pressure. The problem could be a faulty relief valve or other system malfunction. The inner tube may show signs of swelling. Sections may be washed out. This usually means that the inner tube is not compatible with the fluid. Select a hose with a compatible inner tube material.

When a coupling blows off, it can mean that the wrong coupling was used, or that the coupling was not assembled or crimped properly. Remember to check the threads on reusable couplings. They can wear down after several uses and will not be gripping the hose properly. When a hose is stiff and the inner tube is cracked and hardened, this generally means that it was caused by excessive heat. Select a hose that can handle this heat or check the system to see why it is running so hot. In a case where the inner tube has broken loose from the reinforcement and has piled up at the end of the hose, the wrong hose is used in a vacuum application.

9.6 HOSE ASSEMBLY FAILURES

Everyone in maintenance encounters hose failures. Normally, there is no problem. The hose is replaced and the equipment goes back on line. Occasionally, though, the failures come too frequently—the same equipment with the same problems keep occurring. At this point, the task of maintenance personnel is to determine and correct the cause of these repeated failures.

Failure to look into the problem, if the fault lies with the hose, will simply result in repeated loss of hose lines. If the problem lies with the equipment, failure to determine the cause could eventually result in loss of the equipment. A little effort in the beginning can avoid a big headache in the long run.

Every failure should be analyzed, even if that analysis is as basic as deciding that the failure was normal and acceptable. In that case, simply replace the hose line. However, if the failure rate is unacceptable, probc a little deeper to determine the cause of the failure and correct the situation. Hose failures fall into five major categories:

1. Improper application
2. Improper assembly and installation
3. External damage
4. Faulty equipment
5. Faulty hose

9.6.1 Improper Application

To investigate the most common cause of hose failures, improper application, compare the hose specifications with the requirements of the application. The following areas must be considered: (1) maximum operation pressure of the hose, (2) the recommended temperature range of the hose, and (3) the fluid compatibility of the hose. Check all of these areas against the requirements of the application. If they do not match up, another type of hose must be selected.

When contacting a distributor or manufacturer for help, provide all of the information needed to solve the problem. Use the following checklist when describing the problem.

1. Either send in a sample of the hose or provide the part number and date of manufacture. This information is supplied on the lay line of the hose.
2. Describe the type of equipment on which the hose is used and the location of the hose on this equipment.
3. Provide the brand name and type of the fluid used with the hose.
4. Give the maximum and minimum temperature, both internal and external, at which the hose operates. Remember that temperatures can vary widely from one part of the equipment to another. Try to get a reading as close to the failed hose as possible.
5. If the hose is bent, provide the bend radius along the inside of the curve or send along a tracing of the curve on a piece of paper. If the hose bends in more than one plane, say so.

6. Provide the flow rate (gpm) through the hose.
7. List the maximum pressures, both static and transient, to which the hose is subjected.
8. Describe the environment in which the hose operates.

By providing complete information, better and faster answers will result.

9.6.2 Improper Assembly and Installation

The second major cause of premature hose failure are improper assembly and installation procedures. This can involve anything from using the wrong fitting on a hose, to poor routing of the hose. The solution to this cause of failure is to become thoroughly familiar with the proper assembly and installation of hoses and fittings.

9.6.3 External Damage

External damage can range from abrasion and corrosion, to a hose that is crushed by a lift truck. These are problems that can normally be solved simply once the cause is identified. The hose can be rerouted or clamped, or a fire sleeve or abrasion guard can be used. In the case of corrosion, the answer may be as simple as changing to a hose with a more corrosion-resistant cover or rerouting the hose to avoid the corrosive element.

9.6.4 Faulty Equipment

Too-frequent or premature hose failure can be the symptom of a malfunction in your equipment. This is a factor that should be considered, since prompt corrective action can sometimes avoid serious and costly equipment breakdown. The corrective action here would be to review the operation of the equipment to establish that all is proper.

9.6.5 Faulty Hose

Occasionally, a failure problem will lie in the hose itself. The most likely cause of a faulty hose is old age. Check the lay line on the hose to determine the date of manufacture. The hose may have exceeded its recommended shelf life. If it is suspected that the problems lies in the manufacture of the hose, contact the supplier.

9.6.6 Analyzing Hose Failures

A physical examination of the failed hose can often offer a clue to the cause of the failure. Following is a list of problems to look for together with the conditions that could cause them.

Problem	Cause
1. The hose inner liner is very hard and has cracked.	Heat has a tendency to leach the plasticizers out of the tube. This is a material that gives the hose its flexibility or plasticity (see Fig. 9.18). Aerated oil causes oxidation to occur in the inner tube. This reaction of oxygen on a rubber product will cause it to harden. Any combination of oxygen and heat will greatly accelerate the hardening of the inner liner. Cavitation occurring inside the inner liner would have the same effect.
2. The hose is cracked both externally and internally, but the elastomeric materials are soft and flexible at room temperature.	The probable reason is intense cold ambient conditions while the hose was flexed (see Fig. 9.19).
3. The hose has burst and examination of the wire reinforcement after stripping back the outer cover reveals random broken wires the entire length of the hose.	This would indicate a high-frequency pressure impulse condition. If the extrapolated impulses in a system amount to over 1 million in a relatively short time, a spiral reinforced hose should be chosen (see Fig. 9.20).
4. The hose has burst, but there is no indication of multiple broken wires the entire length of the hose. The hose may have burst in more than one place.	This would indicate that the pressure has exceeded the minimum burst strength of the hose. Either a stronger hose is needed or the hydraulic circuit has a malfunction (see Fig. 9.21).
5. The hose has burst. An examinations indicates that the wire braid is rusted and that the outer cover has been cut, abraded, or deteriorated badly.	The only function that the outer cover has is to protect the reinforcement. Elements that may destroy or remove the outer cover are abrasion, chemicals, temperature, and cutting or pinching (see Fig. 9.22). Once the cover protection is gone, the reinforcement is susceptible to attack from moisture or other corrosive matter.
6. The hose has burst on the outside bend and appears to be elliptical in the bent section. In the case of a pump supply line, the pump is noisy and very	Violation of the minimum bend radius is probably the problem in both cases. Check the minimum bend radius and make sure that the application is

(continued)

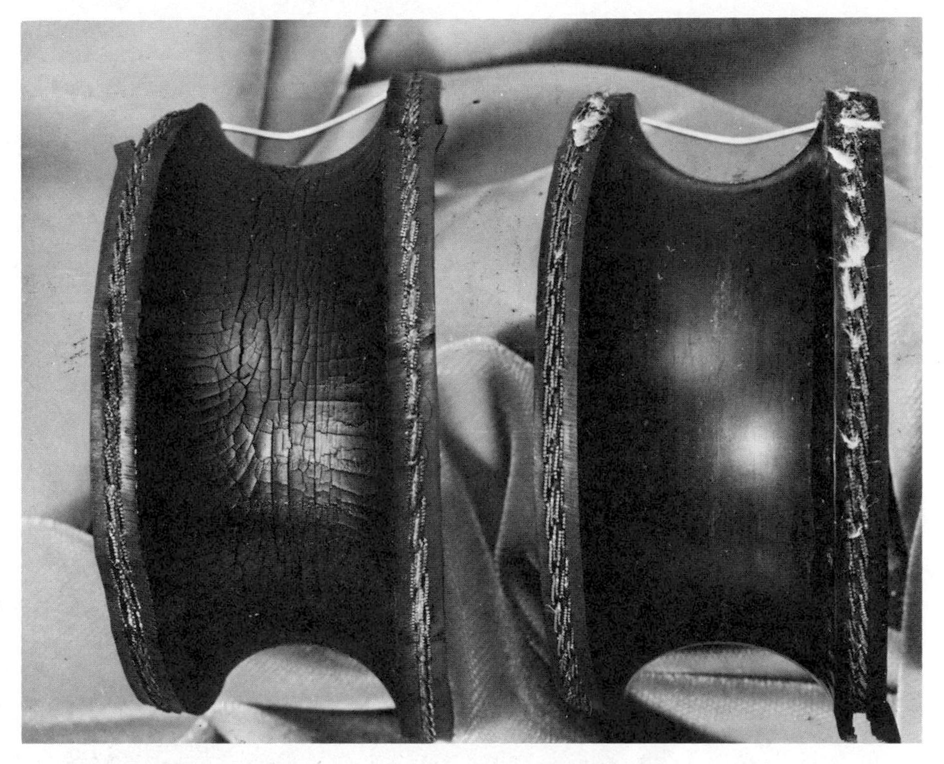

Figure 9.18 Hose failure due to high fluid temperature. (Courtesy of Aeroquip Corp., Jackson, Michigan.)

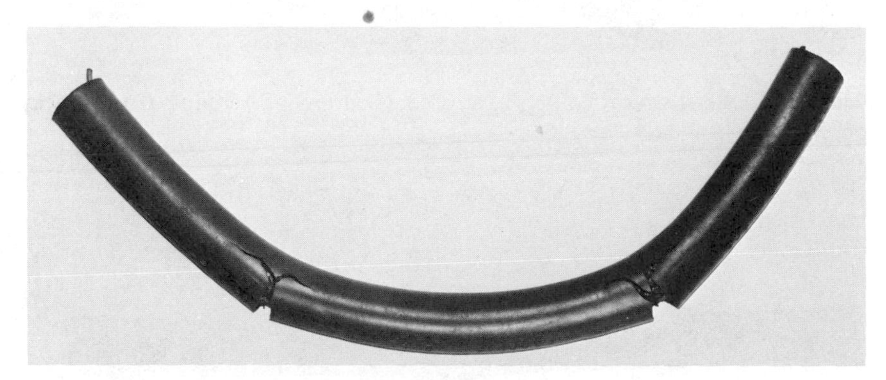

Figure 9.19 Cracked hose due to cold temperature. (Courtesy of Aeroquip Corp., Jackson, Michigan.)

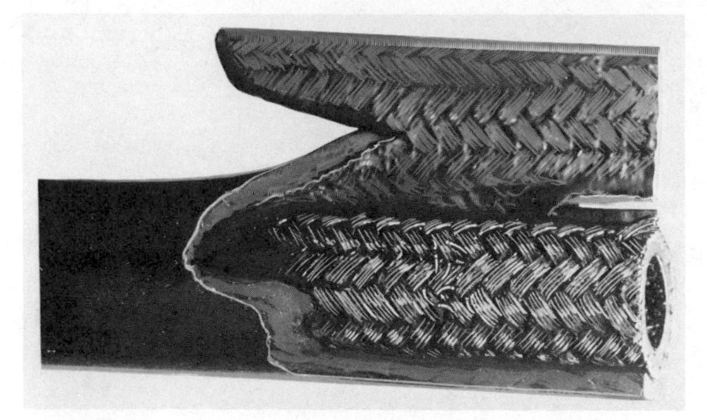

Figure 9.20 Broken wire reinforcement due to pressure pulsations. (Courtesy of Aeroquip Corp., Jackson, Michigan.)

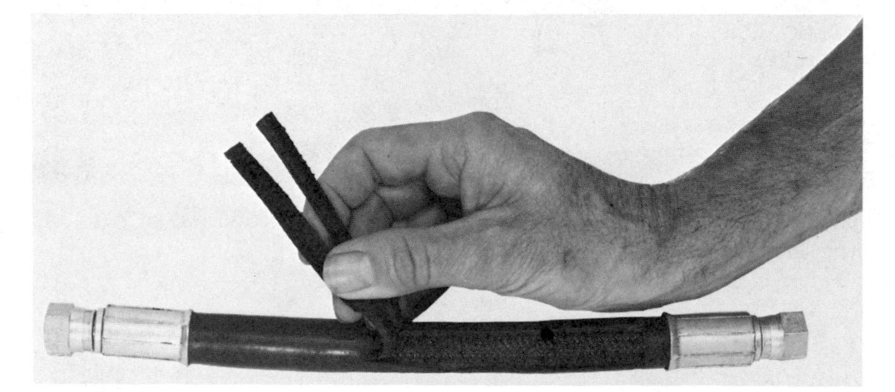

Figure 9.21 Burst strength of hose exceeded. (Courtesy of Aeroquip Corp., Jackson, Michigan.)

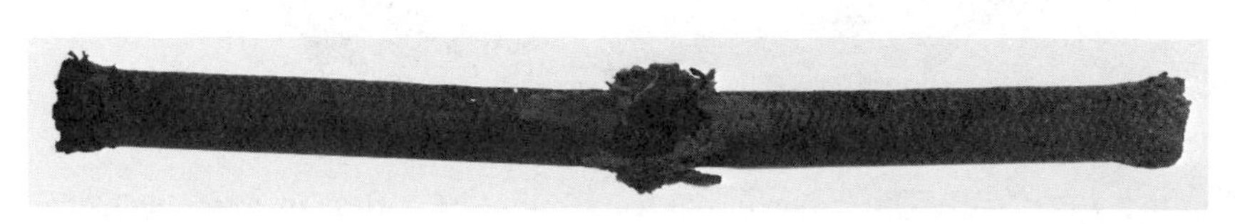

Figure 9.22 Hose failure due to outer-cover deterioration. (Courtesy of Aeroquip Corp., Jackson, Michigan.)

Problem	Cause
hot. The exhaust line on the pump is hard and brittle.	within specifications. In the case of the pump supply line, partial collapse of the hose is causing the pump to cavitate, creating both noise and heat. This is a most serious situation and will result in catastrophic pump failure if not corrected.
7. The hose appears to be flattened out in one or two areas and appears to be kinked. It has burst in this area and also appears to be twisted.	Torquing of a hydraulic control hose will tear loose the reinforcement layers and allow the hose to burst through the enlarged gaps between the braided wire strands. Be sure that there is never any twisting force on a hydraulic hose (see Fig. 9.23).
8. The hose has burst about 6 to 8 in. away from the end fitting. The wire braid is rusted. There are no cuts or abrasions of the outer cover.	Improper assembly of the hose and fitting, allowing moisture to enter around the edge of the outer shell. The moisture will wick through the reinforcement. The heat generated by the system will drive it out around the fitting area, but 6 to 8 in. away it will be trapped between the inner liner and outer cover, causing rusting of the wire reinforcement.
9. There are blisters in the outer cover of the hose. If one punctures a blister, oil will be found.	A minute pinhole in the inner liner is allowing the high-pressure oil to seep between it and the outer cover. Eventually, it will form a blister wherever the cover adhesion is weakest. In the case of a screw-type reusable fitting, insufficient lubrication of the hose and fitting can cause this condition because the dry inner liner will adhere to the rotating nipple and tear enough to allow seepage. A faulty hose can also cause this condition (see Fig. 9.24).
10. The fitting blew off the end of the hose.	It may be that the wrong fitting has been put on the hose. Recheck the manufacturer's specifications and part numbers. In the case of a crimped fitting, the wrong machine setting may have been

(continued)

Problem	Cause
	used, resulting in over- or undercrimping. The outer socket of a screw-together fitting for multiple wire-braided hose may be worn beyond its tolerance. Generally, these sockets should be discarded after being reused about six times. The swaging dies in a swaged hose assembly may be worn beyond tolerance.
	The fitting may have been improperly applied to the hose. The hose may have been installed without leaving enough slack to compensate for the possible 4% shortening that may occur when the hose is pressurized. This will impose a great force on the fitting. The hose itself may be out of tolerance.
11. The inner liner of the hose is badly deteriorated, with evidence of extreme swelling.	Indications are that the hose inner liner is not compatible with the fluid being carried. Consult the hose supplier for a compatibility list or forward for analysis a sample of the fluid being conducted by the hose. Be sure that the operating temperatures, both internal and external, do not exceed recommendations (see Fig. 9.25).
12. The hose has burst. The hose cover is badly deteriorated and the surface of the rubber is crazed.	This could simply be old age. The crazed appearance is the effect of weathering and ozone over a period of time. Try to determine the age of the hose.
13. A spiral reinforced hose has burst and literally split open, with the wire exploded out and badly entangled	The hose is too short to accommodate the change in length occurring while it is pressurized (see Fig. 9.26).
14. The hose is badly flattened out in the burst area. The inner tube is very hard downstream of the burst but appears normal upstream of the burst.	The hose has been kinked either by bending it too sharply or by squashing it such that a major restriction was created. As the velocity of the fluid increases through the restriction, the pressure decreases to the vaporization point of the fluid. This condition causes heat and rapid oxidization to take place, which hardens the inner liner of the hose

(continued)

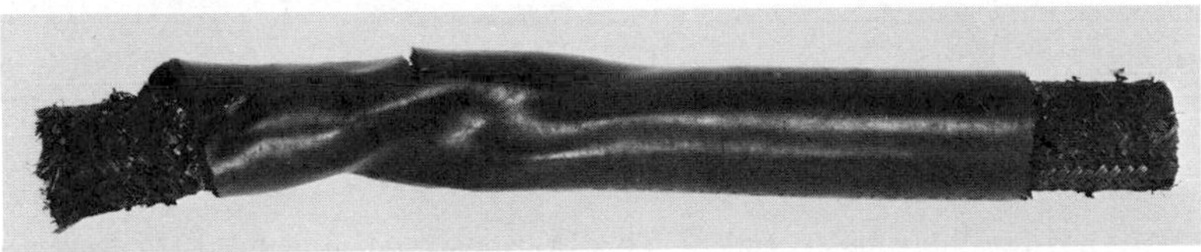

Figure 9.23 Hose failure due to torquing (twisting). (Courtesy of Aeroquip Corp., Jackson, Michigan.)

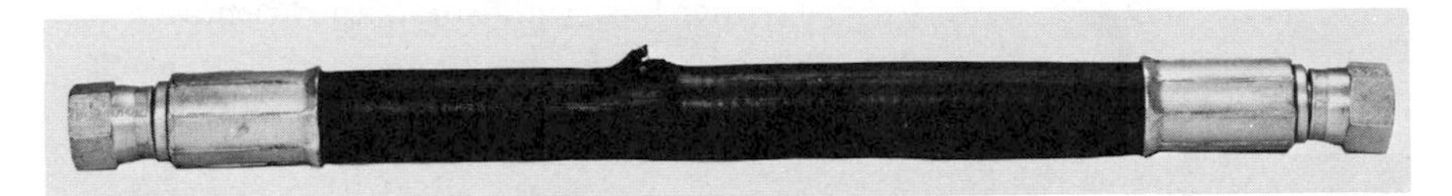

Figure 9.24 Blistering of the outer cover. (Courtesy of Aeroquip Corp., Jackson, Michigan.)

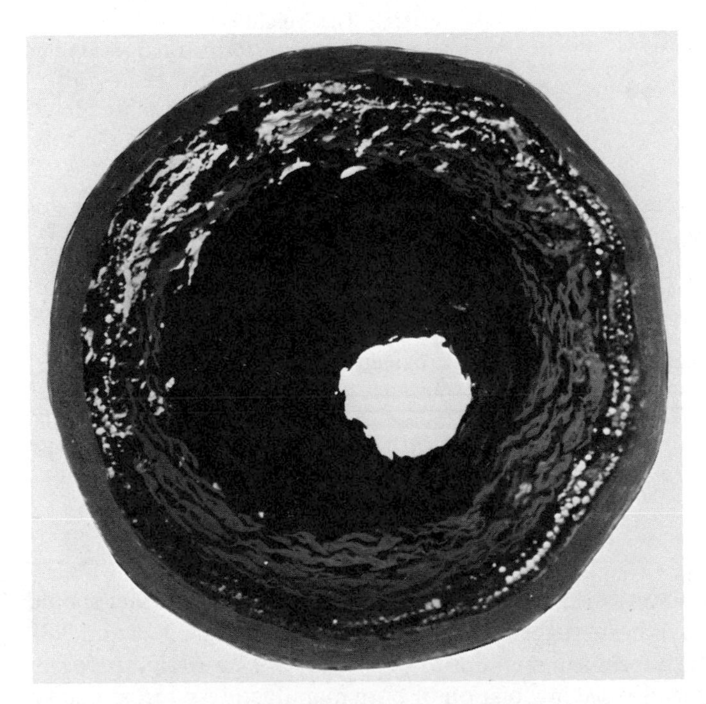

Figure 9.25 Incompatibility between hose and fluid. (Courtesy of Aeroquip Corp., Jackson, Michigan.)

Problem	Cause
	downstream of the restrictions (see Fig. 9.27).
15. The hose has not burst but is leaking profusely. A bisection of the hose reveals that the inner liner has been gouged through to the wire braid for a distance of approximately 2 in.	This failure would indicate that erosion of the inner liner has taken place. A high-velocity needle-like fluid stream being emitted from an orifice and impinging at a single point on the hose inner liner will hydraulically remove a section of it. Be sure that the hose is not bent close to a port that is orificed. In some cases where high velocities are encountered, particles in the fluid can cause considerable erosion in bent sections of the hose assembly.
16. The hose fitting has been pulled out of the hose. The hose has been considerably stretched out in length.	Insufficient support of the hose. It is mandatory to support a very long lengths of hose, especially if they are vertical. The weight of the hose, together with the weight of the fluid inside the hose, is being imposed on the hose fitting. This force can be transmitted to a wire rope or chain by clamping the hose to it.
17. The Teflon hose assembly has collapsed internally in one or more places.	One of the most common causes for this is improper handling of the Teflon assembly. Teflon is a thermoplastic material which is not rubber-like. When bent sharply it simply collapses.
18. The hose fittings keep blowing off nylon hose.	Provided that the right fittings for the hose has been selected and properly attached, the probable cause of failure is heat. Nylons under compression have a tendency to soften and flow out of the compression area when subjected to temperatures bordering 200°F (93°C).

To provide some assistance with hose assembly applications that should be avoided and to indicate suggested means for doing so, Figs. 9.28 and 9.29 are presented. The attachment methods to be avoided appear straightforward. Yet they are frequently used in violation of both manufacturers' recommendations and common sense. Each of these avoidance methods contributes significantly to hose assembly life.

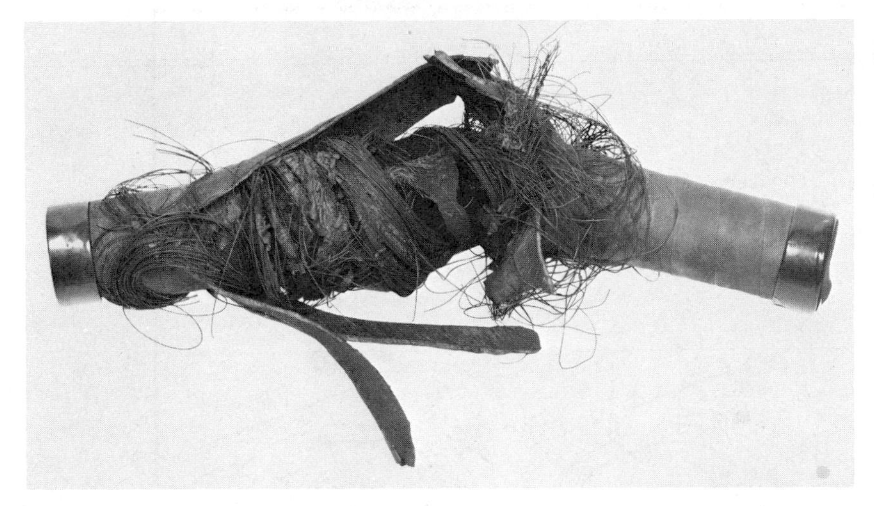

Figure 9.26 Hose too short for application. (Courtesy of Aeroquip Corp., Jackson, Michigan.)

9.7 MODULAR MANIFOLD HYDRAULIC SYSTEMS

In recent years, the area of component and system maintenance has received more attention than any other facet of the fluid power industry. It has led to modularized systems, plug-in components, integrated manifold circuits, and building-block spool cartridges. The reason for this activity is simply economic. Equipment downtime is becoming intolerably expensive. Also, maintenance time has increased in cost to a point where hydraulic equipment users are willing to pay extra for components that exhibit higher reliability, are easier to repair or replace, and can be serviced without dismantling the machine system.

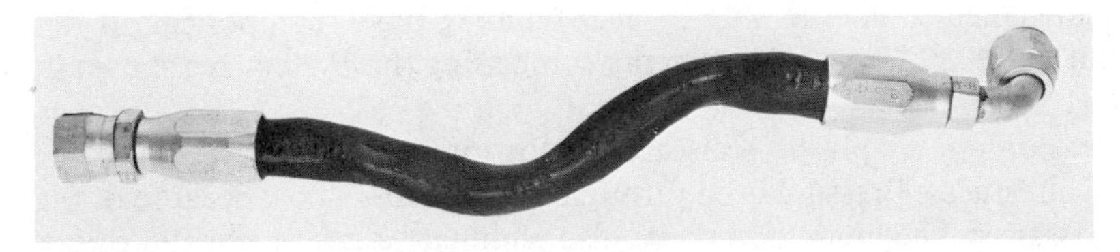

Figure 9.27 Hose failure due to kinking. (Courtesy of Aeroquip Corp., Jackson, Michigan.)

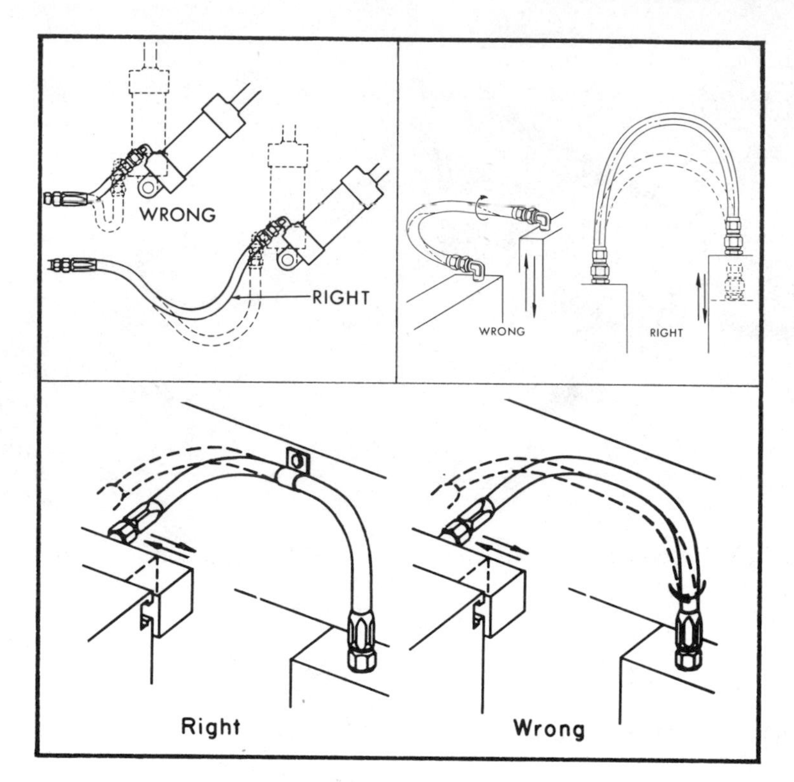

Figure 9.28 Flexing applications. (Courtesy of Aeroquip Corp., Jackson, Michigan.)

A new method of assembling and piping hydraulic systems through manifold modules has been developed and is presently being marketed by several manufacturers. The key to the circuits evolved with this system is the basic building-block module that can perform any and all hydraulic modulating functions with a standard valve body. It incorporates standard interface mounting and sealing arrangements. The primary element of the system is a center block. This center block is a special, predrilled, multiported modular element which can accept directional valves and port adapters. Standard valves can be used in the circuits without special modifications. A typical circuit arrangement made up with typical building-block components is depicted in Fig. 9.30. The basic elements comprising this system are shown in Fig. 9.31.

Elastomeric or plastic seals are used to prevent leakage between the module interfaces. Pressure drop through each module is low because of the short passages. The modular circuits also eliminate a major portion of hydraulic interconnecting piping and fittings, lessen system complexity, and permit versatility in circuit additions and alterations on field-operated hydraulic equipment.

Parallel systems permit mounting of integrated manifold subcircuits with associated valves to the main pressure and return lines. Individual subcircuits can be removed for servicing, replacement, change, and so on, without interference from other functional elements.

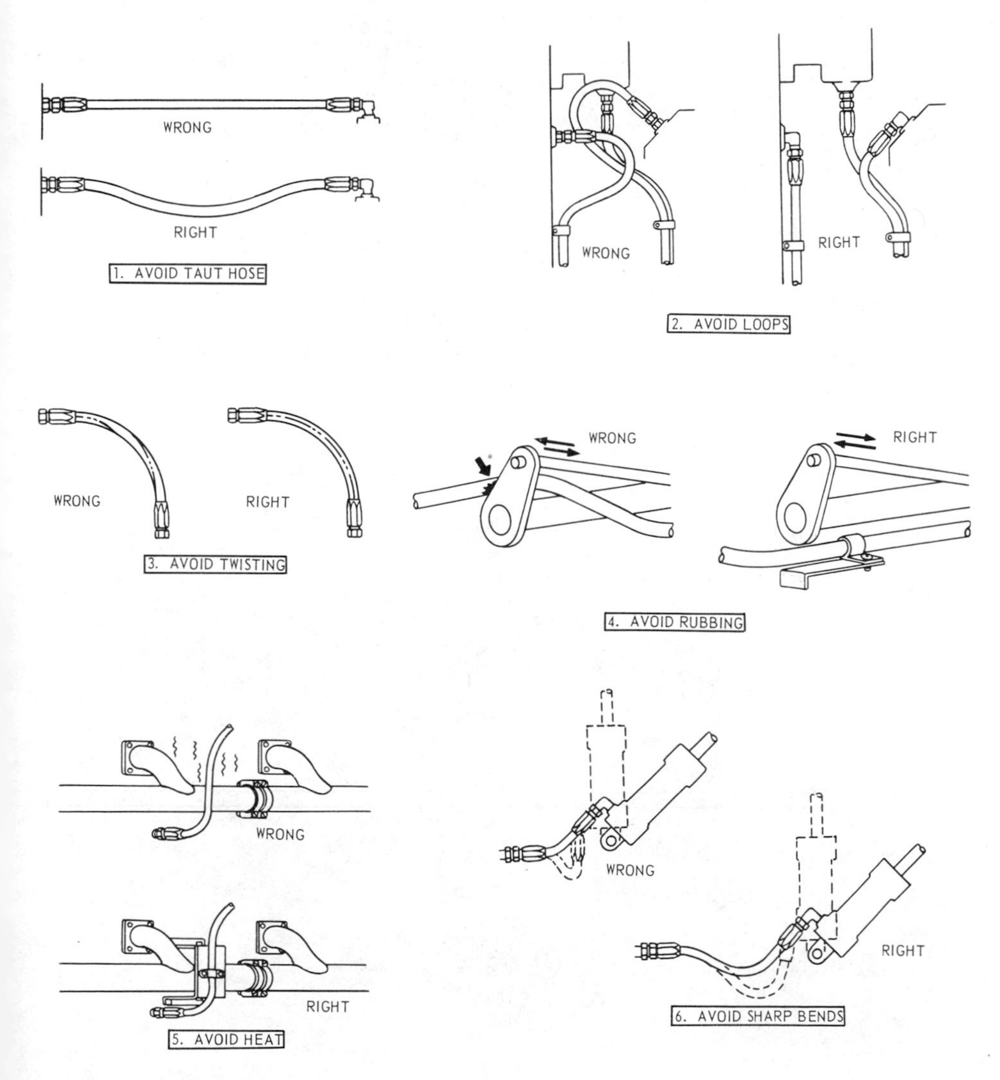

Figure 9.29 Hose assembly installation tips. (Courtesy of John Deere and Co., Waterloo, Iowa.)

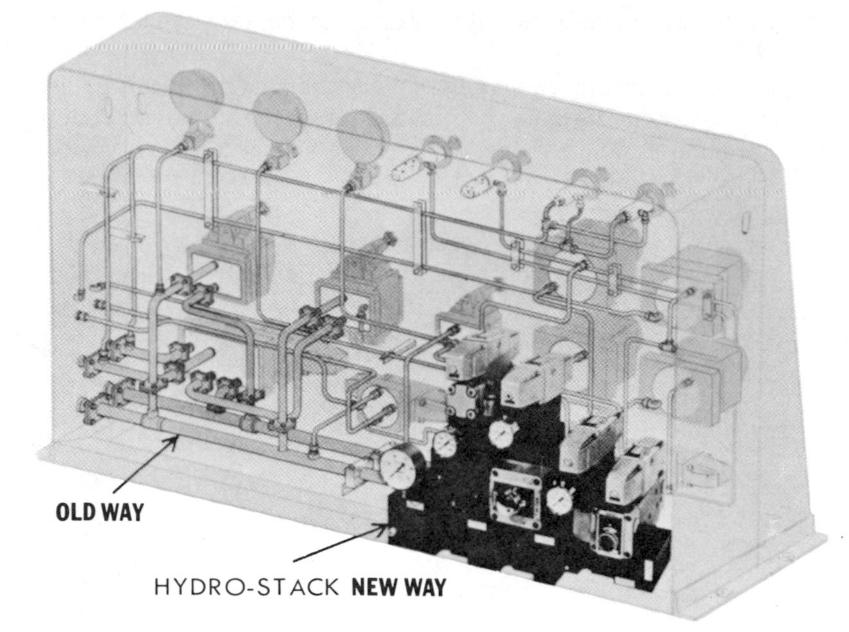

Figure 9.30 Building-block-style, color-coded, hydraulic circuit manifold arrangement. (Courtesy of Hydro-Stack Mfg. Corp., Milwaukee, Wisconsin.)

Figure 9.31 Modular manifold hydraulic control system. (Courtesy of Hydro-Stack Mfg. Corp., Milwaukee, Wisconsin.)

The practical development of a circuit follows a logical design procedure. First, a functional schematic circuit is developed for the hydraulic control system. This schematic is then divided as efficiently as possible into bus, branch, and subcircuits.

Bus circuits include all valves and accessories that are associated with the power supply, pressure, and return flow paths connected to a number of branch circuits. *Branch circuits* include all valves and accessories that directly control an output motor such as a cylinder. Branch circuits are served by bus circuits. *Subcircuits* cover any grouping of two or more valves into suitable modules to permit a workable combination as needed. A circuit diagram of the modular style system can then be prepared which will be equivalent to the functional schematic circuit.

10
Seals for Fluid Power Equipment

Any hydraulic system requires seals to prevent fluid from escaping and causing a leak. The seals hold the fluid under pressure in the system and keep dirt and grime out of the system. Visually, hydraulic seals appear to be simple in construction and design. However, in use they are complex, precision components that must be treated carefully and installed correctly if they are to do their job properly.

Hydraulic seals can be classified into two general categories based on their application: (1) *static seals*, used where no motion occurs between the parts, and (2) *dynamic seals*, used to seal moving parts. Seals are also classified in accordance with their form or shape. Following is a brief discussion of each type of seal according to form and its application (see Fig. 10.1).

10.1 SEAL FORMS

O Rings

The simple O ring is the most popular seal in hydraulics. Usually made of synthetic rubber, it is used in both static and dynamic applications. O rings are designed for use in grooves, where they are compressed (about 10%) between two surfaces. In dynamic use, they must have a smooth surface to work against. O rings are not used where they must cross openings or pass corners under pressure. In static use, under high pressure, they are often strengthened by a backup ring to prevent them from squeezing or extruding out of their grooves. The backup ring is usually made of fiber or synthetic plastic. They are generally not used in rotating applications because of wear problems.

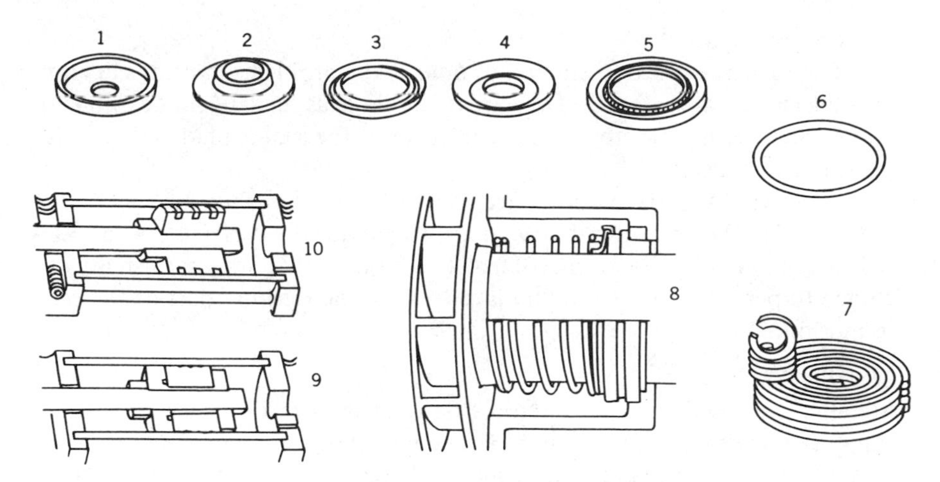

Figure 10.1 Types of hydraulic seals: 1, cup packing; 2, flange packing; 3, U packing; 4, V packing; 5, spring-loaded lip seal; 6, O ring; 7, compression packing; 8 mechanical seal; 9, nonexpanding metallic seal; 10, expanding metallic seal. (Courtesy of Sperry Vickers, Troy, Michigan.)

U and V Packings

U and V packings are dynamic seals for pistons and rod ends of cylinders, and for pump shafts. They are made of leather, synthetic and natural rubber, plastics, and other materials. These packings are installed with the open side, or lip, toward the system pressure so that the pressure will push the lip against the mating surface to form a tight seal. They are pressure-energized seals.

U and V packings are made of several U- or V-shaped elements and are used in packing glands or cases which hold them in one piece. They are popular for sealing rotating shafts, pistons, and the rod ends of cylinders.

Spring-Loaded Lip Seals

These seals are refinements of the simple U or V packing. The rubber lip is ringed by a spring which gives the sealing lips tension against the mating surface. Usually, the seal has a metal case which is pressed into a housing bore and remains fixed. This seal is often used to seal rotary shafts. The lip normally faces in toward the system oil. Double-lip seals are sometimes used to seal in fluids on both sides of an area.

Cup and Flange Packings

Cup and flange packings are dynamic seals and are made of leather, synthetic rubber, plastics, and other material. The surfaces are sealed by expansion of the lip or beveled edge of the packing. They are used to seal cylinder pistons and piston rods.

Mechanical Seals

These seals are designed to eliminate some of the problems in using chevron packings for rotating shafts. They are dynamic seals, usually made of metal and rubber. Sometimes the rotating portion of the seal is made of carbon, backed up with steel.

The seat has a fixed outer part attached to the housing. An inner part is attached to the revolving shaft and a spring holds the two parts of the seal tightly together. A rubber ring (flange-shaped) or a diaphragm is usually included to permit lateral flexibility and to keep the rotating part of the seal in motion.

Metallic Seals

Metallic seals used on pistons and piston rods are very similar to the piston rings used on engines. They may be either expanding or nonexpanding. Used as dynamic seals, they are usually made of steel. Unless fitted very closely, nonexpanding seals will leak excessively. Expanding seals (for use on pistons) and contracting seals (for use on piston rods) are subjected to moderate friction and leakage losses. Precision metallic seals, however, are not as subject to leakage, and are especially well adapted for use in extremely high temperatures. Since metallic seals are more subject to leakage than others, fluid wiper seals with external drains are often used.

Compression Packings

Compression packings (jam packings) are used in dynamic applications. They are made of plastics, asbestos cloth, rubber-laminated cotton, or flexible metals. Compression packings are generally suitable only for low-pressure use. Lubrication is very important, since they will score or scratch moving parts if allowed to run dry.

10.2 DYNAMIC SEAL APPLICATION AND FAILURES

Dynamic seals prevent or control leakage between surfaces that move past each other. Since these seals contact moving surfaces, they will eventually wear out or fail. Periodic replacement of the seal is required. With proper installation and maintenance, however, dynamic seals may last several hundred to several thousand hours. High pressure, high temperature, high speed, and surface roughness work to reduce seal life.

Typical locations for dynamic seals are pump and hydraulic motor drive shafts, pintles on variable-displacement pumps, directional valve pushpins, and actuator rods. In brief, dynamic seals prevent leakage where the action is taking place. Three types of dynamic seals commonly used are lip seals, face seals, and packings.

10.2.1 Radial Lip Seals

Radial lip seals, commonly called *oil seals* or *shaft seals*, are used to retain fluids in or keep dirt out of equipment with reciprocating or rotating shafts. The simplest type, the single-lip seal, is used only for low speeds and low pressures (Fig. 10.2).

Sealing is normally a result of an interference fit between the flexible sealing lip and a shaft. However, as seals age and temperatures change, the interference fit or lip pressure falls off. To maintain a more constant load on the shaft, a garter or finger spring is used. This permits operation at higher speeds and moderate pressures. It should be noted that the seal lip does not act like a squeegee to wipe the shaft dry. The lip must ride on a thin film of lubricant to be successful in its function. If the film gets too thick, the seal leaks. If it gets too thin, the seal wears and gets hard. The harder the seal, the more difficulty the lip has in following the shaft movement (Fig. 10.3).

Tests have shown that maximum seal life is obtained when the shaft sealing surface is 8 to 20 μin. in roughness. If the shaft is too smooth, it will not support a film. If too rough, it wears the seal lip. In either case, premature seal failure may occur. Finish marks should be circumferential rather than axial to retain the fluid.

Seal performance can be impaired by two types of eccentricity:

1. *Shaft-to-bore misalignment:* the distance from the center of shaft rotation to the center of the bore. It can be measured by attaching a dial indicator to the shaft to read off the seal bore while the shaft is rotated. This misalignment can cause uneven wear of the sealing lip and shorter life.

2. *Dynamic runout:* the distance that the center of the shaft is dis-

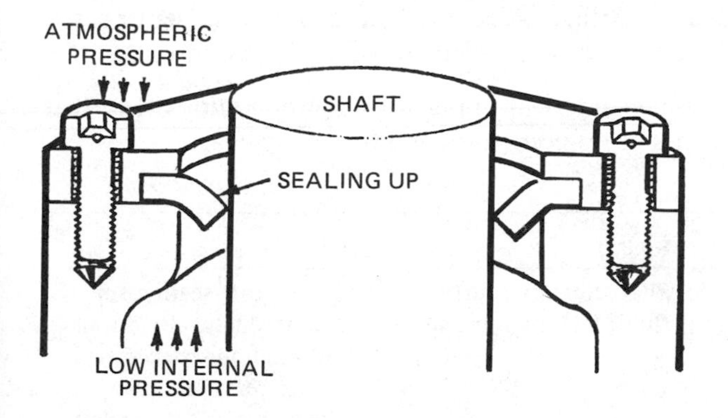

Figure 10.2 Simple lip seal. (Courtesy of Sperry Vickers, Troy, Michigan.)

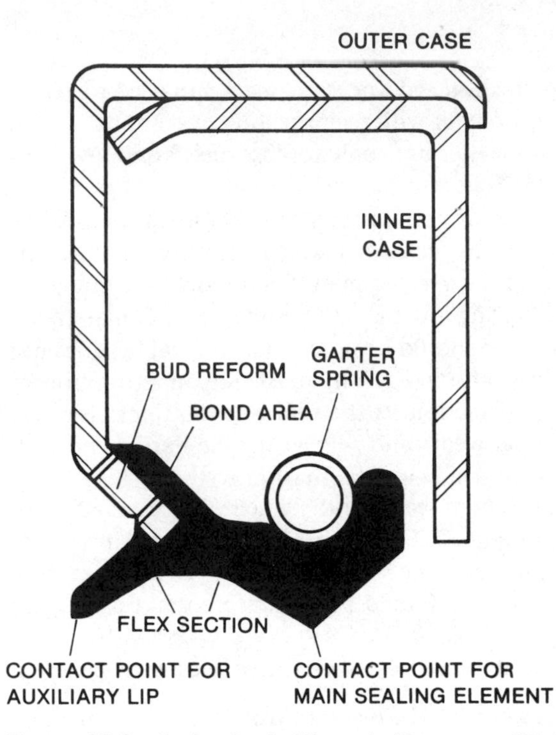

Figure 10.3 Spring-loaded lip seal. (Courtesy of Federal Mogul Co., Philadelphia, Pennsylvania.)

placed from the actual center of rotation. It is usually the result of misalignment, shaft imbalance, or actual flexing of the shaft. Dynamic runout causes wear around the entire sealing element. A spring-loaded seal with an adequate flex section will operate satisfactorily if the total runout is not too great (beyond 0.035 in.).

Figure 10.4 tabulates some shaft seal application considerations. Lip seal troubleshooting guidelines are presented in the following chart.

Problem	Cause	Solution
1. Worn shaft	Shaft too soft; dirty fluid; lack of lubrication	Check shaft sealing surface hardness—Rockwell C30 minimum necessary; replace seal; use wear sleeve on shaft if available, otherwise replace shaft; lubricate parts

Problem	Cause	Solution
2. Rough finish on shaft	Poor machining	Finish shaft surface to 8- to 20-μin. surface finish
3. Damaged shaft	Improper handling	Replace shaft; protect sealing surfaces during handling and assembly
4. Adhesive or paint on sealing surface	Lack of care in assembly	Clean with crocus cloth; mask shaft during seal assembly into bore or painting of unit
5. Seal cocked in bore	Improper installation	Use proper driving tool; install seal at right angle to shaft surface
6. Seal lip reversed	Improper installation	Before replacement, check stock seal to see if it has double lip: one lip faces inward to retain fluid, one faces outward to exclude dirt
7. Seal lip cut or torn	Installed over sharp edges; damaged shaft	Replace seal; lubricate seal and shaft; use thimble to carry lip over keyways, splines, and sharp edges; make sure that lip ID is not stretched
8. Seal lip worn, glazed, or hardened; shaft is intact	Overheated system; incorrect seal size; case pressure too high	Check for hot oil, high case pressure, and correct seal size; also check to see whether seal lubricant is good
9. Seal spring damaged	Excessive pressure or clearance	Replace seal, avoiding excessive spreading of sealing lip and spring; check for proper storage and handling of seals
10. Excessive eccentricity or misalignment; seal lip cannot follow shaft movement	Excessive side load	Align shaft, eliminate shaft side load, or use a better flexible coupling
11. "Built-in" seal flaw such as contamination, poor rubber bond to metal, or flash on seal lip	Manufacturing error	Replace seal

Parameter	Comments	Limits	
Shaft speed	Important factor in seal selection (shown in surface feet per minute)	Slow Moderate High	0–1000 1001–2000 2001–3000
Temperature	Limits shown are points where seal material or medium sealed becomes ineffective; for extreme temperatures, special compounds can be used	Continuous Intermittent	– 50 to +225°F – 65 to +250°F
Pressure	Oils seals are not pressure seals; where pressures above those shown exist, special seals should be used or pressure against sealing lip relieved	Slow Moderate High	7 psi 5 psi 3 psi
Shaft finish	Surface finish, direction, and spiral of finishing marks and leads as well as surface finish value affect sealing; polished or ground surfaces with concentric finish marks are preferred	Slow Moderate High	10–25 μin. 10–20 μin. 10–20 μin.
Shaft hardness	Although shafts as soft as cold-rolled steel can be sealed successfully, hardness of Rockwell C30 or greater is preferred; fluid starvation, abrasives, and high surface speeds require hard shafts	Abrasives No abrasives	Above C-45 Above B-80
Shaft to bore misalignment	Fixed misalignment of center of shaft rotation with bore center; concentrates wear at one side of seal; becomes more severe as speed increases	Slow Moderate High	0.015 TIR 0.010 TIR 0.010 TIR
Shaft runout	Runout (eccentricity or shaft whip) should be kept to an absolute minimum; creates a difficult sealing problem	0–800 rpm 800–2200 rpm 2200–4200 rpm	0.025 TIR 0.020 TIR 0.015 TIR

Figure 10.4 Shaft seal application considerations. (Courtesy of Federal Mogul Co., Philadelphia, Pennsylvania.)

10.2.2 Face Seals

The mechanical face seal is one of the most effective devices in preventing leakage along a rotating shaft which passes in or out of an area of pressurized oil. Two ultra-flat sealing faces are mounted perpendicular to the shaft. The seal seat is attached to and rotates with the shaft, while the spring-loaded seal head is stationary.

The usual seal face materials in hydraulic applications are hard carbon and bronze for the seal head. Many shaft seals now use bronze elements rather than carbon. They are stronger and less susceptible to chipping and similar problems. Also, they are not attacked by synthetic fluids such as the phosphate esters. Steel or cast iron is usually used for the seal seat. The two are separated by an oil film. With an excellent matching of sealing forces and seal flatness, oil surface tension can complete the seal and there is no leakage. Elevated pressures can induce seal wear, but with proper balancing, pressure-induced sealing forces can be kept low.

Only a properly trained person should attempt to repair the sealing surfaces of face seals. The condition of the seal surfaces is so critical that one company provides 40 hr of training to its personnel on face seal operation, repair, and installation. With new replacement parts, do not touch the sealing surfaces with fingers or an old wiping rag. Make sure that the seal seat is perpendicular to the shaft within 0.001 in. TIR (TIR is the total change in indicator reading during one complete rotation of the shaft). Lubricate the sealing surfaces well with the fluid to be sealed before installation.

Examine the old parts for signs of the failure mode that was experienced by the seal. Abrasive wear of the sealing faces means that contaminated oil is in the system. Burned faces indicate dry running of the seal. Heavy wear may mean either excessive operating pressure or a hung-up spring. A cracked carbon ring leaks badly. Worn bearings should be replaced if end play exceeds 0.002 in. or radial looseness is greater than 0.004 in. Replace the shaft with a new one if runout exceeds 0.002 in. TIR. Polish the new shaft to remove burrs or scratches that might damage the seals. Figures 10.5 to 10.17 depict typical seal damage and their cause. The following chart lists some face seal symptoms and causes of leakage.

Problem	Cause	Solution
1. Seal leaking steadily despite apparent full contact of mating rings	Secondary seal leakage caused by: a. Nicked, scratched, or porous seal surfaces b. O-ring compression set c. Chemical attack	Replace seal assembly; check compatibility of oil and seal materials

(*continued*)

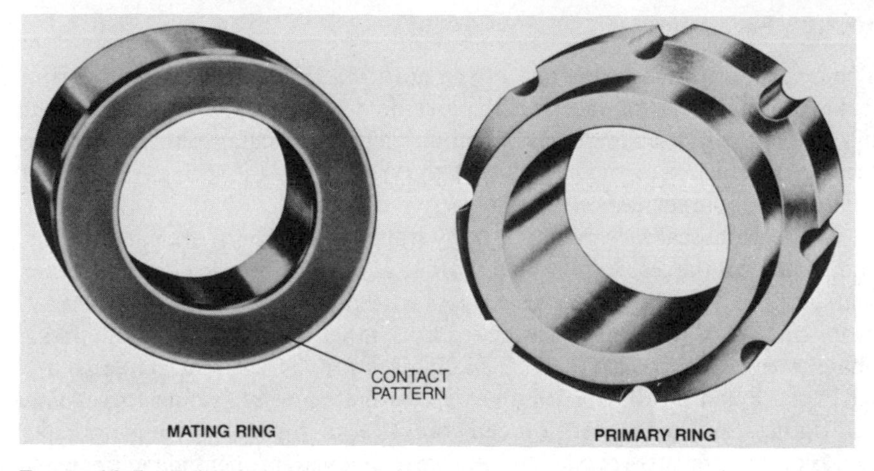

Figure 10.5 Full contact pattern. (Courtesy of John Crane Packing Co., Morton Grove, Illinois.)

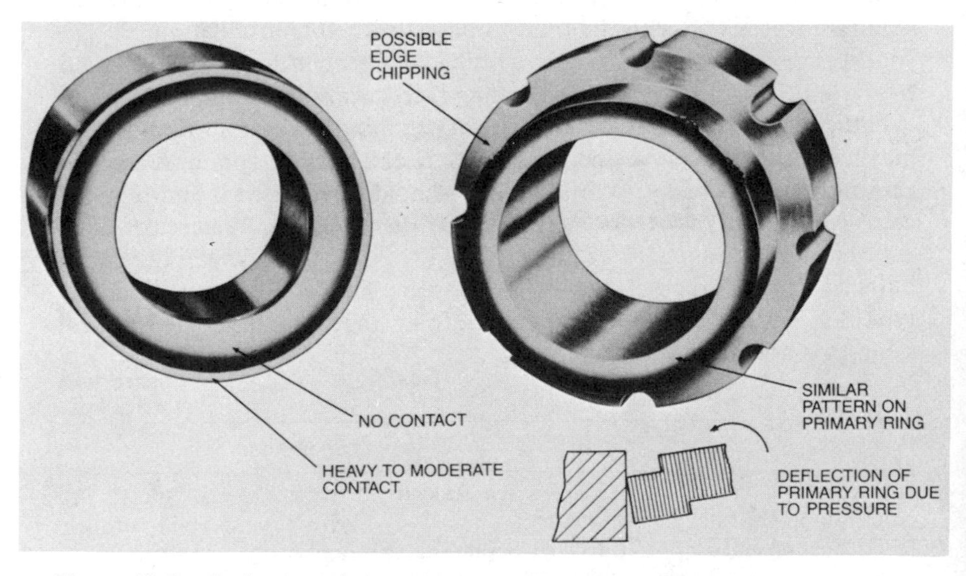

Figure 10.6 Coning (negative rotation). (Courtesy of John Crane Packing Co., Morton Grove, Illinois.)

Problem	Cause	Solution
2. Steady leakage at low pressure; little or no leakage at high pressure	Deflection of primary ring from overpressurization; seal faces not flat because of improper lapping	Replace seal
3. Seal leaks steadily when shaft is rotating; little or no leakage when shaft is stationary	Thermal distortion of seal faces; seal faces not flat because of improper lapping	Replace seal
4. Seal leaks steadily whether shaft is stationary or turning	Mechanical distortion caused by: a. Overtorqued bolts b. Out-of-square clamping parts c. Out-of-flat stuffing box faces d. Nicked or burred gland surface e. Hard gasket	Check torque on mounting bolts Check flatness and squareness of parts Check for nicks and burrs; remove these with fine crocus cloth
5. Seal leaks steadily whether shaft is stationary or rotating; noise from flashing or face popping	Sealed liquid vaporizing at seal interface, caused by: a. Low suction or stuffing box pressure b. Improper running clearance between shaft and primary ring c. Insufficient cooling d. Improper bushing clearance e. Circumferential flush groove in gland plate missing or blocked	Check for vacuum by dripping fluid on seal to see if noise disappears temporarily Check clearance between seal and shaft Replace seal
6. Seal leaks steadily whether shaft is stationary or rotating; high wear and grooving	Poor lubrication from sealed fluid; abrasives in fluid	Clean fluid with filters and/or change fluid Check seal material and fluid compatibility
7. Steady leakage when shaft is rotating; no leakage when shaft is stationary	Out-of-square mating surface, caused by: a. Nicked or burred gland surfaces b. Improper drive pin extension	Check alignment Check for pump housing distortion Check pump bearings Replace seal

(continued)

Problem	Cause	Solution
	c. Misaligned shaft d. Piping strain on pump casing e. Bearing failure f. Shaft whirl	
8. Damaged mating ring; seal leaks whether shaft is rotating or stationary	Misaligned mating ring, caused by: a. Improper clearance between gland plate and stuffing box b. Lack of concentricity between shaft OD and stuffing box ID	Check concentricity between shaft and seal mounting bore; correct if excessive
9. Seal squeals during operation	Inadequate amount of liquid to lubricate seal faces	Bypass flush line may be needed if not in use Enlarge bypass flush line and/or orifices in gland plate
10. Carbon dust accumulating on outside of gland ring	Inadequate amount of liquid to lubricate seal faces Liquid film evaporating between seal faces	Bypass flush line may be needed if not in use Enlarge bypass flush line and/or orifices in gland plate Check for proper seal design with seal manufacturer if pressure in stuffing box is excessively high
11. Short seal life	Abrasive fluid	Prevent abrasives from accumulating at seal faces Bypass flush line will be needed if not in use Use abrasive separator or filter
	Seal running too hot	Increase cooling of seal faces Increase bypass flush line flow Check for obstructed flow in cooling lines
	Equipment mechanically out of line	Align this equipment Check for rubbing of seal on shaft

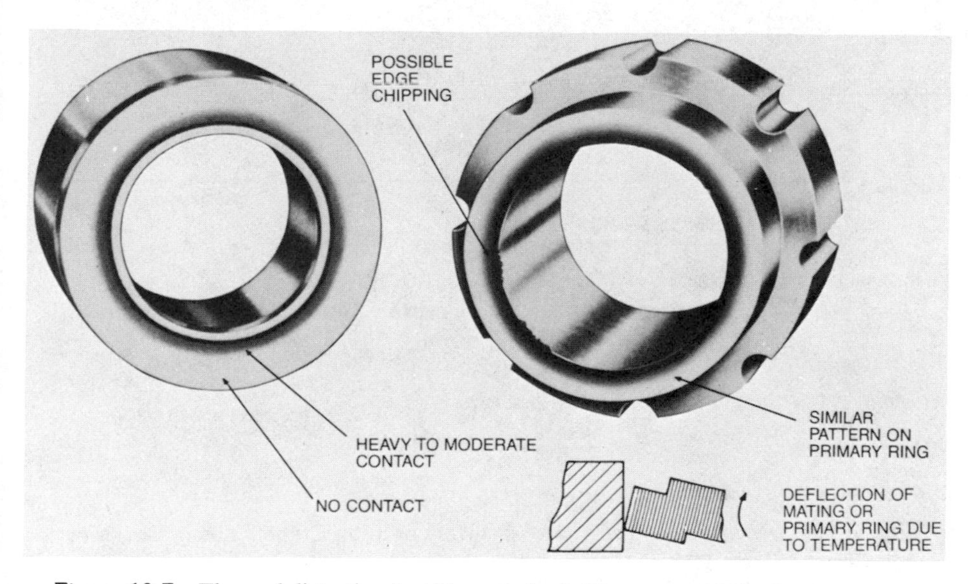

Figure 10.7 Thermal distortion (positive rotation). (Courtesy of John Crane Packing Co., Morton Grove, Illinois.)

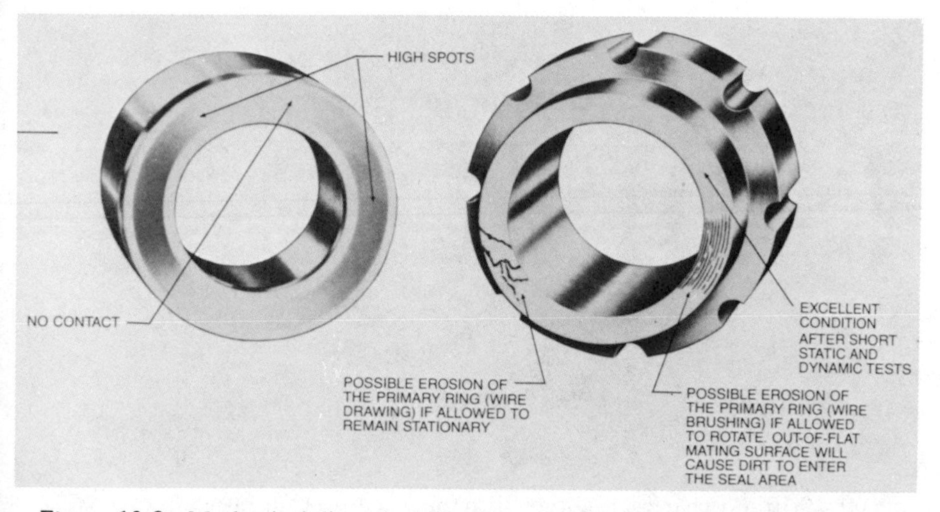

Figure 10.8 Mechanical distortion. (Courtesy of John Crane Packing Co., Morton Grove, Illinois.)

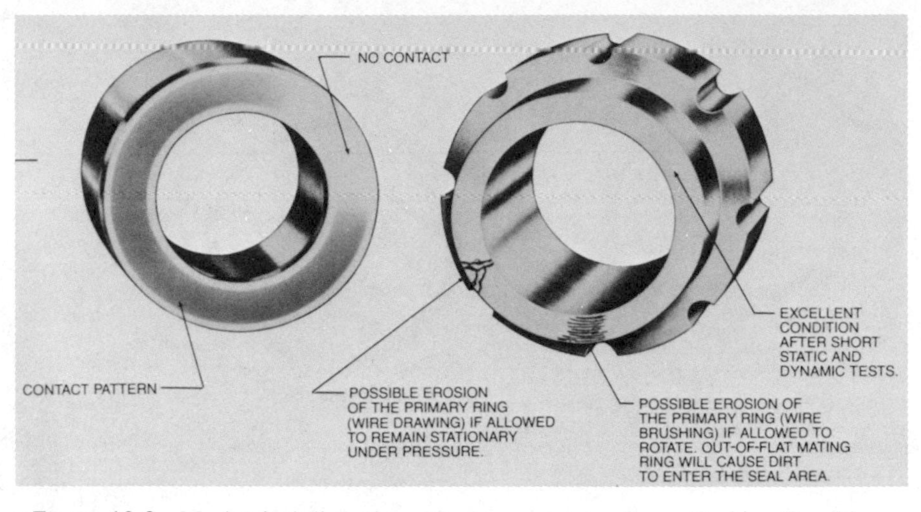

Figure 10.9 Mechanical distortion. (Courtesy of John Crane Packing Co., Morton Grove, Illinois.)

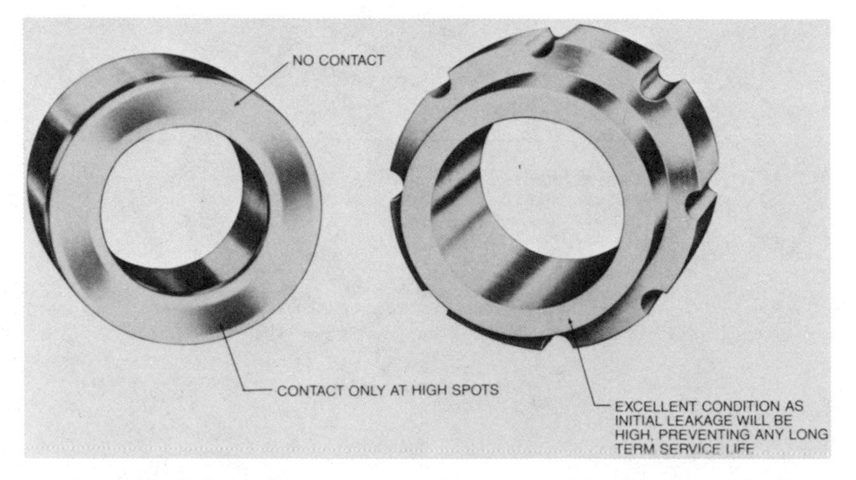

Figure 10.10 Mechanical distortion. (Courtesy of John Crane Packing Co., Morton Grove, Illinois.)

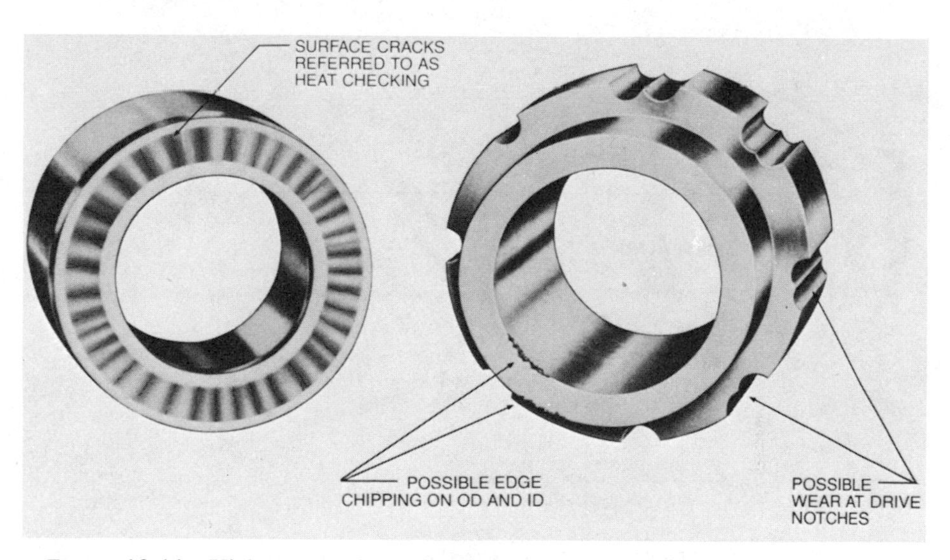

Figure 10.11 High wear or thermally distressed surface. (Courtesy of John Crane Packing Co., Morton Grove, Illinois.)

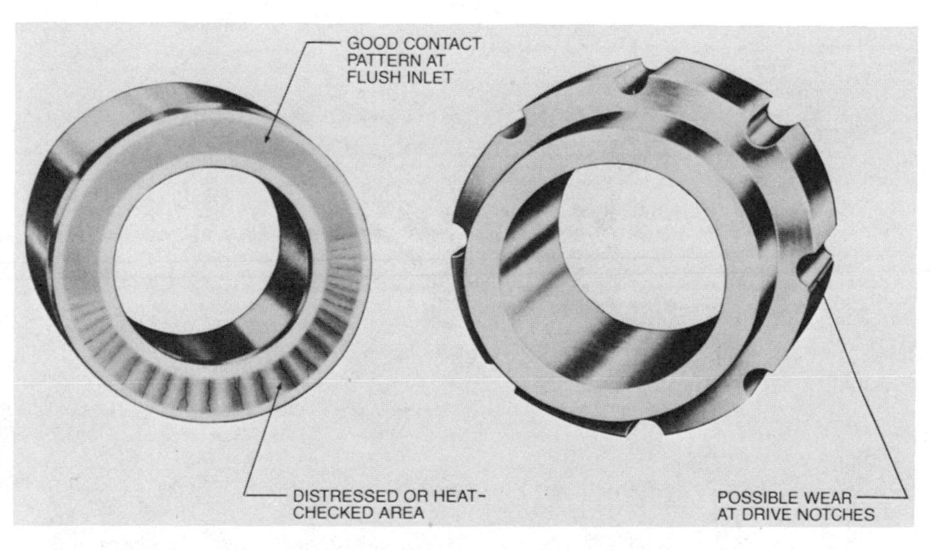

Figure 10.12 Section of thermally distressed surface. (Courtesy of John Crane Packing Co., Morton Grove, Illinois.)

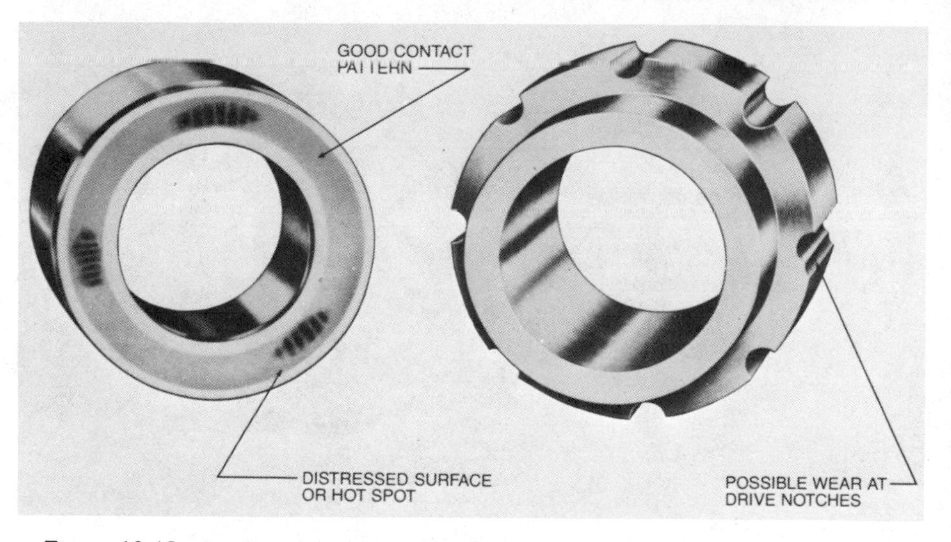

Figure 10.13 Patches of thermally distressed surface. (Courtesy of John Crane Packing Co., Morton Grove, Illinois.)

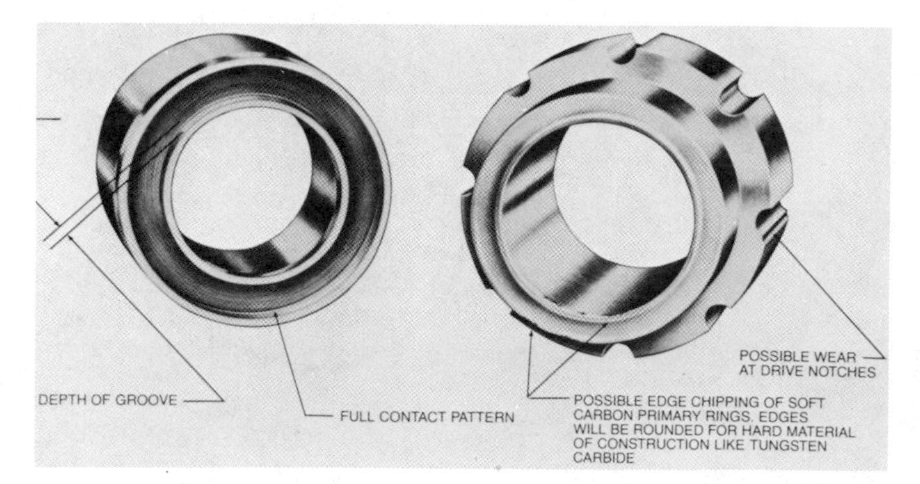

Figure 10.14 High wear and grooving. (Courtesy of John Crane Packing Co., Morton Grove, Illinois.)

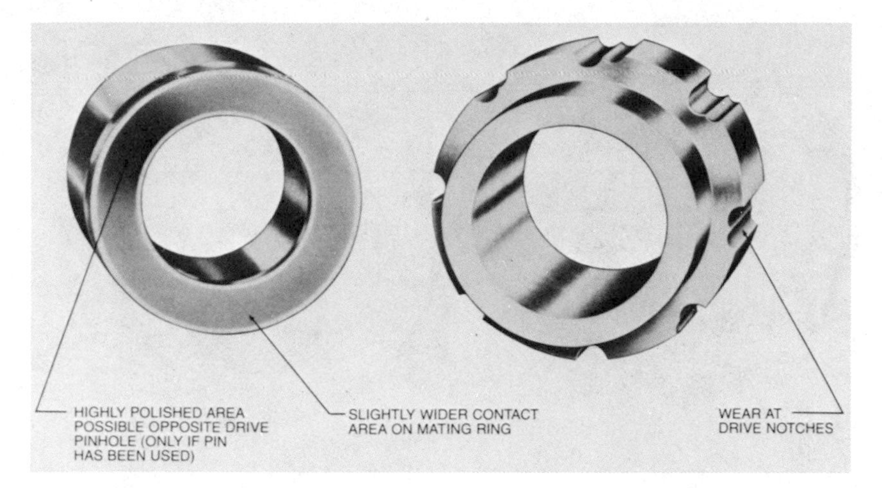

Figure 10.15 Out-of-square mating ring. (Courtesy of John Crane Packing Co., Morton Grove, Illinois.)

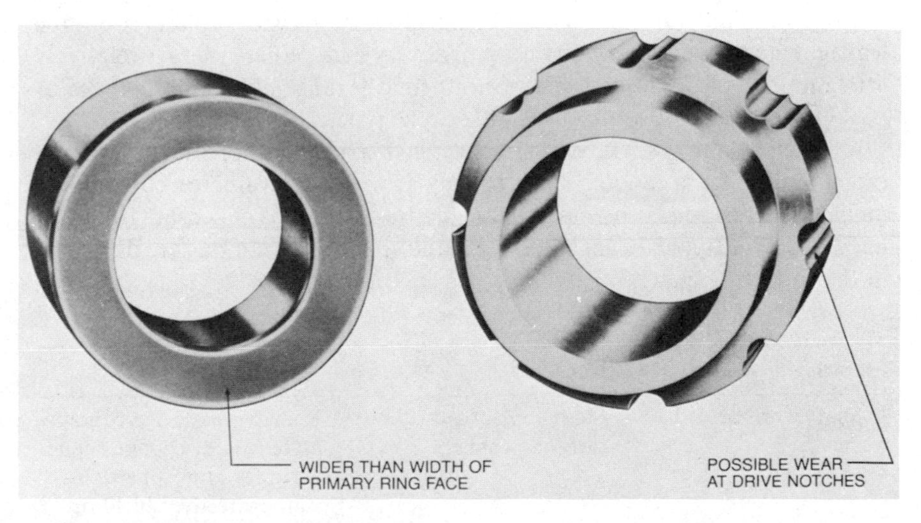

Figure 10.16 Wide contact pattern. (Courtesy of John Crane Packing Co., Morton Grove, Illinois.)

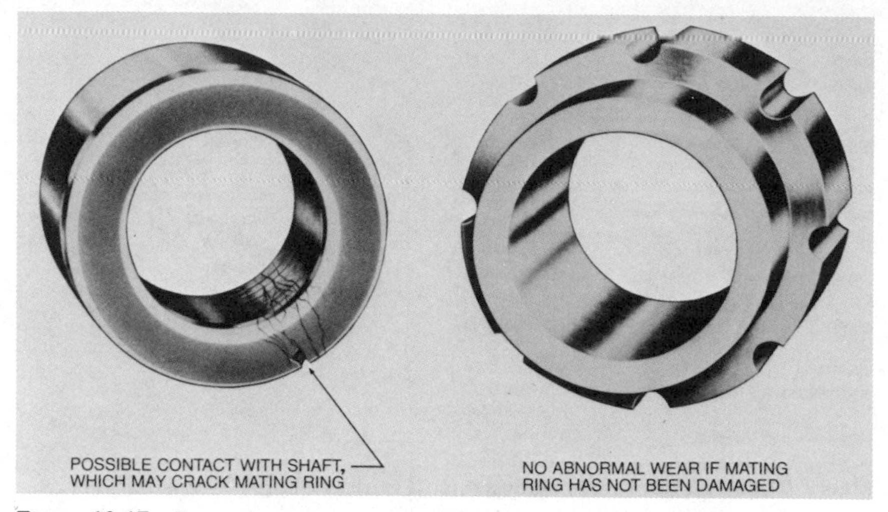

Figure 10.17 Eccentric contact pattern. (Courtesy of John Crane Packing Co., Morton Grove, Illinois.)

10.2.3 Packings

A packing is a material deformed so as to throttle leakage between a moving or rotating part and a stationary one. With rapid motion, there must be enough leakage to lubricate and cool the packing. On some large applications using compressed packings, the desired leakage rate may be as high as 10 drops per minute. On some small O-ring applications with rapid motion, the leakage rate may be as low as 1 drop per every 2 hr. Where there is relatively little motion, packings can seal without fluid leakage. Three basic types of packings are: compression, lip and squeeze type.

The packing is sufficiently pliable when axially compressed to provide radial sealing for a moving shaft or rod. It will not scratch or corrode the moving shaft or rod. It requires frequent adjustment to compensate for packing wear. Some typical packing failures and suggested solutions are described in the following chart.

Problem	Cause	Solution
1. Shaft worn or rod worn	Shaft too soft; dirty oil; dirty atmosphere	Replace shaft; for contaminated fluid, change filter; for dirty atmosphere, install protective shield or boot; check shaft hardness (Rockwell C30 minimum)

(continued)

Problem	Cause	Solution
2. Sealing surfaces are scratched; seals damaged	Improper handling and assembly	Use proper assembly and disassembly tools on overhaul; replace damaged parts
3. Dynamic runout of shaft or eccentric motion is excessive	Excessive shaft runout; excessive side load	Inspect bearings, replace if too loose; check side loads on shaft or rod; seals are not to be used as bearings
4. Rapid wear-out of seal	Seal compressed too much; incorrect seal size	Loosen if adjustment is available; otherwise, check to be certain of correct seal size
5. Glazed or hardened seal	High oil temperature; lack of lubrication	Check for high oil temperature; if seal lubrication is inadequate, correct
6. Seal edges are extruded	Too much clearance between mating parts	Check parts for too much clearance; replace faulty parts; use antiextrusion rings on low-pressure side of seal

10.3 STATIC SEAL APPLICATION AND FAILURES

Static seals prevent leakage between stationary surfaces. To contain pressure, the seal and its mating parts must be in contact at a pressure level higher than the pressure to be sealed. This pressure level may be obtained by parts installation. In other applications, such as when O ring, V ring, and X ring, the initial sealing pressure from installation alone is sufficient to contain only low-pressure oil. High pressure deforms or changes the seal shape and increases the sealing pressure level to complete the seal.

10.3.1 Gaskets

A gasket is an installation-activated seal made of relatively soft material. It must be deformed or compressed to fill surface irregularities and close the gasket structure to fluid leakage. The O ring is replacing the gasket in many of the new hydraulic designs because of its greater reliability and ease of application (Fig. 10.18).

In general, the more compressible (softer) materials are used for low-

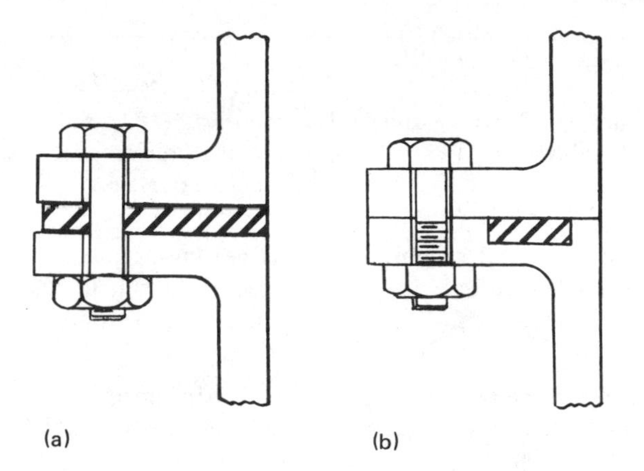

Figure 10.18 Gasket configurations. (a) Flat face. (b) Grooved face.

pressure gasket applications. Common materials are asbestos, cork, paper, plastic, rubber or a combination of these materials. The majority of gasket applications can be represented by two basic flange designs: (1) the flat face or (2) the grooved face. For a given gasket cross section, each material has a minimum clamping load necessary to close its structure to fluid leakage and cause surface irregularities to be filled. Most gasket materials relax and creep after being clamped. A loss in loading then occurs, most of it within the first 18 hr. For critical applications, wait a day after initial assembly and retighten to the original loading, preferably at system operating temperature but with no internal pressure.

As pressures rise and operating conditions become more severe, metal or metal-with-soft-core gaskets are necessary. The required clamping force and surface finish to obtain a good seal vary widely, depending on the type of gasket.

10.3.2 O Rings

One of the most common static seals is the O ring. It has found increasing use as a high-pressure seal. Recommended surface roughness is 32 to 63 μin. With high pressure, the sealing surfaces may either slide or separate. Sliding causes wear. The rougher the surface, the greater the rate of seal wear. If sliding cannot be prevented, a surface roughness of 16 μin. may be necessary to obtain satisfactory seal life.

Separation of the sealing surfaces permits the O ring seal to be extruded into the clearance gap. A sudden pressure drop will trap the extruded edges of the seal. With pressure pulsations, the extruded edges will be nibbled away. Eventually, the seal will leak (Fig. 10.19).

It has been found that as O-ring hardness (called *Durometer*) increases, its resistance to extrusion damage also increases. For example, in laboratory tests with 160° oil, 100,000 pressure pulsations at 1500 psi caused significant O-ring extrusion damage when the diametral extrusion gap was greater than 0.004 in. with 70-Durometer seals, 0.008 in. with 80-Durometer seals, and 0.014 in. with 90-Durometer seals. Do not jump at the first chance to put in high-Durometer O rings. As pressure levels rise, they resist extrusion better than softer seals. But they leak more readily against rougher surfaces (Fig. 10.20).

In all work with O rings, they should be protected as follows: (1) lubricate with light grease or fluid to be sealed; (2) avoid use of sharp tools during assembly or parts removal; and (3) use a brass, paper, or plastic cone to pass the O ring over threads.

Some O-ring problems that are caused by improper installation or application are summarized in Fig. 10.21.

Following is a list of some possible sources of trouble with static seals and some suggested solutions.

Problem	Cause	Solution
1. Seal has been extruded or been nibbled to death	Improper installation; excessive pressure pulsations; parts not flat or smooth	Replace seal and check the following: a. Sealing surfaces must be flat within 0.0005 in.; replace part if out of limits b. Initial bolt torque may have been too low; check manual for proper torque c. Check for proper relief valve setting d. If normal operating

(*continued*)

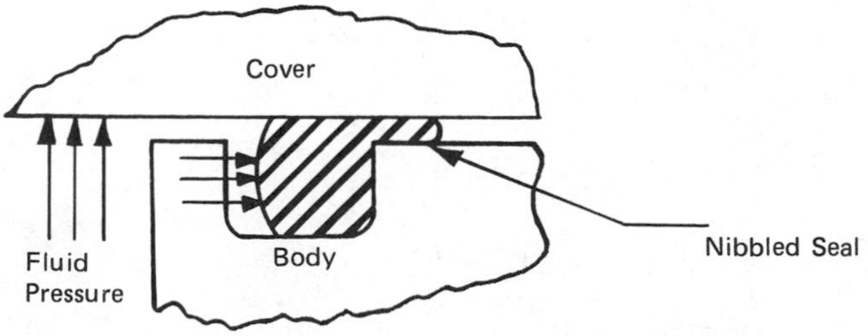

Figure 10.19 O-ring extrusion.

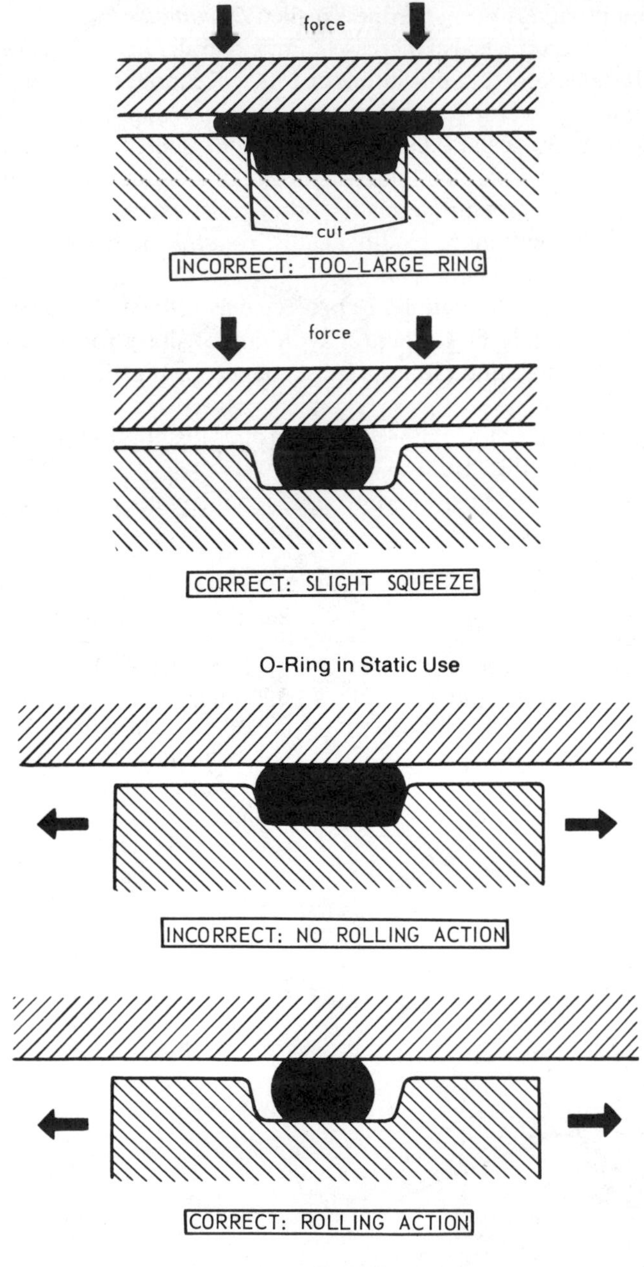

Figure 10.20 O-ring application.

Problem	Cause	Solution
		pressures exceed 1500 psi, backup rings are required
2. Seal is badly worn	Surfaces too rough; seal too soft; dirty oil	Replace seal and check the following: a. Sealing surface too rough; polish to 16 μin. if possible or replace part b. Undertorqued bolts permit movement; check manual for correct setting c. Seal material or Durometer may be wrong; check manual if in doubt
3. Seal is hard or has taken excessive permanent set	Oil too hot; wrong seal material	Replace seal and check the following: a. Determine what normal operating temperature is; check system temperature b. Check manual to determine that correct seal material is being used
4. Sealing surfaces are scratched, gouged, or have spiral tool marks	Careless handling; improper machining	Replace faulty parts if marks cannot be polished out with fine emery paper
5. Seal has been pinched or cut on assembly	Improper assembly	Use the fluid or petrolatum to hold seal in place with blind assembly; use protective shim if seal must pass over sharp threads, keyways, or splines
6. Seal leaks for no apparent reason	Incorrect seal size	Check seal size and parts size; get correct replacement parts

10.4 CHECKING FOR LEAKS

Before disassembling a component, check out the causes of leakage. This may save a repaired item or machine from being returned for repair. The cause may have been something other than the seal. Before cleaning the area around

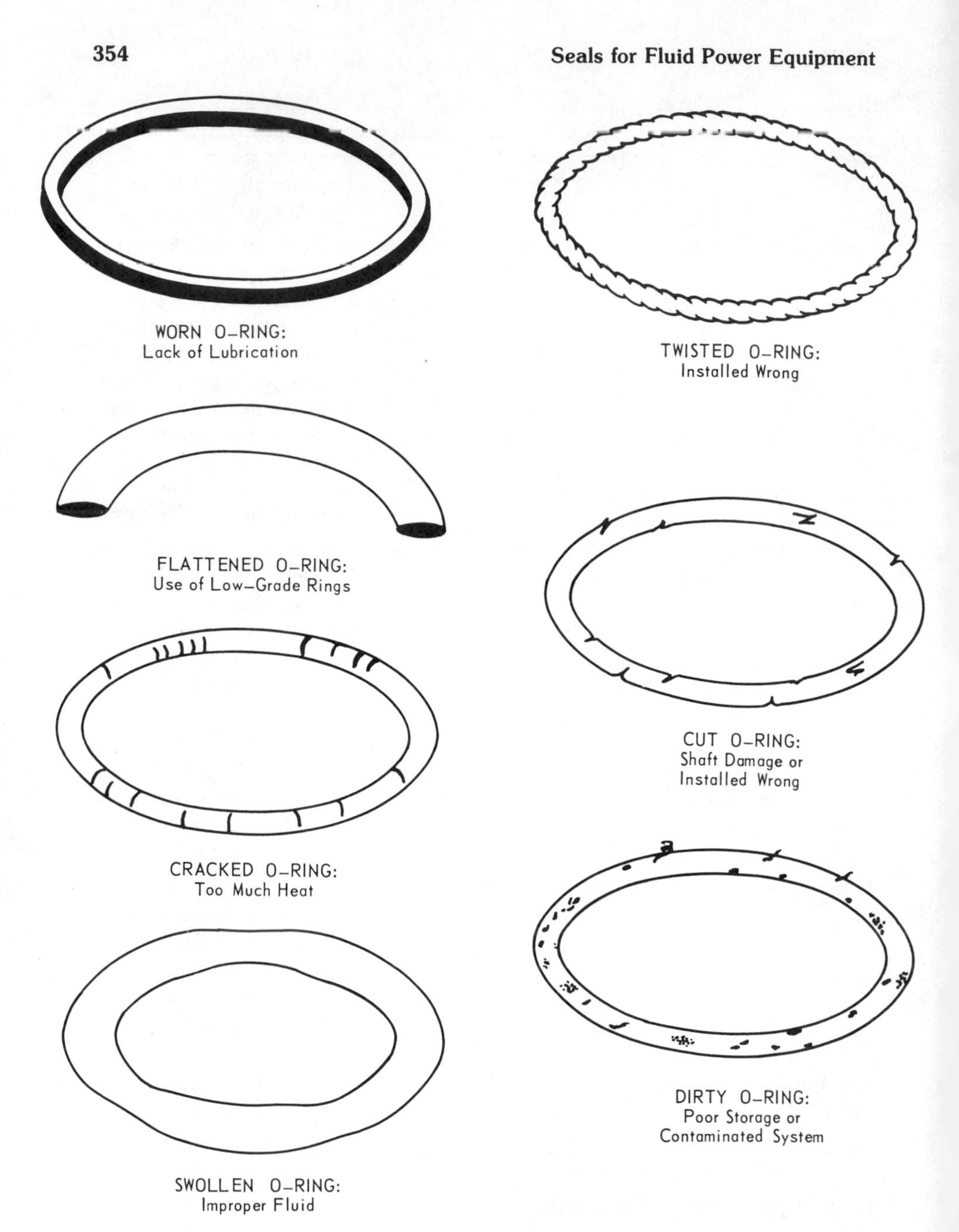

Figure 10.21 O-ring failures. (Courtesy of John Deere and Co., Waterloo, Iowa.)

the seal, find the path of leakage. Sometimes the leakage could be from worn gaskets, loose bolts, cracked housings, or loose line connections.

Inspect the outside sealing area of the seal to see if it is wet or dry. If wet, see whether the oil is running out or merely a lubricating film. During removal, continue to check for causes of leakage. (See Fig. 10.22.)

Check both the inner and outer parts of the seal for wet oil, which indicates leakage. When removing the seal, inspect the sealing surfaces or lips before washing. Look for unusual wear, warping, cuts and gouges, or particles embedded in the seal. On spring-loaded lip seals, be sure that the spring is seated around the lip, and that the lip was not damaged when first installed. Do not disassemble the unit any more than necessary to replace the faulty seals.

Check shafts for roughness at seal contact areas. Look for deep scratches or nicks that could have damaged the seal. Find out if a shaft spline, keyway, or burred end could have caused a nick or cut in the seal lip during installation. (See Fig. 10.23.)

Inspect the bore into which the seal is pressed. Look for nicks and gouges that could create a path of oil leakage. A coarsely machined bore can allow oil to seep out by a spiral path. Sharp corners at the bore edges can score the metal case of the seal when it is pressed in. These scores can make a path for oil leakage. (See Fig. 10.24.)

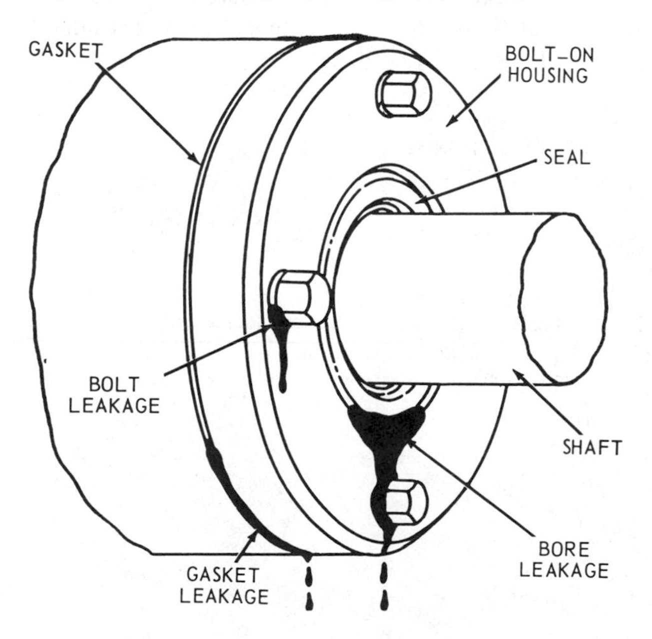

Figure 10.22 Common types of oil seal leaks.

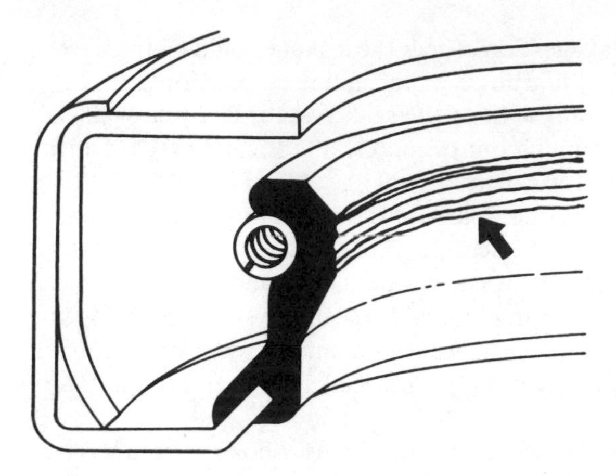

Figure 10.23 Seal worn by rough shaft.

Some hydraulic oils are harmful to certain seals, especially rubber lips. An incorrect oil can either harden or soften the synthetic rubber in seals and so damage them. If the lip is "spongy," this probably means that the seal and the hydraulic fluid are not compatible. If the seal is factory approved, an improper fluid has been used in the system. Hardening of the seal lip can be caused by either heat or chemical reaction with an improper fluid. Hardening of the seal lip on the area of the shaft contact is generally the result of heat from either the shaft or the fluid. (See Fig. 10.25.)

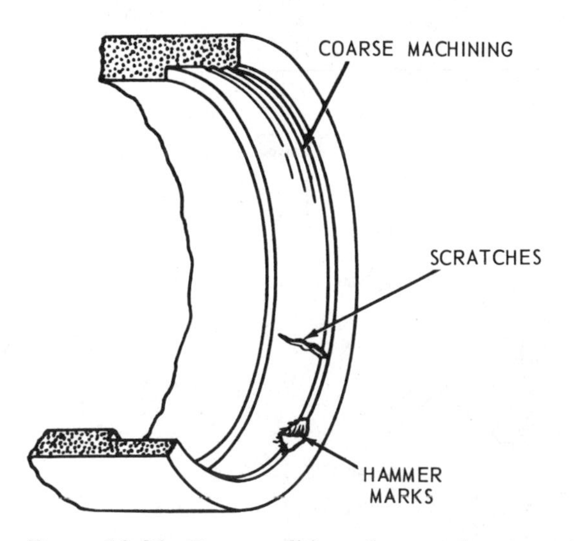

Figure 10.24 Bore conditions that can damage seals and cause leaks.

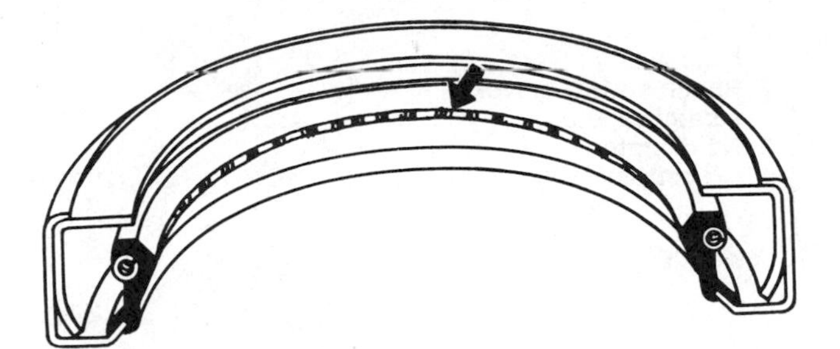

Figure 10.25 Seal lips damaged by heat. (Courtesy of Sperry Vickers, Troy, Michigan.)

10.5 TIPS TO AVOID PREMATURE SEAL FAILURE

Install only seals that are recommended by the manufacturer of the machine or component.

Use only the proper fluids as stated in the machine operator's manual.

Keep the seals and fluids clean and free of dirt.

Before installing seals, clean the shaft or bore area. Inspect these areas for damage. File or stone away any burrs or bad nicks and polish with a fine emery cloth for a ground finish; then clean the area to remove metal particles. In dynamic applications, the sliding surface for the seal should have a mirror finish for best operation. (See Fig. 10.26.)

Lubricate the seal, especially any lips, to ease installation. Use the hydraulic fluid to lubricate the seal. Also, soak packings in the hydraulic fluid before installing them.

With metal-cased seals, coat the seal's outside diameter with a thin film of gasket cement to prevent bore leakage.

Use a factory-recommended tool to install the seal properly. This is very important with press-fit-type seals. If a seal-driving tool is not available, use a circular ring such as an old bearing race that contacts the seal case near the outer diameter, or use a square wooden block. Do not use sharp tools.

Fit packings in snugly without using undue force. Make sure that the packings are not too tight.

Use shim stock to protect seals when installing them over sharp edges such as splines. Place rolled plastic shim stock over the sharp edge; then pull it out when the seal is in place.

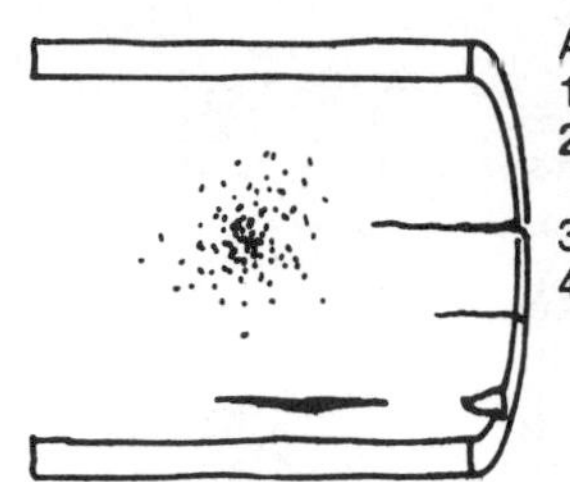

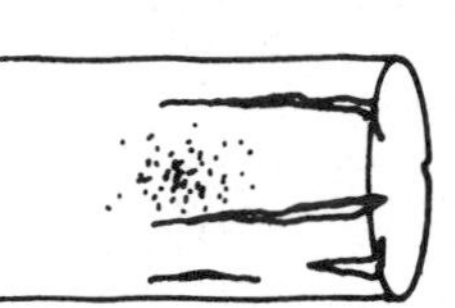

Figure 10.26 Tips for preventing seal damage. (Courtesy of Sperry Vickers, Troy, Michigan.)

Be sure that the seal is driven in evenly to prevent "cocking" of the seal. A cocked seal allows oil to leak out and dirt to enter. Be careful not to bend or "dish" the flat metal area of metal-cased seals. This causes the lips to be distorted. (See Fig. 10.27.)

After assembly, always check the unit by hand for free operation; if possible, before starting up the system.

Try to prevent dirt and grit from falling on piston rods and other parts and being carried into the seal. This material can quickly damage the seal or score metal surfaces.

When a new lip-type seal is installed on a clean shaft, a break-in period of a few hours is required to seat the seal lip with the shaft surface. During this period, the seal polishes a pattern on the shaft and the shaft in turn seats the lip contact, wearing away the knife-edge lip contact to a narrow band. During this period, slight seepage may occur. After seating, the seal should perform without any measurable leakage.

10.6 EXCLUSION RINGS, WIPER RINGS, OR SCRAPER RINGS

Exclusion rings, also known as wiper rings or scraper rings, are devices designed to prevent contaminants from entering the hydraulic system and damaging cylinder walls and other components. These contaminants, including dust, mud, ice, or metal particles, can be abrasive and corrosive. By excluding them, the wiper protects and extends the life of the rod and piston seal of the hydraulic cylinder.

Manufacturers offer exclusion rings in several configurations, cross sections, and materials for a variety of service conditions and operating fluids, Figure 10.28 represents a brief outline of the service conditions shown as light, medium, and heavy duty together with some configurations suitable for each type of service.

Light duty is generally defined as an indoor operation in a low-dust environment, and includes use on machine tools, lift trucks, hydraulic jacks, hydraulic shears, door openers, and so on. Light-duty classifications also includes use in outdoor conditions where the cylinder is in a protected environment such as under a roof, out of the rain, and out of direct wind.

Medium duty is described as an indoor application in a mildly dusty environment. Examples include iron foundries, wood finishing operations, grain mills, concrete mixing, and so on. It also includes outdoor applications, in low- or medium-dust conditions, under semiprotected areas where severe

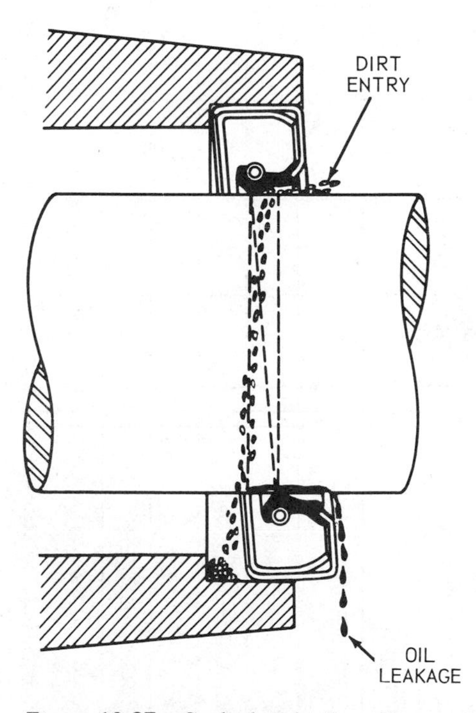

Figure 10.27 Cocked seals allow dirt to enter and oil to leak out. (Courtesy of Sperry Vickers, Troy, Michigan.)

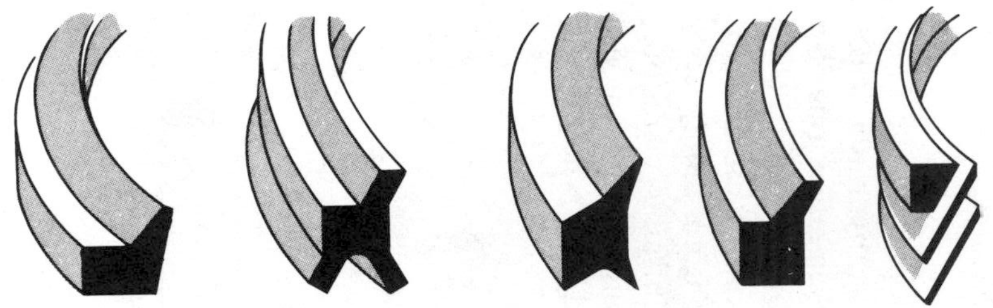

									EPR	FLUORO-CARBON ELASTOMERS
LIGHT-DUTY	X	X	X	X	X	X	X	X	X	X
MEDIUM-DUTY			X	X	X	X	X	X	X	X
HEAVY-DUTY								X	X	X
TEMPERATURE RANGE (°F)	−40 +190	−50 +250	−50 +250	−40 +190	−50 +250	−40 +190	−40 +190	−40 +250	−50 +250	−20 +450
RUBBER		X	X		X			X*	X	X
URETHANE (93 Duro.)	X			X		X	X			
NORMAL HYDRAULIC OIL	X	X	X	X	X	X	X	X		X
OIL-WATER EMULSIONS		X	X		X			X		X
WATER GLYCOL		X	X		X			X		X
PHOSPHATE ESTERS									X	X

*SEAL-GUARD CONSISTS OF A CUSHION RING AND TWO METALLIC RINGS.

Figure 10.28 Exclusion ring configurations and service conditions. (Courtesy of Penton/IPC, Cleveland, Ohio.)

weather and heavy rain are not normally a factor. Light rain or moisture is not detrimental to the cylinder.

Heavy duty is defined as those locations where heavy dust and grit particles are present. Service of this nature includes coal mines, brickyards, steel mills, concrete mills, and so on. Heavy-duty rings are also recommended for off-highway construction equipment used in severe weather conditions, including heavy rain, mud, snow, or ice.

Materials of construction include synthetic rubber, plastic, and metals, depending on the service condition and the fluid used in the hydraulic system. Ethylene propylene rubber is used primarily for fire-resistant fluid in applications such as water glycols and phosphate esters. Fluorocarbon elastomers are gaining increased use where it is desirable to stock one material for a wide variety of applications. They can be used in hydraulic oils and some fire-resistant oils. The material is also applicable over a considerable temperature range, making it ideal for fire-resistant fluids.

11

Fluid Power Systems: Operation, Maintenance, and Troubleshooting

Years ago, the maintenance of hydraulic systems was often performed on a hit-and-miss basis—a "fix-it-when-it-breaks-down" type of operation. With today's sophisticated machinery, high production schedules, and the prohibitive cost of downtime, few companies can afford to operate in this manner. That many of them still do, however, is evidenced by the kind of problems encountered by service personnel in answering customer complaints. Here are several that occur most frequently (Fig. 11.1):

Insufficient oil in the reservoir

Clogged and dirty oil filters

Loose intake lines

Pump shaft turning in the wrong direction

Improper grades of oil (too heavy or too light)

Operating pressures set too high or too low

It is obvious that all of these problems could have been solved (or prevented) with only a basic knowledge of hydraulics and attention to simple maintenance procedures. The expense incurred in carrying out a planned maintenance program is regained many times over in reduced operating costs and savings in machine downtime. Such a program begins with the system designer's selection of good-quality components and adequately sized pipe, tubing, or hydraulic hose.

Maintenance personnel and machine operators can easily be trained to recognize the signs and symptoms of impending hydraulic troubles. The operator can hear any unusual noises, feel excessive operating temperature, and see leaking lines, leaking fittings, or oil deposits on or around the machine. In following up on such a complaint, a trained maintenance mechanic can, at a glance, note the color of the fluid and its level in the reservoir sight glass. Dirty or discolored oil would indicate the need for changing the filter cartridge and possibly the fluid itself.

It has been estimated that as much as 70% of all hydraulic problems may be traced directly to the fluid. Sampling and testing it periodically is a major factor in obtaining reliable performance. Proper maintenance will keep hydraulic troubles to a minimum. However, it is obvious that not all problems will be eliminated. Troubleshooting will be a regular part of the maintenance routine.

To benefit fully from maintenance experience, a system of good reports and records is essential. Such records should include:

A description of the symptoms and the date detected

Description of the preliminary investigation and its results

Explanation of corrective action taken, replacement parts required, date repairs were made, and length of downtime

A record of when fluid is added or changed, filter cartridges replaced, or strainer elements cleaned

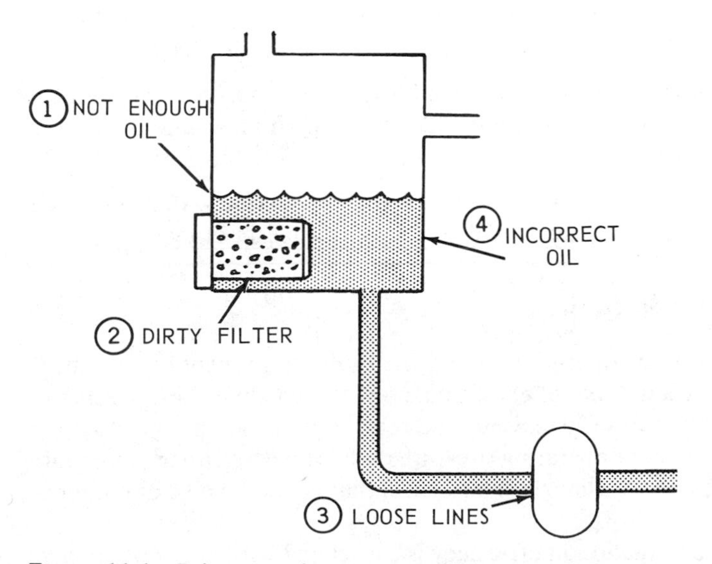

Figure 11.1 Primary maintenance problems.

These reports and records, if analyzed frequently, will indicate areas that require special attention as well as recurring troubles that may be anticipated and corrected before a breakdown occurs.

To provide some basic guidelines for the use and care of hydraulic systems, the following section describes some general guidelines for operation, maintenance, and troubleshooting. These must be coupled with a detailed understanding of the specific system under consideration in order to arrive at a maintenance and troubleshooting plan that will work. There is no substitute for thoroughly understanding the contents of the operation and maintenance manual for a hydraulic system. Thus it is imperative that it be studied by the mechanic prior to working on a given system.

11.1 SYSTEM DESIGN CONSIDERATIONS

There is, of course, little point in discussing the design of a system that has been operating satisfactorily for a period of time. However, a seemingly uncomplicated procedure, such as relocating a system or changing a component part, can cause problems. Because of this, the following points should be considered:

Each component in the system must be compatible with and form an integral part of the system. For example, an inadequate-size filter on the inlet of a pump can cause cavitation and subsequent damage to the pump.

All lines must be of proper size and free of restrictive bends. Undersize or restricted line results in a pressure drop in the line itself.

Some components must be mounted in a specific position with respect to other components or the lines. The housing of an in-line pump, for example, must remain filled with fluid to provide lubrication.

The inclusion of adequate test points for pressure readings, although not essential for operation, will expedite troubleshooting.

11.2 SYSTEM OPERATION

Periodically during the operation of any hydraulic equipment or system, it is prudent to make a number of checks on a regular schedule. Routine checks of the operating pressures will indicate whether or not the pump is maintaining its efficiency. A drop in operating pressure could indicate either a drop-off in pump performance or changes in pressure, that is, relief valve or compensator settings.

Because the reduction in efficiency is associated with a loss of energy, a check of the operating or system temperatures will indicate the presence of such a loss. Normally, the reservoir of a standard design will have a thermometer for checking temperatures. Another way to check temperatures is the use of the touch method, where the normal tolerance level of a human being is approximately 130°F. If one can touch a hot component and hold his or her hand there, it would be expected that the temperature would range somewhere between 100 and 125°F. If the material were beyond the upper ranges of the comfortable, it may be concluded that the system needs attention.

A check of the apparent noise level of the equipment is also another sign to be cognizant of as wear tends to make components noisier. A stethoscope with a metallic pickup can often be used to detect the location of faulty component bearings.

Finally, during the operation of the equipment it is wise to keep a constant eye on the oil level of the equipment. In any hydraulic system the oil not only is a medium of energy transfer but is also the means by which the equipment is lubricated. Lack of oil for lubrication will result in immediate failure of the equipment. Particularly in the case of components featuring hydrodynamic bearings, the maximum period of time that hydrodynamic bearings could be expected to operate without catastrophic failure would be about 30 sec.

11.3 KNOWING THE SYSTEM

Probably the greatest aid to troubleshooting is the confidence of knowing the system. Every component has a purpose in the system. The construction and operating characteristics of each should be understood. For example, knowing that a solenoid-controlled directional valve can be manually actuated will save considerable time in isolating a defective solenoid. Some additional practices that will increase your ability and also the useful life of the system follow:

Know the capabilities of the system: Each component in the system has a maximum rated speed, torque, or pressure. Loading the system beyond the specifications simply increases the possibility of failure.

Know the correct operating pressures: Always set and check pressures with a gauge. How else can you know if the operating pressure is above the maximum rating of the components? A question may arise as to what the correct operating pressure is. If it is not correctly specified on the hydraulic schematic, the following rule should be applied:

> The correct operating pressure is the lowest pressure that will allow adequate performance of the system function and still remain below

the maximum rating of the components and machine. Once the correct pressures have been established, note them on the hydraulic schematic for future reference.

Know the proper signal levels, feedback levels, and dither and gain setting in servo control systems: If they are not specified, check them when the system is functioning correctly and mark them on the schematic for future reference.

Analyze the system and develop a logical sequence for setting valves, mechanical stops, interlocks, and electrical controls. Tracing of flow paths can often be accomplished by listening for flow in the lines or feeling them for warmth. Develop a cause-and-effect troubleshooting guide similar to the charts appearing in Sec. 11.7. The initial time spent on such a project could save hours of system downtime.

The ability to recognize trouble indications in a specific system is usually acquired with experience. However, a few general trouble indications can be discussed.

Excessive heat means trouble: A misaligned coupling places an excessive load on bearings and can be readily identified by the heat generated. A warmer-than-normal tank return line on a relief valve indicates operation at relief valve setting. Hydraulic fluids that have a low viscosity will increase the internal leakage of components, resulting in a heat rise. Cavitation and slippage in a pump will also generate heat.

Excessive noise means wear, misalignment, cavitation, or air in the fluid: Contaminated fluid can cause a relief valve to stick and chatter. These noises may be the result of dirty filters or fluid, high fluid viscosity, excessive drive speed, low reservoir level, loose intake lines, or worn couplings.

11.4 STARTUP AND MAINTENANCE OF HYDRAULIC SYSTEMS

To ensure efficient operation of system components, it is essential, if repairs become necessary, to adhere to the operating instructions supplied with the component or the engineering specifications given. Whether assembling or disassembling, it is important that the internal parts of the components be kept clean. The operating fluid should be of a type indicated by the manufacturer, and must be maintained at a suitable degree of cleanliness. Maintenance frequency and operational guidelines are dependent on the conditions under which the equipment is working.

11.4.1 Startup

Before filling with hydraulic fluid, the tank and pipes should be thoroughly checked to make sure that they are clean. It is important that the check be made immediately prior to the filling and, if necessary, the whole system should be drained.

If reservoirs are painted internally, a careful check should be made that the hydraulic fluid to be used is compatible with the paint. When using highly flammable fluids it is essential, before filling with oil, to check that all hydraulic components used in the system are fitted with seals that are compatible with such fluids, including pipe connections and flanges.

A careful inspection of system piping and wiring should be made, by reference to the appropriate circuit and associated cycle charts, before filling and pressurization. The system should be carefully inspected to ensure proper component alignment. Misalignment between electric motor and pumps can lead to premature damage to the pump.

If not supplied already charged for use, hydraulic accumulators should be charged with the specified volume of nitrogen before installation in the system. Gas charging pressure should be clearly indicated on the accumulator (by sticker) and also in the circuit, so that checks can later be made if required.

A filter must always be used when filling oil into the system. While the degree of filtration required varies, 40 μm is sufficient for most components with the exception of servo equipment. At startup, the setting of the pressure relief valves should be kept as low as possible. An exception to this is in the case of the special fixed-setting pressure relief valves for use in accumulator systems, which do not allow any adjustment. It is recommended, in view of existing safety regulations, that only those personnel directly concerned with startup procedures be present.

Electric motors should be cycled briefly (approximately 5 to 10 sec). The direction of rotation should be verified. It should be determined whether couplings or other connecting elements are loose before startup. Instructions supplied by individual pump manufacturers must also be adhered to.

Before operating pressure setting can be gradually increased, ensure that the pump is running quietly and continuously. Any leakage occurring at this time should be rectified immediately. When running the system at low pressure, the system should be properly vented. Check fluid level in the reservoir and refill if necessary with the same fluid. When operating pressure has been reached and function is satisfactory, pressure switches, float switches, thermostats, and so on, should be set. All component settings should be recorded on a written schedule after successful startup of the system.

11.4.2 Maintenance

Preplanning of maintenance, and in particular, preventative routine servicing, should be considered during the design stage. Often the necessity of completely emptying a reservoir during repairs can be avoided by the simple addition of a shutoff valve in front of a pump or a manifold block. It is recommended that a maintenance manual be prepared during startup, and later be passed on to maintenance personnel to indicate at what intervals certain parts must be inspected.

The fluid level must be checked constantly during the startup period, after startup, then daily, and weekly. During startup, the filters should be checked approximately every 2 to 3 hr and cleaned if necessary. Thereafter, a daily check should be made and after approximately 1 week cleaning may take place as required. Particularly careful attention must be paid to suction filters. After the "running-in" time they must be checked and/or cleaned at least once per week.

Renewing of the system fluid is dependent on several operating factors and is undertaken according to the degree of aging or contamination of the oil. On systems where the ratio of pump flow to tank volume is 1:3 or greater, the first fluid change should take place 50 to 100 hr after startup. On larger systems a fluid change must be made, however, after 10,000 hr of operation at the latest, and on smaller systems after approximately 5000 hr. To keep a running check on the condition of the fluid, take a sample of the oil and pass it through a filter paper or clean cloth. The coloring of the residue will indicate the degree of aging. If the color is blue-black, an oil change is required immediately. It must be cautioned that scheduling oil changes based strictly on operating time is dangerous. The oil condition should be routinely and constantly (at least weekly) monitored. The fluid should be replaced depending on the degree of aging/degradation and contamination as noted by these inspections, irrespective of the operating time period.

Hydraulic accumulator charging pressure should be checked periodically. It is essential that the oil side of the accumulator is not under pressure at the time. The operating temperature should be measured not only in the fluid tank, but also in the region of the pump bearings. An increase in temperature indicates wear. (Increasing friction and leakage while converting hydraulic energy into heat.) The temperature of the fluid near the bearings will increase when the pump is worn, because this wear will reflect itself as higher internal leakage. This internal leakage will be seen in the case drain lines. The bearings are also exposed to the case drain oil. Thus an increase in oil temperature in this area reflects pump wear.

The piping system should be inspected at regular intervals for leakage, particularly for underfloor pipework. Main pressure and pilot pressure should be checked at least weekly. Pressure adjustments should be noted in the maintenance manual. Frequent readjustment of system pressure indicates possible wear on the pressure relief components. Pressure drops should be investigated to determine cause. Figure 11.2 summarizes the points of importance in checking a hydraulic system completely before considering the maintenance tasks completed and turning the system back to operation.

11.5 MAINTENANCE PROCEDURES AND HINTS

Three simple maintenance procedures have the greatest effect on hydraulic system performance, efficiency, and life. Yet the very simplicity of them may be the reason they are so often overlooked. They are:

1. Maintaining a clean sufficient quantity of hydraulic fluid of the proper type and viscosity.
2. Changing filters and cleaning strainers.
3. Keeping all connections tight, but not to the point of distortion, so that air is excluded from the system.

Once the equipment has been placed in operation, it is rewarding to adopt a timetable of regular maintenance. Regular maintenance will require a systematic routine of going through the system on a regular basis to see that the equipment is given the best of available care.

At this point it is good to reemphasize the use of an approved hydraulic fluid. An approved hydraulic fluid includes or covers not only an oil that has good lubricity over the operating temperature range of the equipment but the oil must also contain a certain number of additives to aid in the prevention of accelerated wear. The oil should contain an antifoaming agent, which aids in the prevention of air entrapment in the oil. There are a number of ways in which air becomes trapped in the oil, such as a leaky suction line or oil splashing back into the reservoir and entrapping air. A good premium-grade hydraulic fluid will also contain antiwear additives.

Depending on the application of the hydraulic equipment, it is a good idea to change the oil in the system regularly. Oil may degrade in one of two ways. First, oil may become contaminated and the contamination may act to produce accelerated wear. Second, oil may change its chemical composition by separation of acids and tars. Oil temperatures may cause this breakdown. Separation of the acids from the oil will cause embrittlement of the seals. The separation of tars and sludges from the oil will clog the filters and introduce an increasing amount of "crud" into the system. The chemical breakdown of the hydraulic oil will also mean a change in the lubricity, which is again another contributing factor in the accelerated wear of hydraulic com-

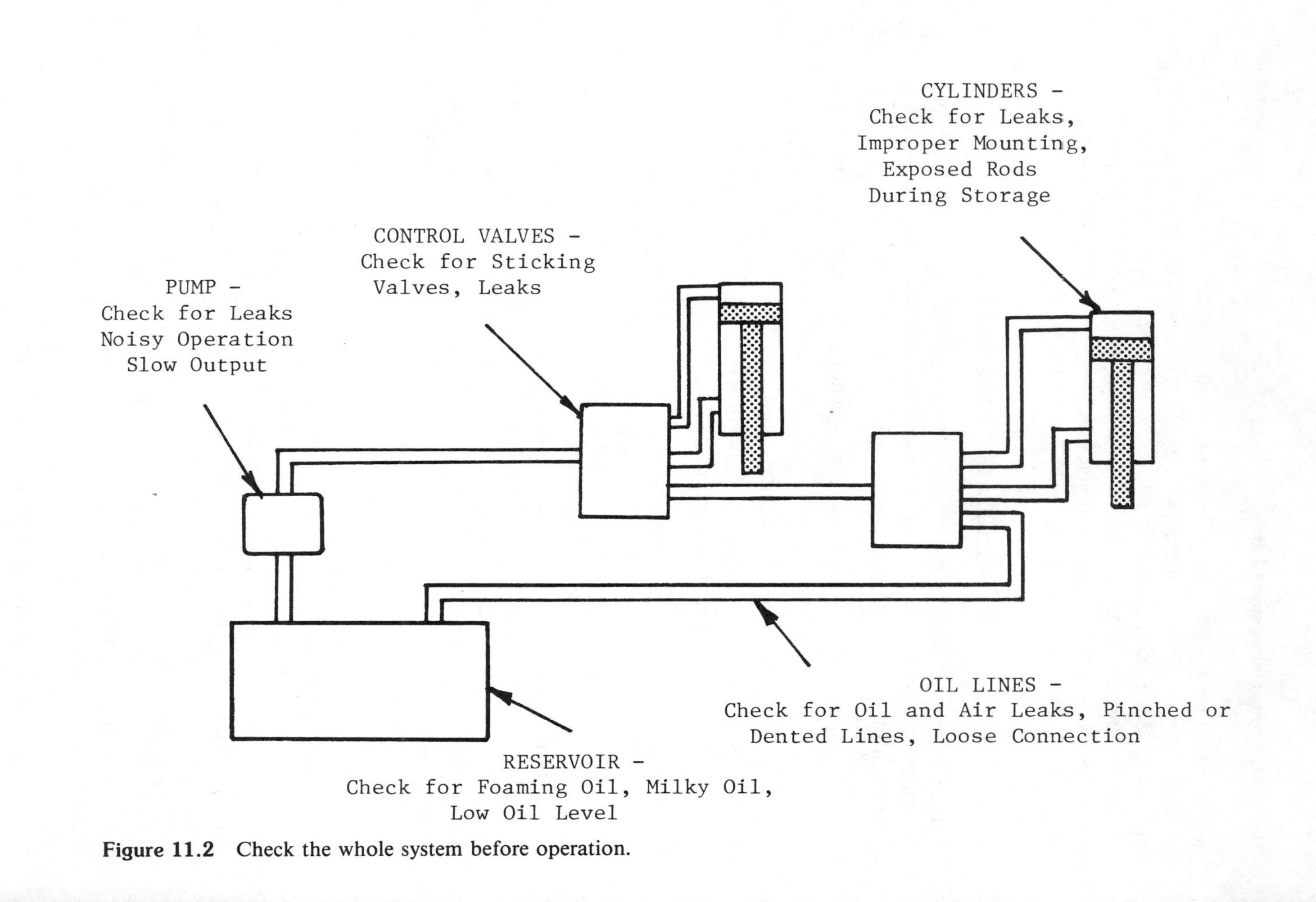

Figure 11.2 Check the whole system before operation.

ponents. Many companies, both hydraulic component manufacturers and oil companies, offer the service of testing by analysis hydraulic oils in both the amount of contamination in and the chemical composition of the oil.

The method of changing oil is as follows. Completely drain the system. This is usually accomplished by removal of the drain plug at the lowest part of the reservoir. Once the oil has been drained from the reservoir, the reservoir cleanout cover is removed to swab and cleanout the bottom of the reservoir. Be sure to use a lint-free rag (a lint type will do more harm than good). At this time it is also appropriate to remove the suction filter and soak it in kerosene, overnight if possible. Before reinstalling in the system, the suction filter should be blown out with an air gun of some type to remove any solvent that is in the filter element and cover or housing.

Where a good amount of dirt is left in the remainder of the hydraulic system, it may be removed by means of a "short fill." If the old oil is still in a good chemical state, it should be retained for cleaning of the system. However, if this oil is chemically bad, it should not be used in the system for any reason. *Short fill* means to introduce sufficient oil into the reservoir to safely operate the system—slightly above the minimum fluid level. The amount of oil needed for this flushing or cleaning procedure should be introduced into the reservoir with a filter cart, if possible. This will ensure that reasonably clean oil is used for this operation.

The oil is then circulated throughout the system for 15 to 20 min to ensure that the new cleaner oil is completely circulated throughout and the dirty oil left in the system is returned to the reservoir. Be sure that the oil level does not go below the recommended minimum. Once the oil has been completely circulated for the 15 to 20 min, it again should be drained from the reservoir and saved in a clean drum for future use. This oil is not contaminated but could contain a relatively high degree of metallic and other particles that have been broken loose from the system. Failing pumps, motors, and so on, create a sea of contamination in the system.

Depending on the type of filter material, filter, or indicator used, it may be wise at this time to change the filter element to prevent bypassing and a partial filtration situation. Remember, however, that even partial filtration is better than no filtration at all. Again, a regular schedule of filter change is strongly recommended. If regular filter changes cannot be made, the alternative method of the bypass indicator can be employed for best economy. As soon as bypassing occurs, the filter should be changed, making the best use of the filter element.

In the case of short-use failures, that is, when a pump or a component has failed soon after installation, the reason for failure is not so obvious and one frequently hears the remark: "Why shouldn't the new pump last for years? The old pump used this oil for years." The answer is the fact that the old pump grew old with increasing contamination and got progressively worse. The new pump, however, with its tighter clearance is not going to be able to tolerate this high level of contamination and is going to go into a state of rapid deterioration. Without a good flush and oil change, a new item may last up to 6 months with luck.

Water (probably condensation, but frequently a water-oil cooler leak) may also become a problem in the system. A needle valve in a low point in the system will provide a check point for the presence of water. Water in the oil will also appear as a milky condition in the hydraulic oil, usually at the oil-level gauge. Water in hydraulic fluids decreases viscosity and increases component wear. Air in the oil makes oil appear foamy or milky. The presence of air may also be observed at the oil-level gauge on the reservoir.

Periodic visual and manual checks are recommended. A couple of areas of interest are: shaft seals, coupling inserts (inserts may disintegrate due to misalignment), and valve spool silting during long periods of nonshifting operation.

11.6 HYDRAULIC SYSTEM MAINTENANCE ACCESSORIES

There are a number of accessories that will make the maintenance of equipment simpler and easier, thereby requiring minimum technical assistance. The use of isolators and snubbers is highly recommended in applying hydraulic gauges. Gauge isolators allow one to remove the system pressure from the gauges when a visual check is not required. A snubber will perform much the same function except that its operation will be manual as in the case of a globe valve.

A solenoid trouble detector on valving is a wise investment if systems are quite involved electrically and downtime during trouble detection should be held to a minimum. A trouble detector is a device applied to a directional control valve for the purpose of detecting and indicating the solenoid's condition. When a solenoid is energized, the trouble detector is provided with a white light which indicates energization. Common reasons for coil burnouts are sticking of the spool due to varnish or silting, under- or overvoltage for the coil itself, or other lumps that may pass the valve, such as dirt or Teflon tape.

When manifolds are used in the hydraulic system it is advisable to include in the manifold design as many test points as possible. These test points will prove invaluable in later days when the equipment is in operation. Omission of these manifold test points will generally make troubleshooting impossible.

The use of a hydraulic filter in the pressure, suction, or return line should be accompanied with appropriate sensors to make efficient use of the elements. Electrical switches are available that will indicate when the filter begins bypassing. The electrical signal may be relayed to control stations or shut off electric motors. Visual indicators may also be employed. Where startup temperatures produce oil viscosities much higher than that recommended for hydraulic pumps, use immersible electric heaters.

Caution must be used when adding electric immersion heaters. Oil must circulate by mechanical means by the heater. Oil is not a good heat conductor and tends to stratify. If the heater cannot be installed directly in the fluid flow, a propellor-type device should be considered to ensure fluid movement.

Where environmental contamination is likely, as in the case of some mobile systems, the use of a pressurized reservoir is a solution. By pressurizing the reservoir one guarantees that contaminated air will not be drawn into the system, provided that a suitable filter is employed in the reservoir pressurizing air system.

An excellent alternative to a filter system employed in a pressurized reservoir is the bladder-type barrier mechanism illustrated in Fig. 2.43.

If system leakage is possible and regular checks of the oil level may be neglected, it is strongly recommended that float-level switches be employed to indicate when the oil level is low. Pump damage is very likely in a short time should the pump run without oil.

System (open-loop) temperatures are best monitored at the reservoir. Therefore, most reservoirs are supplied with temperature gauges. There are some exceptions to this rule, as in the case of closed-loop hydrostatic transmission equipment, where much of the hydraulic oil is usually captive to the system. Without the use of a suitable hot oil shuttle and heat exchanger, it would be prudent to locate a temperature indicator somewhere in this system.

When water-oil heat exchangers are used, a thermostatically controlled water-regulating valve is a wise investment. The sensor is inserted in the reservoir. As the temperature of the oil rises, the water valve opens and modulates to regulate the water flow and control oil temperature.

As mentioned earlier, flexible couplings are available with inserts that disintegrate, indicating misalignment, and thus prevent damage to both the electric motor and the hydraulic pump. All reservoirs should have a magnetic trap (filter). Ten-micrometer air breathers are available for sealing reservoirs. They may be obtained with integral check (relief) valves for controlling under- or overpressurization.

Where it is possible, subplate mounted components should be used; they require the minimum of replacement time. In the case of breakdown these items are extremely handy because all that is required is removal of a few bolts. A four-port in-line-mounted, pipe-threaded valve will require a large amount of time for replacement. By their nature, pipe-threaded fittings become damaged progressively by each separation and require replacement after being used two, three, or four times. The use of SAE straight-threaded O-ring fittings is becoming popular for a good many reasons. Foremost among these reasons are that a new O ring will constitute a new seal, and contamination during disassembly is less likely to occur.

A portable fluid filtering and transfer unit is also a worthwhile investment in that new oil, received in barrels that may not be clean, can be cleaned up. Oil that is used for short fills can also be cleaned. Where it is desirable to pump oil out of the system and back in again this unit is an invaluable piece of equipment. Portable fluid filtering and transfer units may also be connected to systems periodically, and overnight if practical and necessary (see Fig. 11.3).

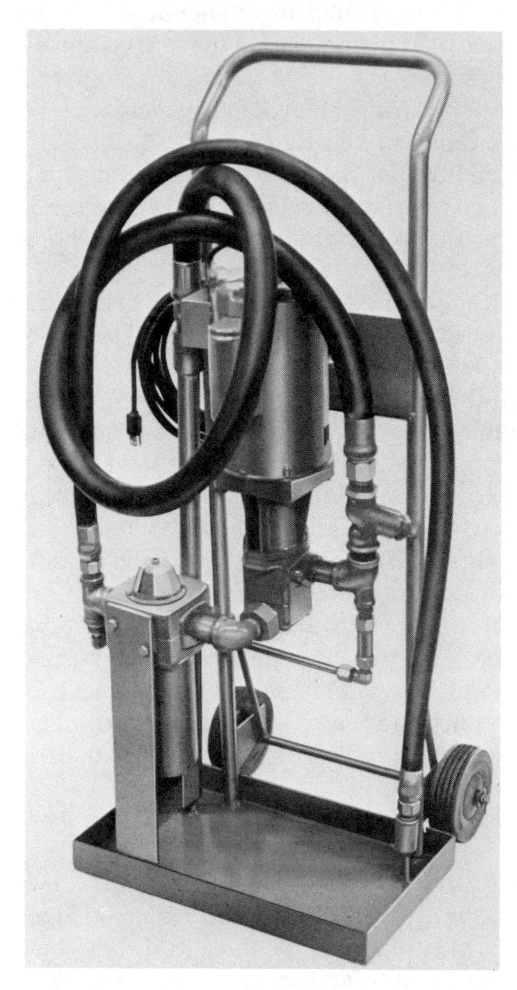

Figure 11.3 Hydraulic oil filtration cart. (Courtesy of Sperry Vickers, Troy, Michigan.)

11.7 TROUBLESHOOTING GUIDES

The troubleshooting charts that follow apply to a general system but should provide an intuitive feeling for a specific system. The charts are arranged in five main categories. The heading of each is an effect which indicates a malfunction in the system. For example, if a pump is exceptionally noisy, refer to Chart I, "Excessive Noise." The noisy pump appears in column A under the main heading. In column A there are four probable causes for a noisy pump. The causes are sequenced according to the likelihood of happening or the ease of checking it. The first cause is cavitation and the remedy is "a." If the first cause does not exist, check for cause 2, and so on.

Chart I

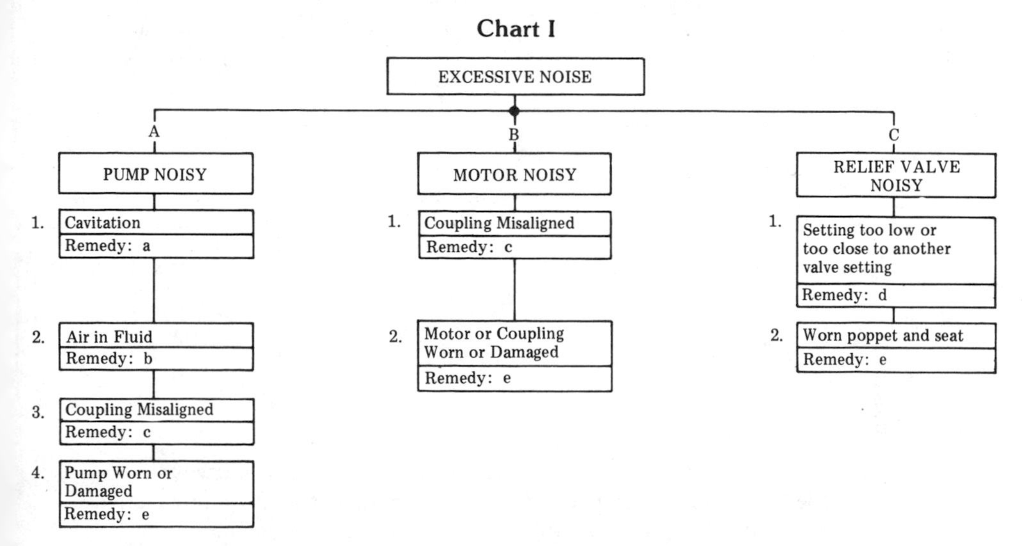

Remedies:

a. Any or all of the following: Replace dirty filters. Wash strainers in solvent compatible with system fluid. Clean clogged inlet line. Clean reservoir breather vent. Change system fluid. Change to proper pump drive motor speed. Overhaul or replace supercharge pump. Fluid may be too cold.

b. Any or all of the following: Tighten leaky inlet connections. Fill reservoir to proper level (with rare exception all return lines should be below fluid level in reservoir). Bleed air from system. Replace pump shaft seal (and shaft if worn at seal journal).

c. Align unit and check condition of seals, bearings, and coupling.

d. Install pressure gauge and adjust to correct pressure.

e. Overhaul or replace.

Courtesy of Sperry Vickers, Troy, Michigan.

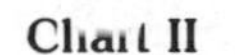

Chart II

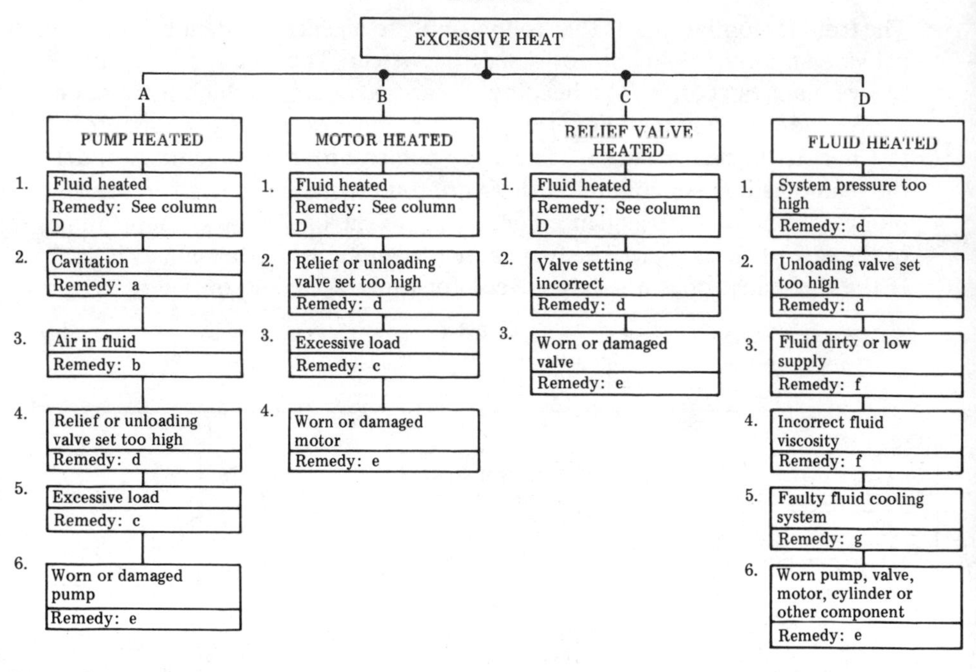

Remedies:

a. Any or all of the following: Replace dirty filters. Clean clogged inlet line. Clean reservoir breather vent. Change system fluid. Change to proper pump drive motor speed. Overhaul or replace supercharge pump.

b. Any or all of the following: Tighten leaky inlet connections. Fill reservoir to proper level (with rare exception all return lines should be below fluid level in reservoir). Bleed air from system. Replace pump shaft seal (and shaft if worn at seal journal).

c. Align unit and check condition of seals and bearings. Locate and correct mechanical binding. Check for work load in excess of circuit design.

d. Install pressure gauge and adjust to correct pressure (keep at least 125 psi difference between valve settings).

e. Overhaul or replace.

f. Change filters and also system fluid if of improper viscosity. Fill reservoir to proper level.

g. Clean cooler and/or cooler strainer. Replace cooler control valve. Repair or replace cooler.

Courtesy of Sperry Vickers, Troy, Michigan.

Chart III

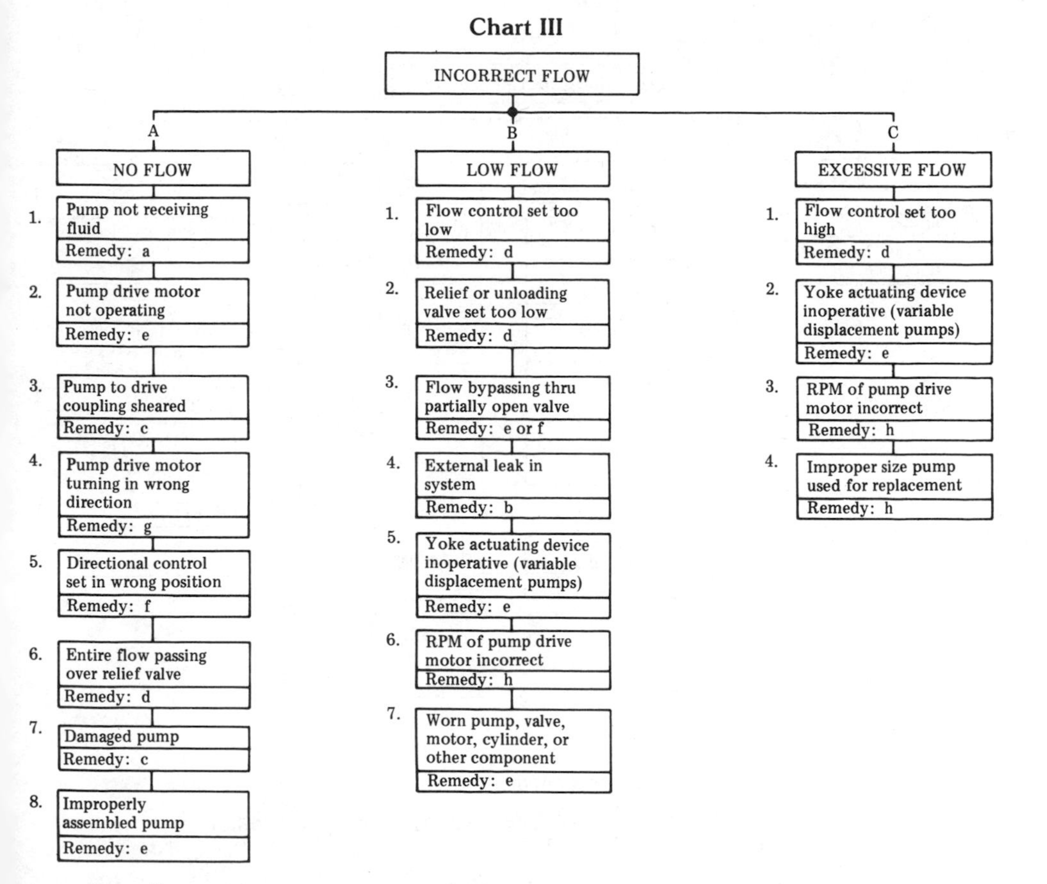

Remedies:

a. Any or all of the following: Replace dirty filters. Clean clogged inlet line. Clean reservoir breather vent. Fill reservoir to proper level. Overhaul or replace supercharge pump.

b. Tighten leaky connections. Bleed air from system.

c. Check for damaged pump or pump drive. Replace and align coupling.

d. Adjust.

e. Overhaul or replace.

f. Check position of manually operated controls. Check electrical circuit on solenoid operated controls. Repair or replace pilot pressure pump.

g. Reverse rotation.

h. Replace with correct unit.

Courtesy of Sperry Vickers, Troy, Michigan.

Chart IV

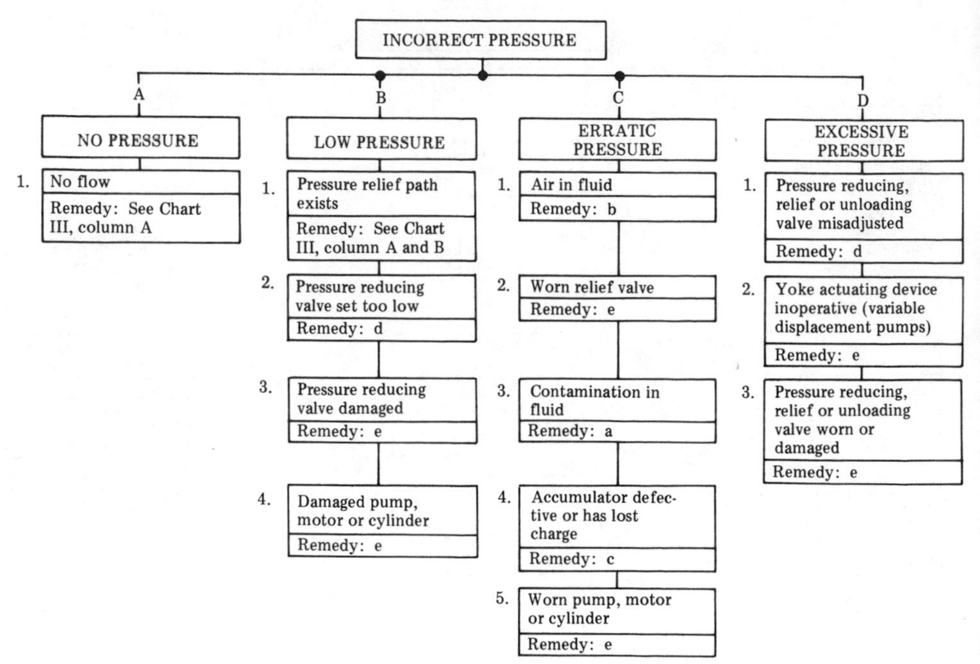

Remedies:

a. Replace dirty filters and system fluid.

b. Tighten leaky connections (fill reservoir to proper level and bleed air from system).

c. Check gas valve for leakage. Charge to correct pressure. Overhaul if defective.

d. Adjust.

e. Overhaul or replace.

Courtesy of Sperry Vickers, Troy, Michigan.

Chart V

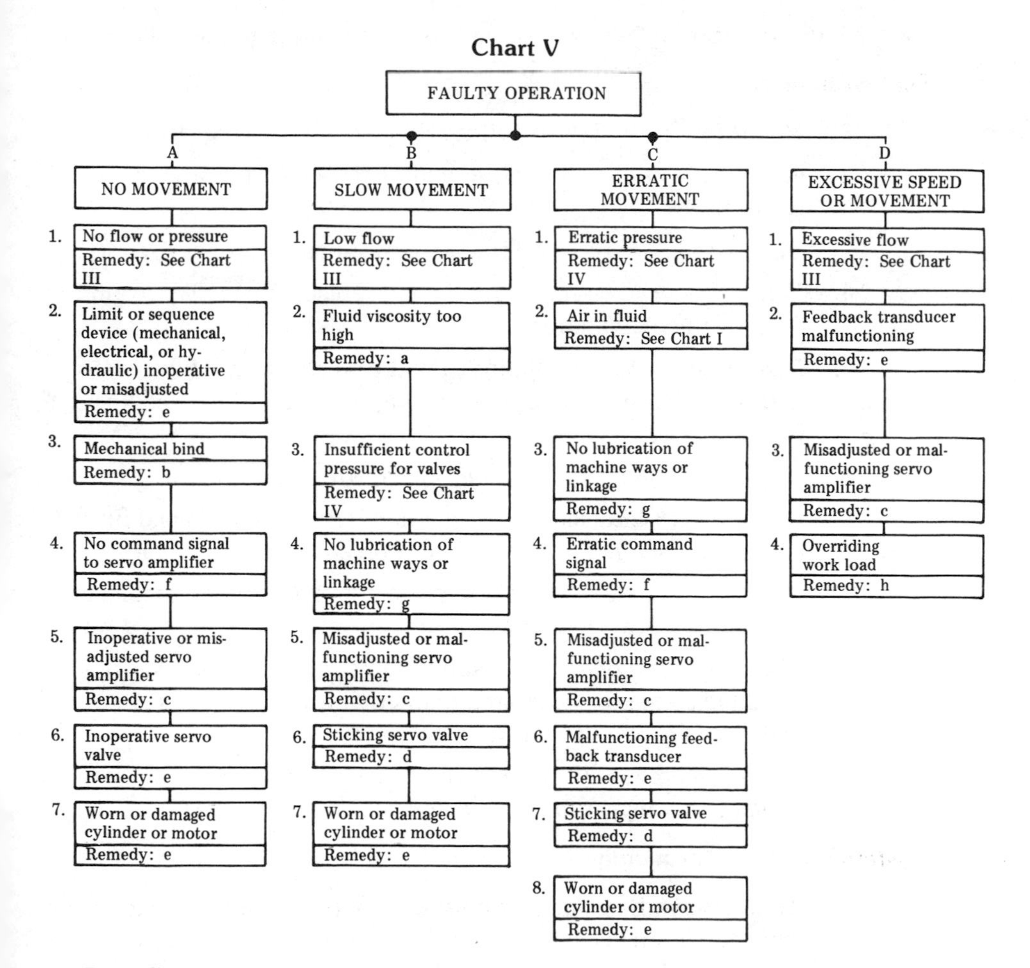

Remedies:

a. Fluid may be too cold or should be changed to clean fluid of correct viscosity.
b. Locate bind and repair.
c. Adjust, repair, or replace.
d. Clean and adjust or replace. Check condition of system fluid and filters.
e. Overhaul or replace.
f. Repair command console or interconnecting wires.
g. Lubricate.
h. Adjust, repair, or replace counterbalance valve.

Courtesy of Sperry Vickers, Troy, Michigan.

Faulty operation (Chart V) can be broken down further as follows.

System Inoperative

1. *No oil in system*: Fill to full mark. Check system for leaks.
2. *Oil low in reservoir*: Check level and fill to full mark. Check system for leaks.
3. *Oil of wrong viscosity*: Refer to specifications for proper viscosity.
4. *Filter dirty or plugged:* Drain oil and replace filters. Try to find source of contamination.
5. *Restriction in system*: Oil lines could be dirty or have inner walls that are collapsing to cut off oil supply. Clean or replace lines. Clean orifices.
6. *Air leaks in pump suction line*: Repair or replace lines.
7. *Dirt in pump*: Clean and repair pump. If necessary, drain and flush hydraulic system. Try to find the source of contamination.
8. *Badly worn pump*: Repair or replace pump. Check for problems causing pump wear, such as misalignment or contaminated oil.
9. *Badly worn components:* Examine and test valves, motors, cylinders, and so on, for external and internal leaks. If wear is abnormal, try to locate the cause.
10. *Oil leak in pressure lines*: Tighten fittings or replace defective lines. Examine mating surfaces on couplers for irregularities.

System Operates Erratically

1. *Air in system*: Examine suction side of system for leaks. Make sure that the oil level is correct. (Oil leak on the pressure side of the system could account for loss of oil.)
2. *Cold oil*: Viscosity of oil may be too high at start of warm-up period. Allow oil to warm up to operating temperature before using hydraulic functions.
3. *Components sticking or binding*: Check for dirt or gummy deposits. If dirt is caused by system internal contamination, try to find the source. Check for worn or bent parts.
4. *Pump damaged*: Check for broken or worn parts. Determine the cause of pump damage.
5. *Dirt in relief valves*: Clean relief valves.

6. *Restriction in filter or suction line*: Suction line could be dirty or have inner walls that are collapsing to cut off oil supply. Clean or replace suction line. Also, check filter line for restrictions.

System Operates Slowly

1. *Cold oil*: Allow oil to warm up before operating machine.
2. *Oil viscosity too heavy*: Use oil recommended by the equipment manufacturer.
3. *Insufficient engine speed*: Refer to operator's manual for recommended speed. If machine has a governor, it may need adjustment.
4. *Low oil supply*: Check reservoir and add oil if necessary. Check system for leaks that could cause loss of oil.
5. *Adjustable orifice restricted too much*: Back out orifice and adjust it. Check machine specifications for proper setting.
6. *Air in system*: Check suction side of the system for leaks.
7. *Badly worn pump*: Repair or replace pump. Check for problems causing pump wear, such as misalignment or contaminated oil.
8. *Restriction in suction line or filter*: Suction line could be dirty or have inner walls that are collapsing to cut off oil supply. Clean or replace suction line. Examine filter for plugging.
9. *Relief valves not properly set or leaking*: Test relief valves to make sure that they are opening at their rated pressure. Examine valves for damaged seats that could leak.
10. *Badly worn components*: Examine and test valves, motors, cylinders, and so on, for external and internal leaks. If wear is abnormal, try to locate the cause.
11. *Valve or regulators plugged*: Clean dirt from components. Clean orifices. Check source of dirt and correct.
12. *Oil leak in pressure lines*: Tighten fittings or replace defective lines. Examine mating surfaces on couplers for irregularities.
13. *Components not properly adjusted*: Refer to machine technical manual for proper adjustment of components.

12

Fluid Power System Malfunction Detection and Diagnosis

12.1 MALFUNCTION DETECTION

If you deal with hydraulically powered equipment for any length of time, you are eventually going to experience some problems with the system. When that time comes, one of your greatest assets will be the ability to diagnose the problem without calling in a hydraulic expert. Often, you will find that the mysterious problem will be a common malfunction that can be spotted and corrected without assistance. All that is required is someone who is aware of obvious clues given off by a malfunctioning hydraulic system. Some of these clues are very subtle and require the use of sophisticated instruments to detect them; most problems make their presence known by sending out warnings that are loud and clear. All you need to troubleshoot a hydraulic system are four tools provided by nature, your "common senses": sight, hearing, touch, and smell.

One good way to forestall problems with a system is to look over your equipment periodically. Watch for any signs of leakage—wet machine parts or hoses, oil stains, low levels in the reservoir, which should be checked regularly, not to check the level but actually to examine the fluid. If the oil has a milky appearance, it is probably saturated with air or water. In either case, your system is in trouble.

Take a look periodically at the conductors in the system. Watch for hose lines that are abrading or for lines that are too short when under pressure. Also, look for kinked or flattened lines, because a kink will restrict flow or, in the case of a pump supply line, you could ruin a pump via cavitation. Look at the system pressure gauges to see whether or not they are within the specified values. Most important, make sure that there is sufficient oil in the hydraulic reservoir *after* the complete system is charged.

A good troubleshooter will listen for unusual sounds coming from a hydraulic system. The loud, shotlike sound you hear when a valve is closed quickly is called *water hammer* and is caused by the sudden stoppage of the moving fluid. A pressure surge may go as high as four times the normal working pressure. A shock wave travels at the speed of sound in a hydraulic fluid, and normal gauges will not record these fast pressure surges.

The main function of a hydraulic pump is to move fluids against resistance. The movement of fluid in a hydraulic circuit is incidental to the basic function of power transmission. Our interest is in the transmission of power in the desired plane with certain motion patterns. We seek to control torque and associated rotary force and motion and/or linear force and motion with a piston and cylinder assembly.

The function of the pump is to accept liquid at the intake and move it through the pumping mechanism. As the fluid passes through the pumping mechanism, it can be pressurized at the output chamber to a value appropriate to the resistance to flow at the terminus.

Several factors are important to service personnel as malfunction of the machine is analyzed.

1. Most hydraulic power pumps used in general industries are positive displacement devices. At each revolution of the shaft of a rotary pump a specific quantity of fluid must enter the pump and be discharged from the pump.

   Those pumps which do not provide a specific quantity of fluid at each revolution of the shaft or reciprocation of a piston are termed hydrodynamic devices as compared to the hydrostatic devices used in usual fluid power systems.

   Centrifugal and propellor pumps are typical hydrodynamic pumps. They are generally used for fluid movement, and power transmission may be a consideration in rare instances. These are the pumps used in water supply systems and the processing industries.

2. As a known predictable flow emanates from the hydraulic pump, it must be dealt with by the circuit components. Thus relief valves, pump volume controls, proper directional valves, and so on, must recognize and provide passage for the known quantity of flow.

3. System pressure level at the pump outlet will be a predictable value created by the resistance to flow within the hydraulic circuit.

4. Intake to the pump is directly related to the ability of the pump to move and pressurize the liquid.

   Restriction of the intake fluid flow to the pump inlet can result in the creation of a vacuum, which can result in a breakdown of the liquid at this intake area with adverse effects on the pumping mechanism. Vanes will not be held in the correct working relationship, pistons will be incorrectly positioned, and severe chatter, hammer, and oscillations can occur with resulting noise and damage. Lubrication can be severely impaired with resulting mechanical abrasion.

   Thus the development of unusual pump noise can be traced to restricted intake flow under certain conditions.

   To remove this restriction to flow it may be necessary to clean suction filters, replace damaged or restricted suction lines, make certain fluid is at the correct temperature to provide adequate flow, and/or increase pressure level in a sealed reservoir to ensure adequate flow to the pump intake.

   Gravity flow or sufficient atmospheric pressure on the fluid in the reservoir should in most instances provide the needed flow.

   At high elevations (Pikes Peak, aircraft installations, etc.) it may be necessary to mechanically assist the flow to the suction of the major power pump.

5. A leak in the suction line may allow entry of air which may pass through the pump and create an undesired resiliency in the output flow. Air passing through the pump can create a different operating sound pattern, which can alert maintenance personnel to this adverse condition, and repair activities can be instituted to correct the air leak in the intake line.

   Low fluid level in the reservoir could result in entry of air into the pump intake line. A float switch installed in the tank may be desirable to monitor fluid level and alert appropriate personnel to low fluid level conditions.

6. A vacuum gauge at the pump inlet may be useful to inform the maintenance personnel of a change in pump intake characteristics and vulnerability to rapid pump wear.

Any significant change in the sound level of a hydraulic pump should be investigated and corrective measures instituted when the cause of the malfunction is determined.

If you should put your hand on a hydraulic pump and wish later that you had not done so, the pump is trying to tell you something. When the maximum temperature of a hydraulic system exceeds 150°F, oil oxidizes. The rate of oxidation approximately doubles with every 18°F increase in temperature according to the characteristics of the fluid. Some fluid power research indicates that the working life of most oil is decreased by 50% for every 15°F rise in temperature above 140°F. Oxidation causes sludge to form, reduces clearance, creates more heat, and causes corrosion. A good troubleshooter will feel the fluid conductors at various sections in the system for hot spots. If any section of a hydraulic system is unusually hot, there is a problem. Remember, high heat can be very damaging to a fluid power system. High heat can be interpreted as operating temperatures above that recommended by the designer and manufacturer of the fluid power system.

Several general observations can be made.

1. Hydraulic systems for machine tools usually work in a temperature range from 100 to 140°F.
2. Construction and agricultural machinery usually operate in a temperature range from 40 to 180°F.
3. Special-purpose power transmission equipment may operate at temperatures other than these values with appropriate consideration of the modification to the system and components to be compatible with the specific range within which the device must operate.

Another problem signal in a system is high-frequency vibration. These vibrations can be felt with your hand. This condition has been known to break welds and other components. If you do not want to use your hands to feel these vibrations, try a small amount of water on the tubing or hose, and watch the results; if the water pops off the surface, you have a serious problem.

Vibration can be the result of sticky pressure control valves caused by contaminated fluid. The cause of the vibration may be a restricted suction line which allows the piston assembly of an axial piston pump to pull away from the ramp surface and pound as the piston reciprocates, with obvious potential for major component damage.

When hydraulic oil is saturated with air, or when a hydraulic pump is cavitating, the air bubbles in the system go to a high-pressure condition which generates a great deal of localized heat. This localized heat may reach as high as 2000 to 4000°F. The heat is caused by the compression of the gases in the oil. Pump cavitation occurs when suction lifts are excessive and local system pressure is below the vapor pressure of the fluid. This condition can be caused by inadequately charged or poorly designed inlet piping, as well as by mismatching the pump to the system, or by dirty inlet filters or restricted inlet lines. Bubbles that form in the low-pressure region and travel through the pump are compressed and collapsed under high pressure. This results in local high fluid velocities and explosive (or implosive) forces which cause erosion of pump parts.

This can be hot enough to scorch the oil. The good troubleshooter can and will smell this burned oil in the reservoir and make necessary corrections.

Nothing should be overlooked. Use any and all factors known to you and check them all out. You should first obtain a drawing of the system together with the current specifications so that you can really know and understand the system. You may have to use pressure, flow, or vacuum gauges or even an oscilloscope from time to time. However, the most important tools you have are your "common sense" and your brains. So use them. Coloring the working lines on a circuit drawing with appropriate colors to indicate pressure and tank lines and suction conductors often helps to quickly identify the energy flow through the circuit.

12.1.1 Using a Hydraulic Tester to Troubleshoot

Troubleshooting involves seven basic steps:

1. Know the system. In short, be familiar with related technical information, bulletins, and the basic machine circuit. Know what each component does and how it functions. Be able to trace the flow of fluid, in all operating modes, from the reservoir through the circuit and back. Review system operation before looking at the machine, and record vital specifications such as operating speeds, pressures, flow rates, and cycle times.
2. Before going to the machine, ask the operator how the machine is malfunctioning. Because the operator works with the machine on a daily basis, he or she knows when service was performed, who works on the machine, its peculiarities, and other functional characteristics. The operator can describe the basic symptoms, such as: pump will not start, system overheats, cylinder will not move load.
3. Operate the machine. Warm up the fluid by cycling the machine and verify the symptoms given by the operator. Be sure to observe gauge readings and operating speeds, listen for noise, inspect the fluid, and note any erratic operation of actuators.
4. Inspect the machine carefully, being sure to observe all safety precautions. Do not operate mobile equipment from the ground; always sit in the operator's seat and have everyone else stand clear. When inspecting the machine, systematically trace the flow of fluid

from the reservoir completely through the circuit, recording observations periodically. Although there is a natural tendency to skip through parts of the circuit and move directly to the area where trouble is suspected, a careful and systematic inspection may identify other problems that can contribute to the malfunction. Thus it is very important to do the inspection in a systematic manner.

Look at the oil and line connections. Check temperatures and look for dirt that could enter the system to restrict cooling. Feel intake and pressure lines for brittleness, softness, leaks, and collapsing, and inspect connections at each valve and component. Brittleness in the hoses is caused by high fluid temperatures; collapse of the suction lines is caused by restrictions in the line due to kinks, dirty suction filter, thickened oil, or lines plugged with rags and other debris. Check for loose housing and mounting bolts. Finally, check shafts and cylinder rods. At every stage of inspection, note any signs of abuse from improper operation or maintenance of the machine.

5. From the information gathered, list in order the probable causes for failure. Keep in mind that more than one symptom may contribute to a failure. Slow actuator speed, for example, may be caused by both worn pump parts and bypass leakage in the cylinder. High oil temperatures can be caused by low oil level and dirt blocking oil cooling surfaces, as well as by high relief valve pressures.
6. Reach a conclusion. From the list of probable causes, select those that, if verified and repaired, would return the machine to normal operation. Keep in mind that some are easier to verify than others. These might be checked first for convenience even though they do not seem to be the primary reason for failure. From the list of items to be checked, decide how to proceed from one to the next to verify the cause of trouble in the shortest time.
7. Test conclusions on the machine. Check pressures and flow rates with appropriate testers, remove housing covers to inspect pump and motor parts, and closely inspect other components and parts of the system. Use these tests as the basis for decisions to replace or repair components.

After the problem has been narrowed to one part of the circuit or one component, quantitative tests should be conducted to determine just what the problem is. Think about where symptoms and probable causes are leading, and perform additional tests to pinpoint the source of the problem. Some tools that might be needed include:

Pressure gauge

Temperature gauge

Flowmeter that operates under system pressure

Hand-held or electric tachometer

Stopwatch

Load valve

Figures 12.1 and 12.2 depict hydraulic circuit analyzers that are available for use in diagnosing system malfunctions. Figure 12.3 shows the placement of a hydraulic analyzer that has a built-in pressure gauge, flowmeter, temperature gauge, and load valve in a circuit. The bypass test is sometimes called a *series test*. The manner in which the bypass test can be used for testing various system component performance is shown on Fig. 12.4.

When testing the hydraulic circuit with the hydraulic tester, first determine how much fluid is flowing in the circuit. Then, by controlling pressure with the load valve, determine the amount of fluid actually flowing through each component. If the test indicates insufficient flow at system pressure, the cause can be pinpointed to such areas as:

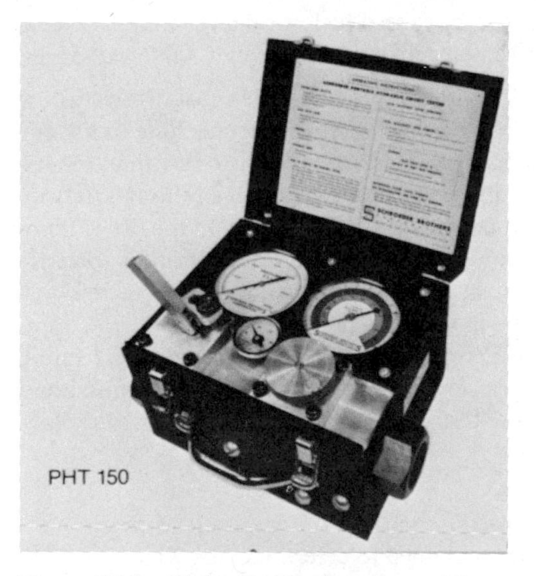

Figure 12.1 Hydraulic circuit analyzer. (Courtesy of Schroeder Bros., Inc., McKees Rocks, Pennsylvania.)

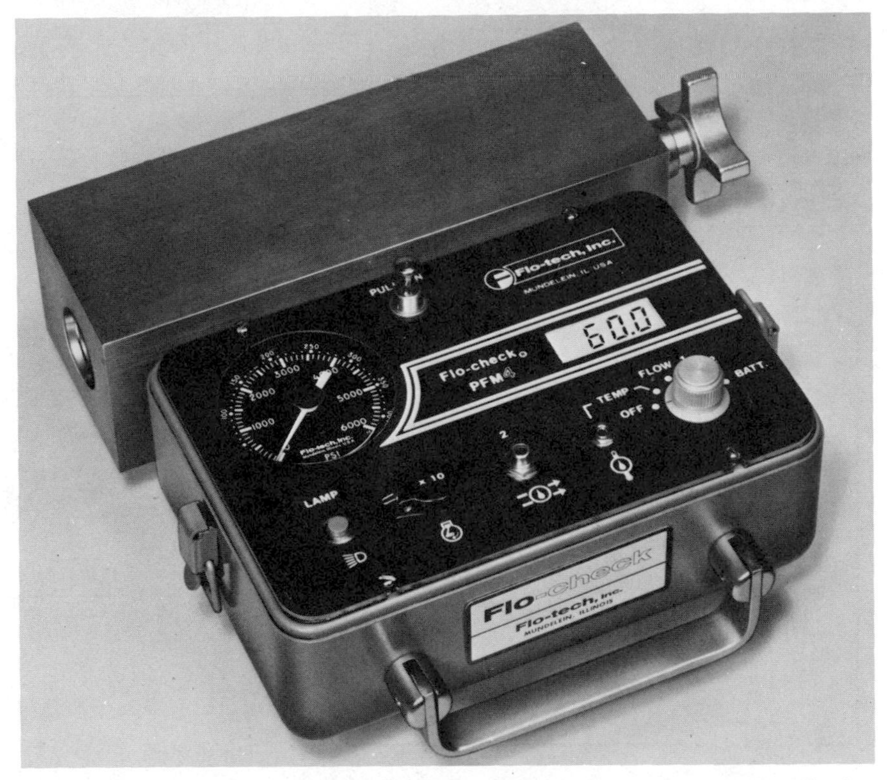

Figure 12.2 Hydraulic analyzer. (Courtesy of Flo-Tech, Inc., Libertyville, Illinois.)

Pump slip

Flow over a faulty relief valve

Leakage past control-valve spools to the reservoir

Leakage past pump or motor parts directly to the return line

Be sure to operate the system long enough to bring the temperature of the fluid within the operating range. If the system is driven by an internal combustion engine, tests should be conducted at constant speed.

12.1.2 Application of a Tester

To provide a more thorough understanding of the application of a hydraulic tester, a more detailed discussion is presented here. As indicated previously, the tester can be installed in the circuit in series with the valves, in parallel, or in a bypass arrangement.

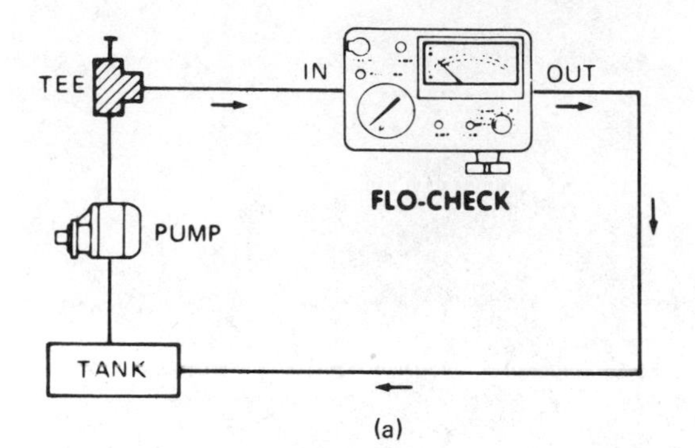

(a)

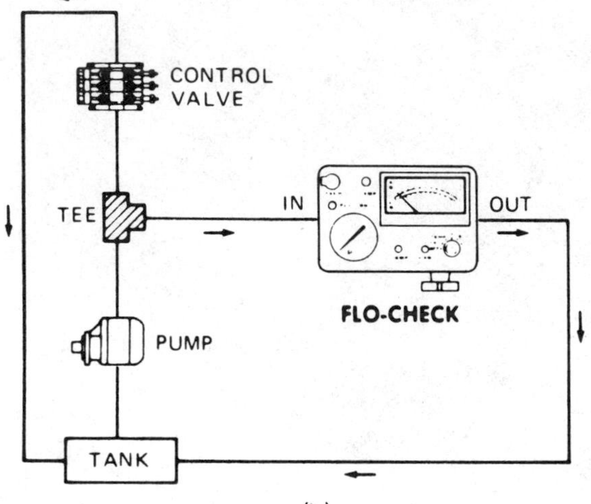

(b)

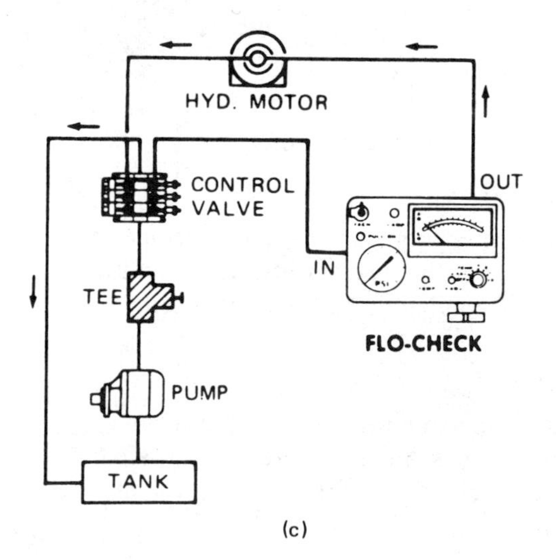

(c)

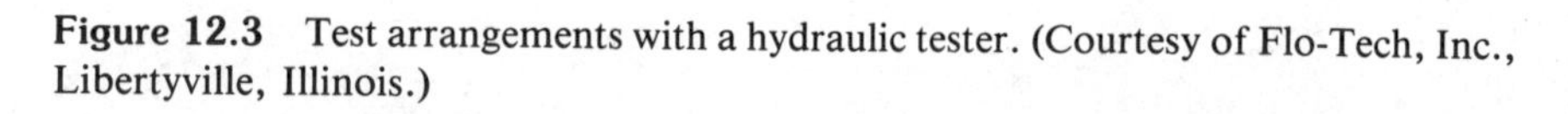

Figure 12.3 Test arrangements with a hydraulic tester. (Courtesy of Flo-Tech, Inc., Libertyville, Illinois.)

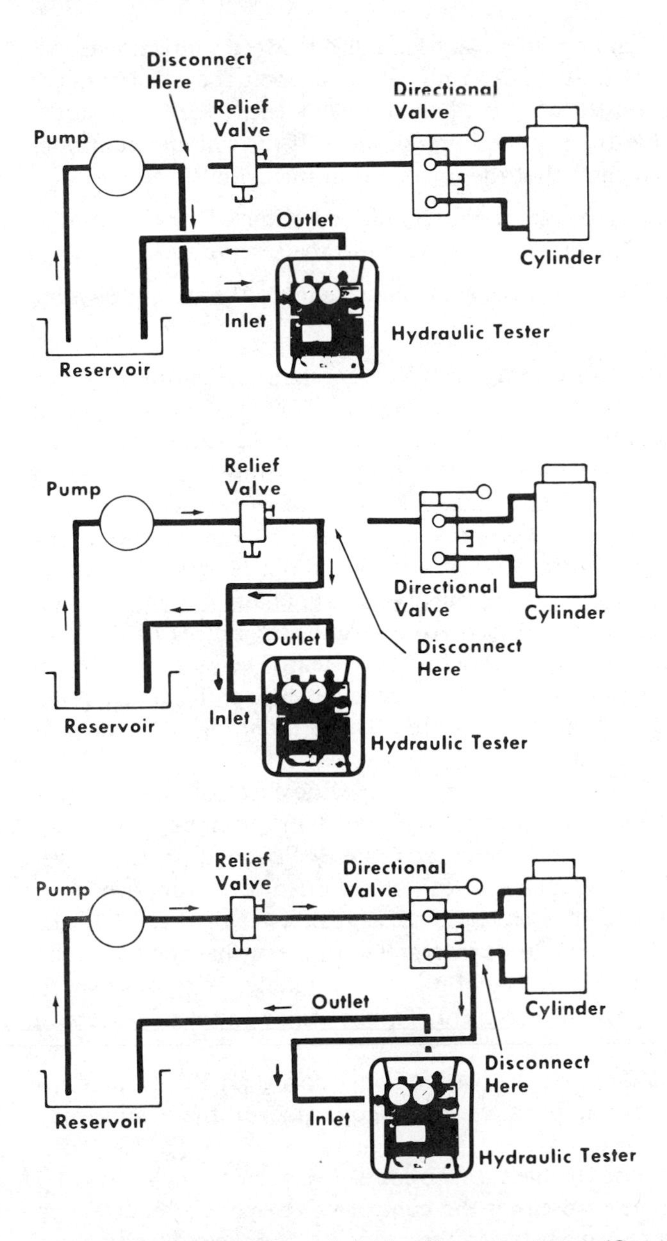

Figure 12.4 Application of a hydraulic system tester. (Courtesy of Owatonna Tool Co., Owatonna, Minnesota.)

In-line test: An in-line tester can be installed between a directional control valve and a cylinder to determine if leakage is in the valve or in the cylinder. If the test valve is closed and the cylinder pressure drop stops, this indicates that the trouble is in the valve. If the cylinder continues to drop, this indicates that the trouble is in the cylinder.

Bypass test: A bypass test connects the tester inlet after the components tested and the outlet is connected to the reservoir.

Tee test: In a tee test the tester is connected into the system by a tee connection.

For example, with the tee test, we may be checking the complete system at one time or selectively as we choose. The tee in Fig. 12.3a is plugged at one side. Pump flow must pass through the tester and the value of pressure and capabilities can be confirmed. If the quantity of flow has degenerated to an unacceptable level, the pump must be repaired or replaced.

The connection of Fig. 12.3b permits a check of the relief valve contained in the directional control valve assembly and the devices operated by the directional control segments. By shifting one spool and moving a piston to the bottomed position, it is possible to determine the setting of the integral relief valve and see if it is correct for the system. Then it is possible to operate each cylinder and see if pressure can be maintained at the needed level. Movement of the actuator can be monitored. The tester of Fig. 12.3 can provide a safety relief function.

The connections of Fig. 12.3c permit a specific check of a component for actual leakage and pressure level in that branch of the circuit. Flow and resistance to motion can be monitored and rate of flow and pressure level can be identified. If it is adequate, it can be recorded for future use.

Should a monitored flow as registered on the tester of Fig. 12.3c not provide proper speed, this could be interpreted to mean that the hydraulic motor has excessive internal leakage if it is not providing the rotative speed corresponding to the motor displacement. Torque will be uniform if excessive internal friction is not encountered.

Generally, if the flow rate drop is too high (more than 25%) and is approximately the same for all positions of the control lever, the trouble may be in the pump. If the flow rate drop is too high but only in one or two positions, the trouble is probably in either the control valve or the cylinders. To tell whether the trouble is in either the control valve or the cylinders, the cylinders can be blocked out between the cylinders and the control valves; then the leakage can be determined for either the cylinder or the control valves.

Testing the Pump

If your system is not working properly and you cannot determine any visual cause, test the pump (Figs. 12.3a or 12.4). Symptoms of trouble in the pump are as follows:

System does not operate. No noise from pump.

System operates too slowly, or is erratic.

System operates too fast.

Pump is too noisy.

General procedures are given here for testing the pump. Service manual procedures for the particular equipment should be followed. To test the pump, proceed as follows:

1. Check the procedures and specifications.
2. Prepare the equipment. Some equipment, such as cranes and loaders, must be tied down when testing the system.
3. Connect an appropriate tester. If the tee test is used, a blocking plate must be installed in the system line so that all the flow will go through the meter (Fig. 12.3a).
4. Slowly open the load control valve of the tester.
5. Turn the pulsation dampener knob connector clockwise several turns.
6. Start the motor and operate the pump at rated speed. Check the operator's manual for proper speed.
7. Warm up to operating temperature. Check the temperature gauge on the instrument.
8. Check gauges. If action is erratic, this indicates low fluid level, clogged filter, or leaking suction line to pump.
9. Increase the pressure by gradually turning the pressure control handle clockwise.
10. Reduce the pressure and record the flow rate at low pressure.
11. Increase the pressure and record the flow rate at different pressures.
12. Open the load control valve. To test other parts of the system in addition to the pump, repeat steps 8, 9 and 11 for different positions of the equipment control valve.

13. Shut off the motor.
14. Check the results against the manufacturer's ratings. Low flow and pressure indicate trouble. Generally, a 15% reduction in flow rate for the pump is acceptable. If the flow rate is within limits, record further tests made at reduced pump speed. If flow rate comparison at different pressures is still within the limits of the lower speed, this indicates that the pump is good. Continuing noise may indicate that the trouble is probably some restriction of fluid to the pump or an air leak in the suction line.

Testing the Valve

Troubles with valves depend on the function of the valve. For example, if the system is operating too slowly, one problem could be the pressure relief valve. If it is barely operating, the relief spring may be broken or improperly adjusted. If the function of the valve is understood, the testing is a simple matter. The biggest job sometimes is to connect the test equipment. Connect the test equipment according to the manufacturer's instructions.

It is obvious that pressure control or regulating valves can be checked with a pressure gauge. Directional control and flow control valves and orifices must be checked with a combination pressure gauge and flow meter. Symptoms that indicate trouble in the pressure regulating valve are as follows:

System unable to lift load.

System operates erratically.

System fails to operate.

Procedures for testing are as follows:

1. Connect the tester. Follow the directions given by the manufacturer.
2. Open the pressure control valve on the tester.
3. Turn the pulsating dampener knob counterclockwise several turns.
4. Start the motor and warm up the system.
5. Move the control valve on the tester to different positions of high and low pressure and volume and record readings.

Note: Some valves require that a cylinder be moved to a completely extended position before pressure can be controlled by the relief valve.

6. Note the pressure when the flow drops to zero. The pressure at which the flow drops to zero is the relief valve setting. If the pres-

sure is below the recommended maximum system pressure, the relief valve should be adjusted.

7. Note the flow rate up to maximum pressure. Assuming that the pump is good, the flow rate should be equal to pump capacity up to 75 to 90% of the maximum pressure. If not, this indicates that the relief valve is leaking. If the pressure regulator valve checks out to be acceptable and the trouble still exists, check the directional control valves.

Internal leaks in the directional control valve may also cause the system to fail to operate or cause the system to operate too slowly. A sticking valve will cause troubles. The procedures for testing the directional control valve are as follows:

1. Connect the tester. Follow the procedures given by the manufacturer.
2. Open the pressure control valve.
3. Turn the pulsating dampener control counterclockwise several turns.
4. Start the motor and warm up the system at rated pump speed.
5. Shift the directional control valve on the system to direct fluid to the tester.
6. Move the tester control handle to the various pressures and flow rates and record. If the volume and pressure obtained in the previous test cannot be obtained, the directional control valve has an internal leak. If the volume obtained in the previous test is obtained, this indicates that the directional control valve is good and the trouble may be in the cylinder.

Testing the Cylinder

If the cylinder is not holding under load, the problem may be an internal leak in either the control valve or the cylinder. Procedures are given here for a bypass hookup and a tee test.

If the *bypass test* is being used (Fig. 12.3c), proceed as follows:

1. Check the equipment manual for flow rate and pressure.
2. Install a safety valve if needed. This may be incorporated in the tester. Some manufacturers recommend the use of an auxiliary safety valve to prevent damage to the system.
3. Connect the tester.

4. Open the load control valve on the tester.
5. Turn the pulsation dampener knob counterclockwise several turns, if installed.
6. Start the motor and operate the pump at rated speed.
7. Warm up to operating temperature. Check the temperature gauge on the instrument if installed. The temperature should not be more than 200° F.
8. Close the hand control valve to get the desired flow and pressure. Open the equipment control valve to extend the cylinder.
9. Read the instruments and record the pressure and rate of flow at various settings of the load control valve on the tester.
10. Compare the flow rate to specifications. If full rated flow at required pressure is obtained, this indicates that the trouble is in the cylinder. *Note:* Cylinder test flow rate should be within 99% of specifications. For double-acting cylinders, you should test in both directions. The seal may leak in one direction but not in the other.

If you use a *tee test* (Fig. 12.3b), proceed as follows:

1. Check the flow rate and pressure for the system.
2. Connect the tester.
3. Install the blocking plate in line to the cylinders being checked. More than one cylinder can be checked at one time if they are fed by the same line. When trouble is detected, a single-cylinder test can be used to determine which cylinder is leaking.
4. Install a safety valve if recommended by the manufacturer of the machine, equipment manufacturer, or manufacturer of the test devices.
5. Open the load control valve on the tester.
6. Start the motor and warm up.
7. Open the equipment control valve to extend the cylinder.
8. Read the instruments and record the readings.
9. Compare with specifications. The results should be the same as with the bypass test. The cylinder can also be checked for a packing leak by disconnecting a loaded cylinder and looking for bypassed fluid resulting from a leak.

12.2 DIAGNOSTIC INSTRUMENTATION

A substantial amount of automatic test equipment for determining the operational condition of hydraulic and pneumatic systems has been developed recently. The results obtained have led industry to adopt the developed techniques and expend further effort to refine them and reduce them to practice. Included in this variety of instruments that are available for hydraulic and pneumatic system troubleshooting are:

Ultrasonic leak detector

Eddy-current probes

Infrared scanners

Spectrometric oil analysis

Ferrographic oil analysis

Photometric particle counters

Acoustic emission detectors

Thermal flowmeters

Vibratory pressure pickup

The operation of the ultrasonic leak detector is described together with its application.

12.2.1 Ultrasonic Leak Detector

The worker hours required for inch-by-inch inspection of pneumatic and hydraulic lines on machine tools, mobile equipment, or industrial machinery have been effectively reduced by more than 80% according to various sources, which have employed a method of checking their systems by ultrasonics. It is now their standard inspection technique at each 200-hr operational check.

Using the ultrasonic system, which consists of a battery-powered, lightweight instrument requiring no setup or auxiliary equipment, a single mechanic can check a complete complex system in 30 min. The portable detector is a passive device; it does not require an acoustic noise source or generator. It responds to ultrasonic energy created by external forces. Air or oil molecules escaping from a leak anywhere in the system, with pressures from 3000 to 100 psi, create ultrasonic energy as the higher-velocity molecules collide with those in the atmosphere (Fig. 12.5).

The detector's hand-held probe microphone responds to ultrasonic energy from 36,000 to 44,000 Hz (cycles per second). Electronics within the

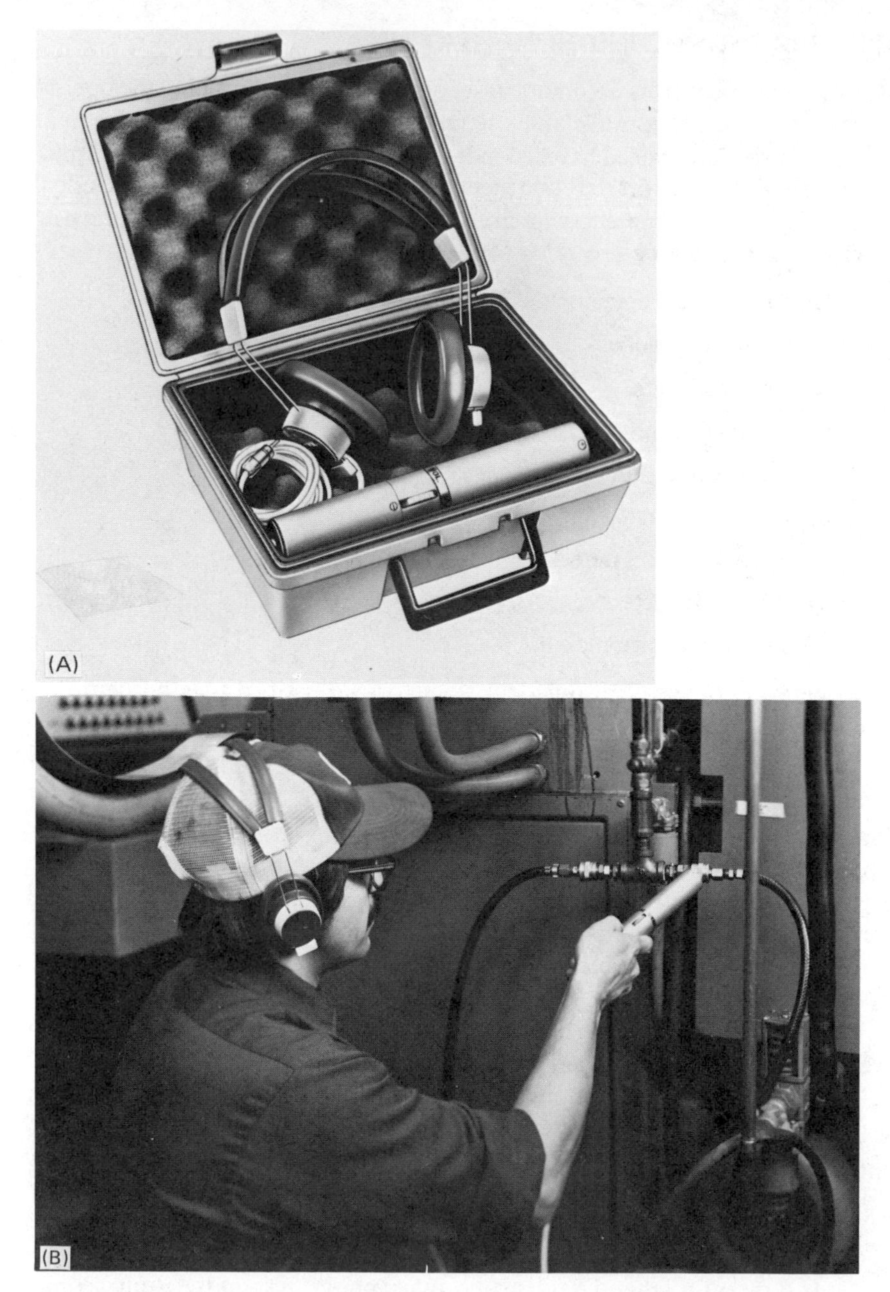

Figure 12.5 Ultrasonic leak detector. (Courtesy of Owatonna Tool Co., Owatonna, Minnesota.)

instrument translates this energy to the range of human hearing and amplifies it through its loudspeaker or earphones.

The characteristic sounds of the leak are preserved. For example, an almost microscopic air leak creates 40,000-Hz energy, which, when translated and amplified, sounds exactly like the hissing of a punctured inner tube.

To inspect a pneumatic system, the mechanic aims the probe along tubing from a distance of about 6 in., paying particular attention to known trouble spots such as swivels, fittings, and other connections. When the detector's speaker emits a hissing sound, he finds the precise location and either adjusts the fitting at the time or marks the part for replacement.

This type of unit can also be used as a vibration sensor or stethoscope, as shown on Fig. 12.6. In this case, the unit is sensitive to lower-frequency signals and these are picked up by the transducer, which must contact the component being diagnosed. The signals are conditioned by the unit to provide a sound or audible-type indication to the person conducting the test. Such items as bad bearings, scratches, cracks, or nicks in the rotating members are readily detected.

12.2.2 Bourdon Tube Pressure Gauge

The Bourdon gauge works on the principle that pressure in a curved tube will tend to straighten it. Thus pressure acts equally on every square inch of area in the tube, but since the surface area on the outside of the curve is greater than the surface area on the shorter radius, the force acting on the outer surface will be greater than the force acting on the inner surface. When pressure is applied, the tube will straighten out until the difference in force is balanced by the elastic resistance of the material composing the tube. If a pointer is attached to the tube and a scale laid out, the main elements of a Bourdon gauge are obtained, as shown in Fig. 12.7.

The working parts of a Bourdon gauge consist of a tube that is bent into a circular arc and is oval in cross section so that it will tend to straighten more easily when under pressure. The open end of the tube passes through a socket which is threaded so that the gauge can be screwed into an opening in the hydraulic system. The closed end of the tube is linked to a pivoted segment gear in mesh with a small rotating gear to which a pointer is attached. Beneath the pointer is a scale reading in pounds per square inch. The gauge is calibrated against known pressures to ensure accurate readings. The working parts are enclosed within a protective case of metal, plastic, or similar material with the dial visible through a clear glass or plastic face. Under pressure the tube tends to straighten and the segment moves about its pivot, rotating the meshing gear and pointer. The pointer assembly is usually pressed

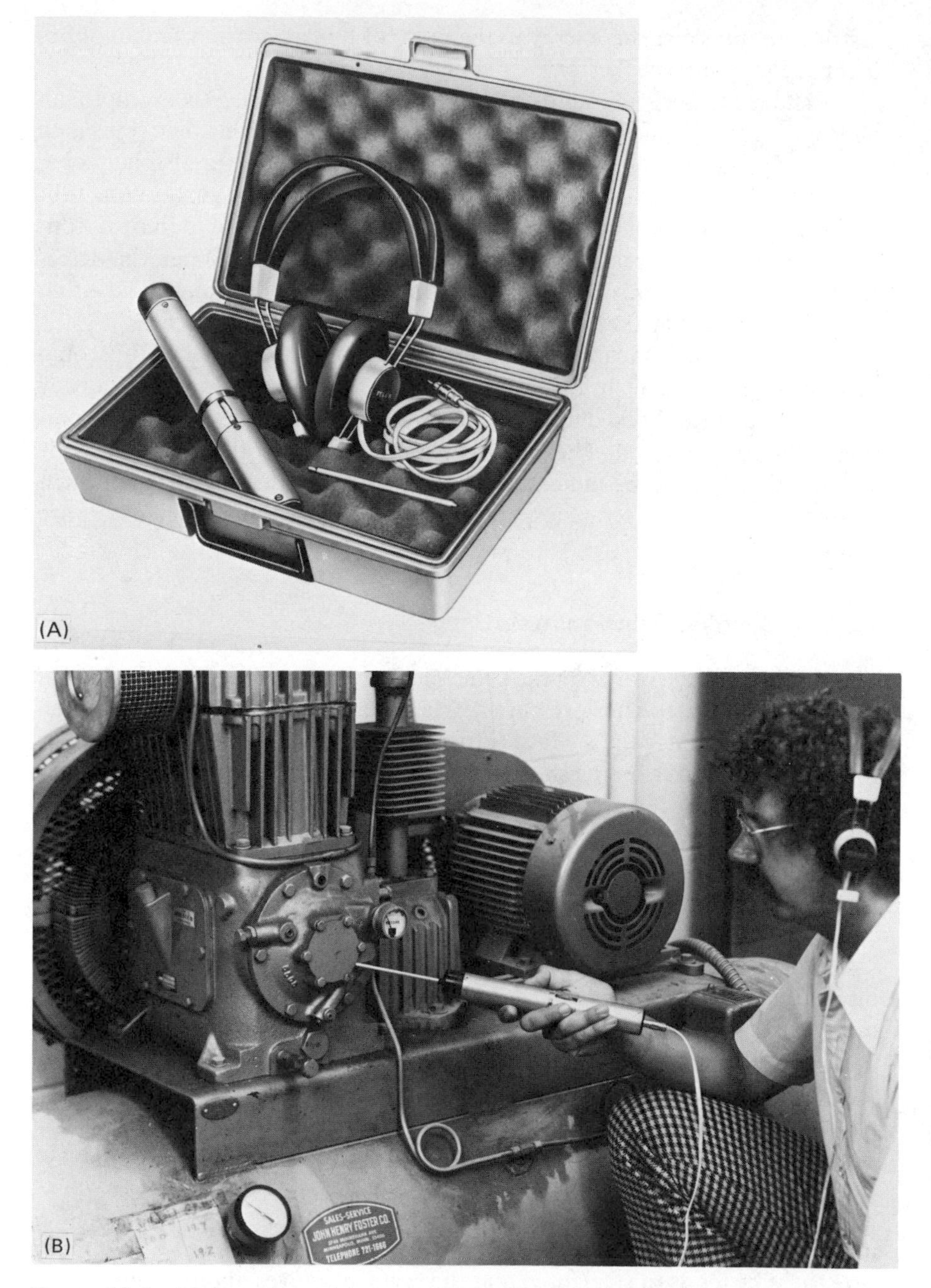

Figure 12.6 Microsonic stethoscope. (Courtesy of Owatonna Tool Co., Owatonna, Minnesota.)

on the shaft in such a manner that it is removable for resetting when the gauge is calibrated against a master unit. The construction details of a typical Bourdon tube pressure gauge are depicted in Fig. 12.8. Possible troubleshooting solutions are given in the following chart.

Problem	Cause	Solution
Incorrect indication	1. Gauge defective	Check zero pressure and remedy or replace broken tube, broken movement, tube spring, broken needle, pegged needle, or gauge pinion gear
	2. Check authenticity of the gauge reading	Check gauge against good gauge
Poor gauge life	1. Gauge subject to mechanical shock	Isolate shock by switching to glycerin-filled gauge
	2. Bourdon tube fatigue	Use gauge isolator to remove continuous pressure on tube
	3. Pegged needles	Add pressure flow snubber to restrict needle movement in addition to glycerin (or fluid-filled gauge)

12.2.3 Viscosity Measurement

The instrument most often used by American engineers and technicians to measure the viscosity of liquids is the Saybolt Universal viscometer or viscosimeter. This instrument measures the number of seconds it takes for a fixed quantity of liquid (60 cm^3) to flow through a small orifice of standard length and diameter at a specified temperature. The viscosity is stated as so many seconds Saybolt Universal (units SSU) at such and such a temperature. For example, a certain oil might have a viscosity of 80 SSU at 130°F.

The Saybolt Universal viscometer consists of a reservoir for the oil surrounded by a bath heated by heating coils to bring the oil to the temperature at which the viscosity is to be measured. The bottom of the reservoir issues into the standard viscometer orifice. Passage through the orifice is blocked by a cork. The reservoir is filled to a marked level, and a container marked at the 60-cm^3 level is placed under the opening. When the oil to be tested is at the desired test temperature, the cork is removed and the number of seconds

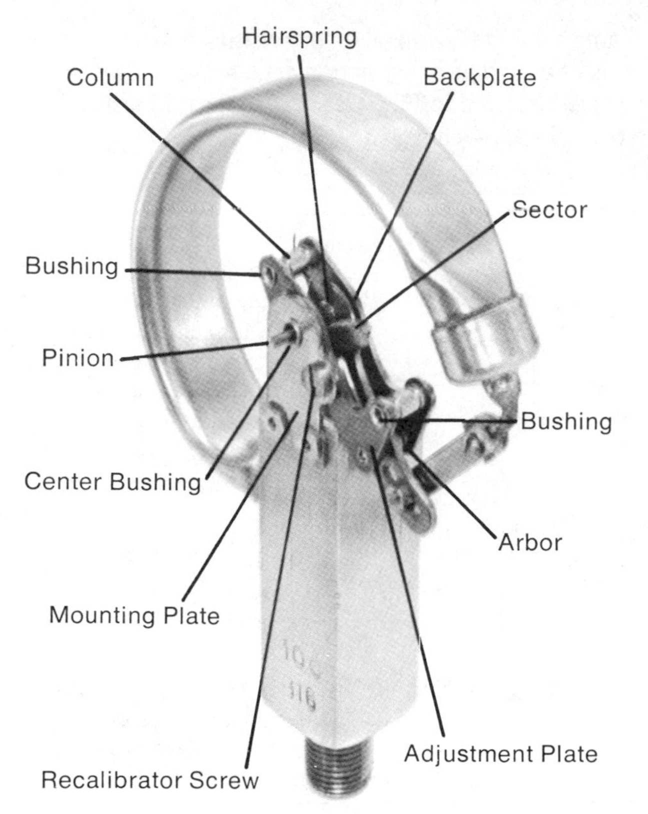

Figure 12.7 Bourdon tube gauge mechanism. (Courtesy of Marsh Instrument Co., Bensenville, Illinois.)

Figure 12.8 Typical gauge construction. (Courtesy of Marsh Instrument Co., Bensenville, Illinois.)

it takes for the liquid to reach the 60-cm^3 level gives the SSU reading. The device is shown in detail in Fig. 12.9.

For quick on-the-job tests it is possible to use comparison devices. Such a device consists of one tube with a known viscosity fluid and a second tube in a parallel plane which is filled with the fluid to be tested. By comparing the fluid under test with the known one, an evaluation is made. Generally, the known fluid is the one recommended for the hydraulic system used.

12.2.4 Liquid-Level and Temperature Indicators

The level and temperature of fluid in the reservoir provide important information on the status of the hydraulic system. Both characteristics can be monitored inexpensively and even controlled automatically at relatively low cost. The simplest level-monitoring devices are dip sticks and sight gauges. Dip sticks are most common on mobile systems; sight gauges, on industrial systems. Together, these two devices take care of the level-monitoring needs of most fluid power systems. Where fluid level is critical, or where the level is unlikely to be observed periodically by the operator, low-level alarms actuated by float switches should be used, sometimes in conjunction with remote-reading gauges. If potential damage from loss of fluid could be costly, the alarm circuit can be wired to shut down the system.

Temperature monitoring is often omitted from mobile systems, although some newer systems have remote-reading thermometers with a readout on the operator's control panel. Industrial systems typically have a thermometer installed as part of the sight-level gauge or as a probe in the suction or return line. High- and low-temperature alarms operated by thermostats are often used and may automatically shut down the system. An immersion-type temperature gauge often used to indicate fluid temperature in a reservoir is shown on Fig. 12.10.

A word of warning: A thermometer installed in a sight-level gauge does not indicate maximum system temperature. It indicates only the temperature of the fluid adjacent to that wall of the reservoir. Maximum system temperature can easily be as much as 130°F higher than that measured in the reservoir. The advantage of measuring reservoir temperature is that it indicates whether or not the system is operating normally, a substantial benefit considering that the thermometer typically costs less than a dollar.

12.2.5 Flowmeters

Flowmeters vary in construction and operation. Generally, there are three broad categories of meters: variable-area types, turbine types, and noninterference types. Following is a review of each type.

Rotameters are the most common variable-area meters, consisting of a tapered tube in which a float is supported by the fluid flowing up through the tube. As the fluid flow increases, the float is lifted higher in the tapered tube, as greater orifice area is required around the float to transmit the flowing fluid. The reading is taken directly from the float position. Rotameters are typically used for low-pressure flow readings; they are most suitable when used for a single fluid.

Interference meters consist of a turbine vane, or propeller, centered in the flow path, which rotates proportionately to flow velocity. The turbine or vane may be geared directly to a meter; in more sophisticated devices, turbine speed may be sensed by a magnetic element outside of the flow tube. A turbine-type flow meter that can be used for low- or high-pressure lines is shown in Fig. 12.11.

Interference meters are the type most commonly used when a flowmeter is required for commercial hydraulic or pneumatic service. They are capable

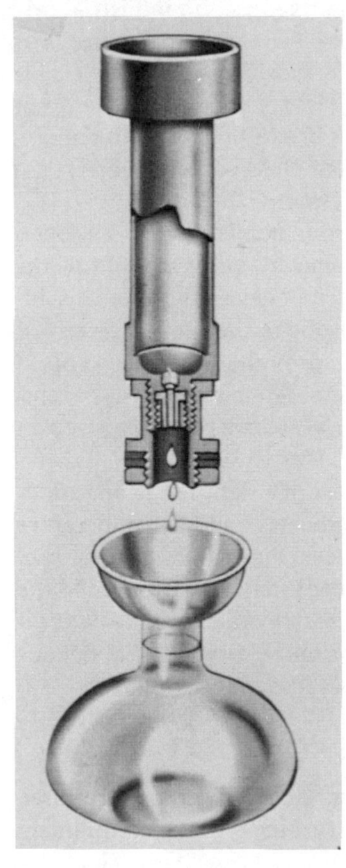

Figure 12.9 Saybolt viscosity testing device.

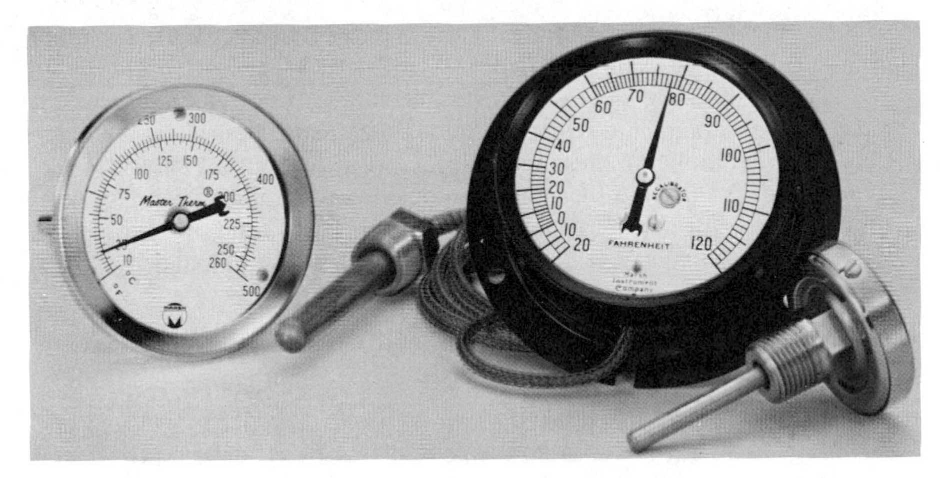

Figure 12.10 Temperature gauges. (Courtesy of Marsh Instrument Co., Bensenville, Illinois.)

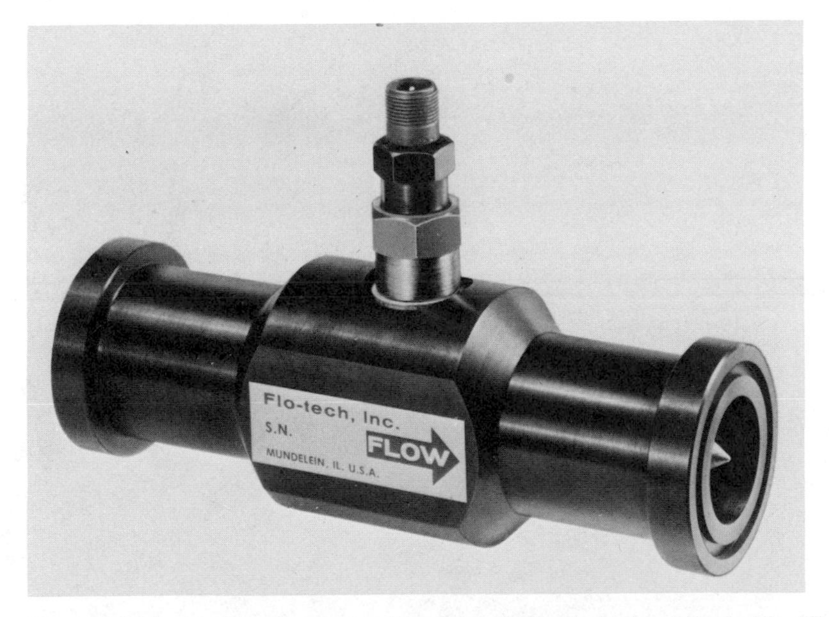

Figure 12.11 Turbine flowmeter. (Courtesy of Flo-Tech, Inc. Libertyville, Illinois.)

of delivering acceptable accuracies at the high pressures normally found in industrial hydraulics and pneumatics.

Positive-displacement meters offer the ultimate in volumetric accuracy but require maximum interference with the flow stream. Basically, they function like a hydraulic or pneumatic motor, with the "output shaft" of the motorlike device driving the gauge readout. Because these meters are very similar to the hydraulic and pneumatic motors used in fluid power applications, they are suitable typically for extremely high pressures and deliver very high accuracies.

Vortex meters use obstructing bodies in the flow path to create vortices, which may be sensed by a variety of devices downstream of the obstruction. In one model, heated thermistors placed in the vortex path are cooled at a faster rate when they encounter the swirling fluid. Another model replaces thermistors with a transverse ultrasonic beam. Vortices interrupt the beam and produce an oscillatory voltage at the receiver. A third model uses vortices to vibrate a pair of metal diaphragms, which act as the plates of a capacitor. Vortex meters can be extremely accurate, typically within 1% of full scale, and are well suited to automatic monitoring and control systems.

13

Fluid Power System Noise Control

"Will the user accept the noise level?" that has been the deciding factor up to now. Designers of fluid-power equipment may soon be faced with standards that set limits on noise. Both equipment designers and users must be concerned. If a plant does federal contract work under the Walsh-Healy Act, and if it has an operation that produces substantial noise levels of 90 dB or more, the noise standards for federal contractors that took effect on May 20, 1969, apply:

> The maximum acceptable noise level for an 8-hr exposure is 90 dB measured on the A scale of a standard sound-level meter at slow response.
>
> Exposure to impulsive or impact noise should not exceed 140 dBA peak sound pressure level.
>
> Where possible, engineering controls should be adopted to reduce sound levels at their source.
>
> You may reduce a worker's cumulative exposures to noise by means of administrative controls. That is, you may schedule a worker so that he or she will not be exposed to excessive noise for an entire 8-hr day.
>
> If engineering controls are not feasible, you must institute a hearing conservation program.

If these regulations are not met, the government can cancel the contract, buy the goods elsewhere, and charge for the difference. The reduction of noise and compliance with the standards necessitates an understanding of the sources of noise in a hydraulic system and the control methods that are available.

13.1 PUMP NOISE

Fast changes in velocity and pressure, and the collapse of bubbles, are the principal sources of hydraulic noise. Pumps and motors are the most common contributors of these factors in hydraulic equipment. All positive-displacement hydraulic power units, including piston, gear, screw, and vane pumps and motors, generate noise in a similar way. The way an in-line piston pump generates noise is typical of most pump types as shown in the vane pump of Fig. 13.1.

Alternating internal loads can deflect the casing of the pump. The motion may be only a few microinches, but it makes noise. Several forces and moments combine to generate the noise of any given unit. The axial forces represents these factors in certain piston pumps. In a nine-cylinder pump, this force varies with the number of pressurized piston cycles, between four and five. The amplitude of the variable portion of this loading is equal to the product of piston area and discharge pressure. The rapid rising and falling between the fixed limits of the load is capable of producing a loud noise.

In each piston in an in-line pump, the flow moves with sinusoidal motion. A half-sine wave describes the discharge of each piston. The basic de-

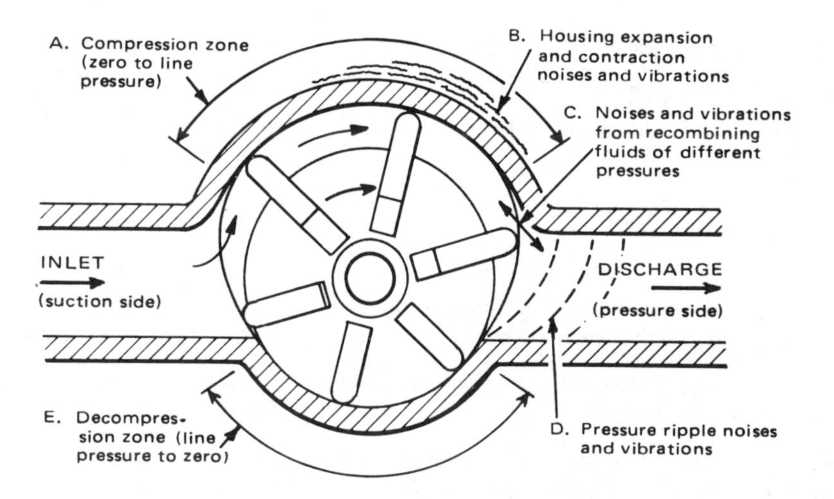

Figure 13.1 Causes of noise in a hydraulic pump. (Courtesy of Parker Hannifin Corp., Cleveland, Ohio.)

liveries of a multipiston pump combine to produce a flow having a small ripple. The way in which the individual fluid elements are compressed determines how much additional flow disturbance is added to the basic pump ripple. If no provision is made for this compression, the fluid will still be at inlet pressure (50 psi or less) when it reaches the discharge port. At the discharge port, the fluid is compressed by a backflow which is at system pressure. This compression interrupts the discharge outflow and can produce a lot of noise if the pump's valving mechanism is not carefully designed. In addition, fluid pressure noise is caused by rough pumping chamber transitions, thereby increasing airborne noise as well.

The internal loading producing sound also produces pump vibration. These loading noises deflect the pump structure, so that component masses are displaced. To balance the relative displacement of its internal masses, the pump housing must shift. The vibrations caused by this shifting transmits to adjacent structures. Figure 13.2a details the noise levels expected from the various pump structures in the early 1970s. Figure 13.2b reflects the significant improvements in the last 10 years resulting from research and development by several major producers of industrial hydraulic pumps.

13.2 REDUCING NOISE AT ITS SOURCE

Good hydraulic circuit design is essential to noise control. This is obvious to anyone familiar with noise of a pump cavitating because of insufficient inlet pressure. Avoiding cavitation and flow noises requires the use of adequate line sizes, suction strainer, input speed, and reservoirs; proper reservoir baffling; and good oil temperature control. In addition, flexible hydraulic lines used in making connections to isolated units can reduce sound levels and dampen vibration.

Fluid pressure pulsation can also produce audible noise; thus it is sometimes desirable to isolate pumps and motors hydraulically. This can be done with acoustic filters which are custom designed for a given application and are tuned for one pump or motor speed. Where speed is variable, gas-charged accumulators may be used. Desurgers, flow-through types, are preferred. A desurger can be a pipe with multiple holes radially located in the wall leading to a metal jacket, with an elastomeric material surrounding the inner pipe and an inert gas under pressure between the outer shell and the elastomeric liner. The gas charge must be constantly monitored to ensure maximum shock-absorbing efficiency.

One equipment user covered enclosed panels with sound-absorbing material and used brick and concrete for mounting machinery too heavy for rubber mounting. These precautions greatly reduced the noise that would have been produced if a metal framework had been used (Fig. 13.3).

a. Noise levels - early 1970s.

Pump type	Decibels[a]
Screw type[b]	72–78
Vane (industrial)	75–82
Axial piston	76–85
Gear (powdered metal)	78–88
Vane (mobile)	84–92
Gear (machined stock)	96–104

[a]The lower range figure is at 500 psi: the higher is at 1000 psi.
[b]1700 gpm, 3500 rpm. (All others 10 gpm at 1200 rpm.)

b. Noise levels - early 1980s.

	1200 rpm	
Pump type	1000 psi	2000 psi
Vane	58	62
Axial piston	64	67
Internal gear	55	–
External gear	65	70

Figure 13.2 (a) Typical noise levels of hydraulic pumps in the early 1970s (courtesy of Parker Hannifin Corp., Cleveland, Ohio). (b) Typical noise levels of hydraulic pumps in the early 1980s (courtesy of Sperry Vickers, Troy, Michigan).

13.2.1 Vibration Mounts

The transmission of machinery vibration through building components generating sound waves in plant areas has been effectively reduced by placing equipment on vibration mounts. This has not only dampened out vibrations, but has also reduced the vibration noise levels. Types of vibration isolators include springs, rubber, or neoprene, or combinations. Vibration isolation has been further improved by mounting equipment on concrete inertia blocks.

13.3 NOISE MEASUREMENT

Noise measurements are usually based on frequency. The decibel is the most common unit used for the measurement of loudness. Zero decibel is the softest sound that the human ear can hear in a very quiet area. Twenty micronewtons per square meter is usually considered as the starting point of the decibel scale. A soft whisper at 5 ft produces 30 dB, while a pneumatic hammer produces 130 dB. The decibel is 20 times the logarithm of the ratio of the measured sound pressure to a selected reference pressure, usually 20 N/in.^{2}.

Some companies have begun to use the *sone* as a method for measuring sound in place of the decibel. The sone is a subjective unit of loudness. One sone is defined as a reference tone of 1000 Hz at 40 dB. the main advantage of the sone is that it varies linearly to the human ear with varied broad-band noises. A hydraulic pump that emits 120 sones of sound noise sounds twice as loud as a pump emitting 60 sones of noise. Overall noise levels, in decibels, of some noise sources are presented in Fig. 13.4.

The basic noise-measuring system is electrical and consists of a microphone, an amplifier, and a meter calibrated to give sound levels over a standard frequency range. This meter usually has the A, B, and the C scales. The A scale is used most often. Fortunately, the number of decibels shown on a sound-level meter makes it easy to determine the source of the noise. Octave-wave band analyzers measure sound pressure levels in bands one octave wide.

Figure 13.3 Noise control. The machine panels have been lined with sound-dampening materials to substantially reduce noise levels. (Courtesy of Farrel Machine Corp., Rochester, New York.)

Decibels	Noise source
140	Hydraulic press at 3 ft
130	Pneumatic riveter at 4 ft
120	Large chipping hammer
110	Loud power motor
100	
90	Shouting
80	
70	Normal talking
60	
50	Light traffic at 100 ft
40	
30	Soft whisper

Figure 13.4 Overall sound levels of some noise sources.

After the sound pressure levels are plotted on a chart, the noise rating can be read directly. Figures 13.5 to 13.7 depict some typical sound- and vibration-measuring equipment.

13.4 RESEARCH TO REDUCE NOISE

Manufacturers of fluid power equipment are investing considerable amounts of money and the time of some of their best engineers and scientists in an effort to find a solution to noise. The solution is not an easy one. It involves the specifics of individual pump design, of system performance requirements, and of the design of the machine itself.

Facilities have and are being devoted to noise control research and development. One manufacturer has a test lab consisting of two separate rooms within a separate building. Each room has its own foundation, roof, and floor. The test room is the heart of the lab. An electric drive motor and other service equipment that might generate interfering noise are in the other room. A control and instrumentation center and a laboratory for auxiliary testing are located in the space outside the two rooms.

This test room has hard-surfaced walls to diffuse sound so that the total acoustic power can be evaluated from a single microphone position. The larger vanes in this room oscillate to assist in this process. Tests are conducted on a heavy, spring-mounted table that permits measuring vibrations generated by a test unit, using accelerometers attached to the table. A 350-hp variable-speed electric motor provides mechanical power to the test room through 11 ft of dynamically balanced shafting. This motor can also drive pumps that supply oil for hydraulic motor tests. There is an oil-conditioning system in the same room as the drive. This system both cools the test oil and maintains its dissolved air at a constant level.

Research to reduce noise includes detailed measurements of internal pressure cycles, basic studies of intrinsic pumping mechanisms, and evaluations of the effects of design variations. The results of much of this work have been used in programming computers that help design quiet fluid power equipment by evaluating harmonic values and providing the most efficient timing for the pumping cycle.

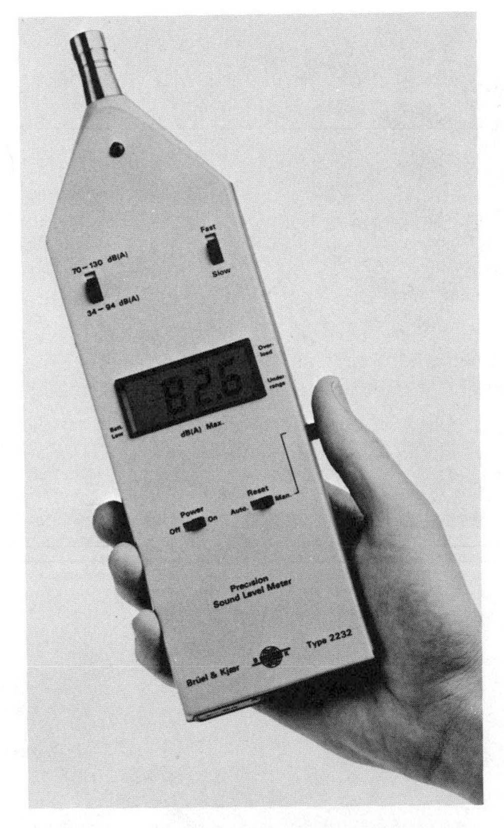

Figure 13.5 Precision sound-level meter. (Courtesy of Bruel & Kjaer Instruments, Inc., Marlborough, Massachusetts.)

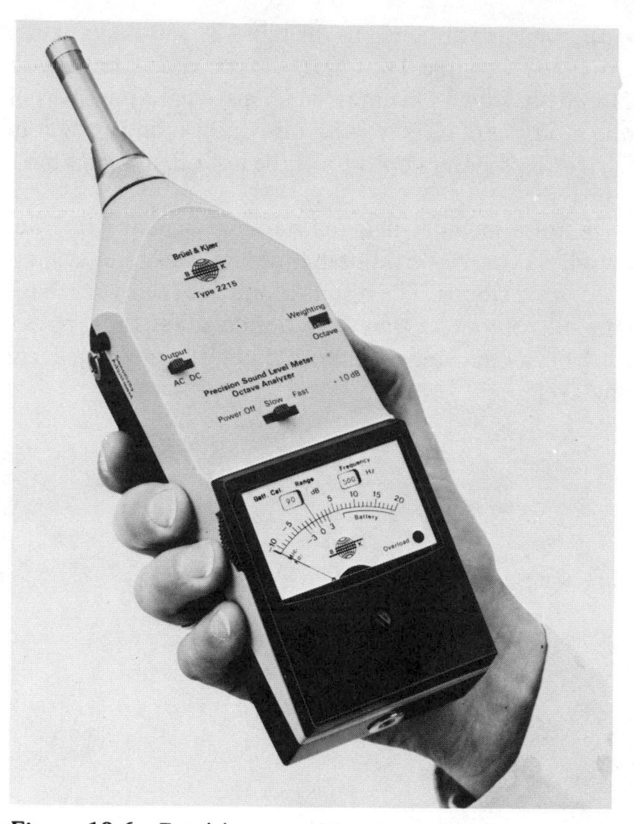

Figure 13.6 Precision sound-level meter and octave analyzer. (Courtesy of Bruel & Kjaer Instruments, Inc., Marlborough, Massachusetts.)

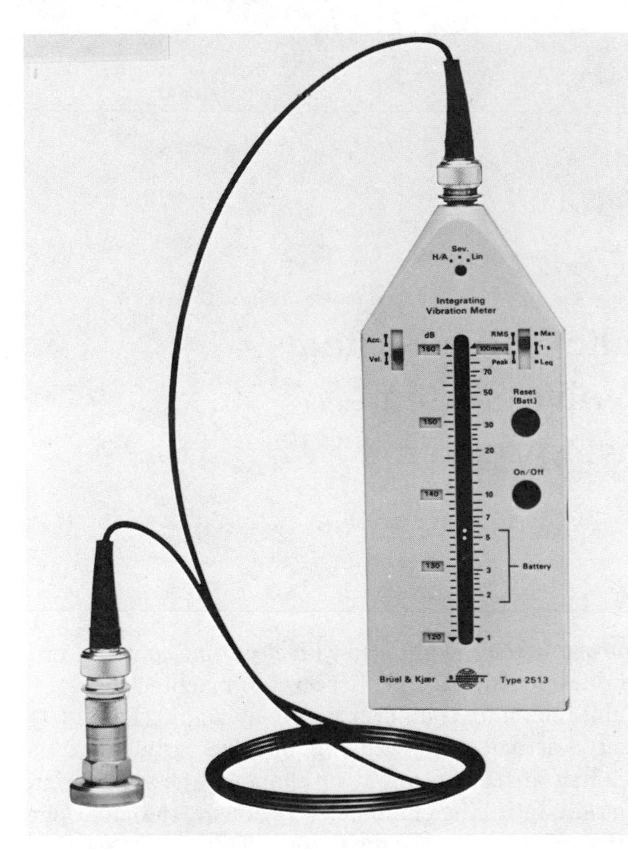

Figure 13.7 Integrating vibration meter. (Courtesy of Bruel & Kjaer Instruments, Inc., Marlborough, Massachusetts.)

14

Electrohydraulic Servo Systems, Proportional Valve Systems, and Load-Sensing Systems

14.1 DESCRIPTION

Electrohydraulic fluid power systems, combined with electronic controls, provide fluid power systems designers with a greater level of machine accuracy, efficiency, versatility, and reliability than ever before available (Fig. 14.1). For example, a new, servo-actuated rug-weaving machine can change the woven pattern produced in a tufted rug simply by changing the appropriate electronic circuit card in the controlling computer. Previously, the cam-operated machine required new cams to alter the rug pattern. This change required almost a complete machine rebuild. As a second example, an automated continuous coal miner can be operated entirely by remote control. This permits the machine operator to remain in a safe location where roof supports ensure the operator's safety. The solution is to use a multichannel digitally coded remote radio control that provides complete isolation with no loss of control. Previously, the operator had to be in the hazardous environment to operate the machine.

On-off electric control of fluid power is as old as modern fluid power technology. Solenoids and relays have been used as fluid power control components for years. Addition of a spring to the solenoid armature provided proportionality a long time ago.

Proportional valving provides control at a level between the capabilities of jet pipe or flapper-type servo valves and systems and conventional alternating current solenoids and/or hydraulic pilot flow control systems.

Direct-current solenoids biased by a spring can actuate a spool-type directional valve in a repetitive pattern to provide power transmission functions in a production machine circuit with acceptable accuracy. This is particularly true when they are biased by an appropriate spring mechanism. Low cost, good performance, and dependability have resulted in expanding use and customer acceptance.

This expanded use of proportional control systems has offered some useful control options to the designer and user of hydraulic power transmission systems. Most of these areas are related to accelerating, deceleration, and blending of the control functions with the machine harmonics.

Load-sensing systems as currently used in compensated flow control valves can be expanded by using the proportional valve as both the directional control and flow orifice mechanism.

The directional control valve can be associated with a normally open pressure level control valve with a downstream load-sensing pilot control as used in a conventional reducing valve. Thus pressure drop across the orifice created by the directional valve can be compensated by an input-reducing-type valve structure or by a variable restrictive-type compensator mechanism on the outflow area.

Some manufacturers manifold the two-valve assemblies together with conventional fasteners or assemble cartridge valve structures in the subbase or interconnecting piping manifold structure.

There is somewhat of a parallel function to that of the jet pipe or flapper-type servo-valve system with a recognition of associated limitations of the proportional mechanism.

Electrohydraulic servocontrol is a child of the military research of the middle and late 1940s. Much of the early development work dealt with the design of a proper feedback control system. The actuators were hydraulic because they alone could deliver required power levels.

Type of system	Accuracy (in.)	Frequency response (Hz)	Comparative efficiency[a]
Valve-motor	0.00001–0.002	50–150	B
Valve-cylinder	0.0005–0.005	30–50	B
Servo pump (packaged)	0.005–0.020	10	A
Servo pump (split)	0.010–0.020	10	A

[a]Rating A is higher.

Figure 14.1 Servo-system comparison.

The control system was first electric and later electronic, to interface with early computers used for rocket guidance or gun fire control. These systems accurately and quickly provided necessary control. The relatively small weight and size of electrohydraulic components was an added requirement. The first industrial applications of these systems were on numerically controlled (NC) machine tools in the late 1950s. (Incidentally, many of these machine tool applications are now electronically controlled, dc electric motor driven.)

Early electrohydraulic systems had many problems: (1) they offered poor reliability, were large in size, and were based on vacuum-tube systems; (2) they were sensitive to oil contamination; and (3) they were very costly.

Servo-valve technology is now widely accepted in industry, and users and manufacturers have gained considerable experience adapting from aerospace to industrial applications. Improvements include:

1. Increased reliability and reduced cost of present electronic control components—solid-state, active devices such as transistors, integrated circuits, and microprocessors.
2. Advances in the state of the art of oil filtration coupled with higher contaminant immunity of hydraulic components.
3. Improved design of electric and hydraulic segments of the servo valve, such as dry force motors, reduction of armature friction, nonspring-centered spools, and mechanical feedback resulting in much higher spool-driving forces. The servo valve itself has become a small servo system.
4. Cost. Although some servo valves still cost as much as they used to, others are quite inexpensive. Increased use in industry in general can be expected to further reduce component costs.

14.1.1 Control System Types

Broadly speaking, three types of electronic control systems are presently available: open loop, simple closed loop, and sophisticated closed loop.

Open-Loop Systems

In an open-loop control system, the operator provides the intelligence to close the control loop: as the operator sees the load approach its desired position, he or she feathers the actuator to place the load smoothly. An electronic link now often replaces the mechanical link, resulting in many advantages. For example, the operator can control the machine remotely for reasons of safety, environmental conditions, and improved visibility.

In addition, electronic inputs may come from several sources simultaneously. For instance, operator command may be the primary control until the load demands start to slow the engine, at which time antistall circuitry will override the operator. In this same manner, other parameters, such as horsepower limiting, overspeed, overtorque, overtilt, and emergency shutdown, can easily be included in the control design.

On occasion, it becomes advantageous to eliminate all mechanical linkage and even electric cables between operator and machine. A radio link can control the hydraulic actuators, so there is no danger of current arc or necessity of dragging a tether cord.

In one system, a radio transmitter produces a low-power, line-of-sight, digitally coded and multiplexed signal to implement electronic control of the hydraulic system. The signal includes address codes making each transmitter and receiver a unique pair, thus eliminating potential interference from nearby transmitters. This system provides multiaccess, on-off, and proportional control.

Simple Closed-Loop System

A closed-loop control system is more accurate because the difference between the actual and desired results is minimized by measuring and feeding back the actual results. The feedback signal may be a measure of speed, torque, force, position, pressure, gravity vector, and so on. Use of feedback can improve accuracy and repeatability, reduce response times, and remove human beings from the system.

Hardware required for simple closed-loop control usually includes off-the-shelf servo valves, proportional valves, linear and rotary servo actuators, and electronic amplifiers and transducers. Some measurement transducers used with this type of system are potentiometers, tachometers, linear variable differential transformers, and pressure and pulse pickoffs. The electronics are proportional; they use integrated circuits and are reliable and economical. Often the electronics are assembled on an easily replaceable plug-in card module which can be quickly replaced and assembled to a test module to verify proper operating conditions.

Sophisticated Closed-Loop Systems

Closed-loop control has arbitrarily been divided into simple and sophisticated to distinguish between analog (i.e., directly proportional) and digital electronic control systems. Another meaning of sophisticated here is the need for extremely high accuracy, at least higher than can be obtained with the analog systems described. Once information has been digitized, the numbers can be processed as desired.

Some systems operate the digital control using feedback-position encoders or pulse pickoffs with digital error comparison. These systems can be programmed manually, using matrix boards or manual-dial-in thumbwheels to control position, feeds, speeds, thicknesses, and so on.

Automatically programmable systems are the next level of complexity. Typically, they use magnetic or paper tape input and can include programmable read-only memory (PROM) for complex sequencing and control. The rug machinc is an cxample of this type of control.

Finally, self-programmable systems have decision-making capability using digital computers. These systems may have adaptive control capability, that is, the selectivity to vary the system configuration depending on performance. If a trend toward unacceptable manufacturing tolerance is noted, for example, the mini- or microcomputers can revise the command so that the output will return to the center of the tolerance band.

Such machines use the most advanced types of electronic hardware, including microprocessors and electronic memories. The electrohydraulic hardware is conventional except that it provides high performance to achieve the higher accuracies and higher response times necessary.

14.1.2 Electrohydraulic Equipment

Servo-controlled hydraulic components include several types of directional control valves, cylinders, fluid motors, and hydrostatic drive-stroke controllers. Electrohydraulic stepping motors also use open-loop control, but their use is limited.

Electrohydraulic proportional control valves used in mobile and industrial applications are available with open- or closed-center configurations. They can provide pressure beyond capability, and some can operate comfortably with filtration to 25 μm. Their response times are about 0.05 to 0.3 sec, depending on size and application. In response to industrial demand, some valves are now surprisingly economical.

Hydrostatic swashplate positioners for axial piston pumps and cylinder barrel positioners for bent-axis assemblies were originally simple levers which moved a yoke to cant the drive angle. Now, bolt-on servo controllers use mechanical or electric feedback control to provide desired pump and/or motor stroke to establish a predetermined power transmission pattern. Two-wire electrical connection may be used to eliminate all mechanical linkages. On some models, dual-coil torque motors permit control from one or more stations or, with a slight wiring change, provide steering control.

On the following pages we review the design and operation of the various components that make up a typical electrohydraulic servo system. This includes hydraulic amplifiers, servo valves, proportional valves, and feedback elements.

14.2 PROPERTIES OF FEEDBACK CONTROL SYSTEMS

14.2.1 Objectives

The purpose of a feedback control system is to monitor an output (controlled variable) in a manner dictated by an input (reference variable) in the presence of spurious disturbances (such as random load changes). The system measures the output, compares the measurement with the desired value of the output as prescribed by the input, and uses the error (difference between actual output and desired output) to change the actual output and bring it into closer correspondence with the desired value of the output (Fig. 14.2). To achieve a more sensitive control means, the error is usually amplified; in general, the higher the gain, the more accurate the system. Thus a feedback control system is characterized by measurement, comparison, and amplification. In brief, a feedback control system is an error-correcting power-amplifying system that produces a high-accuracy output in accordance with the dictates of a prescribed input.

Since arbitrary disturbances (such as amplifier drift, random torques, etc.) can occur at various points in the system, a feedback control system must be able to perform its task with the required accuracy in the presence of these disturbances. Since random noise (unwanted fluctuations) often is present at the input of the system, a feedback control system must be able to reject, or filter out, the noise while producing as faithful a representation of the desired output as is feasible.

14.2.2 Open-Loop Versus Closed-Loop System Characteristics

Because a measure of the output is fed back and compared with the input, any representation of a feedback system contains a closed loop and the system

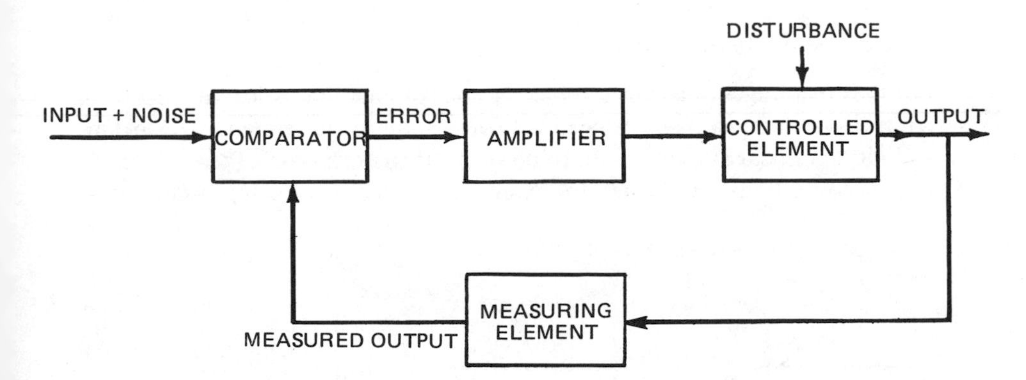

Figure 14.2 Elements of a feedback control system.

is thus called a closed-loop system. Many control systems do not exhibit this closed-loop feature and may be termed open-loop systems (Fig. 14.3). In an open-loop system, the error is reduced by careful calibration.

If open-loop and closed-loop systems are compared, it can be seen that several advantages accrue to the closed-loop system. In a closed-loop system, the percentage change in the response of one of its elements is approximately inversely proportional to the overall amplification of the loop. However, in an open-loop system, the percentage change in the response of the system is approximately proportional to the percentage change in the response of one of its elements. Thus a feedback control system is insensitive to changes in the parameters of its components and can usually be constructed from less accurate and cheaper components than those used in an open-loop system. One exception to the foregoing statement results from an inherent limitation—the closed-loop system can be no more accurate and reliable than its measuring element. The same limitation holds true for an open-loop system.

The error produced in an open-loop system by a given disturbance is much larger than the error produced by the same disturbance in an equivalent closed-loop system, the ratio of errors being approximately proportional to the overall amplification of the loop of the closed-loop system. Thus a feedback control system is relatively insensitive to extraneous disturbances and can be used in situations where severe upsets are expected. One can conclude that the overall amplification (or gain) that can be achieved inside the loop of a feedback control system directly affects the accuracy of the system, the constancy of its characteristics, and the "stiffness" of the system in the face of external upsets or disturbances. In general, it is found that the higher the gain of the system, the better the system. The highest gain that can be used, however, is limited in every case by considerations of stability.

14.2.3 Stability and Dynamic Response

For the advantages of accurcy and constancy of characteristics, the feedback control system must pay a price in the form of a greater tendency toward instability. A linear system is said to be stable if the response of the system to any discontinuous input does not exhibit sustained or growing oscillations.

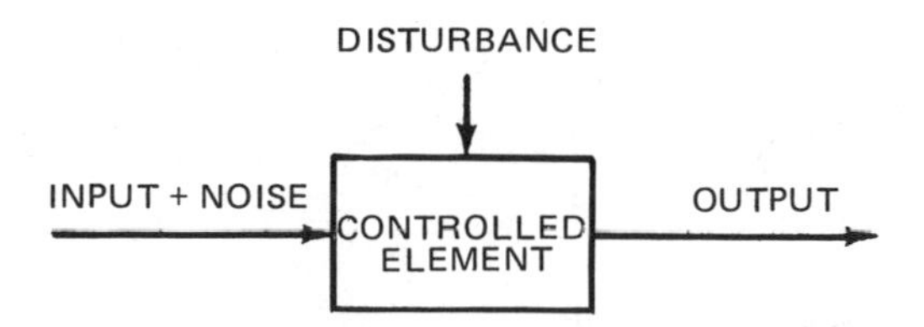

Figure 14.3 Elements of an open-loop control system.

Essentially, this means that the system response will ultimately settle down to some steady value. An unstable system that exhibits steady or runaway oscillations is unacceptable. Unstable behavior must be guarded against in the design, construction, and testing of feedback systems. Because of the possibility of instability, a major portion of control system design is devoted to the task of ensuring that a safe margin of stability exists and can be maintained throughout the operating range of the system.

It can be shown that the cause of instability in a given closed-loop system is due to the fact that no physical device can respond instantaneously to a sudden change at its input. If a sudden change occurs in the error of a feedback system, the output will not correct for the error instantaneously. If the corrective force is great enough (due to a high amplification), the output will accelerate rapidly and cause a reversal of the error. If a high output velocity is attained, the inertia of the output will carry the output past the point where the error is zero. Instability occurs if the maximum magnitude of the error after reversal is equal to, or greater than, the magnitude of the original disturbance in the error. The tendency for a system to become unstable is accentuated as the amplification is increased, since the stored energy in the inertia of the output will be correspondingly increased without any compensating increase in the rate of dissipation of energy in the system. This situation corresponds to an excessive delay in the response of the output. Thus an attempt to increase accuracy by increasing gain or amplification is usually accompanied by an increased tendency toward instability. As a result, design becomes a compromise between accuracy and stability.

14.3 ELECTROHYDRAULIC PROPORTIONAL AND SERVO VALVES

Servo systems use a technique of feeding back to the input all errors that occur within the system during normal operation. The system responds to its feedback and makes proper correction for its error. Of all servo systems, the most responsive and precise is the electrohydraulic system. The heart of the electrohydraulic system is the proportional and/or servo valve.

14.3.1 Valve Construction

Servo valves are either single-stage, two-stage, or in rare instances, three-stage. In a single-stage spool type, the main valve spool is actuated directly by an electric torque motor (Fig. 14.4a). Fluid is ported in a standard four-way configuration. Flows in single-stage valves are limited to about 3 gpm, and their dynamic performance is not as good as two-stage servo valves.

Single-stage proportional valves typically use dc solenoids of conventional design as shown in the double solenoid unit of Fig. 14.4b. This valve also provides a positional transducer to monitor the spool movement.

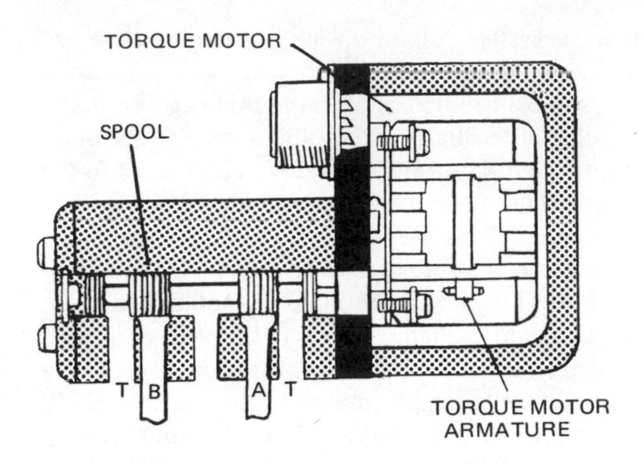

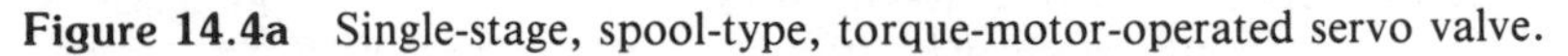

Figure 14.4a Single-stage, spool-type, torque-motor-operated servo valve.

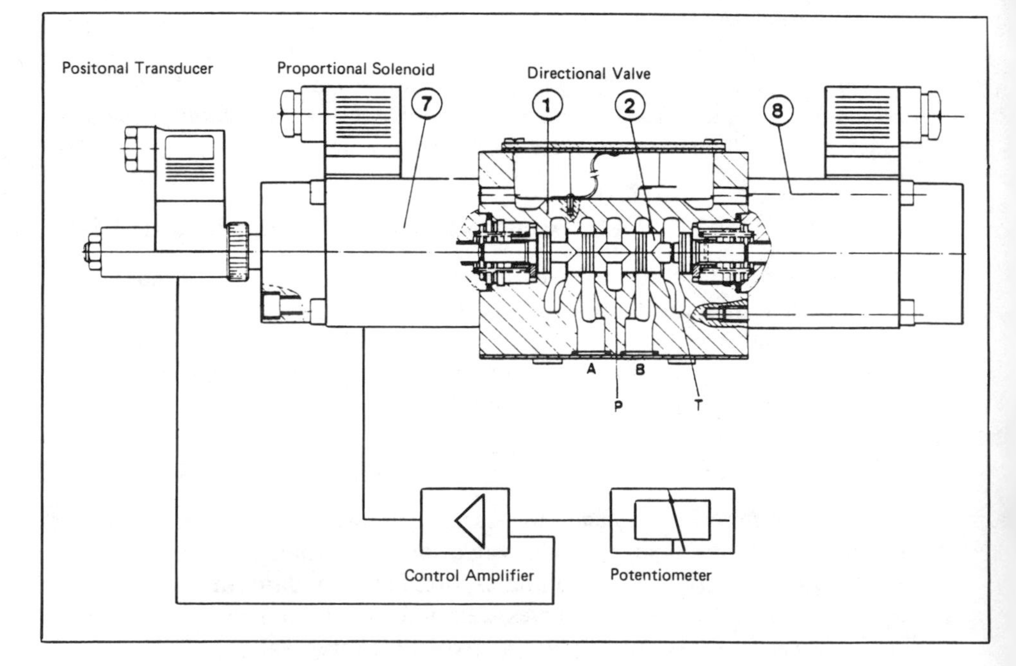

Figure 14.4b Four-way proportional valve with positional transducer. (Courtesy of the Rexroth Corporation, Bethlehem, Pennsylvania.)

In the two-stage valve, the pilot or first stage receives an electromechanical input, amplifies it, and controls the movement of the second (main) stage. In a typical two-stage spool-type valve, the torque motor actuates the pilot spool, which in turn ports fluid to shift the second-stage (main) spool. Most servo valves are two-stage (Fig. 14.5a).

A typical two-stage proportional valve, (Fig. 14.5b) uses a pilot valve which employs two dc solenoids and a spool assembly which provides a feedback function. The solenoid, when energized with a specific input electrical signal within its operating range, will move the pilot spool a proportional amount which directs pressurized pilot fluid to one end of the piloted spring-centered spool. As the piloted spool moves, the centering spring is compressed. This increases the pressure level in the pilot line. The increased pressure level is sensed at piston 10 or 11. One of these piston assemblies will oppose the dc solenoid and force it back to a point where the pilot spool centers and locks fluid in the appropriate end cavity of the piloted spool to hold it in a predetermined shifted position to create the desired direction of movement and the desired rate of flow. The piloted spool will stay at that position until the signal value to the dc solenoid controlling the pilot spool is changed.

Pilot Stages

Four major types of pilot or first stages are used in multistage servo valves: spool, single flapper, double flapper, and jet pipe (Fig. 14.6). All use the principle of varying control pressure as an input electrical signal varies.

In the *spool* arrangement, both orifices change simultaneously. One in-

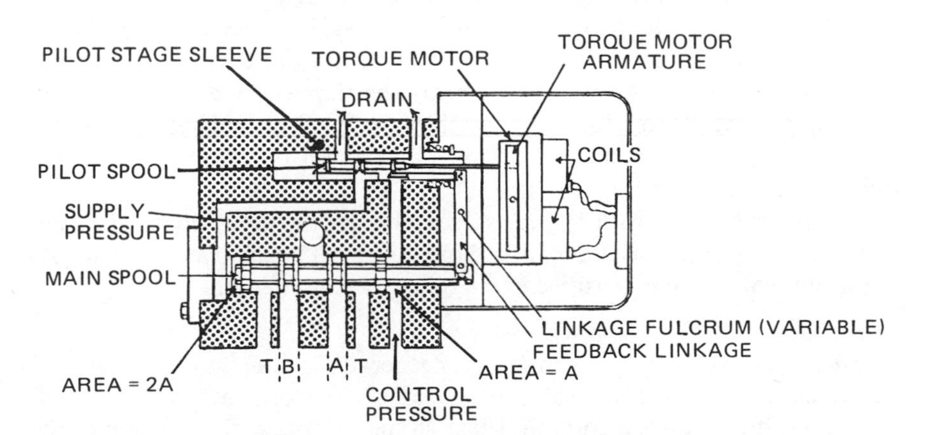

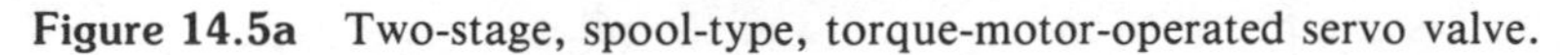
Figure 14.5a Two-stage, spool-type, torque-motor-operated servo valve.

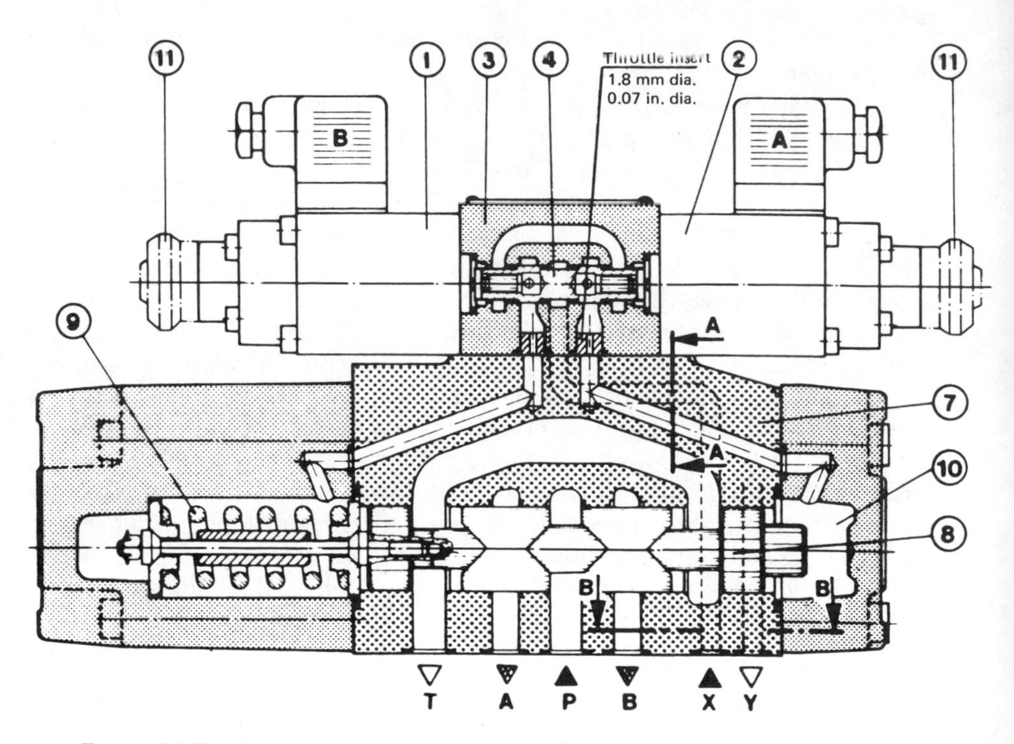

Figure 14.5b Four-way proportional valve. (Courtesy of the Rexroth Corporation, Bethlehem, Pennsylvania.)

creases as the other decreases to create a hydraulic signal. The signal is passed on to the second-stage spool or, in a single-stage valve, directly to the actuator.

In the *single-flapper* type, for each signal line there is one fixed orifice and one variable orifice that is varied by the flapper. A *double flapper* is similar in principle to a four-way spool or an electrical bridge circuit. It has higher pressure gain than a single flapper and is self-compensated against environmental changes.

In servo valves with *jet pipe* stages, the size of the jet nozzle does not change with position as it does in a flapper stage. The relatively large nozzle opening makes it less sensitive to clogging by contaminants.

Feedback

Two-stage servo or proportional valves generally include feedback between the stages. This makes the valve into a small servo system, with all the advantages of closed-loop control. Through the feedback, the position of the main stage is compared to the command of the pilot stage so that any error can be corrected.

Commonly used feedback techniques are the following:

Mechanical-positional feedback maintains a positive positional relationship between the pilot sleeve and the main spool. A spring load is applied to the sleeve to overcome backlash.

Mechanical-force balance designs have a spring that permits adjusting the ratio of motion between the stages.

Electrical feedback using a linear variable differential transformer is a very effective method. However, it requires additional electronic equipment to complete the loop.

Hydraulic-follower feedback permits the simplest valve construction, but the movement ratio between the pilot and second stage cannot be adjusted.

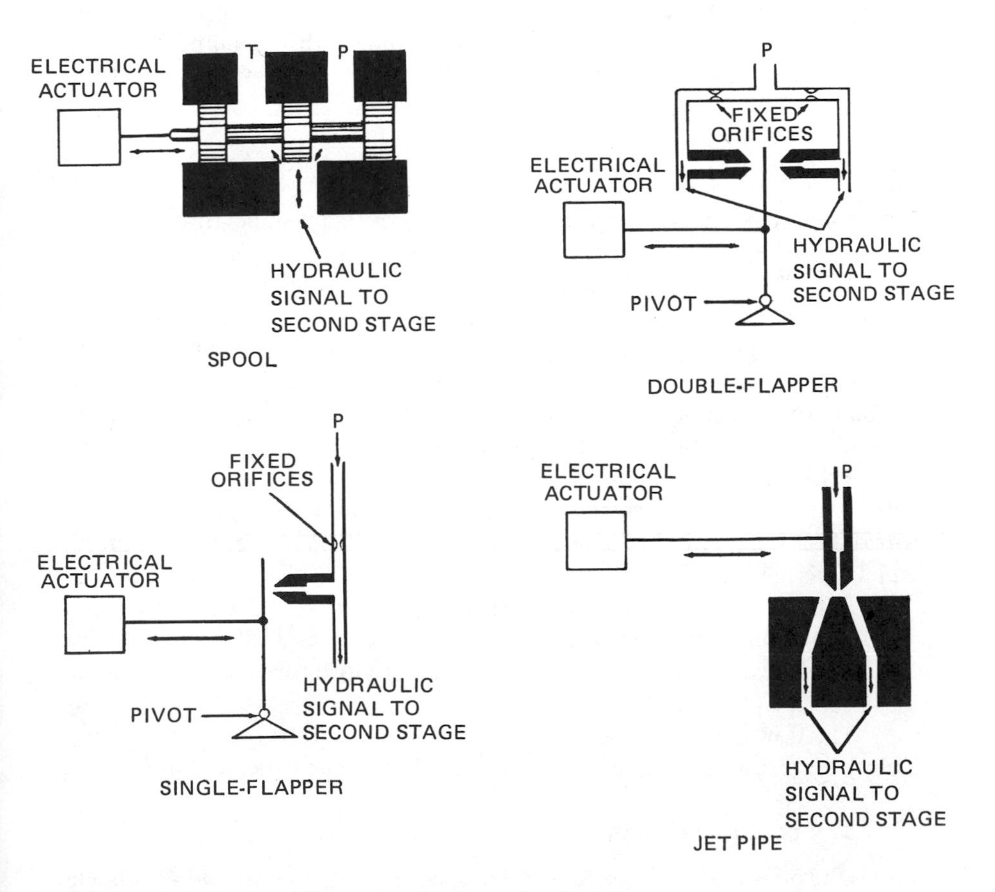

Figure 14.6 Four major types of servo-valve first stages.

Hydraulic-load pressure feedback is the most complex in operation. Basically, main spool movement depends on the pilot differential and the rate of the springs at both ends of the spool. In addition, internal passages connect the spool ends to cylinder port pressures, creating forces that close the valve. The result is a lower flow rate than with springs alonc. As load pressure increases, flow decreases in a reasonably straight line relationship.

Proportional-position arrangement does not necessarily have feedback, but does provide for main spool porting.

In most feedback mechanisms the ratio of movement of the main stage to that of the pilot stage can be adjusted. In mechanical-positioning feedback, the fulcrum of the lever can be changed to provide stepless adjustment. In mechanical force balance, the feedback can be changed by use of different rate springs. Electrical feedback ratio can be changed by adjusting the gain of the amplifier used to compare the input signal to the signal produced by spool displacement. The feedback ratio of hydraulic-follower valves is fixed at 1:1

14.3.2 Main-Stage Porting

Two standard porting arrangement are used in the main stage: three-way and four-way. The four-way version includes a pressure port and two cylinder ports. The three-way configuration has only one cylinder port.

With a three-way valve, the cylinder usually has a piston area ratio of 2:1. The small area end is connected to constant hydraulic pressure. The pressure of the large area end of the cylinder is controlled by the servo valve. At the point of hydraulic balance, the valve maintains the pressure on the large area at one-half that of the small area.

A four-way valve controls pressure on both sides of the actuator. The valve is said to have a bridge-type balance when used on equal-area actuators such as balanced pistons and rotary motors. When the valve is balanced, equal intermediate pressures are supplied to both sides of the actuator.

Port-area configuration is of prime importance to valve performance. The most common porting shapes are square, round, and full annulus. Flow from a valve with square ports is almost directly proportional to the spool displacement.

The flow curve of a valve with the full annulus port shows more gain. This gives more rapid acceleration and deceleration of the actuator.

14.3.3 Spool Lap Condition

Lap is the physical relationship between spool metering lands and port openings. An open-center servo valve refers to an underlap condition in which the lands are slightly narrower than the porting area of the body or sleeve. When the valve is centered, this arrangement permits a constant flow of oil from the pressure side of the pump across the ports to tank. The pressure drop across these restrictions produces an intermediate pressure at the cylinder ports that is normally 40 to 60% of supply pressure.

A closed-center valve has an overlapped arrangement in which the lands are slightly larger than the porting area. This construction is not common in servo valves because it creates a dead zone and makes the valve unresponsive to small signals. Sometimes a slight overlap is used with dither (high-frequency vibration of the spool) to ensure a dynamic line-to-line condition.

A third and most desirable construction is the line-to-line servo valve in which the spool metering lands just coincide with the port openings. With this configuration leakage will provide enough flow to establish the intermediate cylinder port pressures, and the valve is ready to respond to very small movements of the spool.

Pressure gain and flow gain are directly affected by lap conditions. Full system pressure to the cylinder cannot be obtained until the metering land that opens the cylinder port has completely blocked the escape of flow to tank.

In an overlapped valve, once the overlap is passed, pressure gain is higher than with an underlapped valve. Only a slight displacement creates full system pressure in one cylinder line and opens the other line to tank. Line-to-line valves have as good pressure gains as overlapped valves without lost motion of the spool.

14.3.4 Valve Characteristics

Ideally, a servo valve produces zero output flow at zero current. In practice, this ideal condition is seldom attained. The null shift may be due to changes in temperature, supply pressure, or load pressure. It is expressed in terms of null bias or current changes required to restore zero output flow.

In open-loop systems the null shift of the servo valve is so great that the load drifts and wanders from its assigned position. Because of null shift, a servo valve is seldom used to position a load without a closed loop.

Load-flow curves show the effects of load pressure on valve flow (Fig. 14.7). The flow change can be determined for a given load pressure change for any input current. The flow curve is obtained by cycling the valve through its rated input current range and recording the continuous plot of output flow for one cycle of input current. The flow curve measures valve flow gain, hysteresis, linearity, and dead zone.

Flow gain is the change in output flow per unit change in input current. This is taken at zero load pressure unless otherwise specified. Pressure gain is the pressure increase per unit of current increase, psi/A.

Valve dynamics can be expressed either in terms of transient response

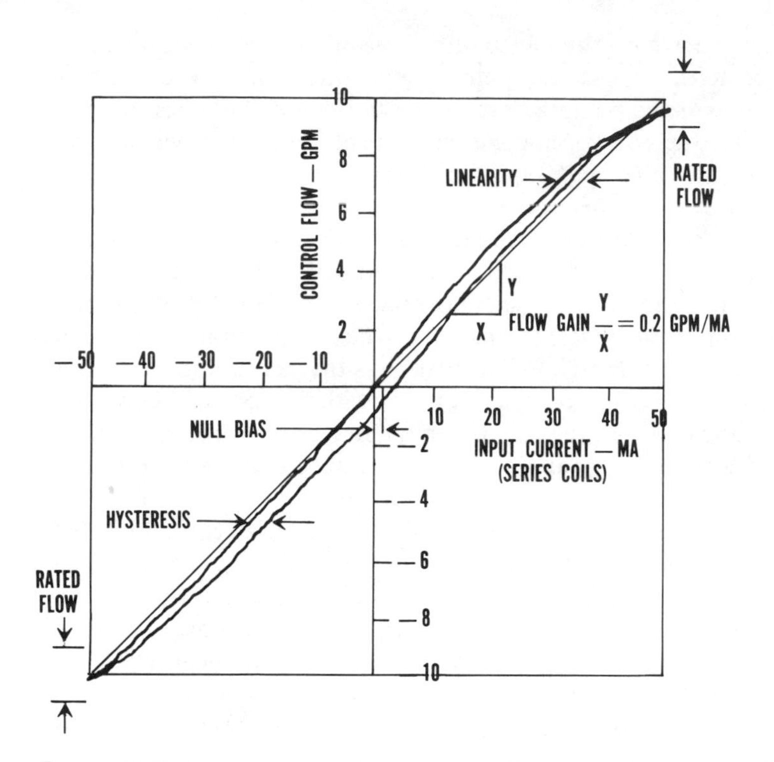

Figure 14.7 Typical servo-valve flow plot. (Courtesy of Moog, Inc., East Aurora, New York.)

or frequency response. A transient response test measures the time required to achieve desired output, the degree of overshoot, and settling time (Fig. 14.8). Frequency-response tests are a measure of the no-load sinusoidal frequency response characteristics.

14.3.4 Flow Control Servo Valves

Servo valves are the most sophisticated and expensive method of controlling actuator speed, but the expense is probably worth the investment if exacting speed control is needed and if the load continually varies (Fig. 14.9a). The valve has a torque motor, which is an armature that is free to rotate within the gap between two permanent magnets. The armature is supported by a thin-walled flexure tube inside of which is a feedback wire and a flapper.

Supply pressure is continuously fed through inlet orifices in the hydraulic amplifier and through nozzles to each side of the flapper. With no movement of the flapper, this control fluid flows through the drain orifice to tank.

When the armature is electrically energized, it rocks in one direction, closing the flow from one of the nozzles of the hydraulic amplificr.

This blockage of the right nozzle causes pressure to build at the right end of the power, or second-stage spool, while left supply fluid continues to flow past the flapper to the tank. Because there is a differential pressure across the spool, it shifts left to permit pressurized fluid to flow to the actuator and the return fluid to flow to tank.

As the spool moves, the ball end of the feedback wire also moves left, trying to move the flapper away from the right nozzle to restore equilibrium. When this feedback torque becomes equal to the magnetic torque of the armature, the armature and flapper move back toward the centered position.

But the spool will stop short of its center position if there is still current to the armature. When these opposing torques are equal, spool position will be proportional to input current. With constant supply pressure, flow to load is proportional to spool position. Thus varying input current controls flow volume and actuator speed.

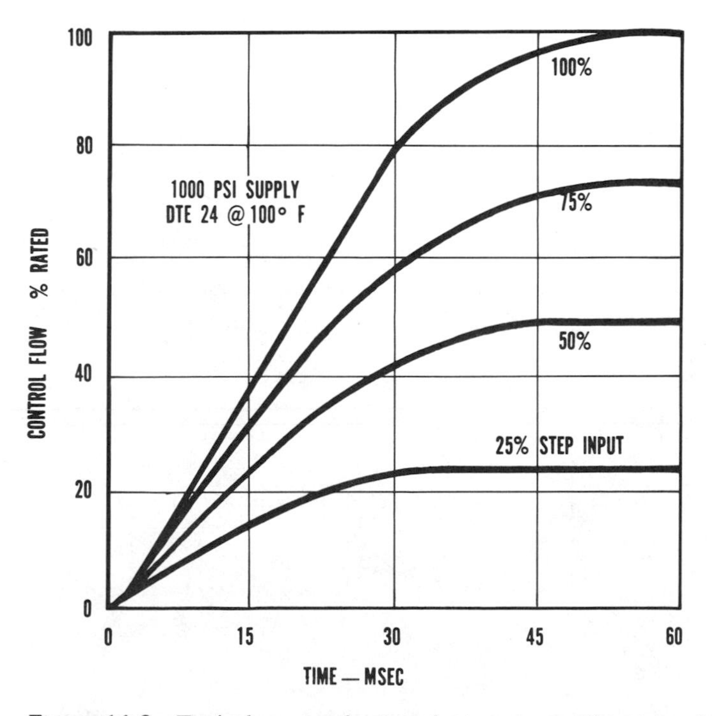

Figure 14.8 Typical servo-valve transient response. (Courtesy of Moog, Inc., East Aurora, New York.)

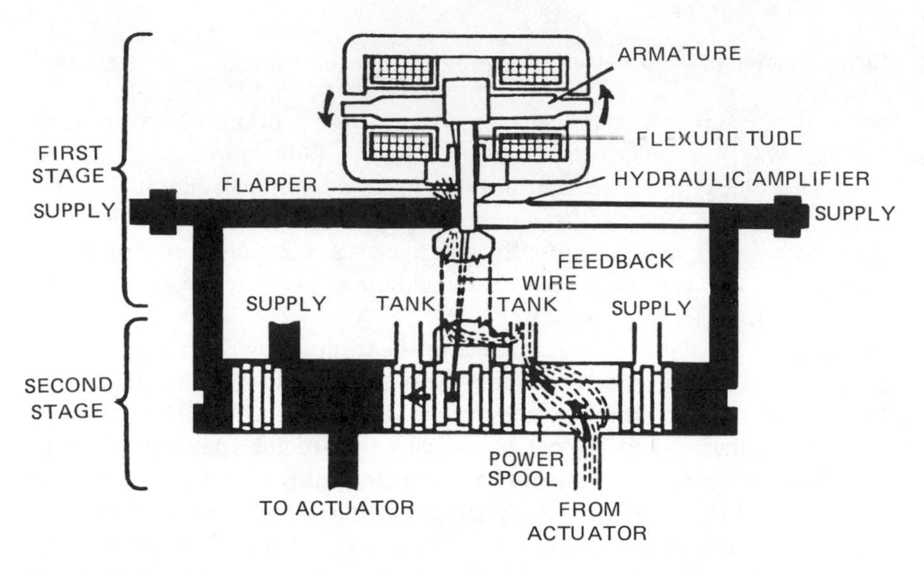

Figure 14.9a Flow control servo valve.

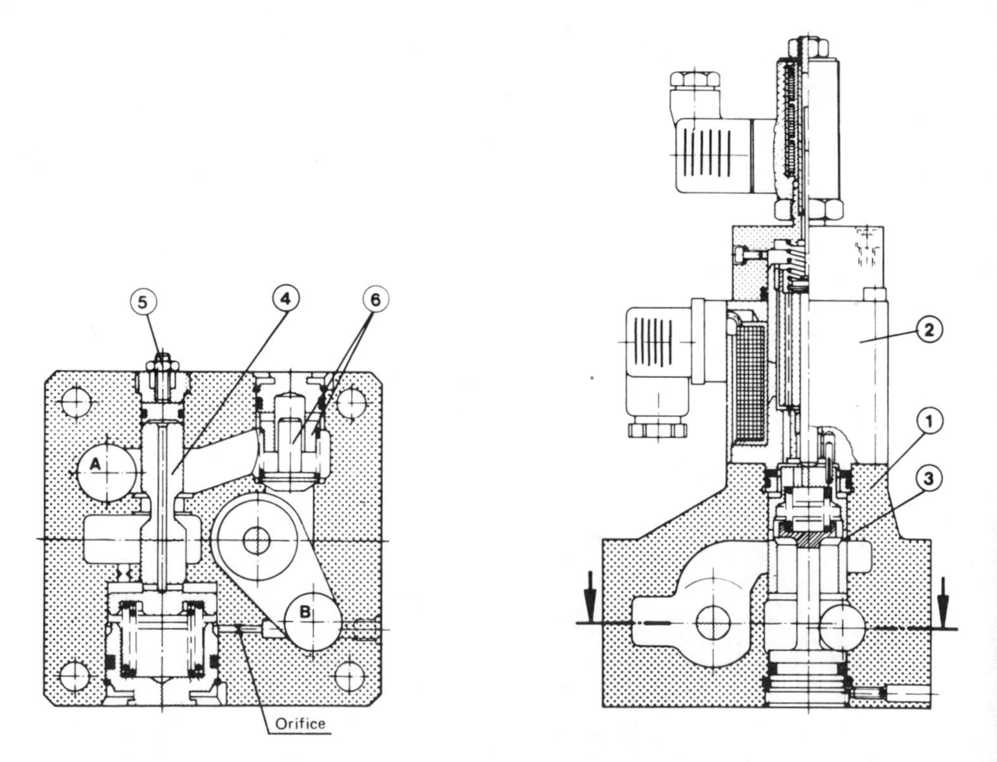

Figure 14.9b Electronic proportional pressure compensated flow control valve. (Courtesy of the Rexroth Corporation, Bethlehem, Pennsylvania.)

Figure 14.9b illustrates an electrohydraulic proportional flow control valve. Pressure-compensated spool 5 and check valve 7 provide conventional reducing valve compensation and return free flow through the valve. Proportional movement of orifice assembly 4 and 8 is provided by dc solenoid 1. Position of orifice device is monitored by LVDT 2. Flow through the valve is varied by signal energy applied at solenoid 1.

14.3.5 Pressure Control Servo Valves

A pressure control servo valve has a polarized electrical torque motor, a double-nozzle hydraulic amplifier, and an output-stage sliding-spool valve (Fig. 14.10a). A stub and sleeve arrangement provide the spool with two drive areas at each end. The differential pressure across one pair of spool end areas is controlled by the flapper and nozzle hydraulic amplifier. The differential pressure across the other pair of spool end areas provides load pressure feedback.

The hydraulic amplifier has two fixed inlet orifices and two variable orifices formed by the circumferential openings between the nozzle tips and the flapper. The flapper connects directly to the armature of the torque motor. Flexing a thin-walled tubular pivot moves the armature and flapper. The tubular pivot also provides a fluid seal between the electromagnetic and hydraulic parts of the servo valve.

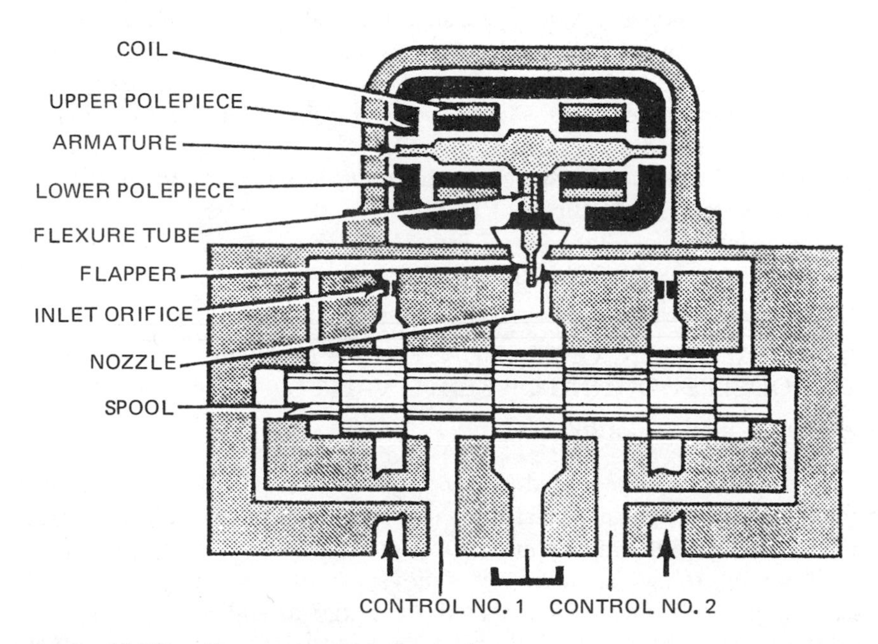

Figure 14.10a Pressure control servo valve.

When electric current is applied to the servo valve, a torque is developed on the armature-flapper. This torque causes the flapper to pivot on the flexure tube support and move between the nozzle tips: one nozzle orifice opens as the other closes, resulting in a differential pressure between the nozzle chambers. The pressure differential creates a force on the second-stage spool that moves the spool. Spool displacement allows flow to the load to develop an output pressure from the reservoir. The load pressure is fed back to the spool end areas so that the spool will move until a balance is achieved between the force due to the hydraulic amplifier and that due to load pressure. One spool end area may connect directly to supply or return if the pressure in a single load line is to be controlled with respect to the supply or return pressure.

A proportional relationship exists between the servo valve output pressure and the electrical input signal due to the force balance at the spool and the proportional action of the torque motor and hydraulic amplifier. Under load flow conditions, output pressure will exhibit some droop due to flow reaction forces acting on the valve spool.

The proportional relief valve of Fig. 14.10b uses conventional piloted relief valve components and adds a proportional function in parallel to the manual pilot adjustment 3. The plunger of solenoid 2 urges the poppet against seat 12. Pilot fluid can be diverted to the tank at minimum pressure when solenoid 2 is not energized. As an increasing electrical signal is applied at solenoid 2, an increasing pressure will be provided up to the point at which control fluid can pass to tank at the value established by parallel manually adjusted pilot control assembly 3. The relief valve of Fig. 14.10b is a normally closed device. The reducing valve of Fig. 14.10c is a normally open device. It responds to the proportional electric signal in a pattern similar to that of the relief valve of Fig. 14.10b.

The proportional pilot valve of Fig. 14.10d can be used to control virtually any pilot controlled pressure device.

14.4 INSTALLATION, OPERATION, MAINTENANCE, AND TROUBLESHOOTING OF SERVO VALVES

14.4.1 Installation and Startup

The servo valve can be mounted in any position. Best performance will be achieved with the servo valve located as close as possible to the load actuator. By minimizing the column of liquid between the control valve and the actuator it is possible to increase accuracy of movement of the controlled device. Possible sponginess resulting from small quantities of entrained gas in the liquid is minimized and deflection of the conductors creating an accumulator action is limited.

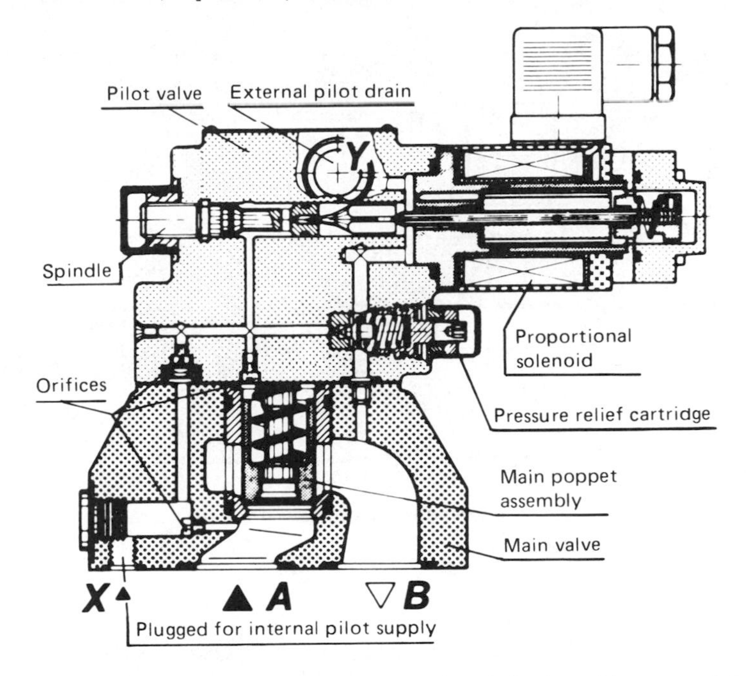

Figure 14.10b Electronic proportional relief valve. (Courtesy of the Rexroth Corporation, Bethlehem, Pennsylvania.)

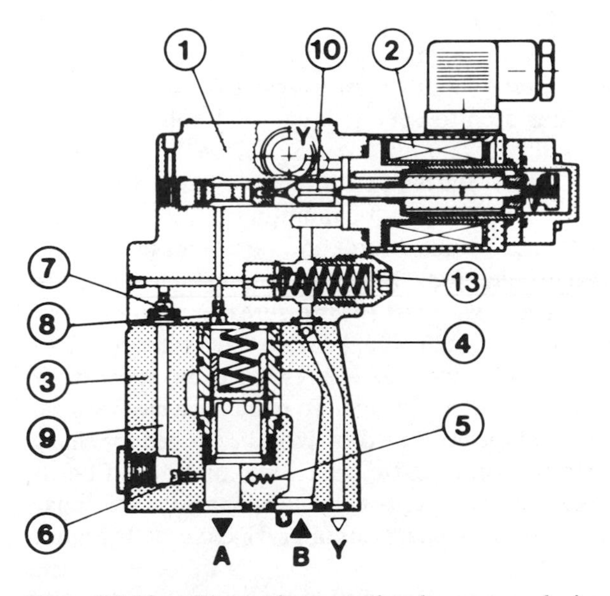

Figure 14.10c Electronic proportional pressure reducing valve. (Courtesy of the Rexroth Corporation, Bethlehem, Pennsylvania.)

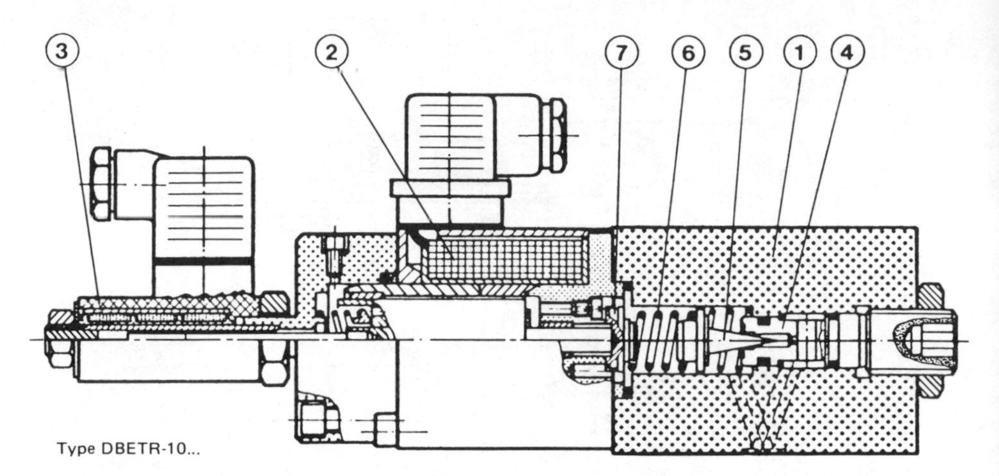

Figure 14.10d Proportional pressure relief valve. (Courtesy of the Rexroth Corporation, Bethlehem, Pennsylvania.)

Prior to initial installation of a servo valve, the hydraulic system must be flushed clean by running the pump to circulate fluid through the system filters. During this flushing process, the load actuator should be manually exercised several times to circulate otherwise trapped oil.

To prolong servo-valve life and to reduce hydraulic system maintenance, it is recommended that a 5- or 10-μm large-capacity filter be installed upstream of each servo valve. Some proportional valve systems can tolerate 25-μm filter systems. Standard practice has been to operate a new hydraulic system for 4 hr with a flushing block at the servo-valve location before the servo valve is installed. However, the period of flushing prior to servo-valve installation varies considerably with the complexity and condition of the system. The flushing is done under conditions of temperature, flow rates, and so on, which reasonably simulate operating conditions. New system filter elements are installed during the flushing process whenever the pressure drop across the filter indicates that the element(s) need changing. When a filter will operate for a period of 2 hr with no perceptible increase in pressure drop, most of the harmful system contamination has been removed. To maintain a clean system, filters must be replaced whenever the pressure drop indicates a need for changing. It is also recommended that an oversize-capacity 5- or 10-μm filter be installed in the return line or in a line that provides constant full-flow capacity. This increases the filter element replacement interval and greatly reduces the system contamination level.

It is often desirable to adjust the flow null of a servo valve independent of other system parameters. The *mechanical null adjustment* on the servo valves allow a ±20% adjustment of flow null (Fig. 14.11). The mechanical null adjustor for one manufacturers valve is an eccentric bushing retainer pin, located above the "return" port designation on the valve body which, when rotated, provides control of the bushing position. Mechanical feedback elements position the spool relative to the valve body for a given input signal. Therefore, a movement of the bushing relative to the body changes the flow null.

Minor differences for adjusting valves will be encountered with different suppliers. Request appropriate service instructions prior to making any adjustment and note recommendations associated with the service and adjustment procedures. Some adjustments require laboratory-type instrumentation, which may be modified for the desired "tuning" of the servo devices.

If the valve is off null, the load will start to move in one direction or the other. The null adjustment for one typical servo valve is accessible by first removing a cap screw located in the valve cover just above the electrical connector. Valve null is adjusted with an Allen wrench. This adjustment should be turned slowly one way or the other to stop load motion. If the load has

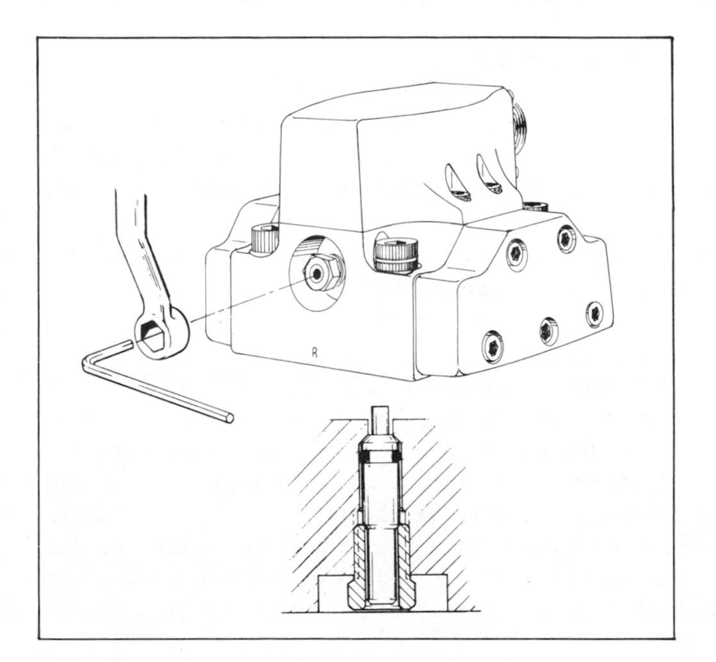

Figure 14.11 Mechanical null adjustment on a typical two-stage servo valve. (Courtesy of Moog, Inc., East Aurora, New York.)

moved against a stop, it may be desirable to reconnect the electrical cable momentarily to bring the load back near its starting point.

After stopping the load, continue to turn the null adjustment until reverse motion occurs. Repeat the sequence of stopping and restarting the load in opposite directions, then leave the null adjustment midway between. The difference in adjustment positions is caused by valve threshold, load friction, and other effects.

Less than ±1 turn of this null adjustment should be sufficient to null the servo valve. If ±2 turns of adjustment fail to achieve null, further turning will not correct the problem.

A review of manufacturers' recommendations may be advisable to determine if other factors are affecting the null adjustment procedures.

14.4.2 Operation

One type of electrohydraulic servo valve consists of an electrical torque motor, a nozzle-flapper pilot stage, and a sliding spool main stage. The torque motor includes coils, pole pieces, magnets, and an armature (Fig. 14.12). The armature is supported for limited movement by a flexure tube. The flexure tube also provides a fluid seal between the hydraulic and electromagnetic portions of the valve.

The flapper attaches to the center of the armature and extends down, inside the flexure tube. A nozzle is located on each side of the flapper so that flapper motion varies the nozzle openings. Pressurized hydraulic fluid is supplied to each nozzle through a filter and inlet orifice. Differential pressures caused by flapper movement between the nozzles are applied to the ends of the valve spool.

The four-way valve spool controls flow from the supply to either control port. The bushing contains flow control ports that are uncovered by spool motion. A feedback wire is deflected by spool movement so that feedback torque is applied to the armature-flapper.

Electrical current in the torque motor coils causes either clockwise or counterclockwise torque on the armature. The torque displaces the flapper between the two nozzles. The differential nozzle pressure moves the spool to either the right or left. The spool continues to move until the feedback torque counteracts the electromagnetic torque. At this point the armature-flapper is returned to center, so the spool stops and remains displaced until the electrical input changes to a new level. The actual flow from the valve to the load will depend on the load pressure. Figure 14.13 depicts a two-stage electrohydraulic servo valve.

Valves utilizing other pilot-stage amplifiers, such as jet pipe types, operate in similar fashion to the nozzle-flapper one described. Motion of the jet pipe causes a force imbalance in the valve and, in turn, spool motion.

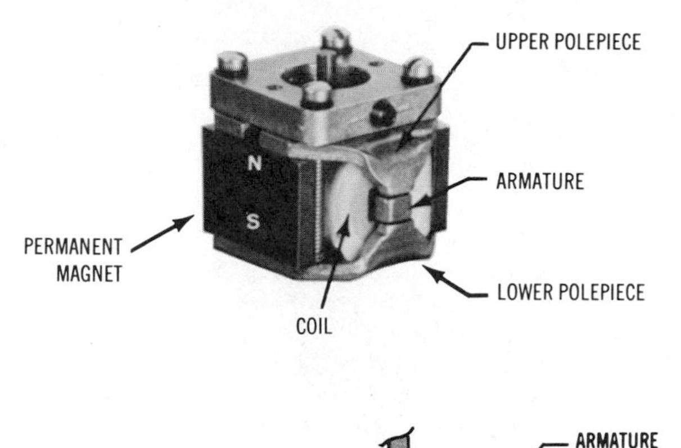

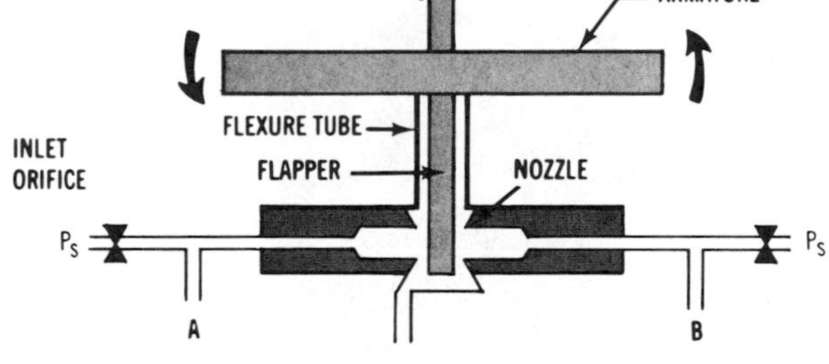

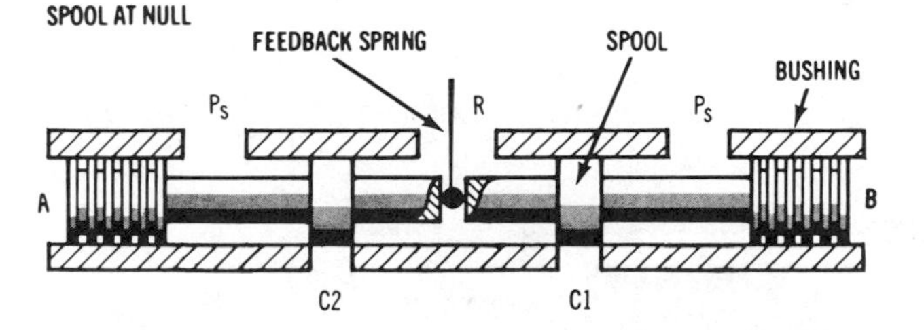

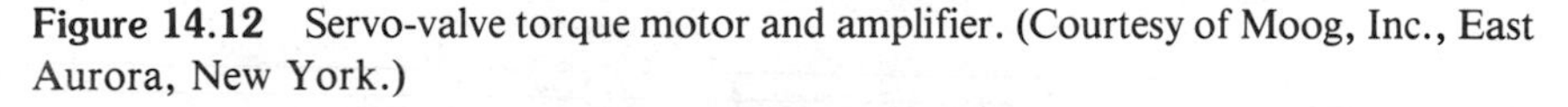

Figure 14.12 Servo-valve torque motor and amplifier. (Courtesy of Moog, Inc., East Aurora, New York.)

Electrical dither is often helpful for improving system resolution. A small-amplitude, relatively high frequency dither signal will continuously cycle the valve spool. This prevents silting of contaminant at the valve spool control lands. Silting can lead to erratic system behavior and poor low-speed control.

Additional dither signal amplitude, or a lower dither frequency, is sometimes used to create continuous load motion. This may help overcome problems of actuator or load breakout friction.

Excessive dither signal can produce needless wear and fatigue. The maximum recommended dither level for standard servo valves is 10 mA peak-to-peak dither (series coils) at a frequency between 50 and 200 Hz. Do not use

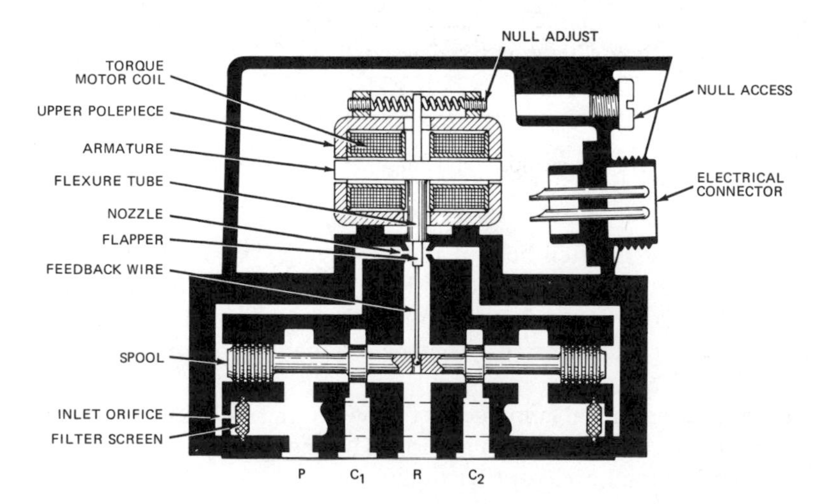

Figure 14.13 Two-stage electrohydraulic servo valve. (Courtesy of Moog, Inc., East Aurora, New York.)

a higher dither frequency as internal valve resonances may be excited that can cause valve failure.

Most servo valves are closed-center types and operate with constant supply pressure. The constant pressure can be provided by a fixed-displacement pump and relief valve, or by an unloading arrangement, or by a variable-displacement pump. In each system, the reservoir should be large enough to avoid fluid foaming at maximum flow. Also, a separate heat exchanger may be required to avoid overheating.

Some valves will operate with supply pressures from 200 to 3000 psi. Performance is considerably better if the supply pressure is above 700 psi. If the supply pressure varies, some valve null shift may occur.

Good fluid filtration will extend the life and improve the performance of servo valves, as well as other system components. The servo valve contains internal filter screens, but these are intended only to protect the pilot stage against inadvertent system contamination. The fluid normally supplied to the servo valve must be filtered to 25 μm absolute or better.

The recommended filtration arrangement is a full-flow, pressure line filter immediately upstream of the servo valve. This filter should *not* contain an automatic bypass valve, as bypass operation can allow quantities of contaminant into the system.

System life can be significantly extended with use of a low-micrometer, full-flow return line filter. Ninety-five percent of the particles in most hydraulic systems are below 10 μm in size. Thus proper silt control such as provided by the return line filter will increase pump, actuator, and valve life by a great many operating hours.

New system filter elements are installed during the flushing process performed during system startup whenever the pressure drop across the filter indicates that the element needs changing. When a filter will operate for a period of 2 hr with no noticeable increase in pressure drop, most of the harmful system contamination has been removed. To maintain a clean system, filters must be replaced whenever the pressure drop indicates a need for changing. Pressure drop can be observed or monitored by installing hydraulic pressure gauges or with mechanical and electrical "dirt alarms." *The filter elements should be changed a minimum of once every year according to some users and experienced maintenance personnel.*

14.4.3 Maintenance and Troubleshooting

Servo valves are delicate, precision-built devices and require the utmost in care, cleanliness, and familiarity with operation for the performance of maintenance tasks. The need for cleanliness was stressed when filtration requirements were reviewed. Servo-valve disassembly and reassembly should be performed in as clean an environment as possible. When performing servicing tasks:

1. Remove electrical leads to the servo valve.
2. Relieve the hydraulic system of any pressure.
3. Then remove the servo valve.

Electrical Checkout

Using an ohmmeter, measure the resistance across the electrical connector pins. Resistance must be within catalog or specification tolerances. If an open circuit (infinite ohms) exists, remove the connector and check for proper solder connections. If a defective connection exists, resolder the coil lead wire to the connector terminal and measure the resistance. If a short circuit still exists, the hydraulic amplifier assembly must be replaced.

Repair or Replacement

1. Replace all O rings, regardless of age or condition.
2. Replace the inlet orifice assemblies if damaged or if foreign matter is present. Always replace the inlet orifice assemblies with ones that have the same rating. The rating, is usually marked on each assembly. Follow manufacturers recommendations when refitting.
3. Replace the filter if the steel mesh is damaged or broken.
4. Clean all parts in a *clean, compatible* commercial solvent. Permit parts to air dry or dry using an air hose with clean, dry air.
5. Lubricate all O rings with clean, filtered hydraulic fluid as used in the hydraulic system prior to installing them on parts, and when installing parts containing O rings.
6. It may be advantageous to return certain defective components, such as the hydraulic amplifier assembly, to the manufacturer for repair.

When making such repairs, note the following:

1. When procuring O rings for replacement, obtain O rings compounded to be compatible with system hydraulic fluid being used, or deterioration may result.
2. Exercise care to avoid cutting or nicking O rings when installing them on parts, and when installing parts containing O rings to prevent subsequent leakage.
3. Most solvents react with O-ring compounds. Remove all O rings from parts before cleaning.

4. Clean all parts in a *clean, compatible* commercial solvent. Permit parts to air-dry or dry using an air hose with *clean, dry air.*
5. Lubricate all O rings with clean filtered hydraulic fluid prior to installing them on parts, and when installing parts containing O rings.
6. It may be advantageous to return certain defective components, such as the hydraulic amplifier assembly to the manufacturer for repair.
7. When returning components, package them so that "rough" handling will not cause further damage.

Troubleshooting Chart

The following troubleshooting chart lists potential problems encountered, probable causes, and solutions.

Problem	Cause	Solution
Servo valve does not respond to command signal	Controller does not function	Replace controller
	Open controller cable	Replace controller cable
	Open coil or open coil lead	Replace hydraulic amplifier assembly
	Contamination wedged in air gap	Clean air gaps
	Jammed spool	Clean valve bushing and spool
Output flow obtained from one control port only (actuator is at limit of stroke or hydraulic motor is rapidly rotating); limited or no response to command signal	Controller not functioning properly	Replace controller
	Filters silted with contamination	Replace inlet orifice assembly
	Plugged inlet orifices	Replace inlet orifice assembly
	Plugged hydraulic amplifier assembly	Replace hydraulic amplifier assembly
	Contamination wedged in air gap	Clean air gaps
	Jammed spool	Clean valve bushing and spool
	Null adjustor erroneously adjusted hardware	Readjust null adjustor

(continued)

Problem	Cause	Solution
High null bias (actuator drifts or hydraulic motor slowly rotates when controller returns to neutral)	Incorrect null adjustment	Readjust null
	Filters silted with contamination	Replace inlet orifice assembly
	Partially plugged inlet orifice	Replace inlet orifice assembly
	Partially plugged hydraulic amplifier assembly	Replace hydraulic amplifier assembly
	Contamination wedged in air gaps	Clean air gaps
Poor response (valve delays in returning to neutral after controller is returned to neutral)	Filters silted with contamination	Replace inlet orifice assembly
Nonrepeatability (valve fails to return to neutral each time the controller is returned to neutral)	Controller not functioning properly	Replace controller
	"Sticky" spool	Clean valve bushing and spool
	Partially plugged hydraulic amplifier assembly	Replace hydraulic amplifier assembly
Servo valve does not follow input command signal (actuator or components are stationary or creeping slowly)	Open coil assembly or open coil leads	Replace hydraulic amplifier assembly
Low flow gain (failure to meet high rate or rapid traverse speeds)	Shorted coil assembly	Replace hydraulic amplifier assembly
High threshold (jerky, possible oscillatory or "hunting" motion in closed-loop system)	"Sticky" spool	Clean bushing and spool assembly

14.5 SERVO-SYSTEM TERMINOLOGY

Coil impedance: the complex ratio of coil voltage to coil current.

Control flow: the flow through the valve control ports to the load expressed in in.^{3}/sec, or gal/min (gpm), or liters/min (lpm).

Dither: an ac signal sometimes superimpossed on the servo-valve input to improve system resolution.

Flow gain: the nominal relationship of control flow to input current, expressed as gpm/mA or lpm/mA.

Frequency response: the relationship of no-load control flow to input current when the current is made to vary sinusoidally at constant amplitude over a range of frequencies.

Hysteresis: the difference in valve input currents required to produce the same valve output as the valve is slowly cycled between plus and minus rated current. Expressed as percent of rated current.

Input current: the electrical current to the valve that commands flow, expressed in milliamperes (mA).

Internal leakage: the total internal valve flow from pressure to return with zero control flow (usually measured with control ports blocked).

Lap: in a sliding spool valve, the relative axial position relationship between the fixed and movable flow-metering edges with the spool at null. Lap is measured as the total separation at zero flow of straight line extensions of the nearly straight portions of the flow curve, drawn separately for each polarity. Expressed as percent of rated current.

Linearity: the maximum deviation of control flow from the best straight line of flow gain. Expressed as percent of rated current.

Load pressure drop: the differential pressure between the control ports (i.e., across the load actuator), expressed in $lb/in.^2$ (psi) or bar.

No-load flow: the control flow with zero load pressure drop.

Null: the condition where the valve supplies zero control flow at zero load pressure drop.

Null bias: the input current required to bring the valve to null, excluding the effects of valve hysteresis. Expressed as percent of rated current.

Null shift: the change in null bias resulting from changes in operating conditions or environment. Expressed as percent of rated current.

Pressure gain: the change of load pressure drop with input current and zero control flow (control ports blocked). Expressed as the nominal psi/mA or bar/mA throughout the range of load pressure between $\pm 40\%$ supply pressure.

Quiescent current: a dc current that is present in each valve coil when using a differential coil connection.

Rated current: the specified input current of either polarity to produce rated flow, expressed in milliamperes (mA).

Rated flow: the specified control flow corresponding to rated current and given supply and load pressure conditions. Rated flow is normally specified as the no-load flow.

Symmetry: the degree of equality between the flow gain of one polarity and that of reversed polarity, measured as the difference in flow gain for each polarity and expressed as percent of the greater.

Threshold: the increment of input current required to produce a change in valve output.

Valve pressure drop: the sum of the differential pressures across the control orifices of the servo-valve spool, expressed in psi or bar.

Appendix A
Fluid Power Symbols and Glossary

A.1 FLUID POWER SYMBOLS

Learning the language of fluid power symbols is important for anyone working with systems. The symbols have proven to be an effective means of communication among engineers, manufacturers, salespeople, and service personnel.

Complex components and entire systems can be presented in simple, easy-to-read symbols. Solving problems is simplified for the troubleshooters because they have a complete representation of a system they can use to think through a problem before beginning the actual service work. Let us take a look at the most commonly used symbols, beginning with those for tanks and filters.

The symbol for a vented tank with an externally mounted filter and the symbol for a pressurized tank with an internally mounted filter are shown in Fig. A.1. Usually, the first item after the tank is the pump. For an oil pump, the triangle in the circle is solid (see Fig. A.2). Notice that the triangle points outward from the center of the circle. The symbol for an air pump or compressor is just the same, except that the triangle is left open. A pump that can pump in either direction is indicated by a circle with two triangles. By adding an arrow, a fixed-displacement pump is changed to a variable-displacement pump.

A directional control valve is symbolized with a series of squares. The number of squares indicates the number of positions a valve has. A two-position valve could have a raise and lower position. A three-position valve could have a neutral, raise, and lower position; a fourth position could be float. The symbol for an open-center directional control valve has the pump flow passing through the valve and back to the tank, as shown in Fig. A.3a. In a

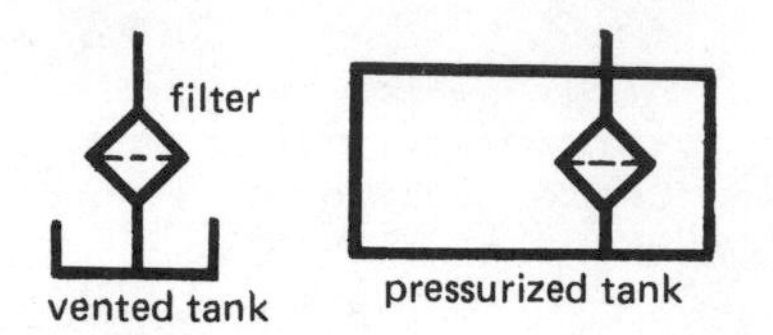

Figure A.1 Vented and pressurized tank symbols.

closed-center valve, pump flow is blocked at the valve (Fig. A.3b). In a closed port valve, the oil contained in the actuator cannot leave the actuator when the valve is in its center position, as can be seen in Fig. A.3c. In the open port valve, the oil can leave the actuator when the valve is centered (Fig. A.3d). Both open- and closed-center valves come in open and closed port versions.

When valves are illustrated in fluid power symbols, they are always shown in the rest position. The arrows in the adjacent squares indicate flow in positions other than center. These arrows never move—you have to imagine that they do.

By adding small symbols to the main symbol, we can show how the valve is actuated. There are many ways a valve can be actuated: for example, solenoids, hand lever, foot pedal, pushbutton, hydraulic pilot, pneumatic pilot, and so on. Further, they can have a combination of actuators, such as a solenoid *and* pilot, solenoid *or* pilot, and so on.

Motors are easy to illustrate. They are just the opposite of pumps (see Fig. A.2). For a motor the triangle points inward. A way to keep it straight is to remember that a pump sends fluid out from itself and so the triangle points outward. The motor receives fluid, so the triangle points inward. Both pumps and motors can be unidirectional or bidirectional (reversible).

The symbol for a cylinder looks just like a cylinder. There are single-acting and double acting cylinders. Also, there are two types of double-acting cylinders; those with equal piston areas and those with unequal piston areas.

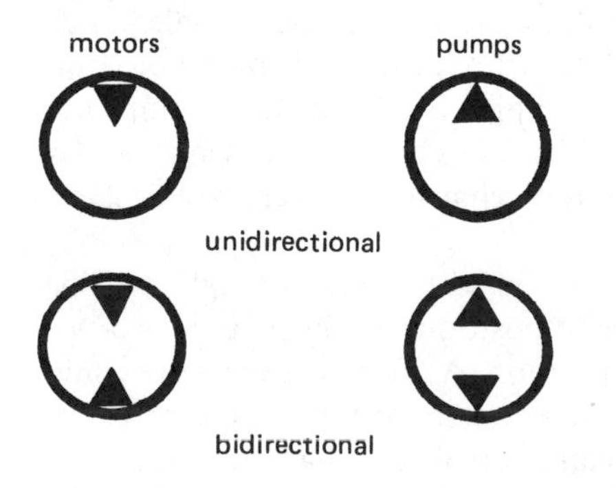

Figure A.2 Hydraulic pump and motor symbols.

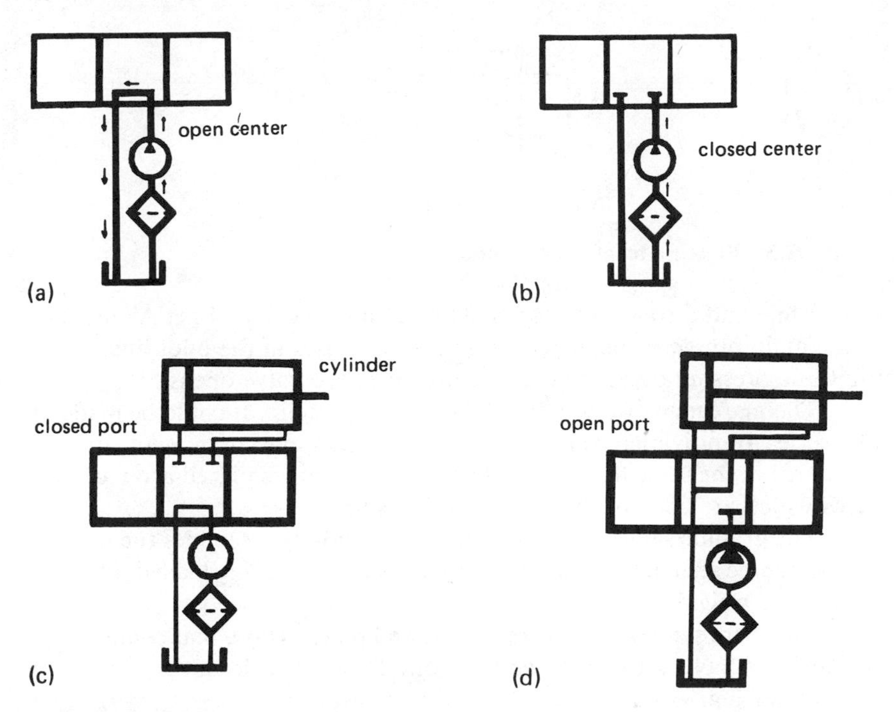

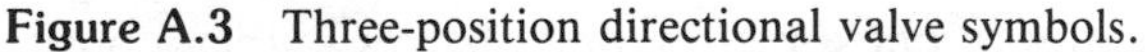
Figure A.3 Three-position directional valve symbols.

The symbol shows each of the configurations, as can be seen in Fig. A.4.

An essential part of the system is the relief valve. It offers an alternative route for the fluid from the pump to either the tank or to atmosphere. Thus it limits the pressure in a system. Imagine that the arrow in the box of the symbol shown in Fig. A.5 is movable. Spring pressure pushes it down to block flow to the tank and hydraulic oil pressure in the pilot line pushes it up to allow pump flow back to the tank. In a pneumatic system, the pilot line is actuated by air pressure and the valve allows air to vent to the atmosphere when the relief valve is actuated. A dashed line is always a representation of

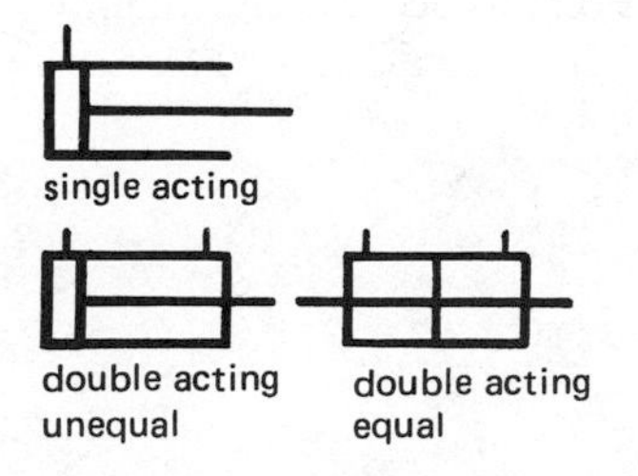

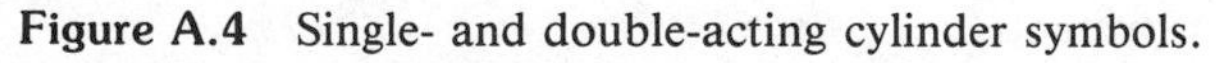
Figure A.4 Single- and double-acting cylinder symbols.

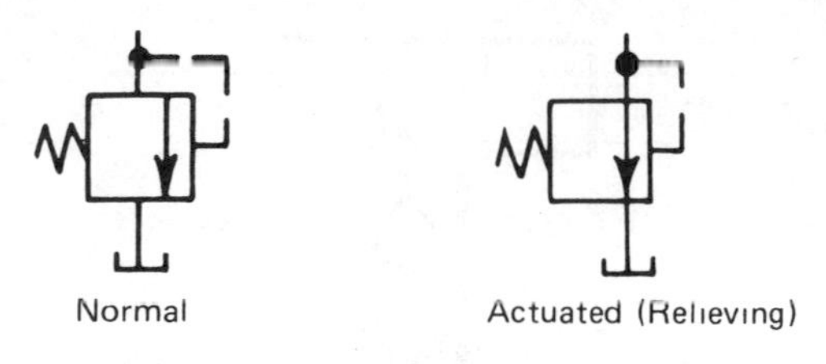

Figure A.5 Pressure relief valve symbols.

a pilot line and a solid line represents a main or working line. As pressure in the main pressure line increases, it also increases in the pilot line. When pilot line pressure overcomes spring pressure, the valve opens.

There are many ways to build a hydraulic system. It could be made up of a series of individual components or more than one component could be contained in the same housing. A dashed line square is placed around those items which are contained in the same housing.

The graphic symbol for an accumulator closely resembles the real object, as can be seen in Fig. A.6. Accumulators may be spring loaded, gas pressurized, or weighted.

There are symbols for hoses, tubes, and pipes. Those that connect with one another have a dot at the connection. Those that do not connect have no dot (see Fig. A.7).

Check valves permit flow in one direction and block it in the opposite direction. They are single-direction valves similar to diodes in electrical systems. Various check valve symbols are shown in Fig. A.8.

Orifices, or restrictors, are often used to control flow in various systems. Some cannot be adjusted and others can. The adjustable one has an arrow drawn through the restrictor symbol to indicate that the size of the orifice can be varied (see Fig. A.9).

Besides the components and their symbols described in the previous paragraphs, there are many others. A complete list of symbols and fluid power terminology is presented in Sec. A.2. You should continually refer to that section if any difficulties are encountered in interpreting any symbols on a fluid power circuit diagram. It is important, however, to become thoroughly

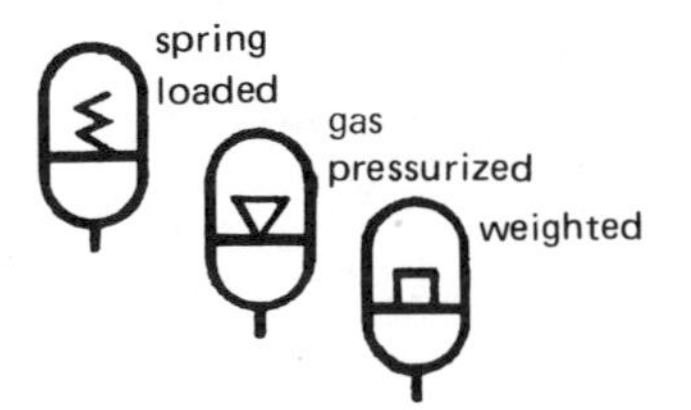

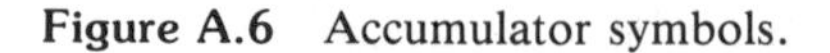

Figure A.6 Accumulator symbols.

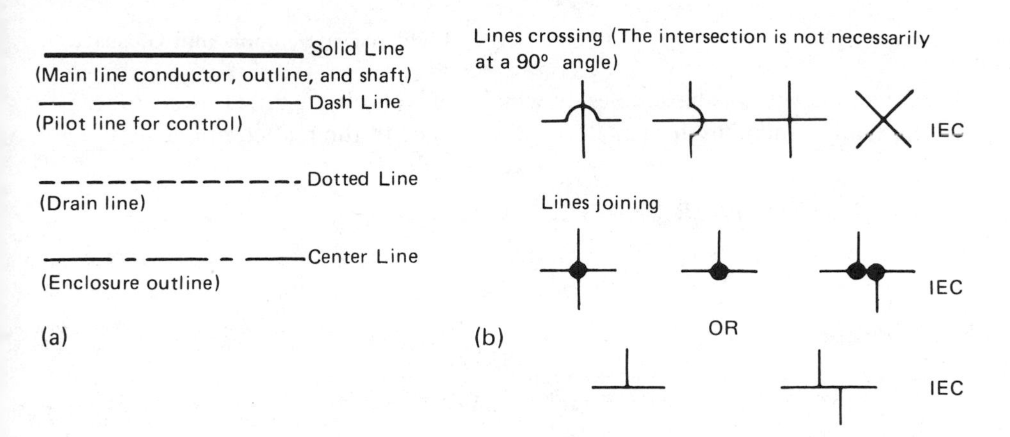

Figure A.7 (a) Lines and (b) intersections on a diagram.

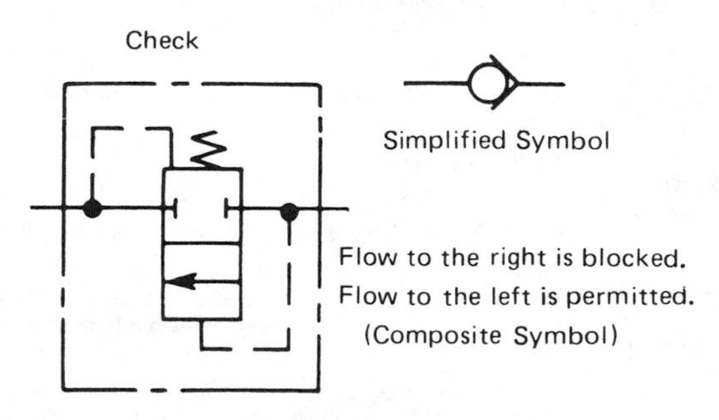

Check, Pilot-Operated to Open

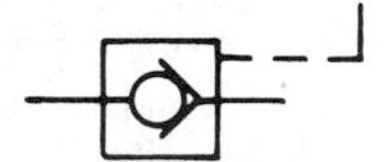

Check, Pilot-Operated to Close

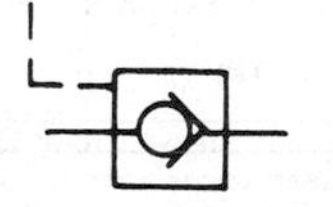

Figure A.8 Check valve symbols.

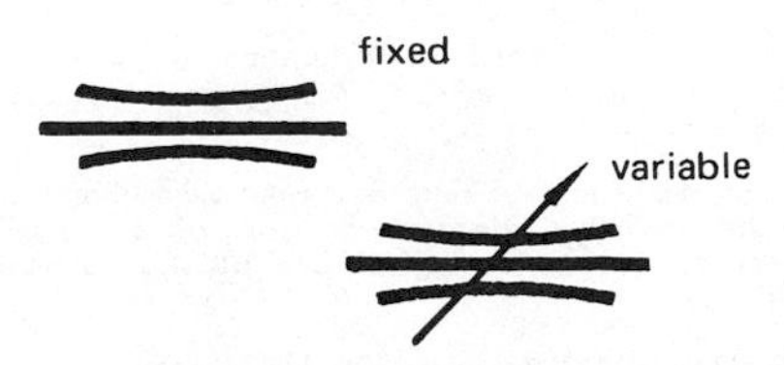

Figure A.9 Symbols for fixed and variable restrictors.

acquainted with the basic rules on which the symbols are based. Remember, the symbols actually bear a close resemblance to the real component.

A.2 FLUID POWER GRAPHIC SYMBOLS

American National Standard / ANS Y 32.10

1. INTRODUCTION

1.1 General Fluid power systems are those that transmit and control power through use of a pressurized fluid (liquid or gas) within an enclosed circuit.
Types of symbols commonly used in drawing circuit diagrams for fluid power systems are Pictorial, Cutaway, and Graphic. These symbols are fully explained in the American Standards Association Drafting Manual (Ref. 2).

1.1.1. *Pictorial symbols* are very useful for showing the interconnection of components. They are difficult to standardize from a functional basis.

1.1.2. *Cutaway symbols* emphasize construction. These symbols are complex to draw and the functions are not readily apparent.

1.1.3. *Graphic symbols* emphasize the function and methods of operation of components. These symbols are simple to draw. Component functions and methods of operation are obvious. Graphic symbols are capable of crossing language barriers, and can promote a universal understanding of fluid power systems.
Graphic symbols for fluid power systems should be used in conjunction with the graphic symbols for other systems published by the American Standards Association. (Ref. 3-7 inclusive).

1.1.3.1 Complete graphic symbols are those which give symbolic representation of the component and all of its features pertinent to the circuit diagram.

1.1.3.2 Simplified graphic symbols are stylized versions of the complete symbols.

1.1.3.3 Composite graphic symbols are an organization of simplified or complete symbols. Composite symbols usually represent a complex component.

1.2 Scope and Purpose

1.2.1 Scope This standard presents a system of graphic symbols for fluid power diagrams.

1.2.1.1. Elementary forms of symbols are:

Circles	Triangles	Lines
Squares	Arcs	Dots
Rectangles	Arrows	Crosses

1.2.1.2 Symbols using words or their abbreviations are avoided. Symbols capable of crossing language barriers are presented herein.

1.2.1.3 Component function rather than construction is emphasized by the symbol.

1.2.1.4 The means of operating fluid power components are shown as part of the symbol (where applicable).

1.2.1.5. This standard shows the basic symbols, describes the principles on which the symbols are based, and illustrates some representative composite symbols. Composite symbols can be devised for any fluid power component by combining basic symbols.
Simplified symbols which are composites of basic symbols are shown for commonly used components.

1.2.1.6. This standard provides basic symbols which differentiate between hydraulic and pneumatic fluid power media.

1.2.2. Purpose

1.2.2.1 The purpose of this standard is to provide a system of fluid power graphic symbols for industrial and educational purposes.

1.2.2.2 The purpose of this standard is to simplify design, fabrication, analysis, and service of fluid power circuits.

1.2.2.3 The purpose of this standard is to provide fluid power graphic symbols which are internationally recognized.

1.2.2.4 The purpose of this standard is to promote universal understanding of fluid power systems.

1.3 Terms and Definitions Terms and corresponding definitions found in this standard are listed starting on page A/62.

2. SYMBOL RULES

(See Section 10)

2.1 Symbols show connections, flow paths, and functions of components represented. They can indicate conditions occurring during transition from one flow path arrangement to another. Symbols do not indicate construction, nor do they indicate values, such as pressure, flow rate, and other component settings.

2.2 Symbols do not indicate locations of ports, direction of shifting of spools, or positions of actuators on actual component.

2.3 Symbols may be rotated or reversed without altering their meaning except in the cases of: a.) Lines to Reservoir, 4.1.1; b.) Vented Manifold, 4.1.2.3; c.) Accumulator, 4.2.

2.4 Line technique (see Ref. 1)
Keep line widths approximately equal. Line width does not alter meaning of symbols.

2.4.1 ———————— Solid Line
(Main line conductor, outline, and shaft)

2.4.2 — — — — — Dash Line
(Pilot line for control)

2.4.3 - - - - - - - - - - - - - Dotted Line
(Drain line)

2.4.4 ——— - ——— - ——— Center Line
(Enclosure outline)

2.4.5 Lines crossing (The intersection is not necessarily at a 90° angle.)

2.4.6 Lines joining

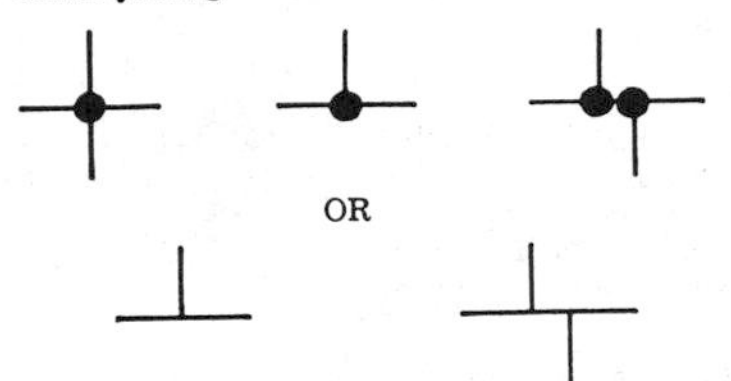

2.5 Basic symbols may be shown any suitable size. Size may be varied for emphasis or clarity. Relative sizes should be maintained. (As in the following example.)

2.5.1 Circle and Semi-Circle

2.5.1.1 Large and small circles may be used to signify that one component is the "main" and the other the auxiliary.

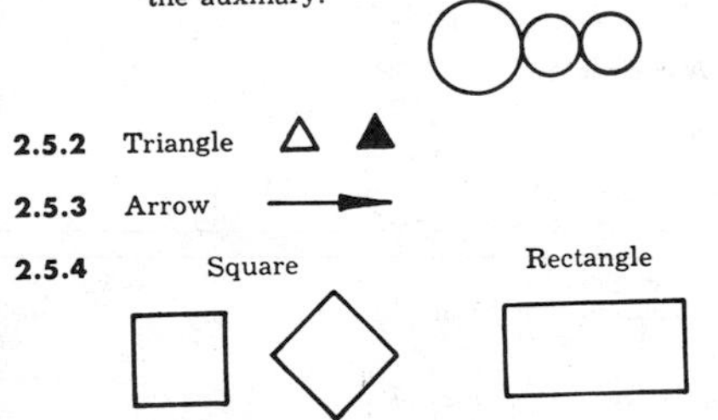

2.5.2 Triangle

2.5.3 Arrow

2.5.4 Square Rectangle

2.6 Letter combinations used as parts of graphic symbols are not necessarily abbreviations.

2.7 In multiple envelope symbols, the flow condition shown nearest an actuator symbol takes place when that control is caused or permitted to actuate.

2.8 Each symbol is drawn to show normal, at-rest, or neutral condition of component unless multiple diagrams are furnished showing various phases of circuit operation. Show an actuator symbol for each flow path condition possessed by the component.

2.9 An arrow through a symbol at approximately 45° indicates that the component can be adjusted or varied.

2.10 An arrow parallel to the short side of a symbol, within the symbol, indicates that the component is pressure compensated.

2.11 A line terminating in a dot to represent a thermometer is the symbol for temperature cause or effect.
See Temperature Controls 7.9, Temperature Indicators and Recorders 9.1.2, and Temperature Compensation 10.16.3 and .4.

2.12 External ports are located where flow lines connect to basic symbol, except where component enclosure symbol is used.
External ports are located at intersections of flow lines and component enclosure symbol when enclosure is used, see SECTION 11.

2.13 Rotating shafts are symbolized by an arrow which indicates direction of rotation (assume arrow on near side of shaft).

3. CONDUCTOR, FLUID

3.1 ———————— Line, Working (main)

3.2 — — — — Line, Pilot (for control)

3.3 - - - - - - - - - Line, Liquid Drain

3.4 Line, sensing, etc. such as gauge lines shall be drawn the same as the line to which it connects.

3.5 Flow, Direction of

3.5.1 Pneumatic

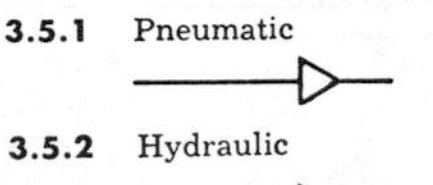

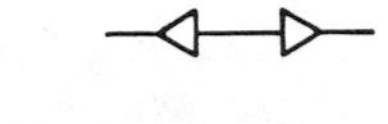

3.5.2 Hydraulic

3.6 Line, Pneumatic Outlet to Atmosphere

3.6.1 Plain orifice, unconnectable

3.6.2 Connectable orifice (e.g. Thread)

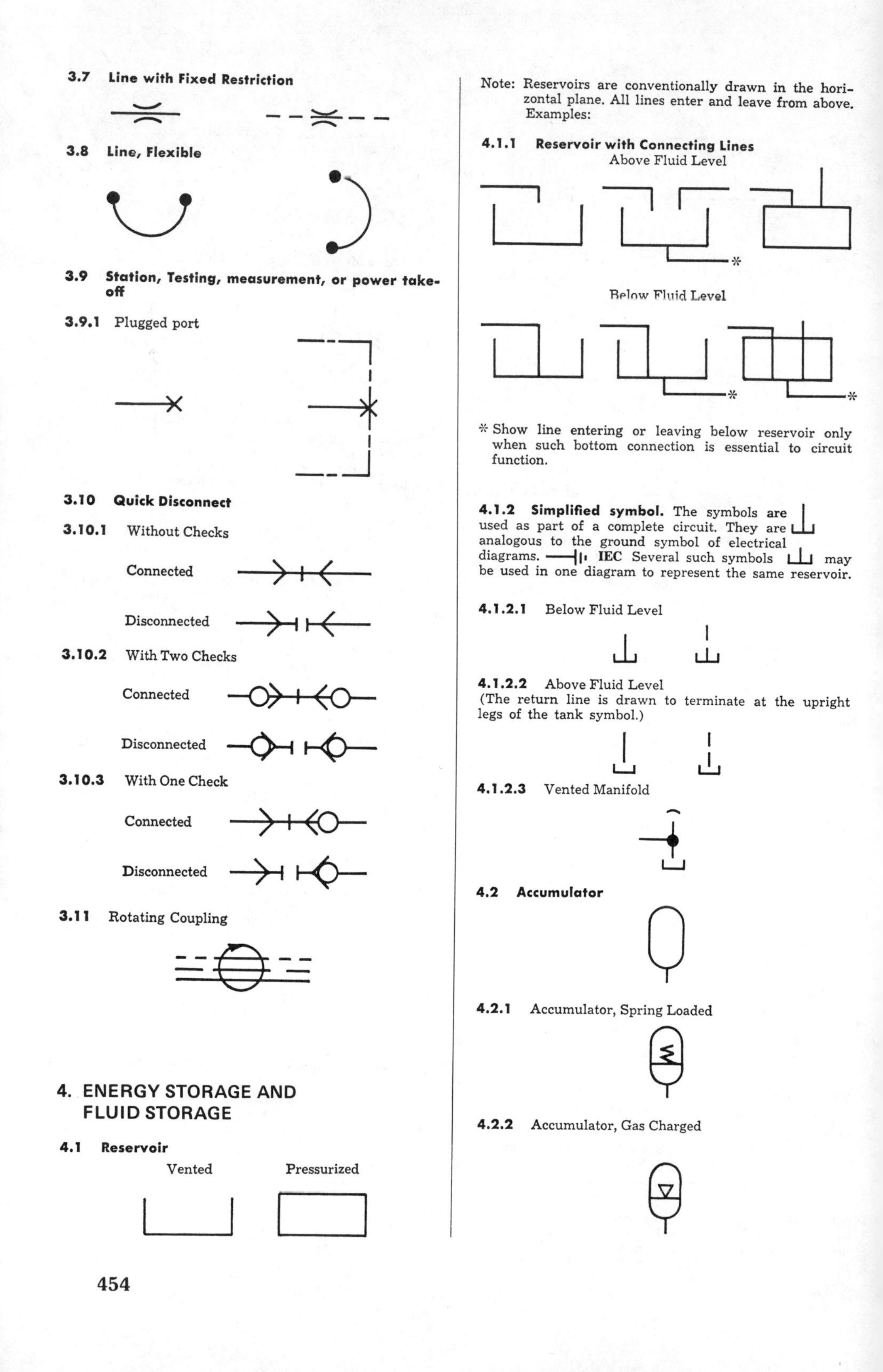

3.7 Line with Fixed Restriction

3.8 Line, Flexible

3.9 Station, Testing, measurement, or power take-off

3.9.1 Plugged port

3.10 Quick Disconnect

3.10.1 Without Checks

Connected

Disconnected

3.10.2 With Two Checks

Connected

Disconnected

3.10.3 With One Check

Connected

Disconnected

3.11 Rotating Coupling

4. ENERGY STORAGE AND FLUID STORAGE

4.1 Reservoir

Vented

Pressurized

Note: Reservoirs are conventionally drawn in the horizontal plane. All lines enter and leave from above. Examples:

4.1.1 Reservoir with Connecting Lines

Above Fluid Level

*

Below Fluid Level

*

*

* Show line entering or leaving below reservoir only when such bottom connection is essential to circuit function.

4.1.2 Simplified symbol. The symbols are used as part of a complete circuit. They are analogous to the ground symbol of electrical diagrams. **IEC** Several such symbols may be used in one diagram to represent the same reservoir.

4.1.2.1 Below Fluid Level

4.1.2.2 Above Fluid Level

(The return line is drawn to terminate at the upright legs of the tank symbol.)

4.1.2.3 Vented Manifold

4.2 Accumulator

4.2.1 Accumulator, Spring Loaded

4.2.2 Accumulator, Gas Charged

4.2.3 Accumulator, Weighted

4.3 Receiver, for Air or other Gases

4.4 Energy Source (Pump, Compressor, Accumulator, etc.) This symbol may be used to represent a fluid power source which may be a pump, compressor, or another associated system.

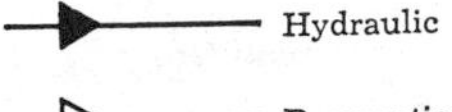

Simplified Symbol

Example:

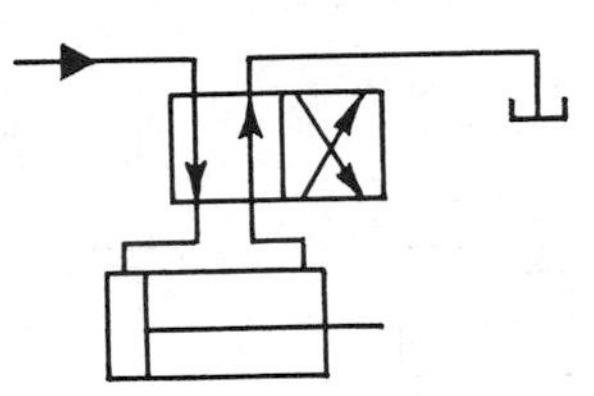

5. FLUID CONDITIONERS

Devices which control the physical characteristics of the fluid.

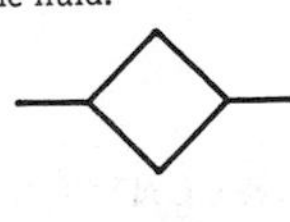

5.1 Heat Exchanger

5.1.1 Heater
Inside triangles indicate the introduction of heat.

Outside triangles show the heating medium is liquid.

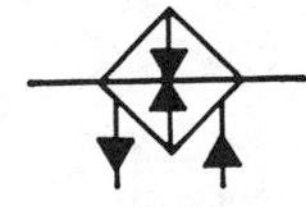

Outside triangles show the heating medium is gaseous.

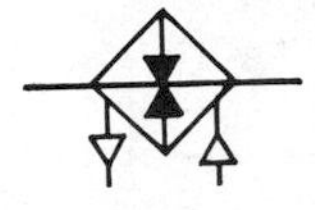

5.1.2 Cooler

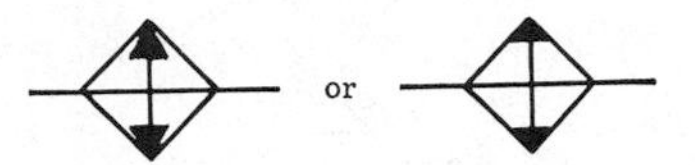

Inside triangles indicate heat dissipation.

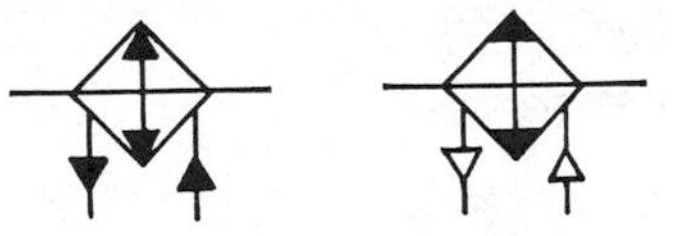

(Corners may be filled in to represent triangles.)

5.1.3 Temperature Controller—(The temperature is to be maintained between two predetermined limits.)

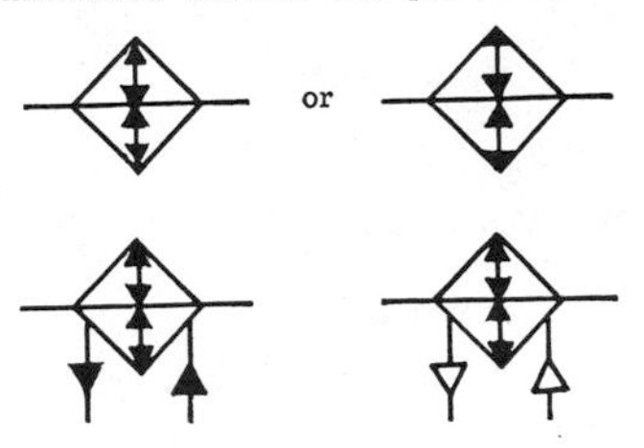

5.2 Filter - Strainer

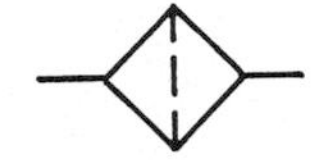

5.3 Separator

5.3.1 With Manual Drain

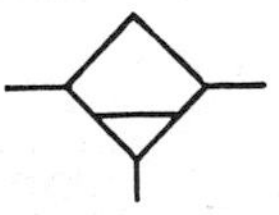

5.3.2 With Automatic Drain

5.4 Filter - Separator

5.4.1 With Manual Drain

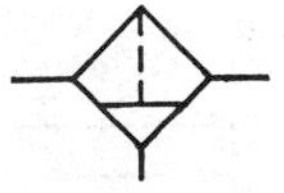

5.4.2 With Automatic Drain

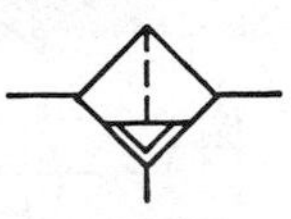

5.5 Dessicator (Chemical Dryer)

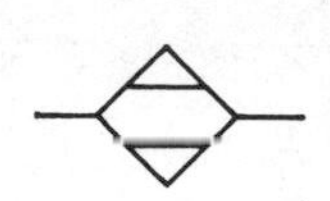

5.6 Lubricator

5.6.1 Less Drain

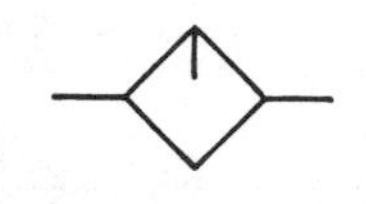

5.6.2 With Manual Drain

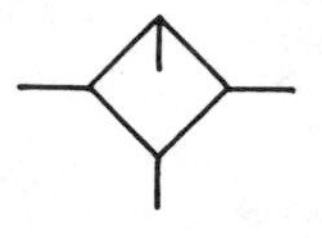

6. LINEAR DEVICES

6.1 Cylinders, Hydraulic & Pneumatic

6.1.1 Single Acting

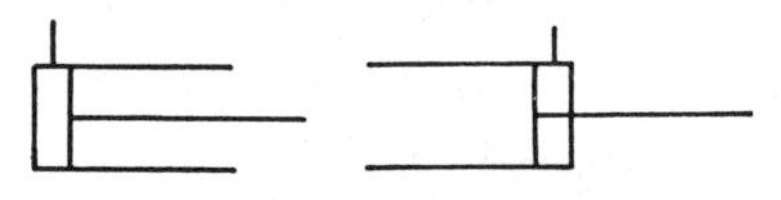

6.1.2 Double Acting

6.1.2.1 Single End Rod

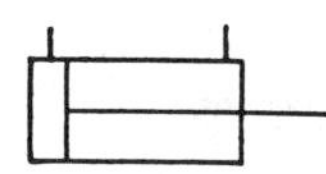

6.1.2.2 Double End Rod

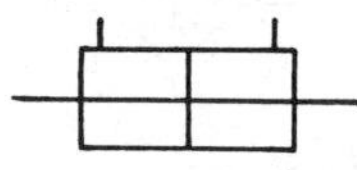

6.1.2.3 Fixed Cushion, Advance & Retract

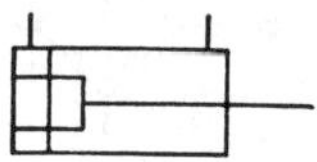

6.1.2.4 Adjustable Cushion, Advance Only

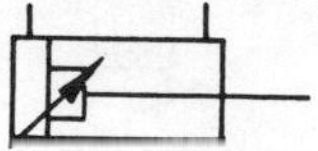

6.1.2.5 Use these symbols when diameter of rod compared to diameter of bore is significant to circuit function.

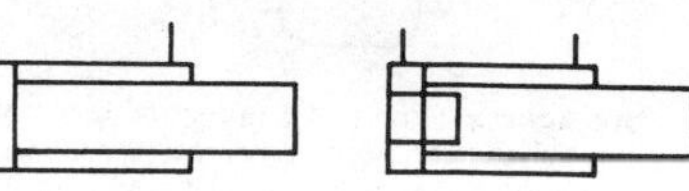

(Non-Cushion) (Cushion, Advance & Retract)

6.2 Pressure Intensifier

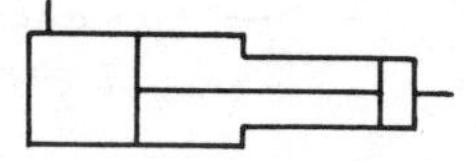

6.3 Servo Positioner (Simplified)

Hydraulic

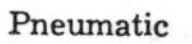

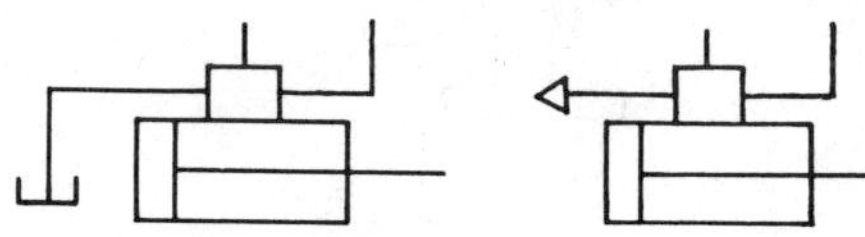

6.4 Discrete Positioner

Combine two or more basic cylinder symbols.

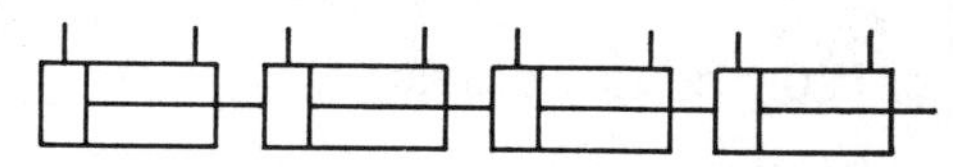

7. ACTUATORS & CONTROLS

7.1 Spring

7.2 Manual

(Use as general symbol without indication of specific type: i.e., foot, hand, leg, arm)

7.2.1 Push Button

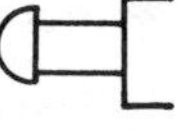

7.2.2 Lever

7.2.3 Pedal or Treadle

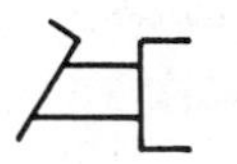

7.3 Mechanical

7.4 Detent

(Show a notch for each detent in the actual component being symbolized. A short line indicates which detent is in use.) Detent may, for convenience, be positioned on either end of symbol.

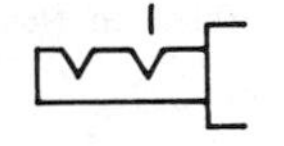

7.5 Pressure Compensated

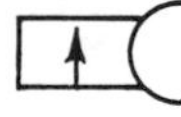

7.6 Electrical

7.6.1 Solenoid (Single Winding)

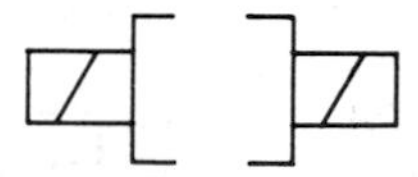

7.6.2 Reversing Motor

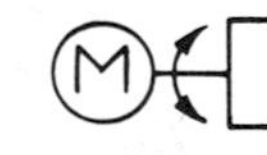

7.7 Pilot Pressure

7.7.1

Remote Supply

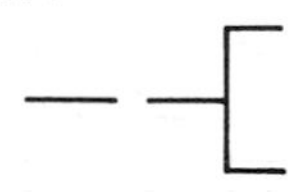

7.7.2

Internal Supply

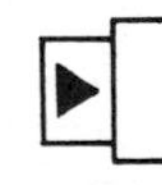

7.7.3 Actuation By Released Pressure

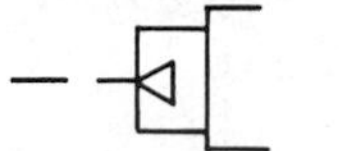

Remote Exhaust

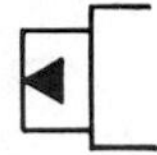

Internal Return

7.7.4 Pilot Controlled, Spring Centered

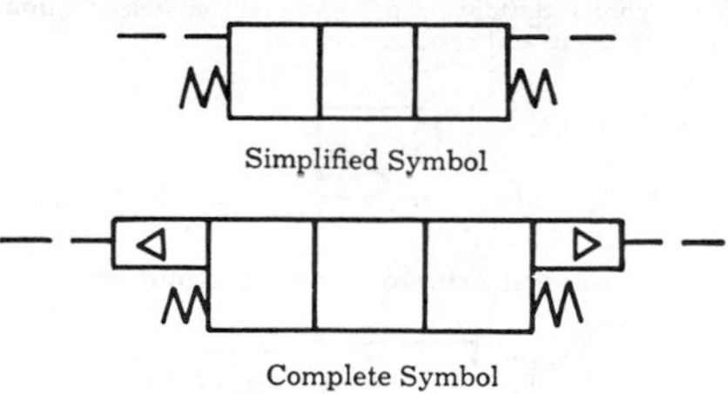

Simplified Symbol

Complete Symbol

7.7.5 Pilot Differential

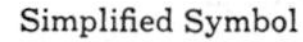

Simplified Symbol

Complete Symbol

7.8 Solenoid Pilot

7.8.1 Solenoid <u>or</u> Pilot

External Pilot Supply

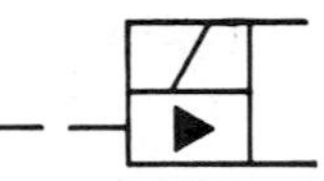

Internal Pilot Supply and Exhaust

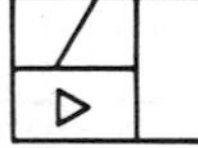

7.8.2 Solenoid <u>and</u> Pilot

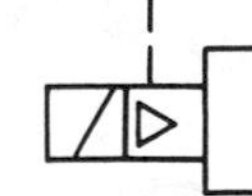

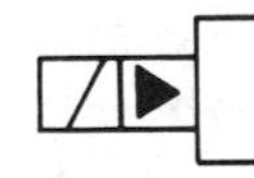

7.9 Thermal—A mechanical device responding to thermal change.

7.9.1 Local Sensing

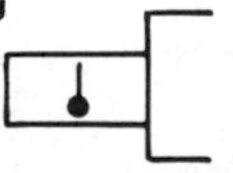

7.9.2 With Bulb for Remote Sensing

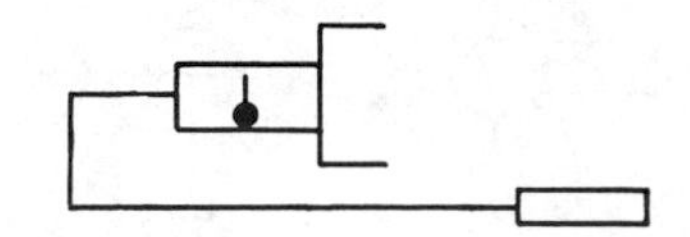

7.10 Servo
(This symbol contains representation for energy input, command input, and resultant output.)

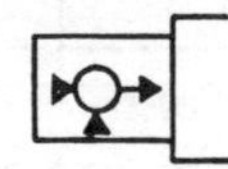

7.11 Composite Actuators (and, or, and/or)

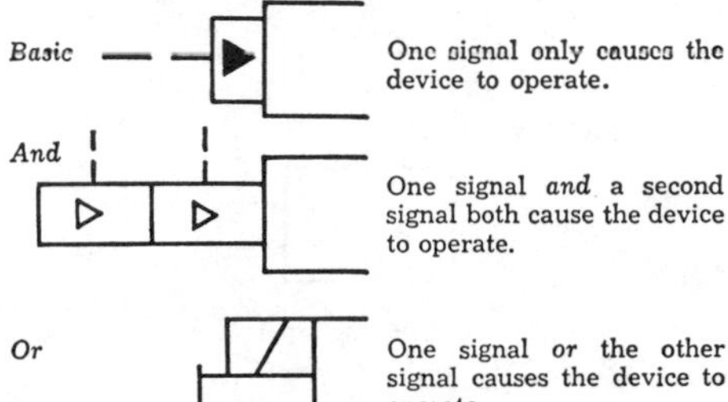

Basic — One signal only causes the device to operate.

And — One signal *and* a second signal both cause the device to operate.

Or — One signal *or* the other signal causes the device to operate.

And/Or

The solenoid *and* the pilot or the manual override alone causes the device to operate.

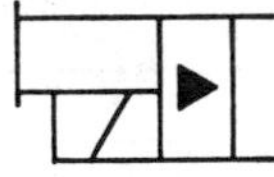

The solenoid *and* the pilot *or* the manual override *and* the pilot.

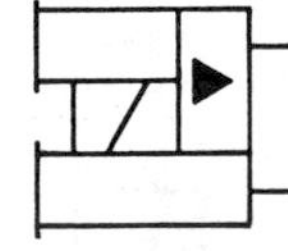

The solenoid *and* the pilot
or
a manual override *and* the pilot
or
a manual override alone.

8. ROTARY DEVICES

8.1 Basic Symbol

8.1.1 With Ports

8.1.2 With Rotating Shaft, with control, and with Drain

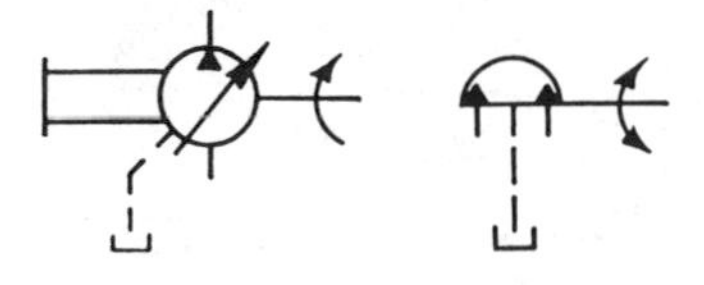

8.2 Hydraulic Pump

8.2.1 Fixed Displacement

8.2.1.1 Unidirectional

8.2.1.2 Bidirectional

8.2.2 Variable Displacement, Non-Compensated

8.2.2.1 Unidirectional

Simplified

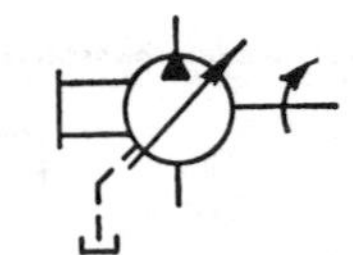

Complete

8.2.2.2 Bidirectional

Simplified

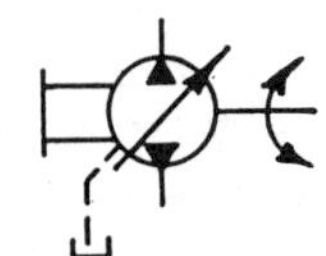

Complete

8.2.3 Variable Displacement, Pressure Compensated

8.2.3.1 Unidirectional

Simplified

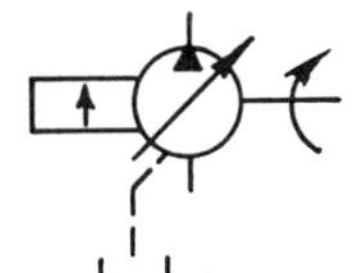

Complete

8.2.3.2 Bidirectional

Simplified

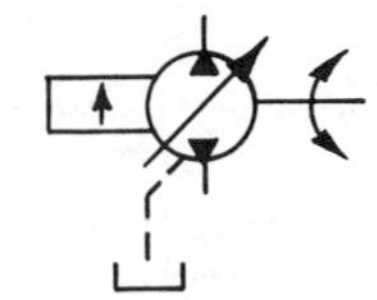

Complete

8.3 Hydraulic Motor

8.3.1 Fixed Displacement

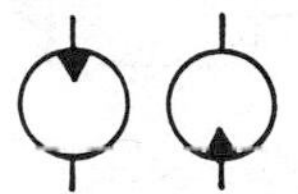

8.3.1.2 Bidirectional

8.3.2 Variable Displacement

8.3.2.1 Unidirectional

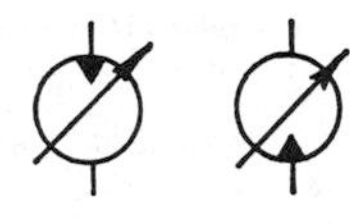

8.3.2.2 Bidirectional

8.4 Pump-Motor, Hydraulic

8.4.1 Operating in one direction as a pump.
Operating in the other direction as a motor.

8.4.1.1 Complete Symbol

8.4.1.2 Simplified Symbol

8.4.2 Operating one direction of flow as either a pump or as a motor.

8.4.2.1 Complete Symbol

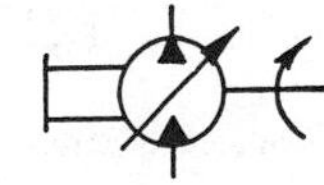

8.4.2.2 Simplified Symbol

8.4.3 Operating in both directions of flow either as a pump or as a motor. (Variable displacement, pressure compensated shown.)

8.4.3.1 Complete Symbol

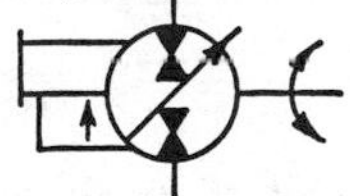

8.4.3.2 Simplified Symbol

8.5 Pump, Pneumatic

8.5.1 Compressor, Fixed Displacement

8.5.2 Vacuum Pump, Fixed Displacement

8.6 Motor, Pneumatic

8.6.1 Unidirectional

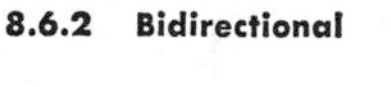

8.6.2 Bidirectional

8.7 Oscillator

8.7.1 Hydraulic

8.7.2 Pneumatic

8.8 Motors, Engines

8.8.1 Electric Motor

8.8.2 **Heat Engine (Internal Combustion Engine)**

9. INSTRUMENTS & ACCESSORIES

9.1 Indicating and Recording

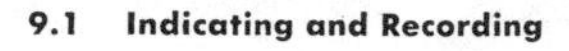

9.1.1 Pressure

9.1.2 Temperature

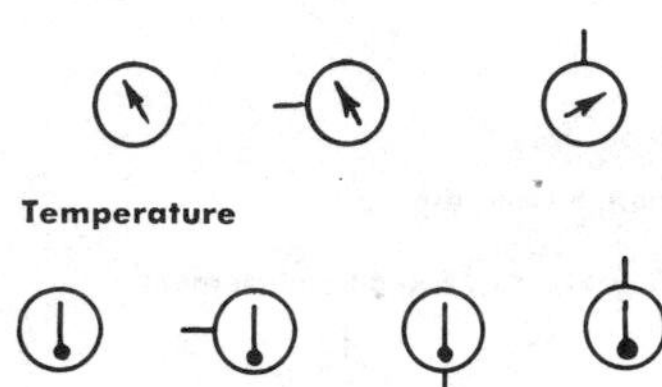

9.1.3 Flow Meter

9.1.3.1 Flow Rate

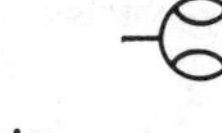

9.1.3.2 Totalizing

9.2 Sensing

9.2.1 Venturi

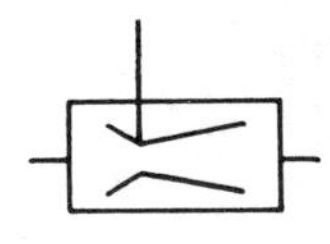

9.2.2 Orifice Plate

9.2.3 Pitot Tube

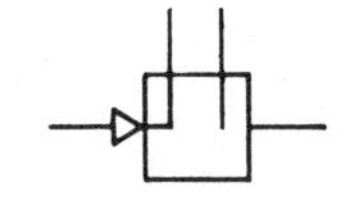

9.2.4 Nozzle

Hydraulic

Pneumatic

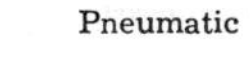

9.3 Accessories

9.3.1 Pressure Switch

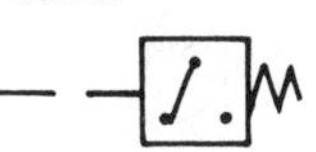

9.3.2 Muffler

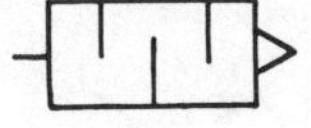

10. VALVES

A basic valve symbol is composed of one or more envelopes with lines inside the envelope to represent flow paths and flow conditions between ports. Three symbol systems are used to represent valve types: single envelope, both finite and infinite position; multiple envelope, finite position; and multiple envelope, infinite position.

10.1 In infinite position single envelope valves, the envelope is imagined to move to illustrate how pressure or flow conditions are controlled as the valve is actuated.

10.2 Multiple envelopes symbolize valves providing more than one finite flow path option for the fluid. The multiple envelope moves to represent how flow paths change when the valving element within the component is shifted to its finite positions.

10.3 Multiple envelope valves capable of infinite positioning between certain limits are symbolized as in 10.2 above with the addition of horizontal bars which are drawn parallel to the envelope. The horizontal bars are the clues to the infinite positioning function possessed by the valve represented.

10.3 Envelopes

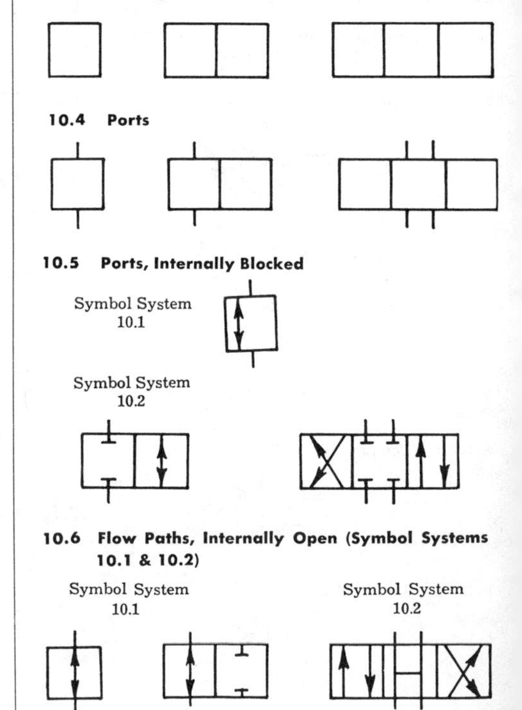

10.4 Ports

10.5 Ports, Internally Blocked

Symbol System 10.1

Symbol System 10.2

10.6 Flow Paths, Internally Open (Symbol Systems 10.1 & 10.2)

Symbol System 10.1

Symbol System 10.2

10.8 Flow Paths, internally Open (Symbol System 10.3)

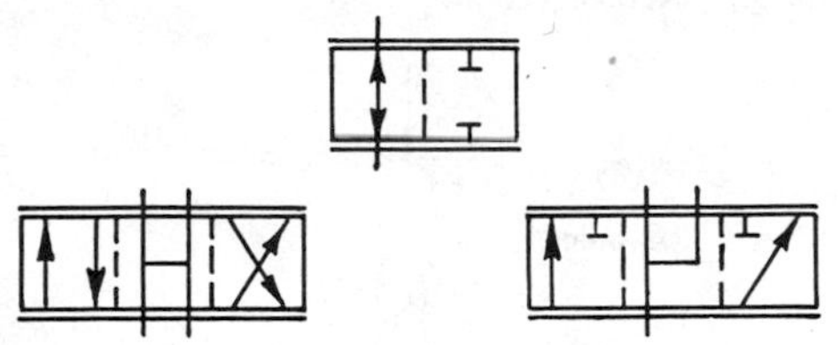

10.9 Two-Way Valves (2 Ported Valves)

10.9.1 On-Off (Manual Shut-Off)

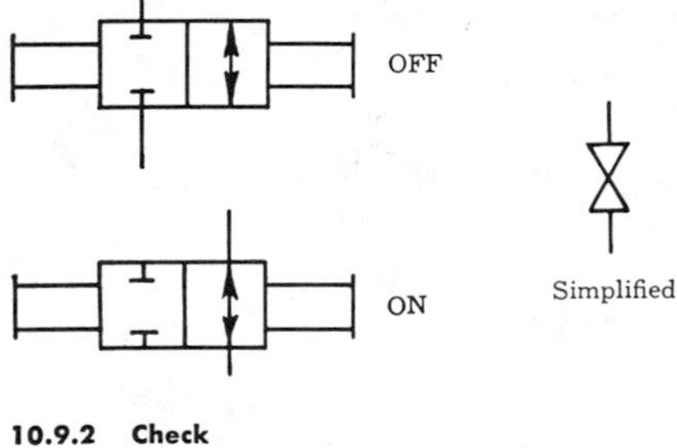

10.9.2 Check

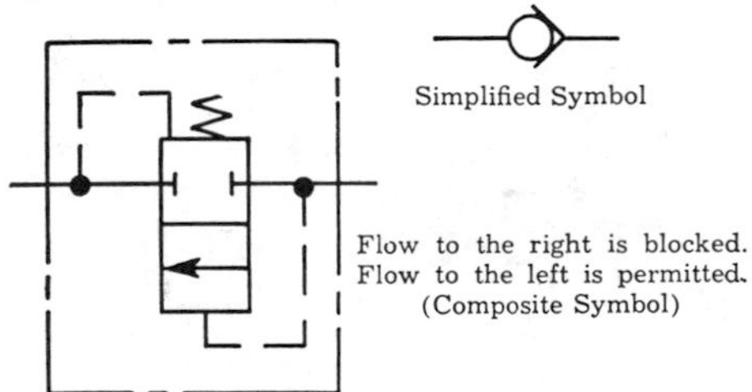

10.9.3 Check, Pilot-Operated to Open

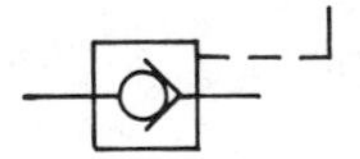

10.9.4 Check, Pilot-Operated to Close

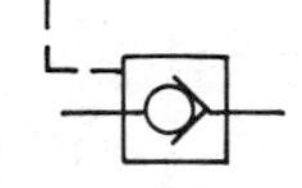

10.9.5 Two-Way Valves

10.9.5.1 Two-Position

Normally Closed Normally Open

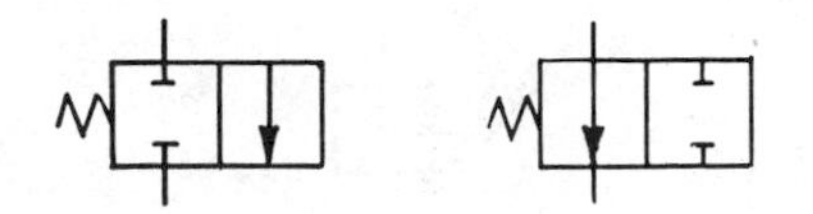

10.9.5.2 Normally Open

10.9.5.2 Infinite Position

Normally Closed Complete Normally Open

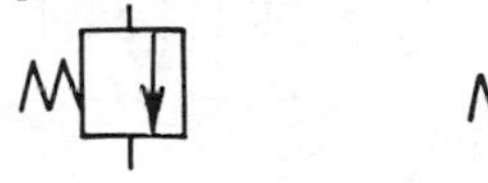

10.10 Three-Way Valves

10.10.1 Two-Position

10.10.1.1 Normally Open

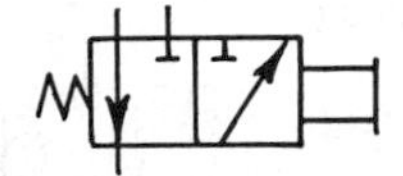

10.10.1.2 Normally Closed

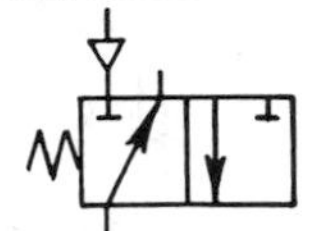

10.10.1.3 Distributor (Pressure is distributed first to one port, then the other)

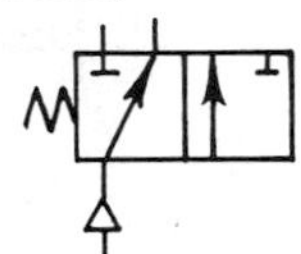

10.10.1.4 Two-Pressure

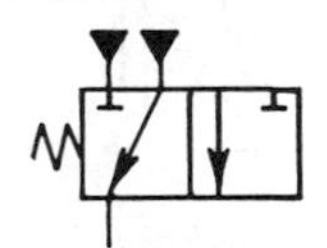

10.10.2 Double Check Valve. Double check valves can be built with and without 'cross bleed.' Such valves with two poppets do not usually allow pressure to momentarily 'cross bleed' to return during transition. Valves with one poppet may allow 'cross bleed' as these symbols illustrate.

10.10.2.1 Without Cross Bleed (One Way Flow)

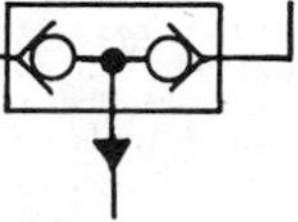

10.10.2.2 With Cross Bleed (Reverse Flow Permitted)

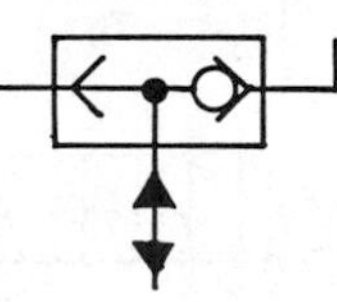

10.11 Four-Way Valves

10.11.1 Two Position

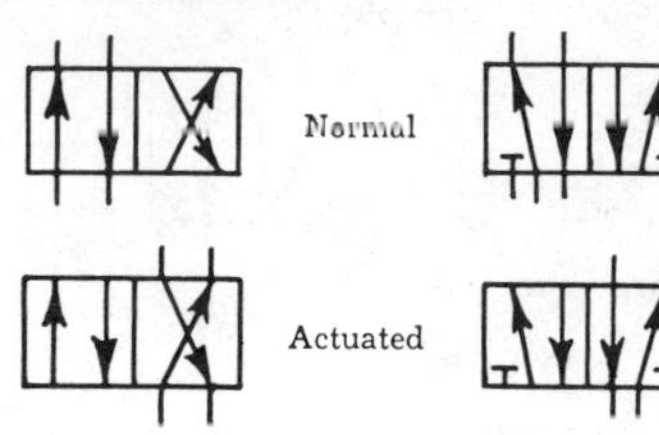

10.11.2 Three Position

(a) Normal

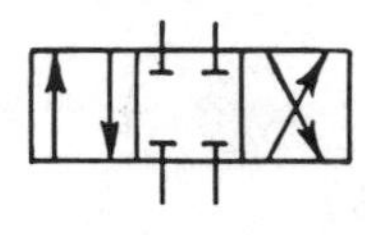

(b) Actuated Left

(c) Actuated Right

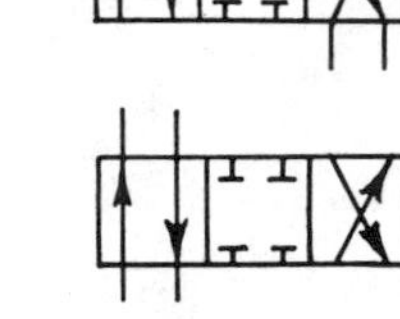

10.11.3 Typical Flow Paths for Center Condition of Three-Position Valves

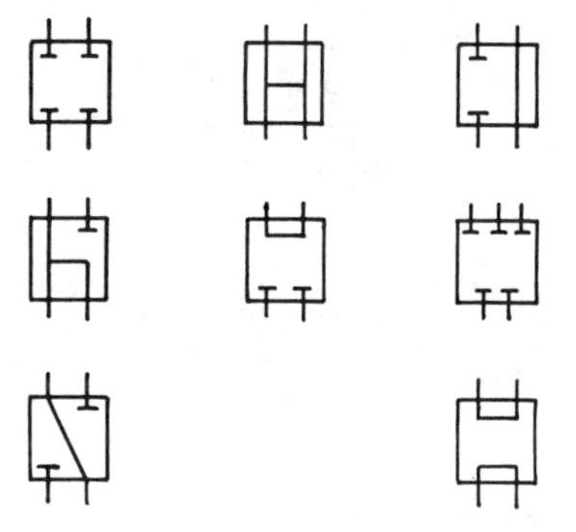

10.11.4 Two-Position, Snap Action with Transition.

As the valve element shifts from one position to the other, it passes through an intermediate position. If it is essential to circuit function to symbolize this "in transit" condition, it can be shown in the center position, enclosed by dashed lines.

Typical Transition Symbol

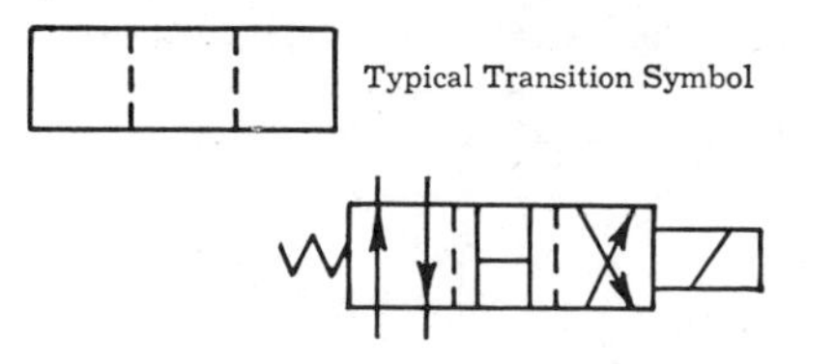

10.12 Infinite Positioning (Between Open & Closed)

10.12.1 Normally Closed

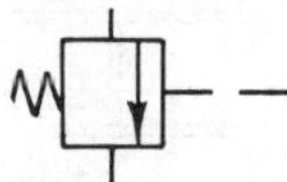

10.12.2 Normally Open

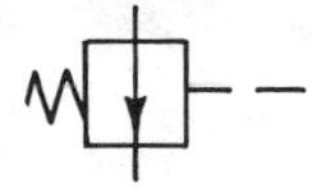

10.13 Pressure Control Valves

10.13.1 Pressure Relief

Simplified Symbol Denotes

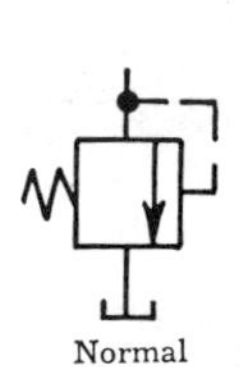

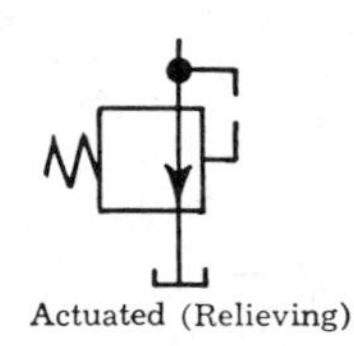

Normal   Actuated (Relieving)

10.13.2 Sequence

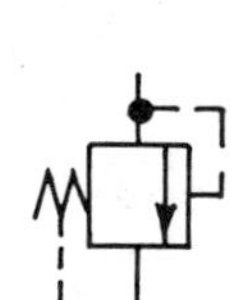

10.13.3 Pressure Reducing

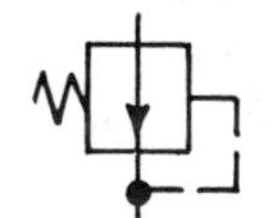

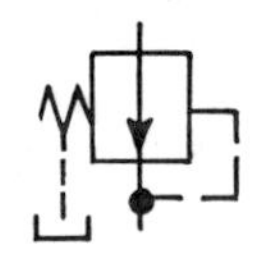

10.13.4 Pressure Reducing and Relieving

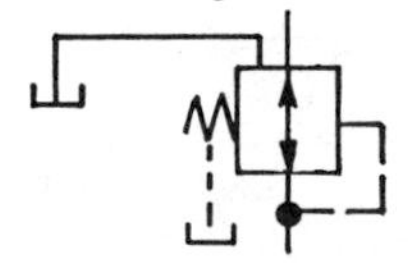

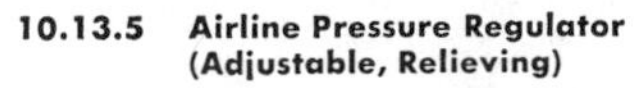

10.13.5 Airline Pressure Regulator (Adjustable, Relieving)

10.14 Infinite Positioning Three-Way Valves

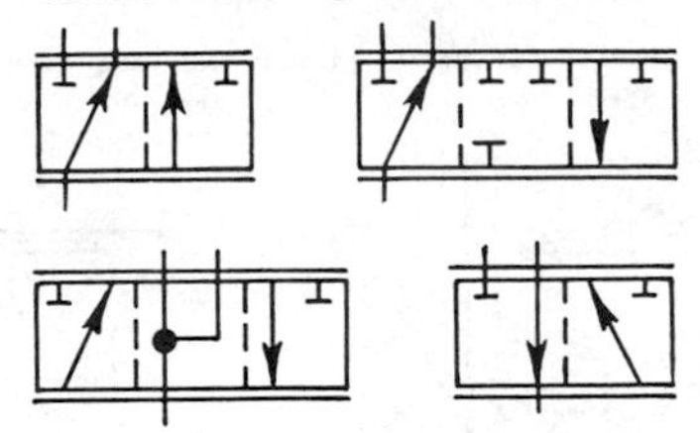

10.15 Infinite Positioning Four-Way Valves

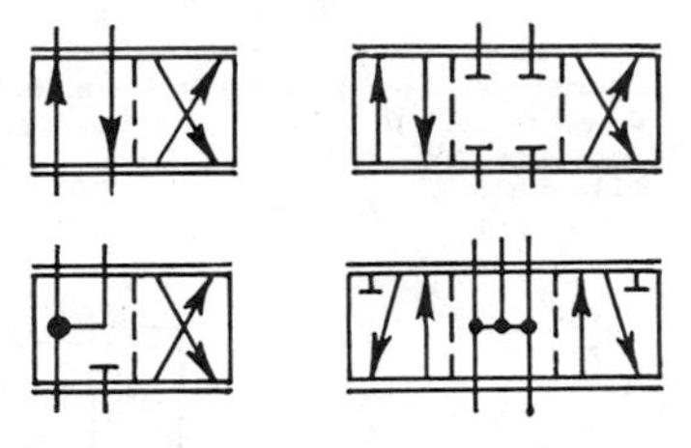

10.16 Flow Control Valves (See 3.7)

10.16.1 Adjustable, Non-Compensated (Flow control in each direction)

10.16.2 Adjustable with Bypass
Flow is controlled to the right.
Flow to the left bypasses control.

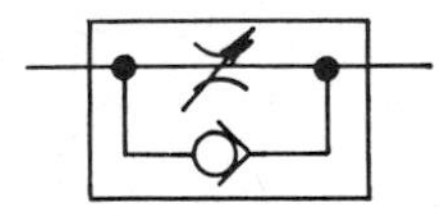

10.16.3 Adjustable and Pressure Compensated with Bypass

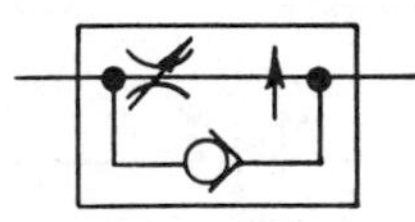

10.16.4 Adjustable, Temperature & Pressure Compensated

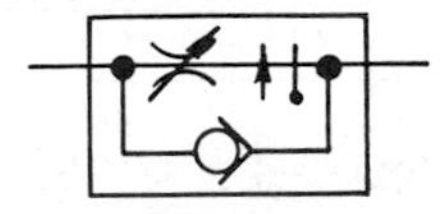

11. REPRESENTATIVE COMPOSITE SYMBOLS

11.1 Component Enclosure. Component enclosure may surround a complete symbol or a group of symbols to represent an assembly. It is used to convey more information about component connections and functions. Enclosure indicates extremity of component or assembly. External ports are assumed to be on enclosure line and indicate connections to component.
Flow lines shall cross enclosure line without loops or dots.

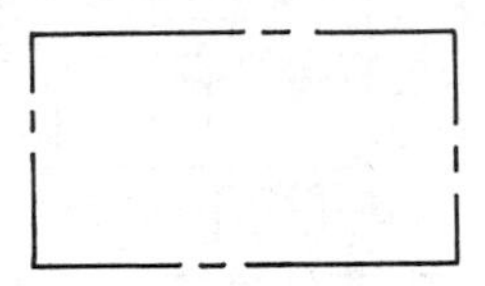

11.2 Airline Accessories
(Filter, Regulator, and Lubricator)

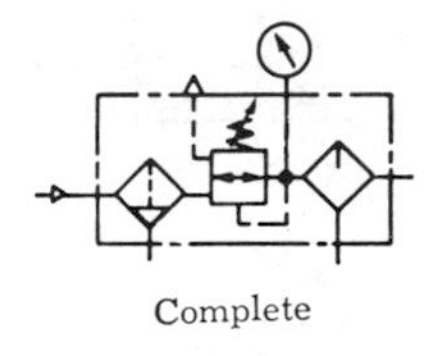

Complete

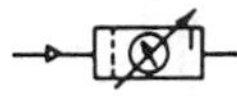

Simplified

11.3 Pumps and Motors

11.3.1 Pumps

11.3.1.1 Double, Fixed Displacement, One Inlet and Two Outlets

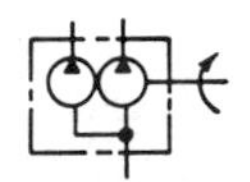

11.3.1.2 Double, with Integral Check Unloading and Two Outlets

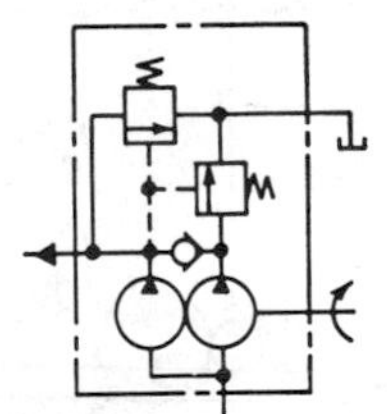

11.3.1.3 Integral Variable Flow Rate Control with Overload Relief

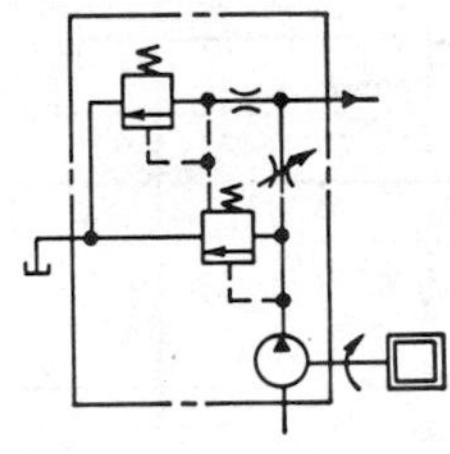

11.3.1.4 Variable Displacement with Integral Replenishing Pump and Control Valves.

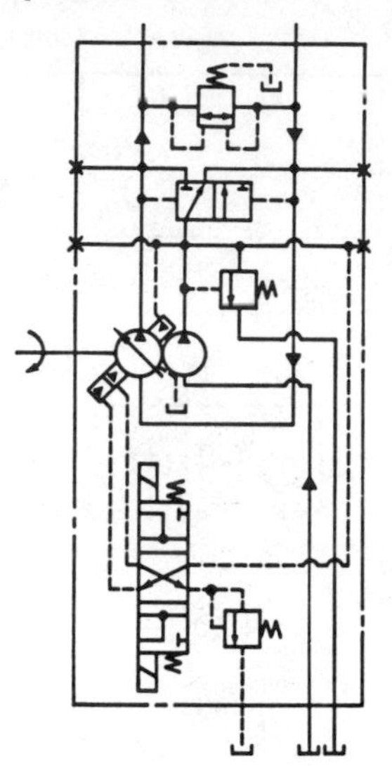

11.3.2 Pump Motor. Variable displacement with manual, electric, pilot, and servo control.

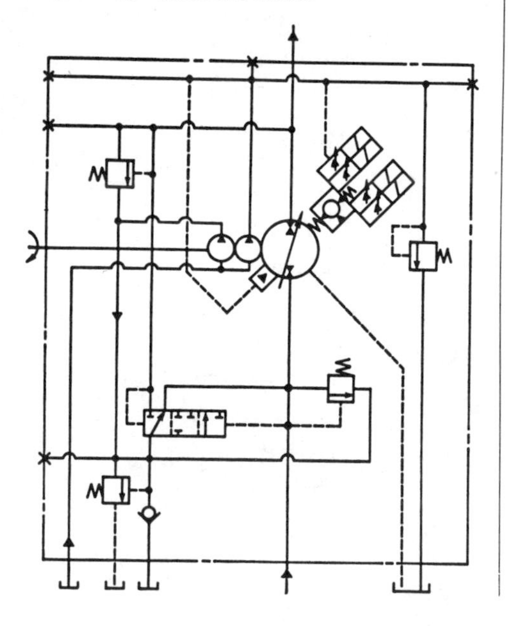

11.4 Valves

11.4.1 Relief, Balanced Type

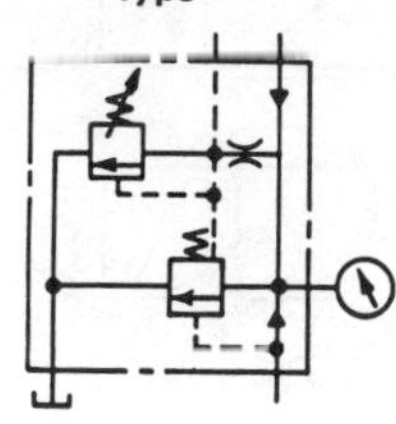

11.4.2 Remote Operated Sequence with Integral Check

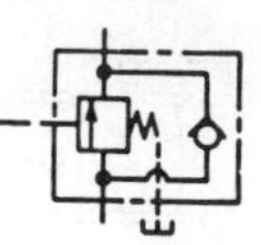

11.4.3 Remote & Direct Operated Sequence with Differential Areas and Integral Check

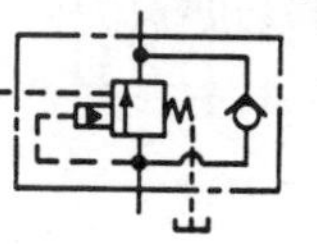

11.4.4 Pressure Reducing with Integral Check

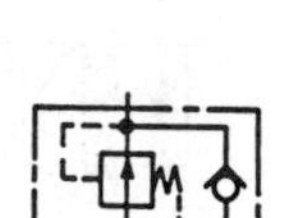

11.4.5 Pilot Operated Check

11.4.5.1 Differential Pilot Opened

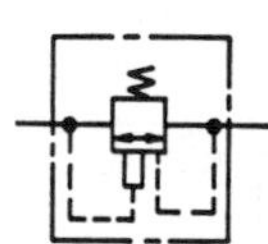

11.4.5.2 Differential Pilot Open and Closed

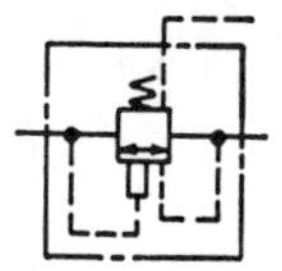

11.4.6 Two Position, Four Connection Solenoid Controlled and Pilot Operated, with Manual Pilot Override

Simplified Symbol

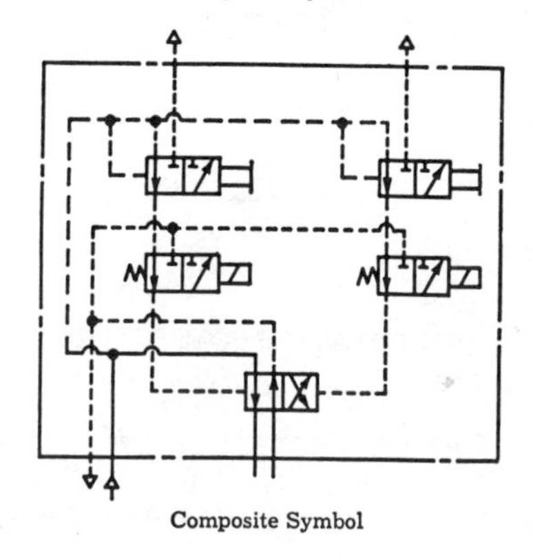

Composite Symbol

11.4.7 Two Position, Five Connection, Solenoid Control Pilot Operated with Detents and Throttle Exhaust

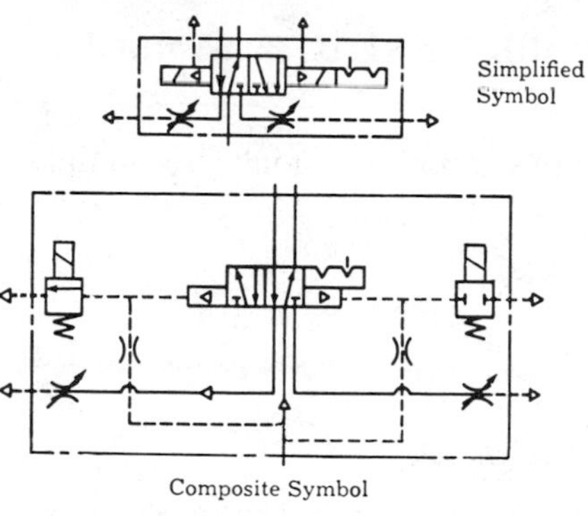

Composite Symbol

11.4.8 Variable Pressure Compensated Flow Control and Overload Relief

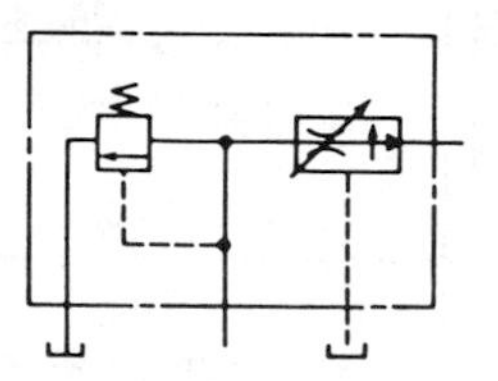

11.4.9 Multiple, Three Position, Manual Directional Control with Integral Check and Relief Valves

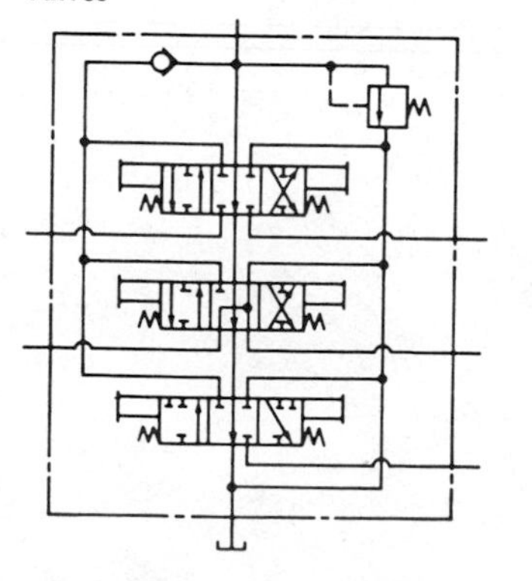

11.4.10 Cycle Control Panel, Five Position

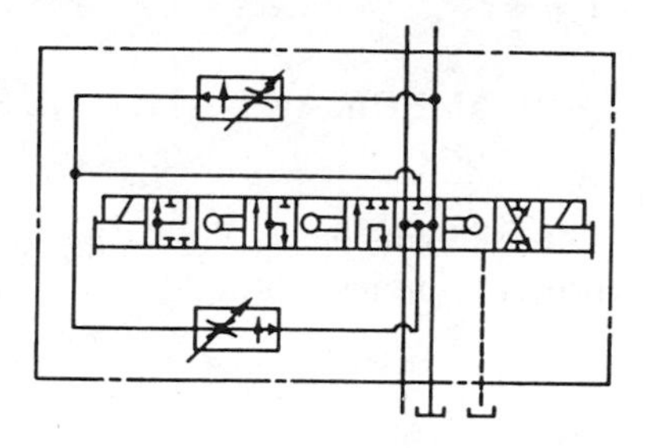

11.4.11 Panel Mounted Separate Units Furnished as a Package (Relief, Two Four-Way, Two Check, and Flow Rate Valves)

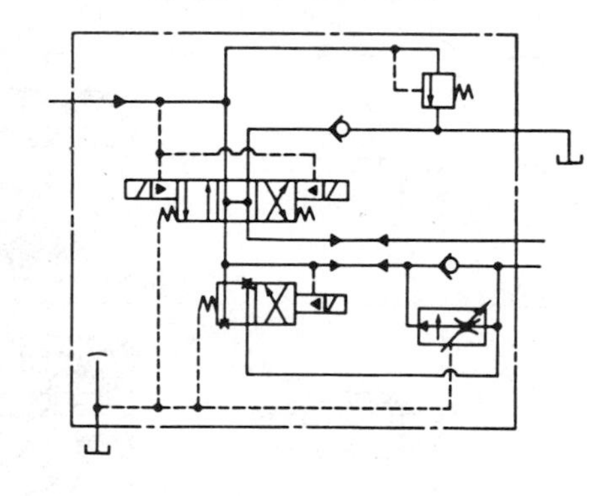

11.4.12 Single Stage Compressor with Electric Motor Drive, Pressure Switch Control of Receiver Tank Pressure

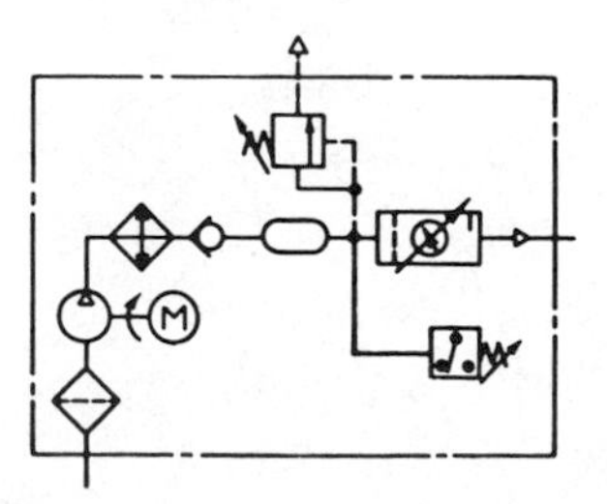

A.3 AMERICAN NATIONAL STANDARDS INSTITUTE SYMBOLS

The following pages provide a compilation of symbols commonly used in fluid power troubleshooting:

DIRECTIONAL CONTROL VALVES

Valves, directional control. Start, stop, and direct fluid flow. They extend and retract cylinders, rotate fluid motors and actuators, and sequence other circuit operations.

Directional control valves are classified according to the number of ports or connecting lines, the number of positions to which they can be actuated, the type of actuator, and the way in which fluid flows through the valve.

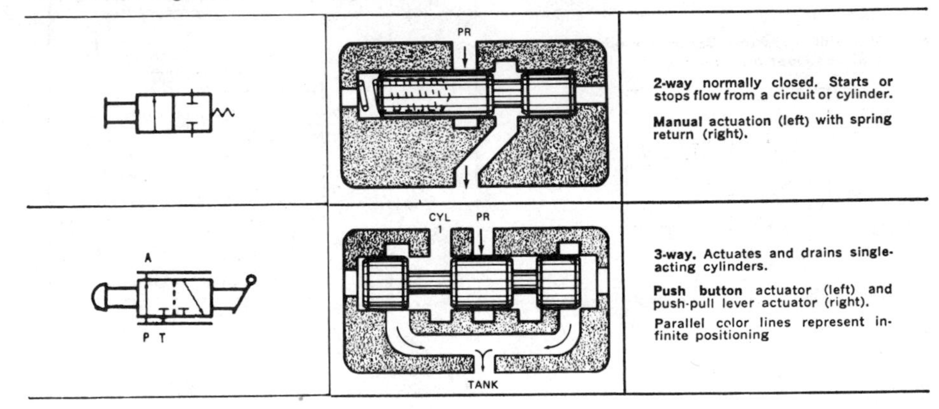

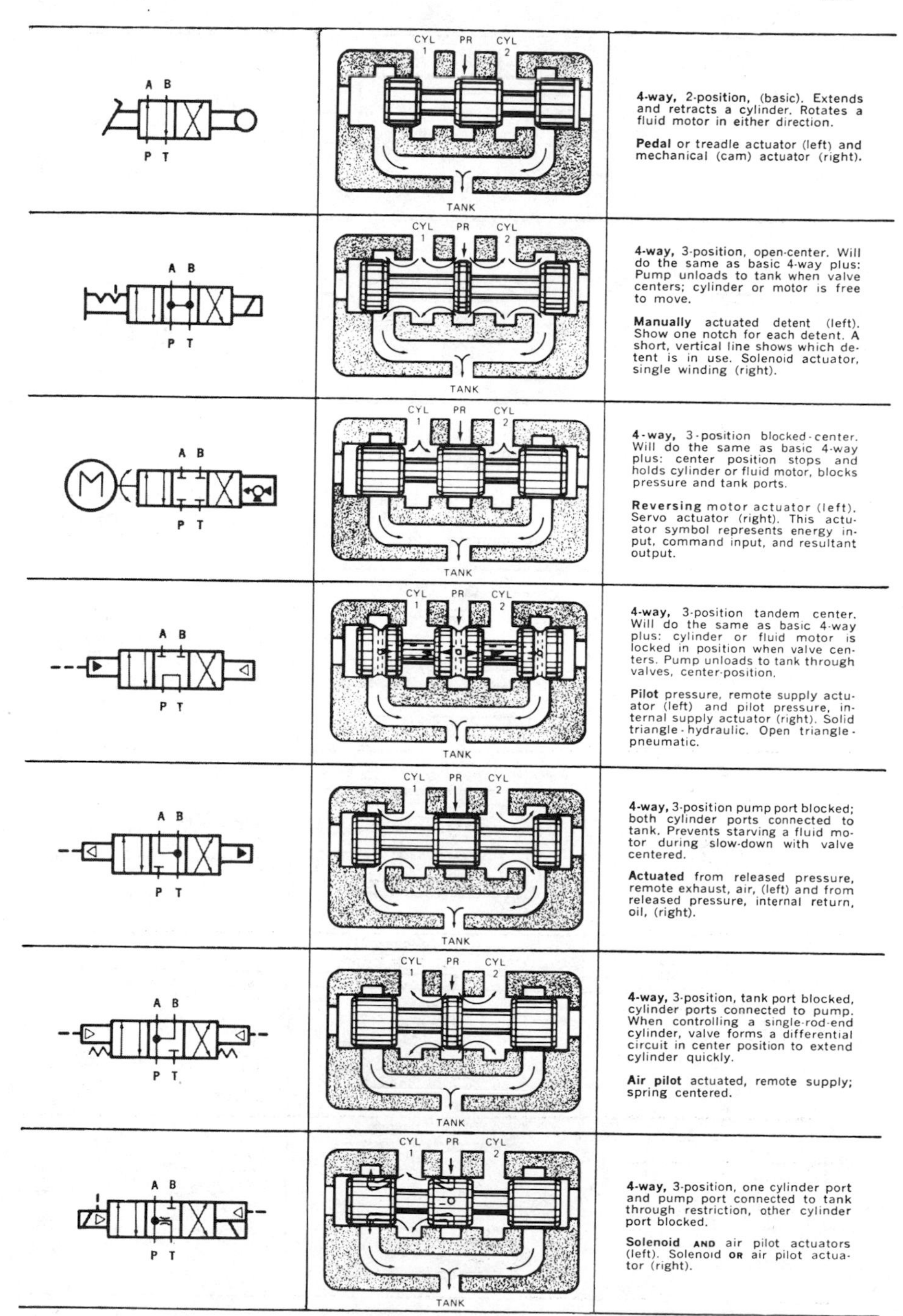

4-way, 2-position, (basic). Extends and retracts a cylinder. Rotates a fluid motor in either direction.

Pedal or treadle actuator (left) and mechanical (cam) actuator (right).

4-way, 3-position, open-center. Will do the same as basic 4-way plus: Pump unloads to tank when valve centers; cylinder or motor is free to move.

Manually actuated detent (left). Show one notch for each detent. A short, vertical line shows which detent is in use. Solenoid actuator, single winding (right).

4-way, 3-position blocked-center. Will do the same as basic 4-way plus: center position stops and holds cylinder or fluid motor, blocks pressure and tank ports.

Reversing motor actuator (left). Servo actuator (right). This actuator symbol represents energy input, command input, and resultant output.

4-way, 3-position tandem center. Will do the same as basic 4-way plus: cylinder or fluid motor is locked in position when valve centers. Pump unloads to tank through valves, center-position.

Pilot pressure, remote supply actuator (left) and pilot pressure, internal supply actuator (right). Solid triangle - hydraulic. Open triangle - pneumatic.

4-way, 3-position pump port blocked; both cylinder ports connected to tank. Prevents starving a fluid motor during slow-down with valve centered.

Actuated from released pressure, remote exhaust, air, (left) and from released pressure, internal return, oil, (right).

4-way, 3-position, tank port blocked, cylinder ports connected to pump. When controlling a single-rod-end cylinder, valve forms a differential circuit in center position to extend cylinder quickly.

Air pilot actuated, remote supply; spring centered.

4-way, 3-position, one cylinder port and pump port connected to tank through restriction, other cylinder port blocked.

Solenoid AND air pilot actuators (left). Solenoid OR air pilot actuator (right).

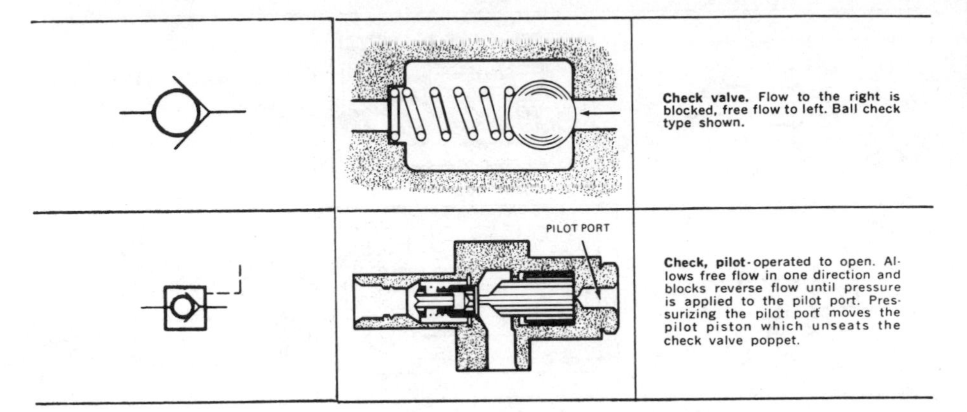

FLOW CONTROL VALVES

Valves, flow control. Control flow by restricting fluid movement in one or both directions. The restriction may be fixed or variable. In some elementary flow control valves, flow varies with changes in fluid line pressure. The more sophisticated ones compensate for fluid pressure and temperature fluctuations.

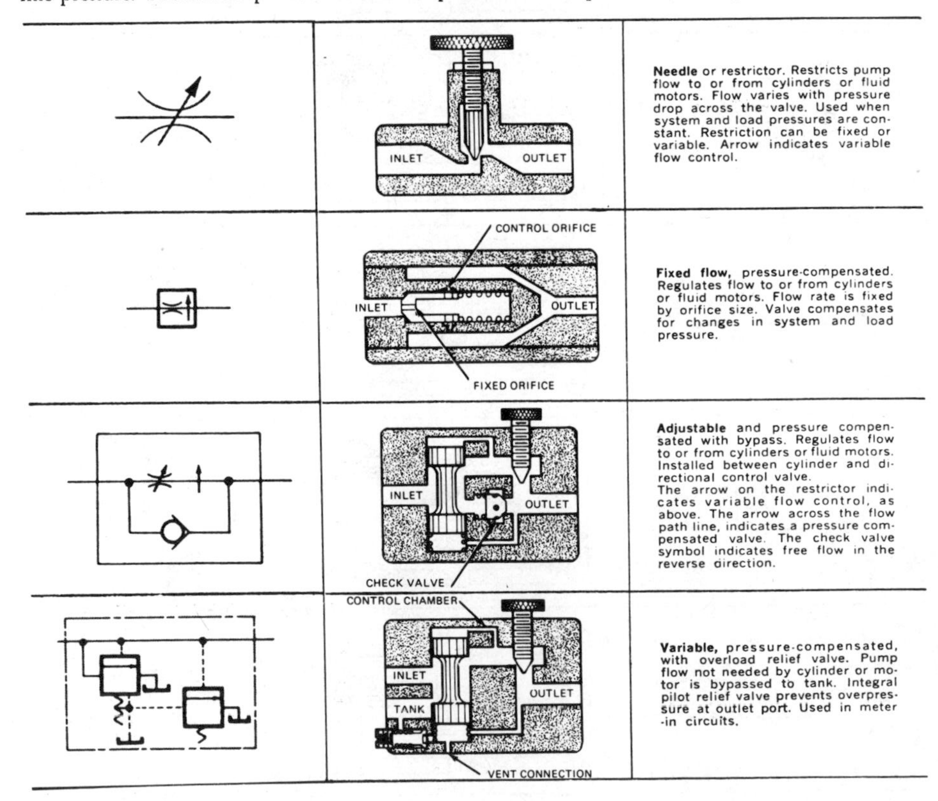

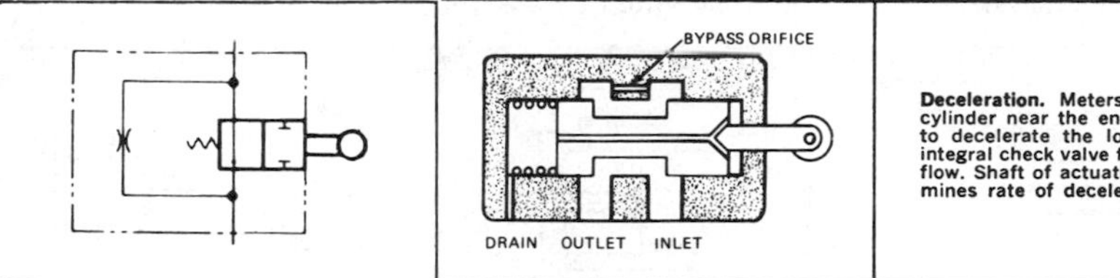

Deceleration. Meters flow from a cylinder near the end of its stroke to decelerate the load. May have integral check valve for free reverse flow. Shaft of actuating cam determines rate of deceleration.

PRESSURE CONTROL VALVES

Valves, pressure control. Limit system pressure, reduce pressure in any part of a circuit, unload a pump during part of a machine's cycle, or determine the pressure at which oil enters part of a circuit. Depending on their function in a circuit, they are called relief, pressure reducing, unloading, sequence, or counterbalance.

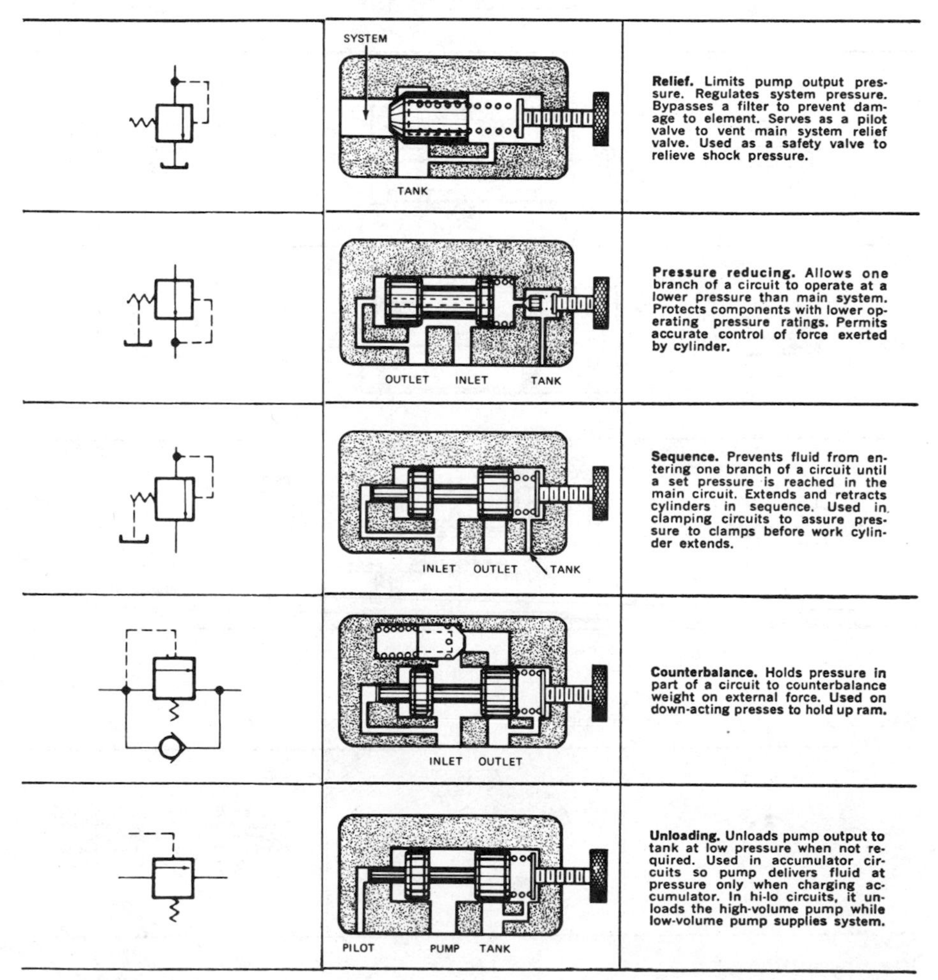

Relief. Limits pump output pressure. Regulates system pressure. Bypasses a filter to prevent damage to element. Serves as a pilot valve to vent main system relief valve. Used as a safety valve to relieve shock pressure.

Pressure reducing. Allows one branch of a circuit to operate at a lower pressure than main system. Protects components with lower operating pressure ratings. Permits accurate control of force exerted by cylinder.

Sequence. Prevents fluid from entering one branch of a circuit until a set pressure is reached in the main circuit. Extends and retracts cylinders in sequence. Used in clamping circuits to assure pressure to clamps before work cylinder extends.

Counterbalance. Holds pressure in part of a circuit to counterbalance weight on external force. Used on down-acting presses to hold up ram.

Unloading. Unloads pump output to tank at low pressure when not required. Used in accumulator circuits so pump delivers fluid at pressure only when charging accumulator. In hi-lo circuits, it unloads the high-volume pump while low-volume pump supplies system.

LINEAR ACTUATORS

These convert fluid energy into linear mechanical force and motion. They usually consist of a movable element such as a piston and piston rod operating within a close-fitting cylindrical bore.

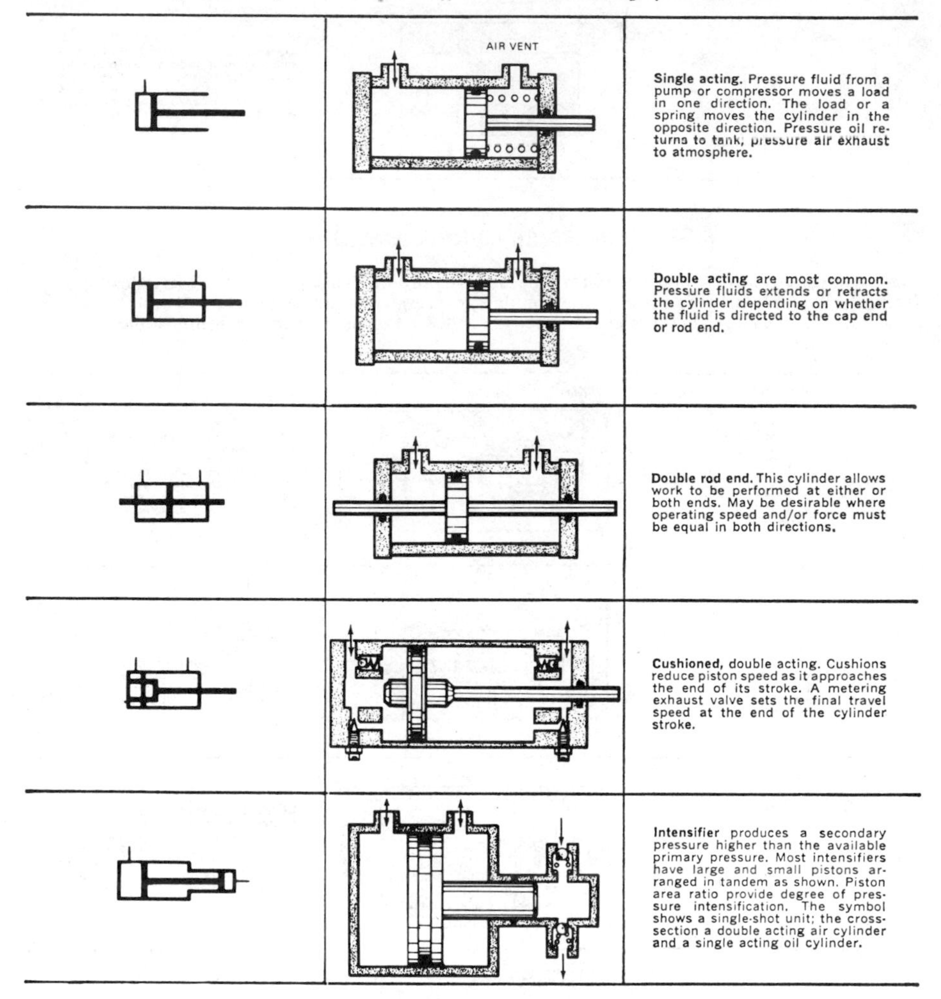

PUMPS

Converts mechanical force and motion to hydraulic power.

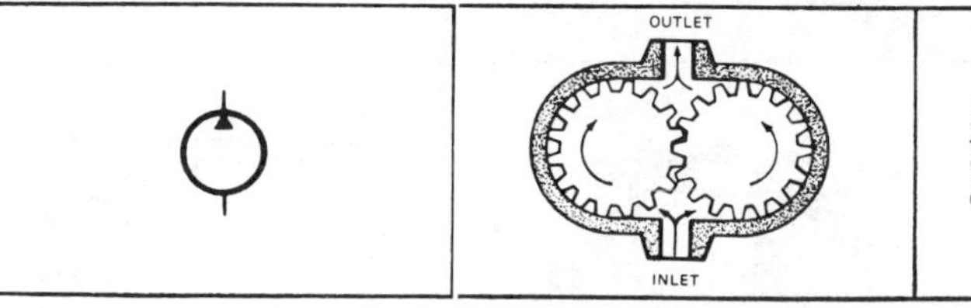

Unidirectional, fixed displacement. Output per revolution is considered to be constant. Total output can be changed by varying drive speed. Pump turns in one direction only. Gear pump shown.

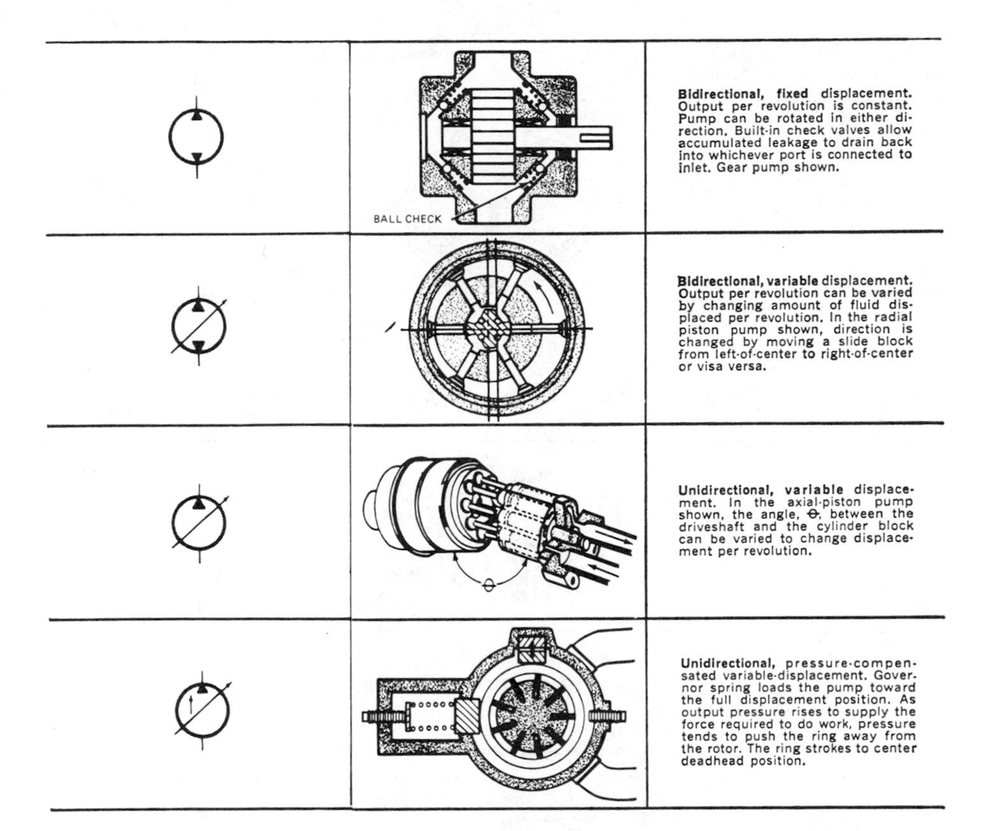

Symbol	Illustration	Description
		Bidirectional, fixed displacement. Output per revolution is constant. Pump can be rotated in either direction. Built-in check valves allow accumulated leakage to drain back into whichever port is connected to inlet. Gear pump shown.
		Bidirectional, variable displacement. Output per revolution can be varied by changing amount of fluid displaced per revolution. In the radial piston pump shown, direction is changed by moving a slide block from left-of-center to right-of-center or visa versa.
		Unidirectional, variable displacement. In the axial-piston pump shown, the angle, θ, between the driveshaft and the cylinder block can be varied to change displacement per revolution.
		Unidirectional, pressure-compensated variable-displacement. Governor spring loads the pump toward the full displacement position. As output pressure rises to supply the force required to do work, pressure tends to push the ring away from the rotor. The ring strokes to center deadhead position.

MOTORS

Converts fluid power into mechanical force and motion.

Symbol	Illustration	Description
	SPRING	**Unidirectional, fixed** displacement motor. Springs behind vanes of vane motor shown keep vanes extended. In the symbols, note that the arrowheads point inward whereas in the pump symbols the arrowheads point outward.
		Bidirectional fixed displacement. Rotation depends on which port the fluid enters. Vanes are kept extended by rocker arms.

COMPRESSORS

A machine that converts mechanical energy to pneumatic energy. Takes air at atmospheric pressure and compresses it to higher pressure.

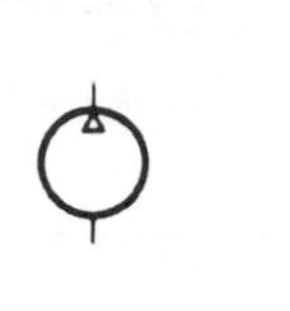

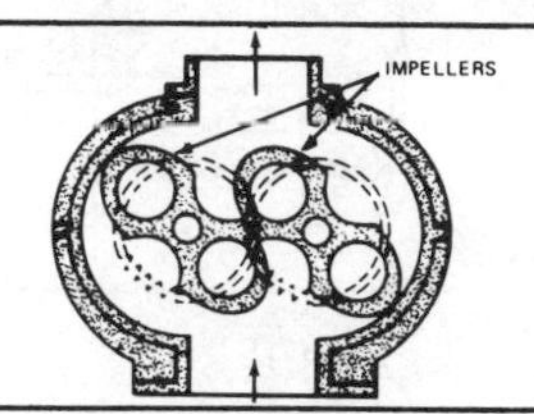

Rotary compressor. A pair of external gears keep two lobes in proper rotative relationship. No lubrication is needed on lobe surfaces because there is no internal contact pressure.

PRESSURE SWITCH

Fluid pressure actuates an electric switch. Pressure switch can sense and react to flow, liquid level or velocity.

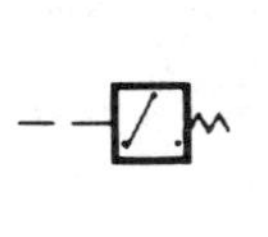

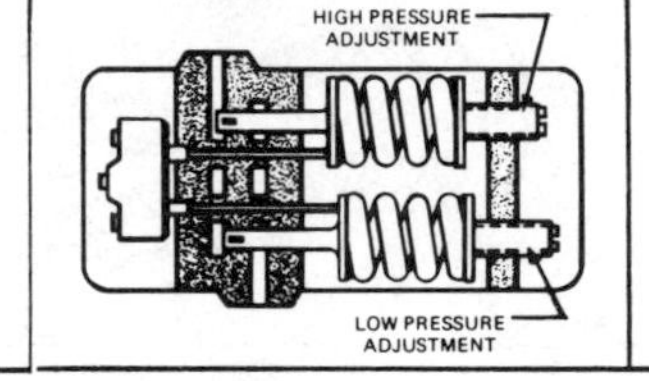

Differential pressure switch. Has a low pressure and a high pressure adjustment.

QUICK DISCONNECT COUPLINGS

Make it possible to separate lines without losing fluid. This eliminates the need for draining a line before disconnecting it. Provide immediate separation or connection with little effort.

CONNECTED DISCONNECTED	**Without checks.** Sleeve valve holds the locking ring locked while the coupling is disconnected. When the nipple is pushed into the socket, it moves the valve sleeves back, opening the valves to permit flow through the valve.
WITH TWO CHECKS CONNECTED DISCONNECTED	**With checks.** Solid poppet valves have captive elastomeric seals for leakproof operation at low pressure. At high pressures the seals compress and the poppets seal on the metal seat.

ROTARY ACTUATOR

Rotates an output shaft through a fixed arc to produce oscillating power. Converts fluid-energy input to mechanical output. Available in a variety of sizes and design types.

	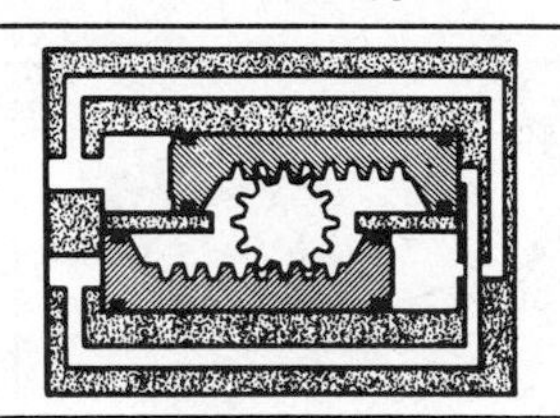	**Double rack** and pinion. Forces rotating the output shaft are balanced. Moving parts operate in a sealed, oil-filled chamber. Angle of rotation is limited. Usually operates at relatively slow speeds.

ACCUMULATOR

A container in which fluid is stored under pressure as a source of pressure fluid. In a hydropneumatic accumulator, compressed gas applies force to the stored fluid. Mechanical accumulators incorporate a mechanical device which applies the force to the stored fluid.

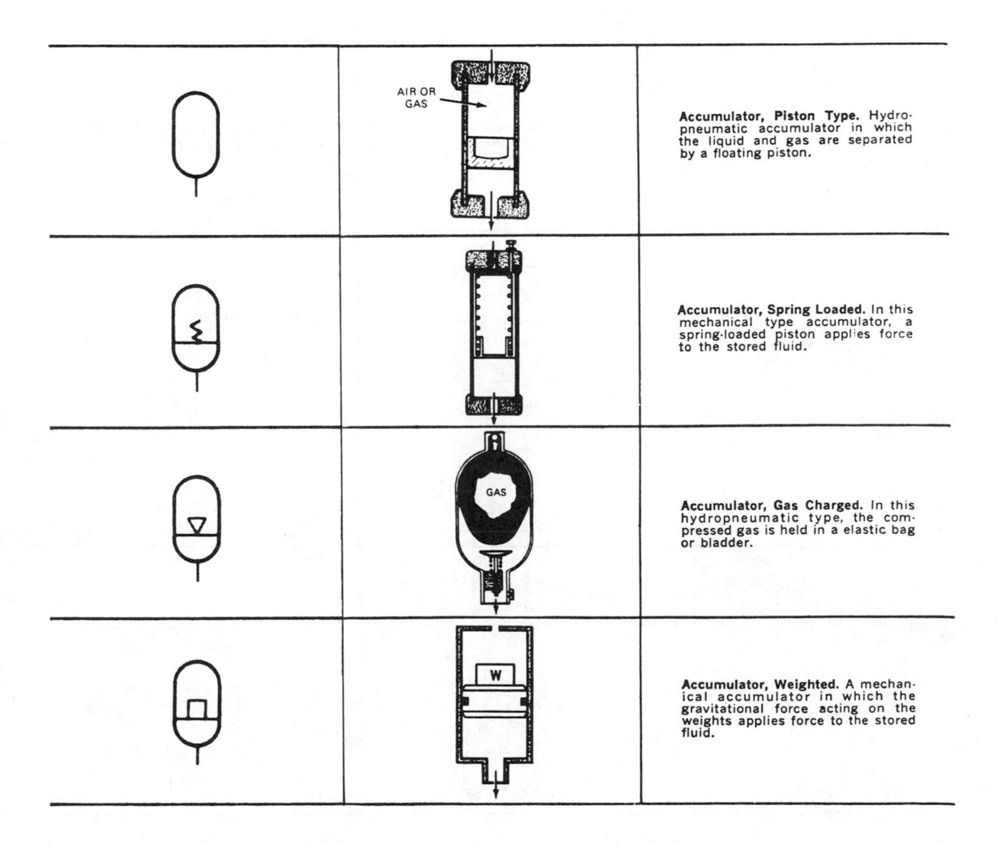

		Accumulator, Piston Type. Hydropneumatic accumulator in which the liquid and gas are separated by a floating piston.
		Accumulator, Spring Loaded. In this mechanical type accumulator, a spring-loaded piston applies force to the stored fluid.
		Accumulator, Gas Charged. In this hydropneumatic type, the compressed gas is held in a elastic bag or bladder.
		Accumulator, Weighted. A mechanical accumulator in which the gravitational force acting on the weights applies force to the stored fluid.

FLUID CONDITIONERS

Devices which control the physical characteristics of the fluid. Individual components condition the fluid to meet specific operating needs.

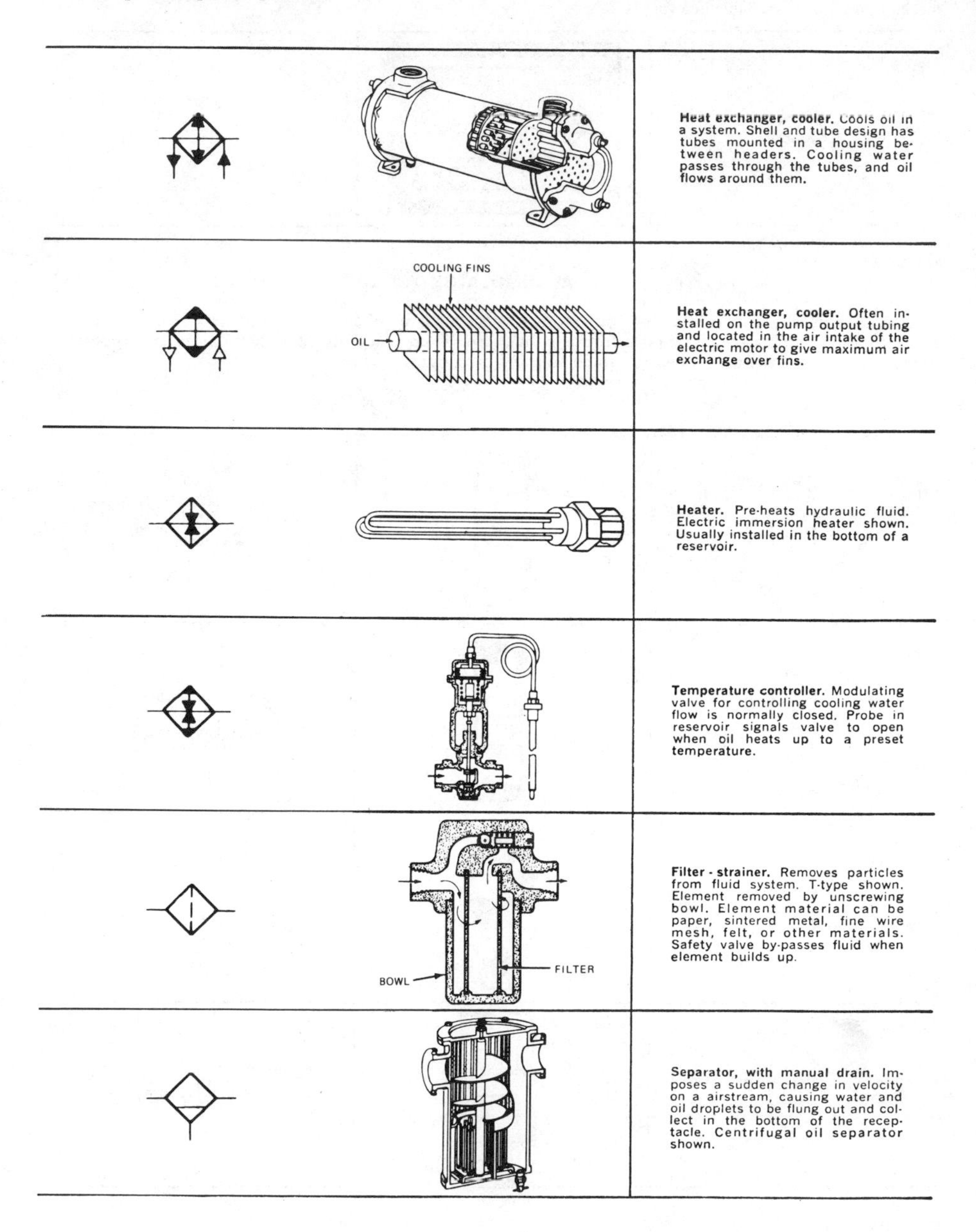

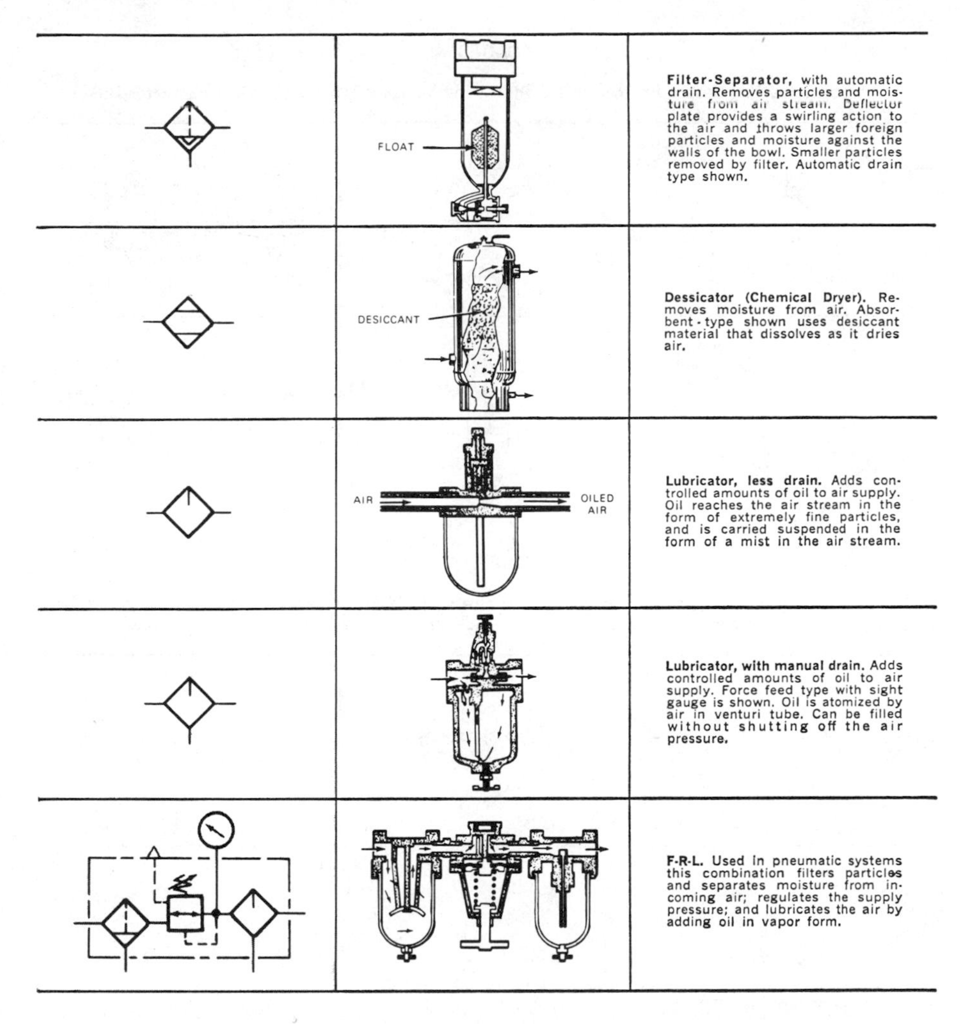

Filter-Separator, with automatic drain. Removes particles and moisture from air stream. Deflector plate provides a swirling action to the air and throws larger foreign particles and moisture against the walls of the bowl. Smaller particles removed by filter. Automatic drain type shown.

Dessicator (Chemical Dryer). Removes moisture from air. Absorbent-type shown uses desiccant material that dissolves as it dries air.

Lubricator, less drain. Adds controlled amounts of oil to air supply. Oil reaches the air stream in the form of extremely fine particles, and is carried suspended in the form of a mist in the air stream.

Lubricator, with manual drain. Adds controlled amounts of oil to air supply. Force feed type with sight gauge is shown. Oil is atomized by air in venturi tube. Can be filled without shutting off the air pressure.

F-R-L. Used in pneumatic systems this combination filters particles and separates moisture from incoming air; regulates the supply pressure; and lubricates the air by adding oil in vapor form.

ROTATING COUPLING

Can turn in one, two, or three planes. Leakproof, and have low torque. Help relieve stresses, caused by shock, misalignment, or vibration.

	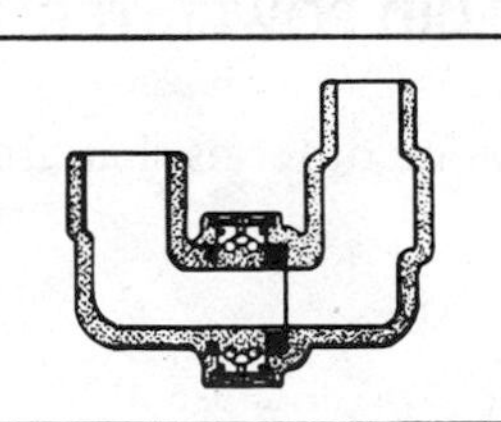	**Swivel joints.** Provide flexible, sealed, movable connections for piping systems which require relative motion between parts.

GAGES

An instrument or device for measuring, indicating, or comparing a physical characteristic.

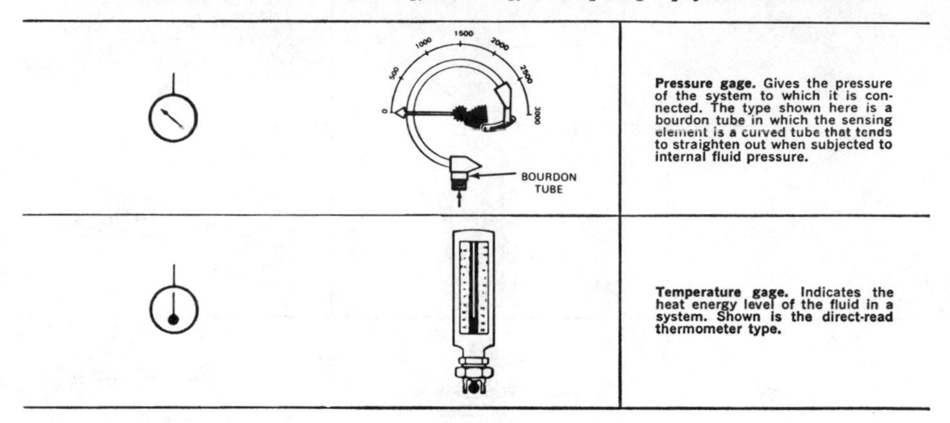

MUFFLER

A device used in pneumatic systems to reduce air noise and blast. Reduce air velocity, change its direction and reduce vibration. Noise is decreased by back pressure control of gas expansion.

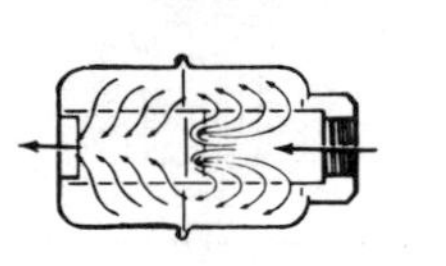

Muffler. Reduces exhaust air velocity by changing air stream direction with mechanical barriers.

A.4 GLOSSARY OF FLUID POWER TERMS*

The following is a compilation of terms that are useful in fluid power troubleshooting:

*From *Fluid Power Handbook & Directory*, 1980–81. Complete glossary available from National Fluid Power Association, 3333 North Mayfair Road, Milwaukee, Wisconsin.

ACCUMULATOR: A container in which fluid is stored under pressure as a source of fluid power.

ADDITIVE: A chemical compound or compounds added to a fluid to change its properties.

AIR, COMPRESSED: Air at any pressure greater than atmospheric pressure.

AIR, FREE: Air under the pressure due to atmospheric conditions at any specific location.

AIR, STANDARD: Air at a temperature of 68°F, a pressure of 14.70 pounds per square inch absolute, and a relative humidity of 36% (0.0750 pounds per cubic foot). In gas industries the temperature of "standard air" is usually given at 60°F.

AIR BREATHER: A device permitting air movement between atmosphere and the component in which it is installed.

AIR RECEIVER: A container in which gas is stored under pressure as a source of pneumatic fluid power.

ANILINE POINT: The lowest temperature at which a liquid is completely miscible with an equal volume of freshly distilled aniline (ASTM Designation D611-55T).

ANTI-EXTRUSION RING: A ring which bridges a clearance to minimize seal extrusion.

BERNOULLI'S LAW: If no work is done on or by a flowing frictionless liquid, its energy due to pressure and velocity remains constant at all points along the streamline.

BLEEDER: A device for removal of pressurized fluid.

BOYLE'S LAW: The absolute pressure of a fixed mass of gas varies inversely as the volume, provided the temperature remains constant.

BULK MODULUS: The measure of resistance to compressibility of a fluid. It is the reciprocal of the compressibility.

CAP: A cylinder end closure which completely covers the bore area.

CAVITATION: A localized gaseous condition within a liquid stream which occurs where the pressure is reduced to the vapor pressure.

CHARLES' LAW: The volume of a fixed mass of gas varies directly with absolute temperature, provided the pressure remains constant.

CIRCUIT, PILOT: A circuit used to control a main circuit or component.

CIRCUIT, REGENERATIVE: A circuit in which pressurized fluid discharged from a component is returned to the system to reduce power input requirements. On single rod end cylinders the discharge from the rod end is often directed to the opposite end to increase rod extension speed.

CIRCUIT, SERVO: A circuit which is controlled by automatic feed back; i.e., the output of the system is sensed or measured and is compared with the input signal. The difference (error) between the actual output and the input controls the circuit. The controls attempt to minimize the error. The system output may be position, velocity, force, pressure, level, flow rate, or temperature.

CIRCUIT, UNLOADING: A circuit in which pump volume is returned to reservoir at near zero gage pressure whenever delivery to the system is not required.

COMPRESSIBILITY: The change in volume of a unit volume of a fluid when subjected to a unit change of pressure.

COMPRESSOR: A device which converts mechanical force and motion into pneumatic fluid power.

COMPRESSOR, MULTIPLE STAGE: A compressor having two or more compressive steps in which the discharge from each supplies the next in series.

CONTAMINANT: Detrimental matter in a fluid.

CONTROL, PRESSURE COMPENSATED: A control in which a pressure signal operates a compensating device.

COUPLING, QUICK DISCONNECT: A coupling which can quickly join or separate lines.

CROSS: A connector with four ports arranged in pairs, each pair on one axis, and the axes at right angles.

CUSHION, CYLINDER: A cushion built into a cylinder to restrict flow at the outlet port thereby arresting the motion of the piston rod.

CYCLE: A single complete operation consisting of progressive phases starting and ending at the neutral position.

CYLINDER: A device which converts fluid power into linear mechanical force and motion. It usually consists of a movable element such as a piston and piston rod, plunger rod, plunger or ram, operating within a cylindrical bore.

CYLINDER, DOUBLE ACTING: A cylinder in which fluid force can be applied to the movable element in either direction.

CYLINDER, PISTON: A cylinder in which the movable element has a greater cross-sectional area than the piston rod.

CYLINDER, PLUNGER: A cylinder in which the movable element has the same cross-sectional area as the piston rod.

CYLINDER, SINGLE ACTING: A cylinder in which the fluid force can be applied to the movable element in only one direction.

CYLINDER, TANDEM: Two or more cylinders with inter-connected piston assemblies.

CYLINDER, TELESCOPING: A cylinder with nested multiple tubular rod segments which provide a long working stroke in a short retracted envelope.

DUROMETER HARDNESS: A comparative indication of elastomer hardness determined by a durometer.

ELBOW: A connector that makes an angle between mating lines. The angle is always 90 degrees unless another angle is specified.

FILTER: A device whose primary function is the retention by a porous media of insoluble contaminants from a fluid.

FILTER ELEMENT: The porous device which performs the actual process of filtration.

FILTER MEDIA, DEPTH: Porous materials which primarily retain contaminant within a tortuous path.

FILTER MEDIA, SURFACE: Porous materials which primarily retain contaminants on the influent face.

FITTING, COMPRESSION: A fitting which seals and grips by manual adjustable deformation.

FITTING, FLANGE: A fitting which utilizes a radially extending collar for sealing and connection.

FITTING, FLARED: A fitting which seals and grips by a preformed flare at the end of the tube.

FITTING, FLARELESS: A fitting which seals and grips by means other than a flare.

FITTING, REUSABLE HOSE: A hose fitting that can be removed from a hose and reused.

FLASH POINT: The temperature to which a liquid must be heated under specified conditions of the test method to give off sufficient vapor to form a mixture with air that can be ignited momentarily by a specified flame.

FLOW, LAMINAR: A flow situation in which fluid moves in aprallel lamina or layers.

FLOW, TURBULENT: A flow situation in which the fluid particles move in a random manner.

FLOW RATE: The volume, mass, or weight of a fluid passing through any conductor per unit of time.

FLOWMETER: A device which indicates either flow rate, total flow, or a combination of both.

FLUID: A liquid or a gas.

FLUID, FIRE RESISTANT: A fluid difficult to ignite which shows little tendency to propagate flame.

FLUID POWER: Energy transmitted and controlled through use of a pressurized fluid.

FLUID POWER SYSTEM: A system that transmits and controls power through use of a pressurized fluid within an enclosed circuit.

FLUIDICS: Engineering science pertaining to the use of fluid dynamic phenomina to sense, control, process information, and/or actuate.

GAGE: An instrument or device for measuring, indicating, or comparing a physical characteristic.

GLAND: The cavity of a stuffing box.

HEAD: The cylinder end closure which covers the differential area between the bore area and the piston rod area.

HEAD: The height of a column or body of fluid above a given point expressed in linear units. Head is often used to indicate gage pressure. Pressure is equal to the height times the density of the fluid.

HEAD, STATIC: The height of a column or body of fluid above a given point.

HEAD, VELOCITY: The equivalent head through which the liquid would have to fall to attain a given velocity. Mathematically is is equal to the square of the velocity (in feet) divided by 64.4 feet per second squared. $h=v2/2g$.

HEAT EXCHANGER: A device which transfers heat through a conducting wall from one fluid to another.

HYDRAULICS: Engineering science pertaining to liquid pressure and flow.

HYDRODYNAMICS: Engineering science pertaining to the energy of liquid flow and pressure.

HYDROKINETICS: Engineering science pertaining to the energy of liquids in motion.

HYDROPNEUMATICS: Pertaining to the combination of hydraulic and pneumatic fluid power.

HYDROSTATICS: Engineering science pertaining to the energy of liquids at rest.

INHIBITOR: Any substance which slows or prevents such chemical reactions as corrosion or oxidation.

INTENSIFIER: A device which converts low pressure fluid power into higher pressure fluid power.

JOINT, SWIVEL: A joint which permits variable operational positioning of lines.

LUBRICATOR: A device which adds controlled or metered amounts of lubricant into a fluid power system.

MANIFOLD: A conductor which provices multiple connection ports.

MANOMETER: A differential pressure gage in which pressure is indicated by the height of a liquid column of known density. Pressure is equal to the difference in vertical height between two connected columns multiplied by the density of the manometer liquid. Some forms of manometers are "U" tube, inclined tube, well, and bell types.

MICRON: A millionth of a meter or about 0.00004 inch.

MOTOR: A device which converts fluid power into mechanical force and motion. It usually provides rotary mechanical motion.

MOTOR, FIXED DISPLACEMENT: A motor in which the displacement per unit of output motion cannot be varied.

MOTOR, ROTARY, LIMITED: A rotary motor having limited motion.

MOTOR, VARIABLE DISPLACEMENT: A motor in which the displacement per unit of output motion can be varied.

MUFFLER: A device for reducing gas flow noise. Noise is decreased by back pressury control of gas expansion.

NEUTRALIZATION NUMBER: A measure of the total acidity or bascity of an oil; this includes organic or inorganic acids or bases or a combination thereof (ASTM Designation D974-58T).

NEWT: The standard unit of kinematic viscosity in the English system. It is expressed in square inches per second.

NIPPLE: A short length of pipe or tube.

PACKING: A sealing device consisting of bulk deformable material or one or more mating deformable elements, reshaped by manually adjustable compression to obtain and maintain effectiveness. It usually uses axial compression to obtain radial sealing.

PASCAL'S LAW: A pressure applied to a confined fluid at rest is transmitted with equal intensity throughout the fluid.

PASSAGE: A machined or cored fluid-conducting path which lies within or passes through a component.

PIPE: A line whose outside diameter is standardized for threading. Pipe is available in Standard, Extra Strong, Double Extra Strong or Schedule wall thicknesses.

PNEUMATICS: Engineering science pertaining to gaseous pressure and flow.

POISE: The standard unit of absolute viscosity in the c.g.s. (centimeter-gram-second) system. It is the ratio of the shearing stress to the shear rate of a fluid and is expressed in dyne seconds per square centimeter; 1 centipoise equals .01 poise.

PORT: An internal or external terminus of a passage in a component.

POUR POINT: The lowest temperature at which a fluid will flow under specified conditions (ASTM Designation D97-57).

POWER UNIT: A combination of pump, pump drive, reservoir, controls and conditioning components which may be required for its application.

PRESSURE: Force per unit area, usually expressed in pounds per square inch.

PRESSURE, ABSOLUTE: The sum of atmospheric and gage pressures.

PRESSURE, ATMOSPHERIC: Pressure exerted by the atmosphere at any specific location. (Sea level pressure is approximately 14.7 pounds per square inch absolute.)

PRESSURE, BACK: The pressure encountered on the return side of a system.

PRESSURE, CRACKING: The pressure at which a pressure operated valve begins to pass fluid.

PRESSURE, DIFFERENTIAL: The difference in pressure between any two points of a system or a component.

PRESSURE, GAGE: Pressure differential above or below atmospheric pressure.

PRESSURE, MAXIMUM INLET: The maximum rated gage pressure applied to the inlet.

PRESSURE, OPERATING: The pressure at which a system is operated.

PRESSURE, OVERRIDE: The difference between the cracking pressure of a valve and the pressure reached when the valve is passing full flow.

PRESSURE, PEAK: The maximum pressure encountered in the operation of a component.

PRESSURE, PROOF: The non-destructive test pressure in excess of the maximum rated operating pressure.

PRESSURE, RATED: The qualified operating pressure which is recommended for a component or a system by the manufacturer.

PRESSURE, SHOCK: The pressure existing in a wave moving at supersonic velocity.

PRESSURE, SYSTEM: The pressure which overcomes the total resistances in a system. It includes all losses as well as useful work.

PRESSURE-SWITCH: An electric switch operated by fluid pressure.

PRESSURE, WORKING: The pressure which overcomes the resistance of the working device.

PUMP: A device which converts mechanical force and motion into hydraulic fluid power.

PUMP, FIXED DISPLACEMENT: A pump in which the displacement per cycle cannot be varied.

PUMP, VARIABLE DISPLACEMENT: A pump in which the displacement per cycle can be varied.

REDUCER: A connector having a smaller line size at one end than the other.

RESERVOIR: A container for storage of liquid in a fluid power system.

RESTRICTOR: A device which reduces the cross-sectional flow area.

RESTRICTOR, CHOKE: A restrictor, the length of which is relatively large with respect to its cross-sectional area.

RESTRICTOR, ORIFICE: A restrictor, the length of which is relatively small with respect to its cross-sectional area. The orifice may be fixed or variable.

Variable types are non-compensated, pressure compensated, or pressure and temperature compensated.

REYN: The standard unit of absolute viscosity in the English system. It is expressed in pound-seconds per square inch.

SEAL, LIP: A sealing device which has a flexible sealing projection.

SEALING DEVICE: A device which prevents or controls the escape of a fluid or entry of a foreign material.

SILENCER: A device for reducing gas flow noise. Noise is decreased by tuned resonant control of gas expansion.

SPECIFIC GRAVITY (LIQUID): The ratio of the weight of a given volume of liquid to the weight of an equal volume of water.

STOKE: The standard unit of kinematic viscosity in the c.g.s. (centimeter-gram-second) system. It is expressed in square centimeters per second; 1 centistoke equals .01 stoke.

STRAINER: A coarse filter.

SURGE: A momentary rise of pressure in a circuit.

TEE: A connector with three ports, a pair on one axis with one side outlet at right angles to this axis.

TIE ROD: An axial external cylinder element which traverses the length of the cylinder. It is pre-stressed at assembly to hold the ends of the cylinder against the tubing. Tie rod extensions can be a mounting device.

TRUNNION: A mounting device consisting of a pair of opposite projecting cylindrical pivots. The cylindrical pivot pins are at right angle or normal to the piston rod centerline to permit the cylinder to swing in a plane.

TUBE: A line whose size is its outside diameter. Tube is available in varied wall thicknesses.

UNION: A connector which permits lines to be joined or separated without requiring the lines to be rotated.

VALVE: A device which controls fluid flow direction, pressure, or flow rate.

VALVE, DIRECTIONAL CONTROL: A valve whose primary function is to direct or prevent flow through selected passages.

VALVE, DIRECTIONAL CONTROL, CHECK: A directional control valve which permits flow of fluid in only one direction.

VALVE, DIRECTIONAL CONTROL, FOUR WAY: A directional control valve whose primary function is to alternately pressurize and exhaust two working ports.

VALVE, DIRECTIONAL CONTROL, SERVO: A directional control valve which modulates flow or pressure as a function of its input signal.

VALVE, DIRECTIONAL CONTROL, STRAIGHTWAY: A two port directional control valve which modulates flow or pressure as a function of its input signal.

VALVE, DIRECTIONAL CONTROL, THREE WAY: A directional control valve whose primary function is to alternately pressurize and exhaust a working port.

VALVE, FLOW CONTROL: A valve whose primary function is to control flow rate.

VALVE, FLOW CONTROL, DECELERATION: A flow control valve which gradually reduces flow rate to provide deceleration.

VALVE, FLOW CONTROL, PRESSURE COMPENSATED: A flow control valve which controls the rate of flow independent of system pressure.

VALVE, FLOW CONTROL, PRESSURE-TEMPERATURE COMPENSATED: A pressure compensated flow control valve which controls the rate of flow independent of fluid temperature.

VALVE, FLOW DIVIDING: A valve which divides the flow from a single source into two or more branches.

VALVE, FLOW DIVIDING, PRESSURE COMPENSATED: A flow dividing valve which divides the flow at constant ratio regardless of the difference in the resistances of the branches.

VALVE, FOUR POSITION: A directional control valve having four positions to give four selections of flow.

VALVE, PREFILL: A valve which permits full flow from a tank to a "working" cylinder during the advance portion of a cycle, permits the operating pressure to be applied to the cylinder during the working portion of the cycle, and permits free flow from the cylinder to the tank during the return portion of the cycle.

VALVE, PRESSURE CONTROL, COUNTERBALANCE: A pressure control valve which maintains back pressure to prevent a load from falling.

VALVE, PRESSURE CONTROL, DECOMPRESSION: A pressure control valve that controls the rate at which the contained energy of the compressed fluid is released.

VALVE, PRESSURE CONTROL, PRESSURE REDUCING: A pressure control valve whose primary function is to limit outlet pressure.

VALVE, PRESSURE CONTROL, RELIEF: A pressure control valve whose primary function is to limit system pressure.

VALVE, SEQUENCE: A valve whose primary function is to direct flow in a pre-determined sequence.

VALVE, SHUTTLE: A connective valve which selects one of two or more circuits because flow or pressure changes between the circuits.

VALVE, SHUTOFF: A valve which operates fully open or fully closed.

VALVE, THREE POSITION: A directional control valve having three positions to give three selections of flow.

VALVE, TWO POSITION: A directional control valve having two positions to give two selections of flow conditions.

VALVE FLOW CONDITION, CLOSED: All ports are closed.

VALVE FLOW CONDITION, FLOAT: Working ports are connected to exhaust or reservoir.

VALVE FLOW CONDITION, HOLD: Working ports are blocked to hold a powered device in a fixed position.

VALVE FLOW CONDITION, OPEN: All ports are open.

VALVE FLOW CONDITION, TANDEM: Working ports are blocked and supply is connected to the reservoir port.

VALVE MOUNTING, BASE: The valve is mounted to a plate which has top and side ports.

VALVE MOUNTING, LINE: The valve is mounted directly to system lines.

VALVE MOUNTING, MANIFOLD: The valve is mounted to a plate which provides multiple connection ports for two or more valves.

VALVE MOUNTING, SUB-PLATE: The valve is mounted to a plate which provides straight-through top and bottom ports.

VALVE POSITION, DETENT: A predetermined position maintained by a holding device acting on the flow-directing elements of a directional control valve.

VALVE POSITION, NORMAL: The valve position when signal or actuating force is not being applied.

VISCOSITY: A measure of the internal friction or the resistance of a fluid to flow.

VISCOSITY, ABSOLUTE: The ratio of the shearing stress to the shear rate of a fluid. It is usually expressed in centipoise.

VISCOSITY, KINEMATIC: The absolute viscosity divided by the density of the fluid. It is usually expressed in centistokes.

VISCOSITY, SUS: Saybolt Universal Seconds (SUS), which is the time in seconds for 60 milliliters of oil to flow through a standard orifice at a given temperature (ASTM Designation D88-56).

VISCOSITY INDEX: A measure of the viscosity-temperature characteristics of a fluid as referred to that of two arbitrary reference fluids (ASTM Designation D567-53).

WYE (Y): A connector with three ports, a pair on one axis with one side outlet at any angle other than right angles to this axis. The side outlet angle is usually 45°, unless another angle is specified.

Appendix B
Hydraulic Facts and Technical Data

B.1 HYDRAULIC FACTS

Here are some key facts that will help you understand hydraulics:

1. Hydraulic power is nearly always generated from mechanical power. Example: A hydraulic pump driven by an engine crankshaft.
2. Hydraulic power output is nearly always achieved by converting back to mechanical energy. Example: A cylinder that raises a heavy plow.
3. There are three types of hydraulic energy:
   a. Potential or pressure energy
   b. Kinetic energy, the energy of moving liquids
   c. Heat energy, the energy of resistance to flow, or friction
4. Hydraulic energy is neither created nor destroyed, only converted to another form.
5. All energy put into a hydraulic system must come out either as work (gain) or as heat (loss).
6. When a moving liquid is restricted, heat is created and there is a loss of potential energy (pressure) for doing work. Example: A tube or hose that is too small or is restricted. Orifices and relief valves are also restrictions but they are purposely designed into systems.
7. Flow through an orifice or restriction causes a pressure drop.

8. Oil must be confined to create pressure for work. A tightly sealed system is a necessity in hydraulics.
9. Oil takes the course of least resistance.
10. Oil is normally pushed into a pump, not drawn into it. (Atmospheric pressure supplies this push. For this reason, an air vent is needed in the top of the reservoir.)
11. A pump does not pump pressure; it creates flow. Pressure is caused by resistance to flow.
12. Two hydraulic systems may produce the same power output: one at high pressure and low flow, the other at low pressure and high flow.
13. A basic hydraulic system must include four components: a reservoir to store the oil, a pump to push the oil through the system, valves to control oil pressure, flow, and direction and a cylinder (or motor) to convert the fluid movement into work.
14. Compare the two major hydraulic systems:
    a. In the open-center system, pressure is varied but flow is constant.
    b. In the closed-center system, flow is varied but pressure is constant.
15. There are two basic types of hydraulics:
    a. Hydrodynamics is the use of fluids at high speeds "on impact" to supply power. *Example:* a torque converter.
    b. Hydrostatics is the use of fluids at relatively low speeds but at high pressures to supply power. *Example:* most hydraulic systems.

B.2 MISCELLANEOUS TECHNICAL DATA

Flow Rate

1. The number of units of volume of a fluid passing through any channel in one unit of time.
2. Normally expressed as gpm (gallons per minute) or cipm (cubic inches per minute).
3. The flow rate determines speed of operation for hydraulic actuators such as cylinders and fluid motors.

Flow

1. There must be a pressure drop (pressure difference) across an ori-

fice or restriction to cause flow through it. Conversely, if there is no flow, there will be no pressure drop.

2. In any fluid system or unit, whenever there is no flow, pressure is the same at all points of the same elevation.
3. In any fluid system or unit, whenever there is a pressure difference that is not absolutely blocked, flow results.
4. Pressure by gravity helps the atmosphere feed a pump when the reservoir is above the pump. This is called a "flooded suction" pumping system.

Specific Gravity

1. The ratio of the weight of a given volume of fluid to the weight of any equal volume of water.
2. Hydraulic fluid, which is normally mineral oil, usually has a specific gravity between 0.8 and 0.9, which means that it weighs between 80 and 90% of water.

Energy

1. The ability or capacity to perform work.
2. Energy put into a system but not taken out in work at the end of a cylinder or motor shaft is converted to heat.

Force

1. Force is an expression used to define the amount of load or resistance to movement that has to be overcome.
2. Force (lb) = area (in.^{2}) × pressure (psi)
3. Force is not possible without resistance.
4. Force exerted by a cylinder is dependent on pressure applied and piston area. (To find the area, square the diameter and multiply by 0.7854.)

Work

Work = force × distance

Horsepower

Horsepower (a measure of the rate of work) involves three factors: force, distance, and time. (In hydraulics: pressure, flow, and time.)

1 hp = 1 gpm at 1500 psi (approx.) (1 hp moves 33,000 lb 1 ft in 1 min.)
= gpm × pressure × 0.00058 (100% efficiency)

$= \text{gpm} \times \text{pressure} \times 0.0007$ (85% efficiency)

$$\text{gpm} = \frac{\text{hp} \times 1725}{\text{pressure}}$$

1 hp = 746 W

1 hp = 42.4 Btu/min

Pressure

1. The force per unit area exerted by a fluid.
2. Normally expressed as psi (pounds per square inch).
3. The pressure determines the load that can be moved in conjunction with the area to which it is exposed.
4. In hydraulics, pressure is caused by resistance to flow. A drop in pressure is a signal that the resistance is gone.

Helpful Hints

1. To gain speed, increase flow.
2. An increase in pressure does not increase the speed of the piston. Increased flow rate increases speed.
3. 2 in. Hg = 1 psi (approx.)
4. With increase in speed, pressure drops.
5. Speed of a cylinder is dependent on its size (piston area) and the rate of oil flow into it.

Formulas

1. Flow velocity through a pipe varies inversely as the square of the inside diameter. Doubling the inside diameter increases the area four times.
2. To find the actual area of a pipe needed to handle a given flow, use the formula

   $$\text{area (in.}^2) = \frac{\text{gpm} \times 0.3208}{\text{velocity}}$$

   or

   $$\text{velocity (ft/sec)} = \frac{\text{gpm}}{3.117 \times \text{area}}$$

3. In standard pipe, the actual inside diameter is larger than the nominal size quoted. A standard conversion chart should be consulted when selecting pipe.

4. Steel and copper tubing size indicates the outside diameter. To find the actual inside diameter, subtract two times the wall thickness from the tube size.
5. The relationship between torque and horsepower is

$$\text{torque (lb-in.)} = \frac{63{,}025 \times \text{hp}}{\text{rpm}}$$

or

$$\text{hp} = \frac{\text{torque (lb-in.)} \times \text{rpm}}{63{,}025}$$

6. To find the amount of oil required to move a piston through a given distance, multiply the piston area (in inches) by the stroke length (in inches). The product is the cylinder capacity in cubic inches. Divide this product by 231 to determine the capacity in gallons. *Note:* Volume displaced by rod must be deducted if the piston is being retracted.
7. To find the pressure, volume, or weight of a gas in a confined volume, the following formula, known as the Ideal Gas Law, can be used.

$$PV = \frac{MRT}{144}$$

where
- P = absolute pressure, psia
- V = volume, ft^3
- M = weight of air, lb
- T = absolute temperature, °R
- R = universal gas constant (air = 53.3 psf)

8. To find the density of water vapor in a cubic foot of compressed gas, use the following formula:

$$dv = \frac{(Pv)(144)(1)}{(85.78)(T)}$$

where dv = pounds of water vapor/ft^3
Pv = water vapor pressure psi @ dew point temperature
85.78 = water vapor constant

$$dv = \frac{(Pv)(144)(1.004)}{(85.78)(T)}$$

where 1.004 = correction factor for deviation from Ideal Gas Law

$$dv = \frac{(Pv)(1.6854)}{(T)}$$

Figure B.1 details the weight of water in a cubic foot of air at various temperatures, and relative humidity.

Rule of Thumb Estimates for Heat Generation

1. Hydrostatic systems: 20 to 40% of input horsepower goes to heat.
2. Torque converter systems: 25 to 30% of input horsepower goes to heat (no retarder).
3. Pump and cylinder systems: 15 to 25% of input horsepower, plus any heat restriction. Heat of restriction: gpm × psi (drop across restriction) × .00058 = amount of horsepower going directly into heat.
4. Duty cycles: percent of the hour each hydraulic system operates, and/or percent of the hour each component operates. Heat load = (input hp × % input hp to heat × % duty cycle + any heat of restriction) × 2545 = Btu/hr heat load to be removed.

Conversion Factors

1 ft^3 of air weighs 0.07648 lb at sea level and 56°F.
Specific heat of oil = 0.5, oil weighs 7 lb/gal.
3.5 Btu/hr are required to raise 1 gal of oil 1°F, in 1 hr.

Useful Formulas for Hydraulic System Heating/Cooling

1. To find oil temperature drop through cooler for a given cooling capacity at a given gpm, use:

$$\frac{\text{Btu/hr removed}}{210 \times \text{gpm}} = \text{temperature drop of oil, °F}$$

2. To find capacity of a cooler when oil temperature drop across cooler and gpm are known:

gpm × temperature drop × 210 = Btu/hr capacity

3. To find air velocity through cooler core from known air flow rate:

$$\frac{\text{Air flow rate, cfm}}{\text{core area, ft}^2} = \text{air velocity through finned core (ft/min)}$$

4. To find the air flow rate required to obtain the necessary air velocity:

Flow rate (ft^3/min) × core area of fins (ft^2) = required air flow rate from fan, cfm

Weights shown in grains.									7,000 grains = 1 lb.	
Temp.	Relative Humidity									Temp.
°C.	10%	20%	30%	40%	50%	60%	70%	80%	90%	100%	°F.
−23	.028	.057	.086	.114	.142	.171	.200	.228	.256	.285	−10
−18	.048	.096	.144	.192	.240	.289	.337	.385	.433	.481	0
−12	.078	.155	.233	.310	.388	.466	.543	.621	.698	.776	10
− 6.7	.124	.247	.370	.494	.618	.741	.864	.988	1.11	1.24	20
0	.211	.422	.634	.845	1.06	1.27	1.48	1.69	1.90	2.11	32
1.6	.237	.473	.710	.946	1.18	1.42	1.66	1.89	2.13	2.37	35
4.4	.285	.570	.855	1.14	1.42	1.71	1.99	2.28	2.56	2.85	40
7.2	.341	.683	1.02	1.37	1.71	2.05	2.39	2.73	3.07	3.41	45
10	.408	.815	1.22	1.63	2.04	2.45	2.85	3.26	3.67	4.08	50
12.7	.485	.970	1.46	1.94	2.42	2.91	3.39	3.88	4.36	4.85	55
15.6	.574	1.15	1.72	2.30	2.87	3.45	4.02	4.60	5.17	5.75	60
18.4	.678	1.36	2.03	2.71	3.39	4.07	4.75	5.42	6.10	6.78	65
21	.798	1.60	2.39	3.19	3.99	4.79	5.59	6.38	7.18	7.98	70
24.9	.936	1.87	2.81	3.74	4.68	5.62	6.55	7.49	8.42	9.36	75
26.7	1.09	2.19	3.28	4.37	5.47	6.56	7.65	8.75	9.84	10.93	80
29.5	1.27	2.54	3.81	5.08	6.35	7.62	8.89	10.16	11.43	12.73	85
32.2	1.48	2.96	4.44	5.92	7.40	8.87	10.35	11.83	13.31	14.78	90
35	1.72	3.44	5.16	6.88	8.60	10.32	12.04	13.76	15.48	17.15	95
37.8	1.98	3.95	5.93	7.91	9.88	11.86	13.84	15.81	17.79	19.77	100
43.3	2.63	5.26	7.89	10.52	13.15	15.78	18.41	21.04	23.67	26.33	110
48.9	3.45	6.90	10.35	13.80	17.25	20.70	24.15	27.60	31.05	34.48	120
54.4	4.44	8.88	13.32	17.76	22.20	26.64	31.08	35.52	39.96	44.42	130

Figure B.1 Weight of water vapor per cubic foot of air.

5. To find air temperature rise through cooler:

$$\frac{\text{Btu/hr removed by cooler}}{1.05 \times (\text{air flow rate across cooler fins, ft}^3/\text{min})} = \text{air temperature rise, °F}$$

6. *Procedure to find inlet oil temperature* with a standard rated cooler that is not quite big enough to remove the heat load. Use the following formula:

$$\frac{\text{Heat load, Btu/hr} \times 100\text{°F}}{\text{in oil temp, °F} - \text{in air temp, °F}} = \text{size of standard cooler required}$$

Example: Given conditions: 266,500 Btu/hr heat load and 54 gpm oil flow, 150°F maximum inlet oil temperature, 90°F maximum inlet air temperature

$$\frac{266{,}500 \text{ Btu/hr} \times 100°\text{F}}{150°\text{F} - 90°\text{F}} = 444{,}166 \text{ Btu/hr standard cooler required}$$

Thus you need a 444,166 Btu/hr rated cooler to remove 266,500 BtU/hr when you only have a 60°F temperature difference between the inlet air and the inlet oil.

B.3 HELPFUL TABLES FOR FLUID POWER SYSTEMS

To provide some detailed information on the flow capabilities, pressure rating, and pressure loss due to fluid friction, Figs. B.2 to B.8 are presented. The pressure ratings contained in the figures were calculated using the American National Standards Association (ASA) Code ASA B31.1-1955, Section 3, for pressure piping; Code ASA B93.4-1966 for steel tubing; and Society of Automotive Engineers (SAE) standards for hose pressure ratings. Figure B.9 details the compatibility between a variety of commonly used seal compounds and various fluids. The compatibility between the seal material and the fluid is given as a relative rating ranging from "Recommended" to "Unsatisfactory." Since many fluids are referred to by their trade name, a code number is included to help identify the manufacturer.

PIPE		PRESSURE–PSI		WATER HAMMER FACTOR	PIPE		PRESSURE–PSI		WATER HAMMER FACTOR
NOM. SIZE INCHES	SCH. NO.	WORKING	BURST		NOM. SIZE INCHES	SCH. NO.	WORKING	BURST	
1/8	40	3500	20,200		2½	160	4200	15,700	5.43
1/8	80	4800	28,000		2½	XXS	6900	23,000	7.82
1/4	40	2100	19,500		3	40	1600	7,400	2.60
1/4	80	4350	26,400		3	80	2600	10,300	2.92
3/8	40	1700	16,200		3	160	4100	15,000	3.56
3/8	80	3800	22,500		3	XXS	6100	20,500	4.64
1/2	40	2300	15,600	63.4	3½	40	1500	6,800	1.94
1/2	80	4100	21,000		3½	80	2400	9,500	2.17
1/2	160	7300	26,700		4	40	1400	6,300	1.51
1/2	XXS	12300	42,100		4	80	2300	9,000	1.67
3/4	40	2000	12,900	36.1	4	160	4000	14,200	2.08
3/4	80	3500	17,600	44.5	4	XXS	5300	18,000	2.47
3/4	160	8500	25,000		5	40	1300	5,500	.960
3/4	XXS	10000	35,000		5	80	2090	8,100	1.06
1	40	2100	12,100	22.3	5	160	3850	13,500	1.32
1	80	3500	15,900	26.8	5	XXS	4780	16,200	1.49
1	160	5700	22,300	36.9	6	40	1210	5,100	.666
1	XXS	9500	32,700	68.3	6	80	2070	7,800	.738
1¼	40	1800	10,100	12.9	6	160	3760	13,000	.912
1¼	80	3000	13,900	15.0	6	XXS	4660	15,000	1.02
1¼	160	4400	18,100	18.2	8	40	1100	4,500	.385
1¼	XXS	7900	27,700	30.5	8	80	1870	6,900	.422
1½	40	1700	9,100	9.46	8	160	3700	12,600	.529
1½	80	2800	12,600	10.9	8	XXS	3560	12,200	.519
1½	160	4500	17,700	13.7	10	40	1030	4,100	.244
1½	XXS	7200	25,300	20.3	10	*80	1800	6,600	
2	40	1500	7,800	5.74	10	160	3740	12,500	.340
2	80	2500	11,000	6.52	10	XXS	3300	11,200	
2	160	4600	17,500	8.60	12	@40	1000	3,800	
2	XXS	6300	22,100	10.9	12	**80	1800	6,500	
2½	40	1900	8,500	4.02	12	160	3700	12,300	.239
2½	80	2800	11,500	4.54	12	XXS	2700	9,400	

Figure B.2 Pressure ratings of steel pipe (based on ASTM A53 grade B or A106 grade B seamless).

TUBING			INTERNAL	FLOW RATE	MINIMUM	PRESSURE–PSI	
OD	Wall–Inches	ID-INCHES	Area Sq In.	GPM@10 FPS	BEND Radius-Inches	WORKING	BURST
1/4	.035	.180	.0255	.797	9/16	3850	15,400
1/4	.049	.152	.0182	.568	9/16	5400	21,600
5/16	.035	.243	.0460	1.45	3/4	3070	12,300
5/16	.049	.215	.0362	1.13	3/4	430[illegible]	17,200
3/8	.035	.305	.0731	2.29	15/16	2580	10,300
3/8	.049	.277	.0603	1.89	15/16	3600	14,400
3/8	.058	.259	.0527	1.65	15/16	4380	17,500
3/8	.065	.245	.0472	1.47	15/16	4750	19,100
1/2	.035	.430	.1452	4.55	1 ¼	1930	7,700
1/2	.049	.402	.1269	3.98	1 ¼	2700	10,800
1/2	.065	.370	.1075	3.37	1 ¼	3580	14,300
1/2	.083	.334	.0876	2.74	1 ¼	4560	18,250
5/8	.049	.527	.2181	6.83	1 ½	2160	8,650
5/8	.065	.495	.1924	6.03	1 ½	2850	11,400
5/8	.083	.459	.1662	5.18	1 ½	3650	14,600
5/8	.095	.435	.1493	4.66	1 ½	4190	16,750
5/8	.109	.411	.1327	4.16	1 ½	4800	19,200
3/4	.049	.652	.3339	10.5	1 ¾	1800	7,200
3/4	.065	.620	.3019	9.46	1 ¾	2390	9,550
3/4	.083	.584	.2679	8.39	1 ¾	3050	12,200
3/4	.095	.560	.2463	7.71	1 ¾	3440	13,950
3/4	.109	.532	.2223	6.96	1 ¾	4000	16,000
7/8	.049	.777	.4742	14.9	2 ¼	1540	6,160
7/8	.065	.745	.4359	13.7	2 ¼	2050	8,200
7/8	.095	.685	.3685	11.5	2 ¼	2980	11,900
1	.049	.902	.6390	20.0	3	1350	5,400
1	.065	.870	.5945	18.6	3	1790	7,150
1	.083	.834	.5463	17.1	3	2280	9,150
1	.095	.810	.5153	16.1	3	2620	10,450
1	.109	.782	.4803	15.0	3	3000	11,900
1	.120	.760	.4537	14.2	3	3300	13,200
1 ⅛	.083	.959	.7223	22.5	3 ½	2030	8,100
1 ⅛	.095	.935	.6866	21.4	3 ½	2330	9,300
1 ⅛	.109	.907	.6461	20.1	3 ½	2660	10,650
1 ⅛	.120	.885	.6151	19.2	3 ½	2940	11,750
1 ¼	.083	1.084	.9229	28.9	3 ¾	1830	7,300
1 ¼	.095	1.060	.8825	27.6	3 ¾	2080	8,350
1 ¼	.109	1.032	.8365	26.2	3 ¾	2400	9,570
1 ¼	.120	1.010	.8012	25.1	3 ¾	2650	10,600
1 ½	.095	1.310	1.348	42.2	5	1740	6,970
1 ½	.109	1.282	1.291	40.4	5	2000	7,990
1 ½	.120	1.260	1.247	39.1	5	2200	8,800
2	.095	1.810	2.573	80.6	8	1300	5,200
2	.120	1.760	2.433	76.2	8	1650	6,600
2	.165	1.670	2.190	68.6	8	2280	9,100
2	.250	1.500	1.767	55.4	8	3480	13,750

Figure B.3 Pressure ratings of ASA steel tubing.

HOSE		I D INCHES	O D INCHES	INTERNAL AREA-SQ IN.	FLOW RATE GPM @ 10 FPS	MINIMUM BEND RADIUS – INCHES	PRESSURE–PSI	
SIZE	SAE						WORKING	BURST
3/16	100–R1	3/16	1/2	.0276	.866	4	3000	12000
3/16	100–R2A	3/16	5/8	.0276	.866	4	5000	20000
–4	100–R5	3/16	33/64	.0276	.866	3	3000	12000
1/4	100–R1	1/4	5/8	.0491	1.54	4	2750	11000
1/4	100–R2A	1/4	11/16	.0491	1.54	4	5000	20000
1/4	100–R3	1/4	9/16	.0491	1.54	3	1250	5000
–5	100–R5	1/4	37/64	.0491	1.54	3 3/8	2500	10000
5/16	100–R1	5/16	11/16	.0767	2.40	5	2500	10000
5/16	100–R2A	5/16	3/4	.0767	2.40	4 1/2	4250	17000
–6	100–R5	5/16	43/64	.0767	2.40	4	2250	9000
3/8	100–R1	3/8	25/32	.1104	3.46	5	2250	9000
3/8	100–R2A	3/8	27/32	.1104	3.46	5	4000	16000
3/8	100–R3	3/8	3/4	.1104	3.46	4	1125	4500
13/32	100–R1	13/32	13/16	.1296	4.06	5 1/2	2250	9000
–8	100–R5	13/32	49/64	.1296	4.06	4 1/2	2000	8000
1/2	100–R1	1/2	29/32	.1963	6.15	7	2000	8000
1/2	100–R2A	1/2	31/32	.1963	6.15	7	3500	14000
1/2	100–R3	1/2	15/16	.1963	6.15	5	1000	4000
–10	100–R5	1/2	59/64	.1963	6.15	5	1750	7000
5/8	100–R1	5/8	1 1/32	.3068	9.61	8	1500	6000
5/8	100–R2A	5/8	1 3/32	.3068	9.61	8	2750	11000
–12	100–R5	5/8	1 5/64	.3068	9.61	6 1/2	1500	6000
3/4	100–R1	3/4	1 3/16	.4418	13.8	9 1/2	1250	5000
3/4	100–R2A	3/4	1 1/4	.4418	13.8	9 1/2	2250	9000
3/4	100–R3	3/4	1 1/4	.4418	13.8	6	750	3000
3/4	100–R4	3/4	1 1/4	.4418	13.8	5	300	1200
7/8	100–R1	7/8	1 5/16	.6013	18.8	11	1125	4500
7/8	100–R2A	7/8	1 3/8	.6013	18.8	11	2000	8000
–16	100–R5	7/8	1 15/64	.6013	18.8	7 3/8	800	3200
1	100–R1	1	1 1/2	.7854	24.6	11	1000	4000
1	100–R2A	1	1 9/16	.7854	24.6	11	2000	8000
1	100–R3	1	1 1/2	.7854	24.6	8	550	2250
1	100–R4	1	1 1/2	.7854	24.6	6	250	1000
–20	100–R5	1 1/8	1 1/2	.9940	31.1	9	625	2500
1 1/4	100–R1	1 1/4	1 13/16	1.227	38.4	16	625	2500
1 1/4	100–R2A	1 1/4	2	1.227	38.4	16 1/2	1625	6500
1 1/4	100–R3	1 1/4	1 3/4	1.227	38.4	10	375	1500
1 1/4	100–R4	1 1/4	1 51/64	1.227	38.4	8	200	800
–24	100–R5	1 3/8	1 3/4	1.485	46.5	10 1/2	500	2000
1 1/2	100–R1	1 1/2	2 1/16	1.767	55.4	20	500	2000
1 1/2	100–R2A	1 1/2	2 1/4	1.767	55.4	20	1250	5000
1 1/2	100–R4	1 1/2	2 5/64	1.767	55.4	10	150	600
–32	100–R5	1 13/16	2 7/32	2.580	80.8	13 1/4	350	1400
2	100–R1	2	2 5/8	3.142	98.4	25	375	1500
2	100–R2A	2	2 3/4	3.142	98.4	22	1125	4500
2	100–R4	2	2 1/2	3.142	98.4	12	100	400
2 1/2	100–R4	2 1/2	3 1/8	4.909	154.	14	65	250
3	100–R4	3	3 5/8	7.069	221.	18	55	225
3 1/2	100–R4	3 1/2	4 1/8	9.621	301.	22	50	200
4	100–R4	4	4 3/4	12.566	394.	26	50	200

ALLOWABLE MINIMUM BURST AND WORKING PRESSURES FOR HOSE ARE BASED ON SAE STANDARDS.

Figure B.4 Pressure ratings of SAE hose.

PIPE SIZE	OD	STANDARD PIPE — SCHEDULE 40 WALL	ID	INT AREA	WT/ FT	GPM @ 2 FPS	GPM @ 5 FPS	GPM @ 10 FPS	GPM @ 15 FPS	GPM @ 20 FPS	GPM @ 25 FPS	EXTRA STRONG PIPE — XS — SCHEDULE 80 WALL	ID	INT AREA	WT/ FT	GPM @ 2 FPS	GPM @ 5 FPS	GPM @ 10 FPS	GPM @ 15 FPS	GPM @ 20 FPS	GPM @ 25 FPS	PIPE SIZE
1/8	.405	.068	.269	.057	.245	.35	.89	1.8	2.7	3.5	4.4	.095	.215	.036	.314	.23	.57	1.1	1.7	2.3	2.8	1/8
1/4	.540	.088	.364	.104	.425	.65	1.6	3.2	4.9	6.5	8.1	.119	.302	.072	.535	.45	1.1	2.2	3.4	4.5	5.6	1/4
3/8	.675	.091	.493	.191	.567	1.2	3.0	6.0	9.0	12.0	15.0	.126	.423	.141	.738	.88	2.2	4.4	6.6	8.8	11.0	3/8
1/2	.840	.109	.622	.304	.852	1.9	4.8	9.5	12.0	19.0	23.8	.147	.546	.234	1.087	1.5	3.7	7.3	11.0	14.7	18.3	1/2
3/4	1.050	.113	.824	.533	1.132	3.3	8.4	16.7	25.1	33.4	41.8	.154	.742	.433	1.473	2.7	6.8	13.6	20.3	27.1	33.9	3/4
1"	1.315	.133	1.049	.864	1.679	5.4	13.5	27.0	40.6	54.1	67.7	.179	.957	.719	2.171	4.5	11.3	22.5	33.8	45.0	56.3	1"
1¼	1.660	.140	1.380	1.495	2.273	9.4	23.4	46.8	70.3	93.7	117	.191	1.278	1.283	2.996	8.0	20.0	40.1	60.2	80.3	100	1¼
1½	1.900	.145	1.610	2.036	2.718	12.7	31.9	63.7	95.6	127	159	.200	1.500	1.767	3.631	11.1	27.7	55.3	83.0	110	138	1½
2	2.375	.154	2.067	3.356	3.653	21.0	52.5	105	157	210	263	.218	1.939	2.953	5.022	18.5	46.2	92.5	139	185	231	2
2½	2.875	.203	2.469	4.788	5.793	30.0	75.0	150	225	300	375	.276	2.323	4.238	7.661	26.5	66.4	133	199	265	332	2½
3	3.500	.216	3.068	7.393	7.576	46.3	116	232	347	463	579	.300	2.900	6.605	10.25	41.4	103	207	310	414	517	3
3½	4.000	.226	3.548	9.886	9.109	61.9	155	310	465	619	774	.318	3.364	8.888	12.50	55.7	139	278	418	557	696	3½
4	4.500	.237	4.026	12.73	10.79	79.7	199	399	598	797	997	.337	3.826	11.50	14.98	72.0	180	360	540	720	900	4
4½	5.000	247	4.506	15.95	12.54	99.9	250	499	749	998	1249	.355	4.290	14.45	17.61	90.5	226	453	679	905	1132	4½
5	5.563	.258	5.047	20.01	14.62	125	313	627	940	1253	1567	.375	4.813	18.19	20.78	114	285	570	855	1140	1425	5
6	6.625	.280	6.065	28.89	18.97	181	452	904	1357	1810	2262	.432	5.761	26.07	28.57	163	408	816	1225	1633	2041	6
7	7.625	.301	7.023	38.74	23.54	243	607	1213	1820	2427	3033	.500	6.625	34.47	38 05	216	540	1080	1620	2160	2699	7
8	8.625	.322	7.981	50.03	28.55	313	783	1567	2350	3134	3917	.500	7.625	45.66	43.39	286	715	1430	2145	2861	3576	8
10 ⊕	10.75	.365	10.02	78.85	40.48	494	1235	2470	3705	4940	6175	.594	9.562	71.81	64.40	450	1125	2249	3374	4498	5623	10 ⊕
12 ⊕	12.75	.406	11.94	111.9	53.56	701	1753	3506	5259	7012	8765	.688	11.37	101.61	88.57	636	1591	3182	4774	6365	7956	12

PIPE SIZE	OD	SCHEDULE 160 PIPE WALL	ID	INT AREA	WT/ FT	GPM @ 2 FPS	GPM @ 5 FPS	GPM @ 10 FPS	GPM @ 15 FPS	GPM @ 20 FPS	GPM @ 25 FPS	DOUBLE EXTRA STRONG PIPE WALL	ID	INT AREA	WT/ FT	GPM @ 2 FPS	GPM @ 5 FPS	GPM @ 10 FPS	GPM @ 15 FPS	GPM @ 20 FPS	GPM @ 25 FPS	PIPE SIZE
1/2	.840	.187	.466	.171	1.310	1.07	2.67	5.34	8.01	10.7	13.4	.294	.252	.050	1.714	.32	.79	1.6	2.4	3.1	3.9	1/2
3/4	1.050	.218	.587	.271	1.940	1.70	4.24	8.49	12.7	17.0	21.2	.308	.434	.148	2.440	.93	2.3	4.6	6.9	9.2	11.6	3/4
1"	1.315	.250	.815	.522	2.850	3.27	8.17	16.3	24.5	32.7	40.8	.358	.599	.282	3.659	1.8	4.4	8.8	13.3	17.7	22.1	1"
1¼	1.660	.250	1.160	1.060	3.764	6.62	16.6	33.1	49.7	66.2	82.8	.382	.896	.630	5.214	4.0	9.9	19.8	29.6	39.5	49.4	1¼
1½	1.900	.281	1.338	1.410	4.862	8.81	22.0	44.0	66.1	88.1	110	.400	1.100	.950	6.408	6.0	14.9	29.8	44.6	59.5	74.4	1½
2	2.375	.343	1.689	2.241	7.450	14.0	35.1	70.2	105	140	175	.436	1.503	1.774	9.029	11.1	27.9	55.6	83.4	111	139	2
2½	2.875	.375	2.125	3.542	10.01	22.2	55.5	111	167	222	278	.552	1.771	2.463	13.70	15.4	38.6	77.1	116	154	193	2½
3	3.500	.437	2.626	5.416	14.30	33.9	84.8	170	254	339	424	.600	2.300	4.154	18.58	26.0	65.1	130	195	260	325	3
4	4.500	.531	3.438	9.283	22.52	58.2	145	291	436	582	727	.674	3.152	7.803	27.54	48.9	122	244	367	488	611	4
5	5.563	.625	4.313	14.61	33.00	91.5	229	458	686	915	1144	.750	4.063	12.97	38.55	81.2	203	406	609	812	1015	5
6	6.625	.718	5.189	21.15	45.30	132	331	662	994	1325	1656	.864	4.897	18.83	53.16	118	295	590	885	1180	1475	6
8	8.625	.906	6.813	36.44	74.70	230	571	1142	1713	2384	2855	.875	6.875	37.12	72.42	233	581	1163	1744	2325	2907	8
10	10.75	1.125	8.500	56.75	115.64	355	889	1777	2666	3555	4443	1.000	8.750	60.13	104.1	377	942	1883	2825	3767	4709	10
12	12.75	1.312	10.126	80.53	160.33	504	1261	2523	3784	5045	6306	1.000	10.75	90.76	125.5	569	1421	2843	4264	5686	7107	12

Figure B.5 Characteristics of steel pipe—size, schedule, flow rates.

MATERIAL						PRESSURE LOSS (PSI/FOOT LENGTH) IN PIPES AT AVERAGE FLOW VELOCITY (FT/SEC) OF														EQUIVALENT PIPE LENGTHS (FT.) FOR CIRCUIT COMPONENTS					
SIZE (INCHES)	PIPE / HOSE	O.D. INCHES	I.D. INCHES	WALL INCHES	I.D. AREA SQ. IN.	5		7		10		15		20		25		30		TEE			ELBOW		
						LOSS	GPM	LOSS	GPM	LOSS	GPM	LOSS	GPM	LOSS	GPM	LOSS	GPM	LOSS	GPM						
1/8	PIPE-SCH 40	.405	.269	.068	.057	1.25	.89	1.79	1.24	2.60	1.75	3.16	2.67	5.47	3.56	6.20	4.45	7.07	5.34						
	PIPE-SCH 80	.405	.215	.095	.036	1.89	.56	3.05	.78	4.26	1.12	5.20	1.68	8.38	2.24	11.1	2.80	12.7	3.36						
	HOSE	–	.125	–	.012	5.96	.186	8.37	.260	11.9	.372	18.0	.558	24.0	.744	30.0	.930	35.7	1.11						
1/4	PIPE-SCH 40	.540	.364	.088	.104	.67	1.62	1.05	2.27	1.64	3.24	1.92	4.96	2.97	6.48	3.23	8.10	3.73	9.72						
	PIPE-SCH 80	.540	.302	.119	.072	1.11	1.12	1.49	1.57	2.11	2.24	2.84	3.36	4.15	4.48	5.08	5.60	6.30	6.72						
	HOSE	–	.250	–	.049	1.57	.758	2.17	1.08	3.00	1.49	4.49	2.23	6.04	2.98	7.49	3.72	8.95	4.44						
3/8	PIPE-SCH 40	.675	.493	.091	.191	.39	2.98	.57	4.18	.86	5.96	1.05	8.94	1.69	11.92	4.27	14.9	5.78	16.9	2.7	.8	2.7	1.2	2.7	.6
	PIPE-SCH 80	.675	.423	.126	.140	.54	2.18	.74	3.06	1.10	4.36	1.34	6.54	1.97	8.72	5.19	10.9	7.20	13.1						
	HOSE	–	.375	–	.110	.685	1.71	.97	2.43	1.34	3.35	2.02	5.03	2.68	6.71	3.33	8.36	3.99	10.0						
1/2	PIPE-SCH 40	.840	.622	.109	.304	.24	4.74	.36	6.65	.49	9.48	.68	14.22	2.09	18.98	3.38	23.7	4.28	28.4	3.5	1.05	3.5	1.5	3.5	.75
	PIPE-SCH 80	.840	.546	.147	.234	.30	3.65	.45	5.12	.71	7.30	.78	10.9	2.47	14.6	3.61	18.2	5.00	21.9	2.9	.9	2.9	1.4	2.9	.68
	PIPE-SCH XX	.840	.252	.294	.050	1.54	.78	2.19	1.09	3.08	1.56	3.65	2.34	6.13	3.12	7.48	3.90	9.55	4.68						
	HOSE	–	.500	–	.196	.387	3.03	.547	4.30	.755	5.94	1.13	8.90	2.4	11.9	3.15	15.3	4.5	17.7						
3/4	PIPE-SCH 40	1.050	.824	.113	.533	.14	8.32	.22	11.7	.27	16.6	.78	25.0	1.47	33.3	2.19	41.6	3.00	49.9	4.5	1.4	4.5	2.1	4.5	1.0
	PIPE-SCH 80	1.050	.742	.154	.432	.16	6.74	.26	9.45	.37	13.5	.87	20.2	1.71	27.0	2.48	33.7	3.52	40.4	4.0	1.2	4.0	1.6	4.0	.8
	PIPE-SCH XX	1.050	.434	.308	.148	.53	2.31	.67	3.24	1.05	4.62	1.31	6.93	1.94	9.24	5.06	11.6	7.02	13.9						
	HOSE	–	.750	–	.442	.171	6.82	.248	9.92	.336	13.4	.502	20.1	1.33	26.8	2.02	33.4	2.90	41.3						
1	PIPE-SCH 40	1.315	1.049	.133	.863	.10	13.5	.13	18.9	.34	26.9	.57	40.4	1.42	53.8	1.64	67.3	2.24	80.7	5.7	1.7	5.7	2.6	5.7	1.2
	PIPE-SCH 80	1.315	.957	.179	.719	.11	11.2	.15	15.7	.24	22.4	.62	33.6	1.23	44.8	1.84	56.1	2.93	67.3	5.2	1.6	5.2	2.5	5.2	1.1
	PIPE-SCH XX	1.315	.599	.358	.863	.26	4.39	.37	6.16	.53	8.78	.67	13.2	2.25	17.6	3.29	22.0	4.20	26.3	3.0	1.0	3.0	1.5	3.0	.75
	HOSE	–	1.00	–	.785	.097	12.2	.136	17.1	.194	24.4	.610	36.6	.987	48.8	1.51	61.2	2.02	73.4						
1-1/4	PIPE-SCH 40	1.660	1.380	.140	1.496	.05	23.4	.08	31.7	.25	46.7	.39	70.1	.78	93.4	1.18	117	1.47	140	7.5	2.4	7.5	3.7	7.5	1.6
	PIPE-SCH 80	1.660	1.278	.191	1.280	.07	20.0	.09	28.1	.26	39.9	.44	58.9	.85	79.8	1.27	99.8	1.80	120	7.0	2.1	7.0	3.5	7.0	1.5
	PIPE-SCH XX	1.660	.896	.382	.630	.13	9.83	.16	13.8	.24	19.7	.71	29.5	1.35	39.3	2.01	49.2	2.76	59.0	4.9	1.5	4.9	2.3	4.9	1.1
	HOSE	–	1.25	–	1.23	.062	19.1	.087	26.8	.125	38.2	.436	57.3	.738	76.4	1.08	95.5	1.52	115						
1-1/2	PIPE-SCH 40	1.900	1.610	.145	2.046	.04	31.8	.06	44.5	.19	63.5	.33	95.3	.64	127	.96	159	1.26	191	9.0	2.8	9.0	4.3	9.0	2.0
	PIPE-SCH 80	1.900	1.500	.200	1.767	.04	27.6	.06	38.6	.21	55.1	.42	82.7	.71	110	1.06	138	1.36	166	8.2	2.6	8.2	4.2	8.2	1.8
	PIPE-SCH XX	1.900	1.100	.400	.950	.09	14.8	.09	20.8	.32	29.6	.51	44.4	1.05	59.2	1.51	74.1	2.14	88.9	6.5	2.0	6.5	3.0	6.5	1.4
	HOSE	–	1.50	–	1.77	.044	27.7	.061	38.6	.180	55.1	.353	82.7	.59	110	.86	138	1.21	166						
2	PIPE-SCH 40	2.375	2.067	.154	3.355	.03	52.3	.08	73.4	.14	105	.24	159	.48	209	.69	262	.85	324	11.0	3.5	11.0	5.5	11.0	2.5
	PIPE-SCH 80	2.375	1.939	.218	2.953	.03	46.0	.09	64.6	.15	92.0	.26	138	.52	184	.73	230	.98	275	10.8	3.4	10.8	5.0	10.8	2.4
	PIPE-SCH XX	2.375	1.503	.436	1.773	.04	27.7	.12	38.8	.21	55.3	.36	82.9	.72	111	1.34	138	1.36	166	8.2	2.6	8.2	4.0	8.2	1.8
	HOSE	–	2.00	–	3.14	.024	48.9	.034	68.6	.123	97.8	.256	147	.41	196	.60	245	.80	293						
2-1/2	PIPE-SCH 40	2.875	2.469	.203	4.788	.03	74.8	.07	105	.11	149	.20	224	.37	299	.53	374	.72	449	14.0	4.2	14.0	6.5	14.0	3.0
	PIPE-SCH 80	2.875	2.323	.276	4.238	.03	66.1	.07	92.6	.12	132	.21	198	.39	264	.57	331	.87	397	13.0	4.0	13.0	6.1	13.0	2.9
	PIPE-SCH XX	2.875	1.771	.552	2.464	.03	38.5	.10	53.4	.17	76.9	.30	115	.59	154	.79	193	1.15	231	10.3	3.1	10.3	4.8	10.3	2.2
	HOSE	–	2.50	–	4.91	.016	76.5	.045	107	.09	153	.18	229	.30	306	.43	382	.617	459						

Figure B.6 Pressure loss due to oil friction in pipes.

Pipe size	150 psi	100 psi	80 psi	60 psi	20 in. Hg. vac.	10 in. Hg. vac.
1/8	16	14	12	11	–	–
1/4	30	25	21	20	11	9
3/8	55	45	41	36	21	16
1/2	85	72	65	60	33	25
3/4	150	130	115	100	60	45
1	250	210	190	165	95	72
1 1/4	430	350	315	285	165	130
1 1/2	600	490	440	400	230	180
2	950	800	710	650	370	275

Figure B.7 Recommended maximum air flow in pipes.

For Various Flow Rates, for 1/2-in., 3/4-in., 1-in., 1-1/4-in. Nominal Pipe Diameters, at Initial Pressures of 60, 80, 100, 125 psi

AIR FLOW RATES					PRESSURE LOSSES FOR EACH 100 FEET OF PIPE LENGTH, PSI							
Free Air, cfm	Equivalents cfm of Compressed Air				1/2-in. Pipe Diameter				3/4-in. Pipe Diameter			
	60 psi	80 psi	100 psi	125 psi	60 psi	80 psi	100 psi	125 psi	60 psi	80 psi	100 psi	125 psi
10	1.97	1.55	1.28	1.05	.59	.46	.38	.31	.14	.11	.09	.08
20	3.94	3.10	2.56	2.10	2.23	1.74	1.42	1.17	.53	.41	.34	.28
30	5.90	4.66	3.84	3.16	4.94	3.84	3.13	2.54	1.14	.90	.74	.60
40	7.87	6.21	5.13	4.21	8.90	6.93	5.55	4.53	1.99	1.55	1.28	1.05
50	9.84	7.76	6.41	5.26	14.20	10.70	8.65	7.01	3.08	2.42	2.00	1.62
60	11.81	9.31	7.69	6.31					4.45	3.47	2.84	2.33
70	13.78	10.87	8.97	7.37					6.06	4.73	3.85	3.14
80	15.74	12.42	10.25	8.42					7.96	6.14	5.01	4.08
90	17.71	13.97	11.53	9.47					10.00	7.75	6.40	5.17
100	19.68	15.50	12.82	10.52					12.60	9.62	7.80	6.33
125	24.60	19.40	16.02	13.15					21.00	15.50	12.40	9.80
150	29.51	23.30	19.22	15.78					31.50	23.00	18.10	14.40

Free Air, cfm	Equivalents cfm of Compressed Air				1-in. Pipe Diameter				1-1/4-in. Pipe Diameter			
	60 psi	80 psi	100 psi	125 psi	60 psi	80 psi	100 psi	125 psi	60 psi	80 psi	100 psi	125 psi
10	1.97	1.55	1.28	1.05	.05	.04	.03	.02	.011	.0086	.0071	.0058
20	3.94	3.10	2.56	2.10	.16	.13	.10	.08	.040	.032	.026	.021
30	5.90	4.66	3.84	3.16	.34	.28	.23	.19	.086	.068	.056	.046
40	7.87	6.21	5.13	4.21	.59	.46	.38	.31	.146	.116	.096	.079
50	9.84	7.76	6.41	5.26	.92	.73	.60	.49	.22	.18	.146	.120
60	11.81	9.31	7.69	6.31	1.30	1.02	.84	.69	.32	.25	.21	.17
70	13.78	10.87	8.97	7.37	1.75	1.36	1.12	.92	.42	.34	.28	.23
80	15.74	12.42	10.25	8.42	2.24	1.76	1.44	1.18	.55	.44	.36	.30
90	17.71	13.97	11.53	9.47	2.88	2.23	1.85	1.49	.69	.55	.45	.37
100	19.68	15.50	12.82	10.52	3.45	2.69	2.21	1.81	.84	.66	.55	.45
125	24.60	19.40	16.02	13.15	5.38	4.18	3.41	2.79	1.31	1.03	.85	.69
150	29.51	23.30	19.22	15.78	7.81	5.75	4.91	3.99	1.87	1.47	1.20	.99
175	34.44	27.20	22.43	18.41	10.80	8.10	6.80	5.45	2.58	2.00	1.64	1.32
200	39.36	31.00	25.63	21.05	14.50	10.90	8.79	7.11	3.31	2.58	2.12	1.73
250	49.20	38.80	32.04	26.31					5.30	4.05	3.30	2.67
300	59.00	46.60	38.45	31.57					7.51	5.78	4.71	3.83
350	68.90	45.30	44.86	36.83					10.30	7.90	6.45	5.15
400	78.70	62.10	51.26	42.09					13.70	10.30	8.30	6.74

Figure B.8 Pressure loss due to air friction in pipes.

Fluid Name	Code	Military Specification	Trade Name/Number	Color	TYPE OF SEAL COMPOUND – COMMON NAME								
					Buna-N	Butyl	Corfam	EP	Viton	Silicone	Neoprene	Nat. Rubber	Polyure'ne
Water–Glycol	1		Houghto-Safe 600 Series	red	R	R	R	R	R	S	S	R	U
	1		Houghto-Safe 500 Series	red	R	R	R	R	R	S	S	R	U
	1	MIL-H22072	Houghto-Safe 271	red	R	R	R	R	R	S	S	–	U
	4		Ucon Hydrolube	yel. or red	R	R	R	R	R	R/S	S	R	U
	4		Ucon M1	yellow	R	R	R	R	R	S	S	S	–
	5		Celluguard	red	R	R	R	R	R	S	S	–	U
	10		Safety Fluid 200	bright pink	R	R	R	R	R	S	S	–	U
Water/Oil Emulsion	1		Houghto-Safe 5000 Series	white	R	U	R	U	R	–	S	U	U
	3		FR	creamy	R	U	R	U	R	–	S	U	U
	7		Irus 902	yellow	R	U	R	U	R	U	S	U	M
	8		Pyrogard C & D	pale yel.	R	U	R	U	R	–	S	U	U
Water-Soluble Oil	–		–	milky	R	M	R	–	R	–	S	S	M/U
Water-Fresh	–		–	–	R	R	R	R	R	R	M	R	M/U
Water-Salt	–		–	–	R	R	R	R	R	R	M	R	M/U
Phosphate Ester	1		Houghto-Safe 1000 Series	green	U	R	M/U	R	R	M	U	U	M
	1	MIL-H-19547B	Houghto-Safe 1120	green	U	R	M/U	R	R	M	U	U	M
	2		Pydraul F-9, 150, 625	cloudy bl.	U	R/S	M/U	S	R	R	U	U	S
	5		Fyrquel	lt. green	U	R	M/U	R	R	M	U	U	M
	7		Shell SFR B.C.D.	aqua gr.	U	R	M/U	R	R	M	U	U	M
	8		Pyrogard 42,43,53,55,190,600	pale yel.	U	R	M/U	R	R/S	M	U	U	M
	2		Skydrol 500A	purple	U	S	U	R	U	M	U	U	U
	2		Skydrol 7000	green	U	S	U	R	U	M	U	U	U
	2		Pydraul 312, 135 (2)	blue gr.	U	M	M	M	R	R	U	U	–
	2		Pydraul AC	cloudy bl.	U	S	M/U	S	R	R	U	U	M/U
	2		Pydraul 60	cloudy bl.	U	R	M/U	R	U	S	U	U	M/U
	8		Pyrogard 210 (3)	yellow	U	M	–	M	R	R	U	U	M/U
Diester	–	MIL-H-7808	Lube Oil-Aircraft	amber	S	U	R	U	R	U	U	U	U
Chlorinat. Hydrocarb.	2		Aroclor 1200 Series (1)	clear	M	S	–	S	R	S	U	U	U
	2		Pydraul A-200	cloudy bl.	U	M	M	M	R	R	U	U	M/U
Silicate Ester	2		OS-45 Type 4	clear	S	U	–	S	R	U	R	U	R
	6	MLO-8200	Oronite 8200	clear	S	U	–	U	R	U	R	U	R
	6	MIL-8515	Oronite 8515	clear	S	U	–	U	R	U	R	U	R
	9	MIL-H-8446B	Brayco 846	red brown	S	U	–	U	R	U	R	U	R
Kerosene	–			clear	R	U	R	U	R	U	M/U	U	R
Jet Fuel	–	MIL-J-5624	JP-3, 4, 5 (RP-1)	lt. straw	R	U	R	U	R	U	U	U	S
Diesel Fuel	–	–	–	clear	R	U	R	U	R	U	M/U	U	R
Gasoline	–	–	Gasoline	various	R	U	R/S	U	R	U	U	U	R
Petroleum Base	–	MIL-H-6083	Preservative Oil	red	R	U	R	U	R	U	R	S	R
Petroleum Base	–	MIL-H-5606	Aircraft Hyd. Fluid	red	R	U	R	U	R	U	S	U	R

NOTES: (1) Halogenated (2) Petroleum and halogenated hydrocarbon and phosphate ester mixture (3) Chlorinated phosphate ester

MANUFACTURERS CODE NUMBERS

NO.	MANUFACTURER	NO.	MANUFACTURER	NO.	MANUFACTURER
1	E. F. HOUGHTON	5	STAUFFER CHEMICAL	8	MOBIL OIL
2	MONSANTO	6	STANDARD OIL (ORTHO CHEMICAL)	9	BRAY OIL – ROYAL LUBRICANT
3	GULF	7	SHELL CHEMICAL	10	TEXACO
4	UNION CARBIDE & CHEMICAL				

Figure B.9 Seal compatibility with common fluids. R, recommended; S, satisfactory; M, marginal; U, unsatisfactory; –, insufficient data.

Index